Perspectives in Environmental Chemistry

TOPICS IN ENVIRONMENTAL CHEMISTRY
A SERIES OF ADVANCED TEXTBOOKS AND MONOGRAPHS

Perspectives in Environmental Chemistry

Edited by
DONALD L. MACALADY

Colorado School of Mines

New York Oxford
OXFORD UNIVERSITY PRESS
1998

OXFORD UNIVERSITY PRESS

OXFORD   NEW YORK
ATHENS   AUCKLAND   BANGKOK   BOGOTÁ   BOMBAY
BUENOS AIRES   CALCUTTA   CAPE TOWN   DAR ES SALAAM
DELHI   FLORENCE   HONG KONG   ISTANBUL   KARACHI
KUALA LUMPUR   MADRAS   MADRID   MELBOURNE
MEXICO CITY   NAIROBI   PARIS   SINGAPORE
TAIPEI   TOKYO   TORONTO

and associated companies in

Berlin   Ibadan

Copyright © 1998 by Oxford University Press, Inc.

Published by Oxford University Press, Inc.,
198 Madison Avenue, New York, New York 10016

Library of Congress Cataloging-in-Publication Data
Perspectives in environmental chemistry / edited by
Donald L. Macalady.
p.   cm.—(Topics in environmental chemistry)
Includes bibliographical references and index.
ISBN 0-19-510208-8 (cloth). — ISBN 0-19-510209-6 (pbk.)
1. Environmental chemistry.   I. Macalady, Donald L., 1941–
II. Series.
TD193.P45   1997   97–4123
628.5—dc21   CIP

9 8 7 6 5 4 3 2 1
Printed in the United States of America
on acid-free paper

SIMPLE SYSTEMS ARE NOT

Contents

Preface

Environmental chemistry has become a discipline which is far more than a series of applications of chemistry to environmental problems. The complexity and diversity of environmental systems demands a concomitant diversity in the group of researchers who call themselves environmental chemists. This book is in many ways a reflection of this diversity. It represents an attempt to provide a broad perspective on the immensely difficult scientific problems faced by environmental chemists. The format is a collection of chapters by some of the leading scientists in the field who present their research and their ideas about the most important areas of future research. Though an attempt was made to provide a thorough coverage of the field, no effort to comprehensively treat environmental chemistry is possible in a single volume.

The diversity of academic backgrounds and affiliations of the authors is obvious from the list of contributors, which provides a general picture of the scope of environmental chemistry research. This mixture of backgrounds also provides a rather formidable challenge to editors who seek to provide a reference work/textbook in a consistent, readable format. My goal and that of the authors was to present these perspectives in a style and at a level that allows upper-level undergraduate and beginning graduate students in the environmental sciences to find the book helpful in their studies. The book should also be useful to instructors as a general or supplementary text in environmental chemistry taught at these levels. Finally, it should provide an excellent means by which scientists newly interested in environmental chemistry can obtain an in-depth introduction to a wide variety of modern environmental chemical research areas.

The book is organized into sections describing fundamental work in the environmental chemistry of condensed-phase systems, studies in atmospheric chemistry, and chapters more directly related to field data. The order of chapters within each section is less formally defined. In the first section, separation of inorganic and organic topics was attempted with only moderate consistency. Among the atmospheric chemistry chapters, separation based on the tropopause is the only organizational principle. The final four chapters on more applied topics are presented in alphabetical order by first author.

The writing of a Table of Contents is usually not considered to be a task of great significance. However, in this case it serves as a rather interesting and useful demonstration of the ideas presented above. It is very difficult to find any chapter which is exclusively related to a single traditional area of chemistry (or biology, or geology, etc.). For example, consider the study of the chemistry of colloids in aquatic systems. This requires a rather intimate understanding of inorganic chemistry, organic chemistry, and the physical chemistry of surfaces and colloidal systems, as well as a smattering of microbiology, fundamental physics, mathematics, sedimentary geology, mineralogy, and hydrology. The study of chemical interactions in the stratosphere requires an equally broad set of discipline-based understandings, as well as detailed knowledge of the remaining portions of the atmosphere. All of the chapters presume a broad understanding of analytical chemistry and modern analytical methodology. So, the process of considering the order of the chapters in the Table of Contents provides an additional appreciation of the breadth and tremendous diversity of our discipline.

The efforts to bring this book to publication began several years ago as an attempt to publish written versions of papers presented at a symposium "Frontiers in Environmental Chemistry" at an American Chemical Society Meeting in Denver, Colorado. Subsequently, titles and authors have been added and a few deleted. Most of the chapters have gone through several updated and revised versions. Those presenters at the original symposium who were prompt in the submission of their chapter have shown incredible patience with the editor as seemingly innumerable and interminable changes in the scope and timescale of this project have been made. For this, I am extremely grateful. On the other hand, more-recently added authors have been tolerant of persistent urgings to get chapter revisions accomplished in minimum time. This also is much appreciated. My hope is that the quality and utility of the resulting work justifies this patience and hard work.

There are many others who deserve thanks, a few of whom will be mentioned here. John Birks of the University of Colorado, who is also the Oxford University Press series editor for this book, was especially helpful with several of the atmospheric chapters and final preparations before the chapters were sent to the publisher. This final push was also greatly facilitated by the generous granting of office and technical help by the Chemistry Department of the University of Otago in Dunedin, New Zealand, where I am currently on leave of absence. All of the trout in New Zealand waters that were not caught as a result of this work are also grateful! I would also like to specifically thank the following persons for editorial assistance, review suggestions, computer program deciphering, and general moral support: Bob Rogers of Oxford University Press; Stephen Daniel, Thomas Wildeman, James Ranville, and George Kennedy of the Colorado School of Mines; Esther and Alison Macalady, Alan Elzerman of Clemson University; and Keith Hunter, Diana Evans, Julie Leith, Andrea Krause, Mervyn Thomas, Eng Tan, Dave Larsen, and Michael Crawford of the University of Otago. Finally, the students in Chemistry 302 and 555 at the Colorado School of Mines, as well as those in Chemistry 466 at the University of Otago, provided many helpful suggestions as a result of their study of early versions of many of the book chapters.

DON MACALADY
Dunedin, New Zealand
May 1996

CONTRIBUTORS

Dr. Michael J. Barcelona, Research Professor
Department of Civil and Environmental Engineering
National Center for Integrated Bioremediation Research and Development
University of Michigan
Ann Arbor, MI, 48109

Dr. John W. Birks, Professor
Department of Chemistry and Biochemistry and
Cooperative Institute for Research in Environmental
  Sciences
University of Colorado
Boulder, CO 80309-0216

Dr. Guy P. Brasseur
National Center for Atmospheric Research
Boulder, CO 80307

Dr. William H. Brune, Professor
Department of Meteorology
Pennsylvania State University
University Park, PA 16802,

Dr. Peter L. Croot
Department of Analytical and Marine Chemistry
University of Göteborg
Chalmers Institute of Technology
Göteborg, Sweden

Dr. Jason Geddes
NEL Laboratories
1030 Matley Lane
Reno, NV 89502

Dr. Stefan B. Haderlein
Federal Institute for Environmental Science and
Technology (EAWAG) and Swiss Federal Institute
  of Technology (ETH)
CH-8600 Dübendorf, Switzerland

Dr. Janet G. Hering, Associate Professor
California Institute of Technology
Environmental Engineering Science (138-78)
Pasadena, CA 91125

Dr. Keith A. Hunter, Professor
Department of Chemistry
University of Otago
P. O. Box 56
Dunedin, New Zealand

Dr. Jonathan P. Kim
Department of Chemistry
University of Otago
P. O. Box 56
Dunedin, New Zealand

Stephan Kraemer, Research Associate
California Institute of Technology
Environmental Engineering Science (138-78)
Pasadena, CA 91125

Dr. Sonia M. Kreidenweis, Assistant Professor
Department of Atmospheric Science
Colorado State University
Fort Collins, CO 80523

Dr. Franck Lefèvre
National Center for Atmospheric Research
Boulder, CO 80307

Dr. Donald L. Macalady, Professor
Department of Chemistry and Geochemistry
Colorado School of Mines
Golden, Colorado 80401

Dr. Ann M. Middlebrook
National Oceanic and Atmospheric Administration
Aeronomy Laboratory
Boulder, CO 80303

Dr. Glenn C. Miller, Professor
Department of Environmental and Resource
  Sciences
University of Nevada,
Reno, NV 89557

Dr. Stephan R. Müller
Federal Institute for Environmental Science and
  Technology (EAWAG) and Swiss Federal Institute of Technology (ETH)
CH-8600 Dübendorf, Switzerland

Dr. Petra M. Radehaus
Visiting Professor of Technical Microbiology
Technical University of Chemnitz
Department of Mechanical Engineering & Processing Technology
Reichenhainer Straße 70
D-09126 Chemnitz, Germany

**Dr. James F. Ranville, Research
Assistant Professor**
Department of Chemistry and Geochemistry
Colorado School of Mines
Golden, CO 80401

Dr. Ronald L. Schmiermund
Knight Piesold LLC
1050 17th St., Suite 500
Denver, CO 80265

Dr. René P. Schwarzenbach, Professor
Federal Institute for Environmental Science and
   Technology (EAWAG) and Swiss Federal Insti-
   tute of Technology (ETH)
CH-8600 Dübendorf, Switzerland

Dr. Laura Sigg, Professor
Federal Institute for Environmental Science and
   Technology (EAWAG), and Swiss Federal Insti-
   tute of Technology (ETH)
CH–8600 Dübendorf, Switzerland

Dr. Jerald L. Schnoor, Professor
Environmental Engineering Program
The University of Iowa
Iowa City, IA, 52242

Dr. Anne K. Smith
National Center for Atmospheric Research
Boulder, CO 80307

Dr. Alan T. Stone, Professor
Department of Geography and Environmental
   Engineering
G.W.C. Whiting School of Engineering
The Johns Hopkins University
Baltimore, MD 21218

Dr. Werner Stumm, Professor
Institute for Environmental Science and Technology
   (EAWAG)

Swiss Federal Institute of Technology (ETH Zürich)
CH–8600 Dübendorf, Switzerland

Dr. Pieter P. Tans
Climate Monitoring and Diagnostics Laboratory
National Oceanic and Atmospheric Administration
Boulder, Colorado 80303

Dr. Margaret A. Tolbert, Professor
Department of Chemistry and Biochemistry and
Cooperative Institute for Research in Environmental
   Sciences
University of Colorado
Boulder, CO 80309

Dr. Paul G. Tratnyek, Assistant Professor
Department of Environmental Science and
   Engineering
Oregon Graduate Institute of Science and
   Technology
P. O. Box 91000
Portland, OR 97291-1000

Dr. Markus M. Ulrich
Federal Institute for Environmental Science and
   Technology (EAWAG), and Swiss Federal Insti-
   tute of Technology (ETH)
CH–8600 Dübendorf, Switzerland

Dr. David Updegraff, Professor Emeritus
Department of Chemistry & Geochemistry
Colorado School of Mines
Golden, CO 80401

Dr. Thomas Wildeman, Professor
Department of Chemistry & Geochemistry
Colorado School of Mines
Golden, CO 80401

Dr. James Charles Wilson, Professor
Department of Engineering
University of Denver
Denver, CO 80208-0177

PART I

ENVIRONMENTAL CHEMISTRY OF CONDENSED PHASES

CHAPTER 1

The Solid–Water Interface in Natural Systems

WERNER STUMM

ABSTRACT. Most chemical processes in natural waters occur at the surface of minerals and inorganic and organic particles. Many of the macroscopic processes or even global cycles are rate-controlled by surface reactions. The concept of active surface sites at the solid–water interface is essential in understanding surface-controlled processes. These surface functional groups interact chemically with H^+, OH^-, metal ions, and ligands. Surface complex formation equilibria permit the prediction of the extent of adsorption as a function of pH and solution variables. The resulting array of surface complexes establishes surface reactivities that include (1) dissolution of minerals and weathering of rocks, (2) precipitation, nucleation, crystal growth, and biomineralization, and (3) catalysis of chemical processes—most importantly, redox processes. A better understanding of the bonding between solids and aquatic adsorbates presents new challenges and will push the boundaries of aquatic surface chemistry and, in turn, of aquatic chemistry.

INTRODUCTION

Figure 1.1 is a simplified steady-state model of the earth's surface geochemical system likened to a chemical engineering plant. The various reservoirs of the earth (atmosphere, water, sediments, soils, biota) contain material that is characterized by high area-to-volume ratios. The interaction of rocks with water and the atmosphere in the presence of photosynthesized organic matter continuously produces reactive material of high surface area. There are trillions of square kilometers of surfaces of inorganic, organic, and biological material that cover our sediments and soils and that are dispersed in our waters (Hochella and White, 1990). Weathering imparts solutes to the water, and erosion brings particles into surface waters and oceans. A large flux of settling detrital and biogenic particles continuously runs through the water columns. The steady state conveyor belt of settling particles which are efficient sorbents of heavy metals, nutrients, and organic matter are critical agents in regulating their concentrations in the water column. Much of the hydrospheric chemistry and a significant part

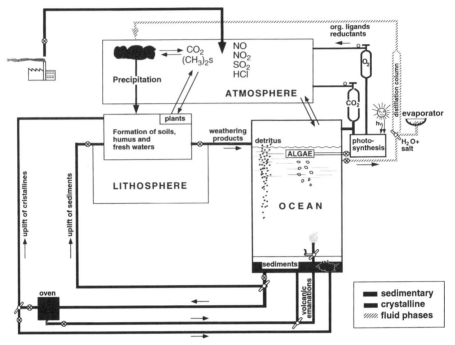

Figure 1.1. Circulation of rocks, water, and biota. Steady-state model for the earth's surface geochemical system likened to a chemical engineering plant. The interaction of water with rocks in the presence of photosynthesized organic matter continuously produces reactive material of high surface area in the surface environment. This process provides nutrient supply to the biosphere and, along with biota, forms the array of small particles (soils). Weathering imparts solutes to the water, and erosion brings particles into surface waters and oceans. A large flux of settling detrital and biogenic particles continuously runs through the water column. The steady-state conveyor belt of settling particles which are efficient sorbents of heavy metals and other trace elements regulates their concentrations in the water column. The sediments are the predominant sink of trace elements. (Inspired by Siever, 1968.)

of the atmospheric chemistry is regulated and maintained by a highly efficient interface chemistry.

Adsorption (accumulation of substances at the solid–water interface):

1. affects the distribution between the aqueous phase and particulate matter;

2. affects the electrostatic properties of the interfaces (surface charge, colloids); and

3. influences the reactivity of surfaces (most importantly, heterogeneous nucleation and precipitation), dissolution of minerals (corrosion of structures), and catalysis of redox and photoredox process, ester hydrolysis, and Lewis-acid (metal) catalysis.

Objectives

This chapter is written to briefly review and exemplify the role of chemical processes at the solid–water interface in regulating the composition of natural waters. Specifically, we will illustrate (1) how adsorption (and incorporation) of heavy metals at the surface of settling particles in oceans and lakes will affect the residence time, residual concentrations, and ultimate fate of the heavy metals, (2) how surface complex formation with protons, metal ions, and ligands at mineral surfaces changes their reactivity and affects their rates of dissolution (weathering), and (3) how surfaces can modify and catalyze redox processes.

THE FORMATION OF COORDINATE BONDS (SURFACE COMPLEX FORMATION) AT THE SOLID–WATER INTERFACE

Interactions at the solid–water interface can be characterized in terms of the chemical and physical properties of the solute, the sorbent, and the water. Of importance are the following processes:

1. *Surface complexation reactions* (the formation of coordinate bonds at the surface with metals and with ligands).

2. The *electric interactions* at surfaces, extending over longer distances than chemical forces.

3. *Hydrophobic expulsion* (hydrophobic substances) which involves nonpolar organic solutes that are usually only sparingly soluble in water and tend to reduce their contact with water. Such molecules seek relatively nonpolar environments and thus may accumulate at solid surfaces and may become adsorbed on organic sorbents.

4. *Adsorption of surfactants* (molecules that contain hydrophobic and hydrophilic moieties). Interfacial tension and adsorption are intimately related through the Gibbs adsorption law; its main message—expressed in a simple way—is that substances that tend to reduce surface tension tend to become adsorbed at interfaces.

5. The *adsorption of polymers* and of polyelectrolytes—most importantly, humic substances and proteins—is a rather general phenomenon in natural waters and soil systems that has far-reaching consequences for the interaction of particles with each other and on the attachment of colloids (and bacteria) to surfaces.

Here we will concentrate on adsorption caused by surface complex formation. Emphasis is on surface chemistry of the oxide–water interface because its coordination chemistry is much better understood than that of other surfaces. Furthermore, oxide surfaces (also including aluminum silicates) are especially important in natural systems.

Inner- and Outer-Sphere Surface Complexes

Surfaces (on minerals and organic particles) are treated as extending structures having active sites (surface functional groups such as $-OH-$, $COOH$, $-SH$, $-SS$, $-NH_2$). These groups interact with protons, hydroxide ions, metal ions and ligands, thereby imparting charge to the surface.

Hydrous oxides are covered with OH-functional groups; for example, an iron oxide contains at its surface $\equiv Fe-OH$ groups that can interact with protons, metals ions, or ligands:

$$\equiv FeOH + H^+ \quad \rightleftarrows \quad \equiv FeOH_2^+$$

$$\equiv FeOH \quad \rightleftarrows \quad \equiv FeO^- + H^+$$

$$\equiv FeOH + Cu^{2+} \quad \rightleftarrows \quad \equiv FeOCu^+ + H^+$$

$$\equiv FeOH + F^- \quad \rightleftarrows \quad \equiv FeF + OH^-$$

Most generally, as illustrated in Figure 1.2, a cation or an anion can associate with a surface as an inner-sphere or outer-sphere complex. An inner-sphere-type surface complex results when a chemical (i.e., largely covalent) bond is formed between the metal and the electron-donating surface oxygen ions (as in an inner-sphere-type solute complex). On the contrary, when an ion of opposite charge approaches the surface group within a critical distance, the adsorbed ion and the surface functional group are separated by one (or more) water molecules, resulting in an outer-sphere complex (as with solute ion pairs). Furthermore, ions may be in the diffuse swarm of the double layer.

In inner-sphere complexes the surface hydroxyl group acts as a σ-donor ligand which increases the electron density of the coordinated metal ion. Cu(II) bound inner-spherically is a different chemical entity than Cu(II) bound outer-spherically; the inner-sphere Cu(II) has different chemical properties—for example, a different redox potential (with regard to Cu(I))—and its equatorial H_2O molecules are expected to exchange faster than in a solute $Cu(H_2O)_n^{2+}$ complex. The reactivity of a surface is primarily affected by inner-sphere surface complexes (Sposito, 1984; Stumm, 1992).

Table 1.1 summarizes schematically the type of surface complex formation equilibria that characterize the adsorption of H^+, OH^-, cations, and ligands at a hydrous oxide surface. The various surface hydroxyl groups at an oxide surface may not be

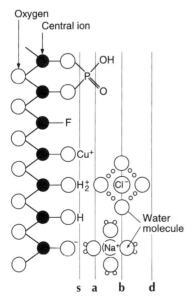

Figure 1.2. Schematic portrayal of the hydrous oxide surface, showing planes associated with surface hydroxyl groups (s), inner-sphere complexes (a), outer-sphere complexes (b), and the diffuse ion swarm (d). In case of an inner-sphere complex with a ligand (e.g., F^-, HPO_4^{2-}) the surface hydroxyl groups are replaced by the ligand (ligand exchange). (Modified from Sposito, 1984.)

TABLE 1.1
Adsorption (Surface Complex Formation Equilibria)[a]

Acid–base equilibria

$$S-OH + H^+ \rightleftharpoons S-OH_2^+$$

$$S-OH (+ OH^-) \rightleftharpoons S-O^- + (+ H_2O)$$

Metal binding

$$S-OH + M^{z+} \rightleftharpoons S-OM^{(z-1)+} + H^+$$

$$2S-OH + M^{z+} \rightleftharpoons (S-O)_2M^{(z-2)+} + 2 H^+$$

$$S-OH + M^{z+} + H_2O \rightleftharpoons S-OMOH^{(z-2)+} + 2 H^+$$

Ligand exchange (L^- = ligand)

$$S-OH + L^- \rightleftharpoons S-L + OH^-$$

$$2S-OH + L^- \rightleftharpoons S_2-L^+ + +2 OH^-$$

Ternary surface complex formation

$$S-OH + L^- + M^{z+} \rightleftharpoons S-L-M^{z+} + OH^-$$

$$S-OH + L^- + M^{z+} \rightleftharpoons S-OM-L^{(z-2)+} + H^+$$

[a]*From Schindler and Stumm (1987, modified).*

Surface hydroxyl groups may structurally not be equivalent. The following groups can be envisaged:

$$\begin{array}{c} S \\ \diagdown \\ \qquad \diagup OH, \quad S-OH, \quad S\diagleft \begin{array}{c} OH \\ OH \end{array}, \quad S\diagleft \begin{array}{c} OH \\ OH \\ OH \end{array} \\ S \diagup \end{array}$$

The surface complexation models used with the equations given above are not sensitive to the detailed structure of the interfacial region; that is, they are only qualitatively correct at the molecular level. Despite this, quantitative descriptions of adsorption isotherms, surface charge, and titration curves can be obtained by curve-fitting techniques. Hiemstra et al. (1989) have elaborated on a multi-site proton adsorption model, taking into account the various types of surface groups. Instead of using a two-step charge buildup ($S-OH_2^+$, $S-OH$, $S-O^-$), these authors modeled successfully the acid-base behavior of oxide surfaces using only one distinct pK value, formally attributed to the equilibrium (e.g., in case of oxides of Al or Fe(III)):

$$S-OH_2^{1/2+} \longleftrightarrow S-OH^{1/2-} + H^+$$

Metal binding and ligand binding (ligand exchange) can then be formulated:

$$S-OH^{1/2-} + M^{z+} \longleftrightarrow S-OM^{(z-3/2)+} + H^+$$

$$S-OH^{1/2-} + L^- \longleftrightarrow S-L^{1/2-} + OH^-$$

fully equivalent, but to facilitate the schematic representation of reactions and equilibria, one usually considers the chemical reaction of "a" surface hydroxyl group, $S-OH$.

These functional groups contain the same donor atoms as found in functional groups of soluble ligands; that is, the surface hydroxyl group on a hydrous oxide has donor properties similar to those of the corresponding counterparts in dissolved solutes, such as hydroxides or carboxylates—for example,

$$RCOOH + Cu^{2+} = RCOOCu^+ + H^+$$

$$S-OH + Cu^{2+} = S-OCu^+ + H^+$$

The adsorption of ligands (anions and weak acids) can also be compared with complex formation reactions in solution—for example,

$$Fe(OH)^{2+} + F^- = FeF^{2+} + OH^-$$

$$S-OH + F^- = S-F + OH^-$$

The central ion of a mineral surface acts as a Lewis acid and exchanges its structural OH against other ligands (ligand exchange).

Spectroscopic Probes of Surface Species

Noninvasive surface spectroscopic methods can be applied in the presence of liquid water. The best-known methods include electron spin resonance, Fourier transform infrared spectroscopy, and X-ray absorption spectroscopy (Brown, 1990). A review by Charlet and Manceau (1993) elaborates on the use of X-ray absorption spectroscopy to gain insight into the structure and reactivity of hydrous oxide particles. A recent discourse by Sposito (1995) illustrates the application of electron spin resonance and extended X-ray absorption fine structure (EXAFS) studies[1] to elucidate the coordination chemistry of adsorption reactions.

Figure 1.3a illustrates the spectroscopic (Fourier transform infrared spectroscopy, FTIR) assessment of the specific (inner-sphere) absorption of oxalate on the surface of TiO_2 (anatase) at different pH values (Hug and Sulzberger, 1994). Figure 1.3b shows schematically the structure of a surface complex of Cu(II) on a δ-Al_2O_3 surface (inferred from electron spin resonance measurements) (Motschi, 1987; see also Sposito, 1995). Figure 1.3c gives a proposed structure for selenite adsorbed on goethite. The structure is modeled on the basis of EXAFS spectra of selenite solutions in presence and absence of suspended goethite.

Mass Laws to Quantify Extent of Adsorption

The following criteria are characteristic for all surface complexation models (Dzombak and Morel, 1990):

1. Sorption takes place at specific surface coordination sites.
2. Sorption reactions can be described by mass law equations.
3. Surface charge results from the sorption reaction itself.
4. The effect of surface charge on sorption can be taken into account by applying a correction factor derived from the electric double layer theory to the mass law constants for surface reactions.

The extent of adsorption (surface coordination) can be predicted as a function of pH and solution variables (Figure 1.4). Dzombak and Hudson (1995) have recently shown how surface complexation models can be extended to permit physicochemical modeling of ion exchange on soils, sediments, and aquatic particles.

---

[1]The mechanism by which an EXAFS spectrum is produced involves the ejection of excess-energy photoelectrons from an atomic absorber of beyond-edge X-ray photons. These electrons are singly scattered from the atoms that are first and second nearest neighbors of the absorber and thereby generate backscattered waves that can interfere with the ejected photoelectron wave to modulate the X-ray absorption spectrum. This modulation feature constitutes the EXAFS portion of the spectrum. Evidently, it contains implicit information about the number and type of near-neighbor scattering atoms, as well as about their positions relative to the absorber atom. The EXAFS spectrum gives information on the time scale of 10^{-4} ps, concerning only the two or three closest shells of neighbors around an absorbing atom (≤ 0.6 nm) because of the small photoelectron mean free path in most materials.

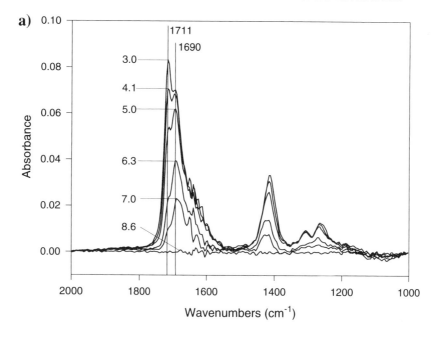

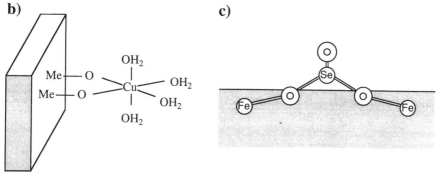

Figure 1.3. Inner-sphere surface complexes. (a) Fourier transform infrared (FTIR) spectra of oxalate adsorbed on TiO_2; oxalate solution concentration 10^{-4} M and pH values between 3.0 and 8.6. Surface complex formation starts below pH 8.6. First, a spectrum with a maximum at 1690 cm^{-1} is observed. With decreasing pH, the amplitude increases and an additional peak at 1711 cm^{-1} appears. These bands are assigned to C=O stretching vibrations. The changing spectral shape is an indication of the formation of at least two different surface complexes. (Adapted from Sulzberger and Hug, 1994.) (b) Surface complex of Cu(II) on δ-Al_2O_3 (structure inferred from electron spin resonance measurements). Adapted from Motschi, H., Aspects of the molecular structure in surface complexes; spectroscopic investigations, in W. Stumm (ed.), *Aquatic Surface Chemistry*. Copyright © 1987 Wiley-Interscience, New York. Reprinted by permission of John Wiley & Sons, Inc. (c) Proposed structure for SeO_3^{2-} coordinated with Fe atoms of goethite based on extended X-ray absorption fine structure (EXAFS) spectroscopy. The spectrum could be modeled by a structure consisting of one Se absorber, three backscattering ions at 0.170 nm from the absorber, and two iron backscattering ions at 0.338 nm from the absorber. Reprinted with permission from Hayes, K. F., et al., In situ x-ray absorption study of surface complexes: selenium oxyanions on α·FeOOH. *Science*, 138, 783–786. Copyright 1987 American Association for the Advancement of Science.

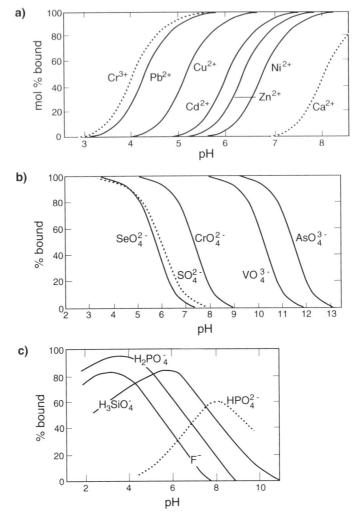

Figure 1.4.   (a) Extent of surface complex formation as a function of pH (measured as mol % of the metal ions in the system adsorbed or surface bound). [TOTFe] = 10^{-3} M (2×10^{-4} mol reactive sites L^{-1}); Metal concentrations in solution = 5×10^{-7} M; I = 0.1 M $NaNO_3$. (The curves are based on data compiled by Dzombak and Morel, 1990.) (b) Surface complex formation with ligands (anions) as a function of pH. Binding of anions from dilute solutions (5×10^{-7} M) to hydrous ferric oxide [TOTFe = 10^{-3} M]. I = 0.1. (Based on data from Dzombak and Morel, 1990.) (c) Binding of phosphate, silicate, and fluoride on goethite (α-FeOOH); the species shown are surface species (6 g FeOOH per liter, $P_T = 10^{-3}$ M, $Si_T = 8 \times 10^{-4}$ M) (Adapted from Sigg and Stumm, 1981). (The curves are calculated with the help of experimentally determined equilibrium constants.)

REGULATION OF METAL IONS BY PARTICLES IN LAKES, RIVERS, AND OCEANS

The concentration of most trace elements (M or mol kg^{-1}) are larger in solid or surface phases than in the water phase. Thus, the capacity of particles to bind trace elements, in competition with complex formation in solution, influences the residual concentration and the speciation of the trace metals in the water. As we have seen, there is one common property of all natural particles; their surfaces contain functional groups which can interact with H^+, OH^-, and metal ions. Alkaline and alkaline-earth ions interact with clays mostly through ion exchange processes. But the adsorption of heavy metals is mostly dominated by surface complex formation with the single coordinated OH groups on the edge surfaces (aluminol and sinalol groups). Biological surfaces play an important role in binding metal ions and protons; they contain $-COOH$, $-NH_2$, and $-OH$ groups. As has been shown by Hudson and Morel (1990), Sigg (1994), and Müller and Sigg (1990), naturally occurring particles, even bacteria and algae, have surfaces that interact with metal ions in a manner similar to oxides.

In Figure 1.1 we envisaged a scheme of some of the chemical, physical, and biological processes which regulate the concentration of trace elements in the water column of lakes and oceans. In the sea as well as in lakes, the ions of metals, metalloids, and other reactive elements interact competitively with soluble ligands (hydroxides, carbonates, and organic solutes such as chelate complex formers including those formed by aquatic humus and those exuded by phytoplankton). Particle surfaces can tie up significant proportions of trace metals even in the presence of solute complex formers. The effects of particles on the regulation of metal ions are enhanced because the continuously settling particles (phytoplankton formed by photosynthesis and particles introduced by rivers) act like a conveyor belt in transporting reactive elements into the sediment.

Indeed, the partitioning of metals and other reactive elements between particles and water is the key parameter in establishing the residence time and, thus in turn, the residual concentrations of these elements in the ocean and in lakes. The residence time of an element, τ_M, is given by the total quantity of this element in the reservoir (Mol) divided by the total input or output (Mol time^{-1})

$$\tau_M = \frac{\text{total no. of moles of M in reservoir}}{\text{rate of addition or removal}} \quad \left[\frac{\text{Mol}}{\text{Mol year}^{-1}} \right]$$

The more reactive an element is, the more will it be bound to particles and the more rapidly will it be removed and the shorter will be its residence time:

$$\frac{C_{\text{part.}}}{C_{\text{total}}} = \frac{\tau_P}{\tau_M}$$

where $C_{\text{part.}}$ and C_{total} are the particulate and total concentrations of an element, τ_P is the residence time of the particles, and τ_M is the residence time of this element with respect to removal via scavenging.

Phytoplankton

The surfaces of biogenic organic particles (algae), as well as organic biomass derived from them, contain functional groups or ligands that are often more efficient in binding bioreactive elements than inorganic surfaces and are thus believed to represent in surface waters the most important scavenging phase and carrier from surface to deep water. In addition to *adsorption* processes, phytoplankton can *absorb* (assimilate) certain nutrient metal ions (or metal ions that are mistaken as nutrients by the organisms). As with other nutrients, this uptake can occur in stoichiometric proportions.

Large spatial and temporal differences in both trace metal concentrations and chemical speciation in the sea have led to wide variations in biological availability of metals and their effects on phytoplankton. The scenario in oceans is similar to that in lakes: The bulk of the trace elements, adsorbed to the biological surfaces (or absorbed into the cells) is carried into the deep sea by settling. As the particles fall to the water column, they become mineralized (or become a food source for successive populations of filter feeders).

For a general review on trace metal interactions with phytoplankton, see Sunda (1991), Bruland et al. (1991), Morel et al. (1991), and Sigg et al. (1994). Whitfield and Turner (1987) have shown that the elements in the ocean can be classified according to their oceanic residence time into accumulated elements (Figure 1.5b), recycled elements (Figure 1.5c), and scavenged elements (Figure 1.5d). The elements that are most sorbable are characterized by short residence times, and those elements that interact little with particles are characterized by long residence times. Bruland et al. (1994) have analyzed the chemical and biological behavior together with global circulation pattern on the distribution of nutrient-type and scavenged-type metals between North Atlantic and North Pacific oceans.

SURFACE STRUCTURE AND SURFACE REACTIVITY

Adsorption (especially inner-sphere coordination) affects surface reactivity by changing both the solid surface and the adsorbate, its coordinating shell, and its electronic structure. This alters the forces between the atoms and changes the energetic requirements for the reaction—that is, reduces its activation energy. To understand the process we need atomic scale theories (electronic structure) and atomic scale images.

Surface-Controlled Dissolution of Minerals

When a mineral dissolves, several successive elementary steps may be involved: (1) mass transport of dissolved reactants from bulk solution to the mineral surfaces, (2) adsorption of solutes, (3) interlattice transfer of reacting species, (4) chemical reactions, and (5) detachment of reactants from the surface, followed by mass transport into the bulk of the solution. Under natural conditions the rates of dissolution of most minerals is controlled by a surface process and the related detachment process. Therefore the surface structure—that is, the speciation of the functional groups at the surface—will influence the dissolution kinetics. Thus, the reaction occurs schematically in two sequences:

$$\text{surface sites} + \text{reactants (H}^+, \text{OH}^-, \text{or ligands)} \xrightarrow{\text{fast}} \text{surface species}$$

$$\text{surface species} \xrightarrow{\text{slow}} \text{Me(aq)}$$

where the second sequence reflects the detachment of the metal ion (Me) from the lattice to the aqueous solution. Although each sequence may consist of a series of smaller reaction steps, the rate law of surface-controlled dissolution is based on the idea that (1) the attachment of reactants to the surface sites is fast and (2) the subsequent de-

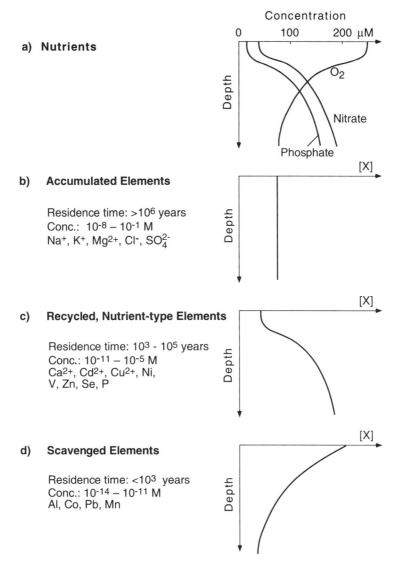

Figure 1.5. Schematic depth ocean profiles for elements. This figure is based on a classification of elements according to their oceanic profiles given by Whitfield and Turner (1987). The mean oceanic residence time of an element can be used to divide the elements into three categories that exhibit characteristically different deep sea profiles resulting from progressively stronger particle–water interactions. Uptake of some of the elements, especially the recycled ones, occurs somewhat analogously to that of nutrients. There are some elements such as Cd that are nonessential but may be taken up (perhaps because they mimic essential elements) the same way as nutrients. The concentration ranges given show significant overlap, since the concentrations of the elements also depend on crustal abundance. Note that the [X] concentration scale is different for each category of elements. (Modified from Whitfield and Turner, 1987.)

tachment of the metal species is slow and thus rate-limiting. Therefore, the rate law becomes very simple; the dissolution rate, R (mol m^{-2} time^{-1}), is proportional to the surface concentration of the surface species (mol m^{-2}):

$$R = k\{\text{surface species}\}$$

We reach the same conclusion if we treat the reaction sequence according to the activated complex theory (or transition state theory). The particular surface species that has formed is the precursor of the activated complex. For a review of the literature see Helgeson et al. (1984), Casey et al. (1988), Brady and Walther (1989), Stumm and Wollast (1990), and Blesa et al. (1994).

Figure 1.6 illustrates various dissolution pathways for an oxide such as iron(III)(hydr)oxide, proton-promoted dissolution, ligand-promoted dissolution, and reductive dissolution. In the *proton-promoted* dissolution (Figure 1.6a), three fast protonation steps are followed by a slow detachment of the Fe(III). The attack of H^+ (or H_3O^+) onto the bridging oxygen site significantly weakens the Fe−O−Fe (or more generally the Me−O−Me) bridging bonds. There is a direct link between the dissolution rate and the amount of adsorbed H^+ species. The latter can be determined by surface titration experiments or calculated from known surface speciation mass law constants (see acid–base equilibria in Table 1.1). A weakening of the critical metal oxygen bonds occurs as consequence of the protonation of the oxygen ions neighboring a surface metal center and imparting charge to the to the mineral lattice. The rate of dissolution, R_H [mol m^2 h^{-1}], is proportional to the third power of the surface protonation

$$R_H = k_H(C_H^s)^3$$

where C_H^s is the surface protonation in mol m^{-2} (Figure 1.7). The surface protonation is accessible experimentally from an alkalimetric titration of the oxide dispersion.

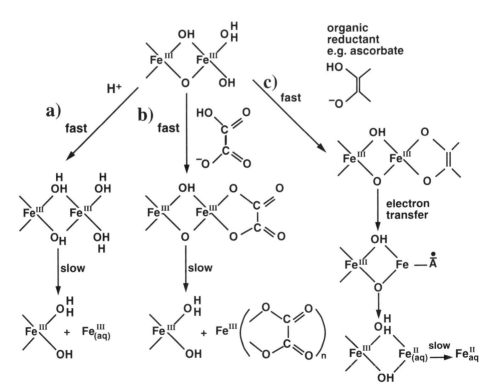

Figure 1.6.   Various pathways for the dissolution of Fe(III)(hydr)oxides: with protons (acids), with ligands, and with reductants. The scheme is also valid for other oxides.

Figure 1.7. The rate of dissolution of δ-Al_2O_3 is third order in the surface density of H^+. The actual slope of the straight line is 3.1. Reprinted from *Geochim. Cosmochim. Acta*, 50, Furrer, G. and Stumm, W., The coordination chemistry of weathering: I. dissolution kinetics of δ·Al_3O_3 and BeO, pp. 1847–1860. Copyright 1986. With kind permission from Elsevier Science Ltd, The Boulevard, Langford Lane, Kidlington OX5 1GB, UK.

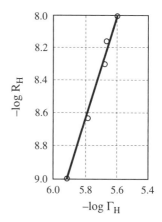

In the *ligand-promoted* dissolution (Figure 1.6b) the formation of a surface complex by a suitable ligand, such as oxalate, citrate, diphenols, or hydroxycarboxylic acids enhances the dissolution of oxides and silicates. These ligands occur in natural waters as a byproduct of biodegradation of organic matter or as exudation products of plants. These ligands bring electron density into the coordination sphere of the metal oxygen bonds, thus facilitating the detachment of the metal ion from the surface. Ligands with two or more donor atoms (bi- or multidentate ligands) can form surface chelates (ring-type complexes). Figure 1.8 illustrates the effects of various ligands on the dissolution rate of δ-Al_2O_3. The dissolution rate, R_L, is proportional to the surface concentration of ligands, C_L^s [mol m^{-2}]:

$$R_L = k_L C_L^s$$

In the proton-promoted dissolution of many aluminum silicates, the critical point of attack by the proton is the oxygen atom that bridges an Si and an Al atom (Hellmann et al., 1990, Wieland and Stumm, 1992). The dissolution reaction can be interpreted as a

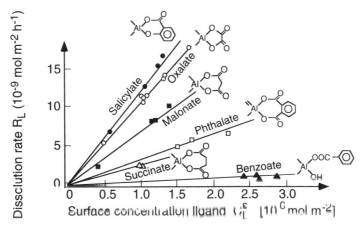

Figure 1.8. In accordance with the reaction scheme given in the text, the rate of ligand-promoted dissolution of δ-Al_2O_3 is in linear dependence on the surface concentration of the surface complexes C_L^s. (Adapted from Furrer and Stumm, 1986.)

coupled release of Al and Si (Figure 1.9). The dissolution of kaolinite and of some other aluminum silicates is also enhanced by oxalate and salicylate through ligand association. Although these ligands form surface complexes only with Al centers, they promote the releases of both Al and Si centers during the dissolution process. This indicates that the detachment of the Al from the lattice structure is the rate-determining step.

Ab initio quantum mechanical studies

Xiao and Lasaga (1994) provide detailed molecular mechanisms which can be used to explain the pH dependence of mineral dissolution rates of aluminum silicates. The *ab initio* results indicate that the attack of H^+ on the bridging oxygen site significantly weakens the $Si-O-Al$ bridging bonds.

The dissolution rate of most oxides at high pH increases with increasing deprotonation, equivalent to the binding of OH ligands:

$$R_{OH} = k'_{OH}(C^s_{OH})^i$$

The overall rate of dissolution is given by

$$R' = k'_H(C^s_H)^j + k'_{OH}(C^s_{OH})^i + k'_L(C^s_L) + k'_{H2O}$$

the sum of the individual reaction rates, assuming that the dissolution occurs in parallel at different metal centers (Furrer and Stumm, 1986). The last term in the equation is due to the effect of hydration and reflects the pH-independent portion of the dissolution rate.

This equation for the dissolution rate is valid for conditions far away from equilibrium (i.e., the back reaction can be neglected). More generally, the validity range can be extended to near solubility equilibrium by correcting the rate expression to

$$R = R'[1 - \exp(\Delta G/RT)]$$

where ΔG is the free energy of the dissolution reaction. For ΔG close to zero, an important limitation to the rate is given by the square brackets; if $\Delta G/RT$ is smaller than about -2, the correction becomes negligible.

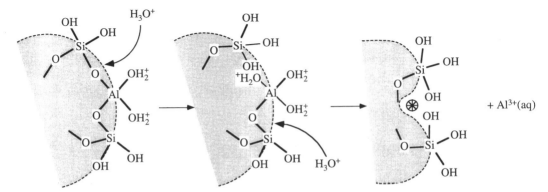

Figure 1.9. Pictorial view of the attack on the more labile Al(III) ions in an aluminosilicate. Water ingress following Al(III) detachment favors further attack deeper in the solid framework. (Adapted from Hellmann et al., 1990.)

CATALYSIS ON SURFACES

The recent advances in modern surface chemistry has been reviewed by Friend (1993). She illustrates how inorganic and metal-organic chemists have been able to produce rapid advances by understanding at the molecular level how solids interact with individual molecules to speed reactions and how such catalytic reactions have been used to improve everything from material synthesis to pollution control (Figure 1.10a). Much of the progress in the study of catalysis has been made with surfaces that are carefully shielded from open air and water. The surfaces of interest to the environmental chemists, on the other hand, are in contact with water. But even in presence of water, there is bonding between solid surfaces and solutes, and chemistry and physics come together at these surfaces to influence chemical reactions.

In Figure 1.10 the catalysis of processes important in the desulfurization of fossil fuel on molybdenum surfaces (absence of open air and water) is compared with the catalysis by a hydrous catalysis. The latter is illustrated by the catalysis of a hydrous oxide surface on the oxidation of Fe(II) by oxygen. A simple qualitative explanation for this catalytic effect can be given as follows: The surface OH groups act as sigma electron donors which increase the electron density of the adsorbed Fe^{2+}, making the surface-bound Fe(II) a much better reductant than Fe^{2+} in solution (Luther, 1990;

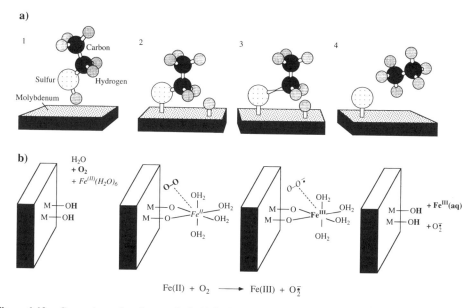

Figure 1.10. Comparison of surface catalysis (a) in the absence of water and (b) at the solid-water interface. (a) The molybdenum surface catalyzes a reaction that removes sulfur from fuels; when a molecule of ethanediol interacts with molybdenum, it loses a hydrogen and binds to the surface, the bond between carbon and sulfur then weakens, and the remaining atoms may recombine to form ethane. (Adapted from Friend, 1993). (b) A hydrous metal oxide surface (TiO_2, Fe_2O_3, Al_2O_3) catalyzes the oxidation of Fe(II) by oxygen. When an Fe(II) binds inner-spherically to the hydrous oxide surface, the functional oxygen donor atoms of the surface hydroxyl groups donate electron density to the Fe(II) and make the Fe(II) a better reductant and enable an electron transfer to the outer spherically bound oxygen. The products of the reaction are superoxide, O_2^- and Fe(II)(aq). In both cases, the interaction of molecules or ions with the surface leads to a slight alteration of the electronic structure of the adsorbate; a reaction is promoted whose activation energy is reduced by the surface.

Wehrli, 1990; Stumm and Morgan, 1996). Adsorption to surface hydroxo groups can be compared with hydrolysis (binding with OH^- in solution) that also catalyzes the oxygenation of Fe(II). The following generalization is possible: The oxidation of transition metal ions (Fe^{2+}, Mn^{2+}, Cu^+, and VO^{2+}) by oxygen is catalyzed by specific adsorption to hydrous oxide surfaces and by hydrolysis (Figures 1.11 and 1.12).

Other redox processes are often mediated by suitable surfaces. The binding of an oxidant or reductant species to a surface often changes the mode and rate of electron transfer processes. Adsorption of a reductant or oxidant, especially in an inner-sphere surface complex, changes the geometry of the reactants coordinating shell and electronic structure and thus influences redox intensity (free energy of redox reaction) and reactivity (rate of electron transfer). The oxidant and reductant have to encounter each other in a suitable structural arrangement to make an electron exchange possible. The surface may assist in various ways to facilitate such an encounter. Furthermore, adsorption may accelerate ligand exchange kinetics. Thus, M(III) cations will exchange their ligands faster once they are adsorbed to an oxide surface.

The hydrous ferric oxide surface often mediates redox reactions of organic substances. A possible example is given by the scheme below.

$$\text{Oxidized S} \quad \overset{-ne^-}{\searrow}\Big(\overset{+ne^-}{\quad} \quad \text{"Fe(III)"} \quad \overset{-ne^-}{\searrow}\Big(\overset{+ne^-}{\quad} \quad \text{Oxidized org.}$$

$$HS^- \qquad\qquad\qquad \text{"Fe(II)"} \qquad\qquad\qquad \text{Reduced org.}$$

Electron transfer mediator

Different electron-mediating systems such as porphyrin complexes, $MnO_2/Mn(II)$ and quinone-like systems are operative in natural systems. Furthermore, the organic carbon oxidation can be coupled by bacteria to the dissimilatory reduction of iron(III)oxides (Lovely and Philips,1988, Nealson and Myers, 1990).

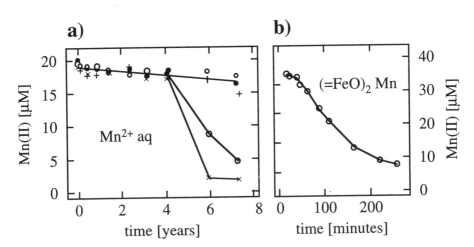

Figure 1.11. Oxygenation of Mn(II). (a) Homogeneous solution at pH 8.4. (Data from Diem and Stumm, 1984.) The curves which show a decrease in [Mn^{2+}] after 4 years are for experiments where small quantities ($5 - 6 \times 10^{-7}$ M) of Mn(III,IV) oxide were added as a catalyst. (b) 10 mM goethite suspension at pH 8.5. (Data from Davies and Morgan, 1988.) The surface complex reacts within hours, whereas the homogeneous Mn(II) solutions are stable for years.

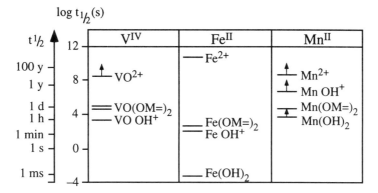

Figure 1.12. Effects of hydrolysis and adsorption on the oxygenation of transition-metal ions. Arrows indicate lower limit. Data represent orders of magnitude. Reprinted from *Geochim. Cosmochim. Acta*, 53, Wehrli, B. and Stumm, W., Vanadyl in natural waters: adsorption and hydrolysis promote oxygenation, pp. 69–77. Copyright 1989. With kind permission from Elsevier Science Ltd, The Boulevard, Langford Lane, Kidlington OX5 1GB, UK.

The solubilization of iron(III)(hydr)oxides is of great importance in the iron cycle, as well as in the general electron cycling which occurs in natural waters, sediments, soils, and atmospheric waters. Electron transfer reactions mediated by Fe(III)(hydr)oxide/Fe(II) or other heterogeneous redox couples often substitute for, or supplement, microbially mediated redox cycling, especially at high redox gradients occurring at oxic–anoxic boundaries (Macalady et al., 1986; Schwarzenbach et al., 1993). The heterogeneous and homogeneous Fe(III)/Fe(II) couple also plays a dominating role in the electron cycling of atmospheric water (Behra and Sigg, 1990; Hoigné et al., 1994).

The Reductive Dissolution of Fe(III)(Hydr)oxides

There is a special type of ligand-promoted dissolution known as *reductive dissolution*. This case involves the transfer of electrons from the environment, through a reducing ligand or reductant, to an oxide surface, followed by its dissolution.

The kinetic scenario of a reductive dissolution of an iron(III)oxide is given in Figure 1.6c. The three basic steps are: adsorption of the reductant and formation of a surface complex (this reaction is relatively fast), electron transfer within the complex; and the release of metal centers into solution. The constant dissolution rate over the observed time interval allows steady-state approximation for the system. The dissolution rate R_R (mol m^{-2} t^{-1}) is proportional to the surface ligand concentration (mol m^{-2}):

$$R_R = k_R \{Fe^{III}-R\}$$

This model was applied to the reduction of hematite and other iron oxides by ascorbate (Suter, et al., 1988) and by H$_2$S (see Figure 1.13) (Dos Santos and Stumm, 1992; Biber, et al., 1994). For general reviews see Stone and Morgan (1987), Sulzberger, et al. (1989), Hering and Stumm (1990), Stumm (1992), and Blesa, et al. (1994).

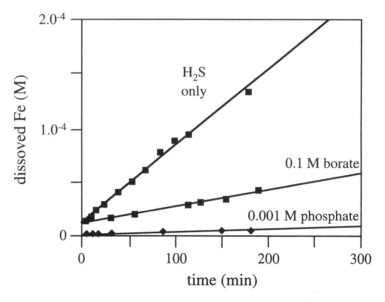

Figure 1.13. Reductive dissolution of goethite by H_2S (constant $p_{H2S} = 10^{-3}$ ATM). The dissolution is strongly inhibited by 0.1 M borate or by 10^{-3} M phosphate (all data at pH = 5). Reprinted from *Geochim. Cosmochim. Acta*, 58(9), Biber, M.V., et al., The coordination chemistry of weathering: IV. Inhibition of the dissolution of oxide minerals, pp. 1999–2010. Copyright 1994. With kind permission from Elsevier Science Ltd, The Boulevard, Langford Lane, Kidlington OX5 1GB, UK.

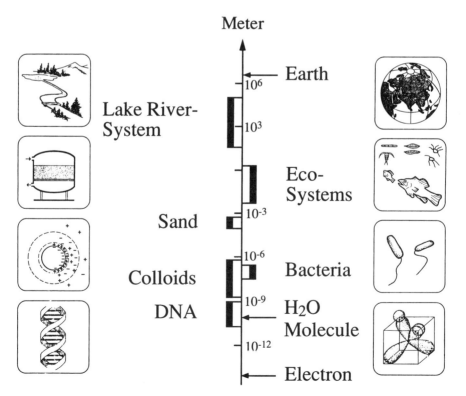

Figure 1.14. The "size spectrum" of the environmental engineer and the environmental chemist.

THE TASK OF THE ENVIRONMENTAL ENGINEER AND ENVIRONMENTAL CHEMIST

Figure 1.14 shows the "size spectrum" of the various tasks in which environmental engineers and scientists are involved. Such a person is concerned with electron transfer and with reactions of atoms, ions, and macromolecules, deals with viruses and other microorganisms, and is involved in problems of water technology involving colloids, porous media, and filters. In addition, the engineer and/or scientist deals with soils and the subsurface environment, studies the ecology of entire river, lake, and ocean systems, and assess pollution from a local point of view but also includes considerations of regional and global cycles of atmosphere, water, and rock interactions. One common denominator in all these systems is the surface and the processes which are influenced or controlled by it. Thus, many macroscopic and even global processes are controlled at the nanoscale.

References

Behra, P., and Sigg, L., 1990, Evidence for redox cycling of iron in atmospheric water. *Nature*, 344, 419–421.

Blesa, M., Morando, B. J., and Regazzoni, A. E., 1994, *Dissolution of Metal Oxides*, CRC Press, Boca Raton FL.

Biber, M. V., Dos Santos-Afonso, M., and Stumm, W., 1994, The coordination chemistry of weathering. IV. Inhibition of the dissolution of oxide minerals. *Geochim. Cosmochim. Acta*, 58(9), 1999–2010.

Brady, P. V., and Walther, J. V., 1989, Controls on Silicate Dissolution Rates in Neutral and Basic pH Solutions at 25°C. *Geochim. Cosmochim. Acta*, 53, 2823–2830.

Brown, G. E., Jr., 1990, Spectroscopic studies of chemisorption reaction mechanisms at oxide water interface. In: M. F. Hochella and A. F. White, eds., *Mineral–Water Interface Geochemistry*, *Reviews in Mineralogy*, Vol. 23, Mineralogical Society of America, Washington, D.C.

Bruland, K. W., Donat, J. R., and Hutchins, D. A., 1991, Interactive influences of bioactive trace metals on biological production in oceanic waters. *Limnol. Oceanogr.*, 36, 1555–1577.

Bruland, K. W., Orians, K. J., and Cower, J.P., 1994, Reactive trace metals in the stratified central North Pacific. *Geochim. Cosmochim. Acta*, 58, 3171–3182.

Casey, W. H., Westrich, H. R. and Arnold, G. W., 1989, The Surface of Labradorite Feldspar Reacted with Aqueous Solutions at pH = 2, 3, and 12. *Geochim. Cosmochim. Acta*, 52, 2795–2807.

Davies, S. H. R., and Morgan, J. J., 1988, Manganese(II) oxidation kinetics on oxide surfaces. *J. Colloid Interface Sci.*, 129(1), 63–77.

Diem, D., and Stumm, W., 1984, Is dissolved Mn^{2+} being oxidized by O_2 in the absence of Mn-bacteria by surface catalysts? *Geochim. Cosmochim. Acta*, 48, 1571–1573.

Dos Santos-Afonso, M., and Stumm, W., 1992, The reductive dissolution of iron(III)(hydr)oxides by hydrogen sulfide. *Langmuir*, 8, 1671.

Dzombak, D., and Morel, F. M. M., 1990, *Surface Complexation Modeling: Hydrous Ferric Oxide*, Wiley-Interscience, New York.

Friend, C. M., 1993, Catalysis on surfaces. *Sci. Am.*, 268(4), 42–47.

Furrer, G., and Stumm, W., 1986, The coordination chemistry of weathering. I. Dissolution kinetics of $\delta \cdot Al_2O_3$ and BeO. *Geochim. Cosmochim. Acta*, 50, 1847–1860.

Hayes, K. F., Roe, A. L., Brown, E., Jr., Hodgson, K. O., Leckie, J. O., and Parks, G. A., 1987, *In situ* x-ray absorption study of surface complexes: selenium oxyanions on $\alpha \cdot FeOOH$. *Science*, 138, 783–786.

Helgeson, H. C., Murphy, W. M. and Aagard, P., 1984, Thermodynamic and Kinetic Constraints on Reaction Rates among Minerals and Aqueous Solutions. *Geochim. Cosmochim. Acta*, 48, 2405–2432.

Hellmann, R., Eggleston, C. M., Hochella, M. F., and Crerar, D. A., 1990, The formation of leached layers on albite surfaces during dissolution under hydrothermal conditions. *Geochim. Cosmochim. Acta*, 54, 1267–1287.

Hering, J., and Stumm, W., 1990, Oxidative and reductive dissolution of minerals. In: M. F. Hochella and A. F. White, eds., *Mineral–Water Interface Geochemistry*, *Reviews in Mineralogy*, Vol. 23, Mineralogical Society of America, Washington, D.C.

Hiemstra, T., de Wit, J. C. M., and van Riemsdijk, W. O., 1989, Multisite proton adsorption modeling at the solid/solution interface of (hydr)oxides, II. Application to various important (hydr)oxides. *J. Colloid Interface Sci.*, 133, 105–117.

Hochella, M. F., and White, A. F., 1990, Interface geochemistry: an overview. In: M. F. Hochella and A. F. White, eds., *Mineral–Water Interface Geochemistry*, *Reviews in Mineralogy*, Vol. 23, Mineralogical Society of America, Washington, D.C.

Hoigné, J., Zuo, Y., and Nowell, L., 1994, Photochemical reactions in atmospheric waters: role of dissolved iron species. In: G. Helz, R. R. Zepp, and D. G. Crosby, eds., *Aquatic and Surface Photochemistry*, Lewis, Boca Raton, FL.

Hudson, R. J. M., and Morel F. M. M., 1990, Iron transport in marine phytoplankton: kinetics of cellular and medium coordination reactions. *Limnol. Oceanogr*, 35, 1002–1020.

Hug, S. J., and Sulzberger, B., 1994, *In situ* Fourier transform infrared spectroscopic evidence for the formation of different surface complexes of oxalate on TiO_2 in the aqueous phase. *Langmuir*, 10, 3587–3597.

Lovley, D. R. and Phillips, E. J. P., 1988, Novel Mode of Microbial Energy Metabolism: Organic Carbon Oxidation Coupled to Dissimilatory Reduction of Iron or Manganese. *Appl. Environ. Microbiol.*, 54(6), 1472–1480.

Luther, G. W., 1990, The frontier-molecular-orbital theory approach in geochemical processes. In: W. Stumm, ed., *Aquatic Chemical Kinetics*, Wiley-Interscience, New York.

Macalady, D. L., Tratnyek, P. G., and Grundl, T. J., 1986, Abiotic reduction reactions of anthropogenic organic chemicals in anaerobic systems: a critical review. *J. Contam. Hydrol.*, 1, 1–28.

Morel, F. M. M., Hudson, R. J. M. and Price, N. L., 1991, Limitations of productivity by trace metals in the sea. *Limnol. Oceanogr.*, 36, 1742–1755.

Motschi, H., 1987, Aspects of the molecular structure in surface complexes; spectroscopic investigations. In: W. Stumm, ed., *Aquatic Surface Chemistry*, Wiley-Interscience, New York.

Müller, B., and Sigg, L.,1990, Interaction of trace metals with natural particle surfaces: comparison between adsorption experiments and field measurements. *Aquatic Sci.*, 52(1), 75–92.

Nealson, K. H. and Myers, C. R., 1990, Iron Reduction by Bacteria: A Potential Role in the Genesis of Brandes Iron Formation. *Am. J. Sci.*, 290, 35–45.

Schindler, P., and Stumm, W., 1987, The surface chemistry of oxides, hydroxides, and oxide minerals. In: W. Stumm, ed., *Aquatic Surface Chemistry*, Wiley-Interscience, New York, pp. 83–100.

Schwarzenbach, R. P., Imboden, D., and Gschwend, P. M., 1993, *Environmental Organic Chemistry*, Wiley-Interscience, New York.

Siever, R., 1968, Sedimentological consequences of a steady-state ocean-atmosphere. *Sedimentology*, 11, 5.

Sigg, L., 1994, Regulation of trace elements in lakes. In: J. Buffle and R. deVitre, eds., *Chemical and Biological Regulation of Aquatic Processes*, Lewis, Boca Raton, FL, pp. 177–197.

Sigg, L., Kuhn, A., Xue, H., Kiefer, E., and Kistler, D., 1994, Cycles of trace elements (copper and zinc) in a eutrophic lake: role of speciation and sedimentation. In: C. P. Huang, C. R. O'Melia, and J. J. Morgan, eds., *Aquatic Chemistry*, Advances in Chemistry Series, No. 244, American Chemical Society, Washington, D.C., pp. 177–194.

Sigg L., and Stumm, W., 1981, The interaction of anions and weak acids with the hydrous surface of goethite (α-FeOOH). *Colloids Surf.*, 2, 101–117.

Sposito, G., 1984, *The Surface Chemistry of Soils*, Oxford University Press, New York.

Sposito, G., 1995, Adsorption as a problem in coordination chemistry: the concept of the surface complex. In: C. P. Huang, C. R. O'Melia, and J. J. Morgan, eds., *Aquatic Chemistry*, ACS Advances in Chemistry Series, No. 244, American Chemical Society, Washington, D.C.

Stone, A. T., and Morgan, J. J., 1987, Reductive dissolution of metal oxides. In: W. Stumm, ed., *Aquatic Surface Chemistry*, Wiley-Interscience, New York.

Stumm, W., 1992, *Chemistry of the Solid–Water Interface*, Wiley-Interscience, New York.

Stumm, W., and Morgan, J. J., 1996, *Aquatic Chemistry*, 3rd ed., Wiley-Interscience, New York.

Stumm, W., and Wollast, R., 1990, Coordination chemistry of weathering: kinetics of the surface-controlled dissolution of oxide minerals. *Rev. of Geophys.*, 28(1), 53–69.

Sulzberger, B., and Hug, S. J., 1994, Light-induced processes in the aquatic environment. In: G. Bidoglio and W. Stumm, eds., *Chemistry of Aquatic Systems: Local and Global Perspectives*, Kluwer, Dordrecht.

Sulzberger, B., Suter, D., Siffert, C. Banwart, S., and Stumm, W., 1989, Dissolution of Fe(III)(hydr)oxides in natural waters: laboratory assessment on the kinetics controlled by surface coordination. *Mar. Chem.*, 28, 127–144.

Sunda, W., 1991, Trace metal interactions with marine phytoplankton, *Biol. Oceanogr.*, 6, 411–442.

Suter, D., Siffert, C., Sulzberger, B., and Stumm, W., 1988, Catalytic dissolution of iron(III)(hydr)oxides by oxalic acid in the presence of Fe(II). *Naturwissenschaften*, 75, 571–573.

Wehrli, B., 1990, Redox reactions of metal ions at mineral surfaces. In: W. Stumm, ed., *Aquatic Chemical Kinetics*, Wiley-Interscience, New York.

Wehrli, B., and Stumm, W., 1989, Vanadyl in natural waters: adsorption and hydrolysis promote oxygenation. *Geochim. Cosmochim. Acta*, 53, 69–77.

Whitfield M , and Turner, D. R., 1987, The role of particles in regulating the composition of natural waters. In: W. Stumm, ed., *Aquatic Surface Chemistry*, Wiley-Interscience, New York. pp. 457–494.

Wieland, E., and Stumm, W., 1992, Dissolution kinetics of kaolinite in acid aqueous solutions at 25°C. *Geochim. Cosmochim. Acta*, 56, 3339–3355.

Xiao, Y., and Lasaga, A. C., 1994, *Ab initio* quantum mechanical studies of the kinetics and mechanisms of silicate dissolution: H^+ catalysis. *Geochim. Cosmochim. Acta*, 58, 5379–5400.

CHAPTER 2

An Overview of Environmental Colloids

JAMES F. RANVILLE AND RONALD L. SCHMIERMUND

ABSTRACT. In most natural waters there exist mobile, nonaqueous phases, called *colloids*, which are "permanently" dispersed. In most cases, colloids consist of solid particles less than a few micrometers in diameter. In this chapter, environmental colloids are described in terms of the features which make them environmentally important, including particle size, composition, origin, surface charge, specific surface area, and stability as a suspension. Specific examples of environmental colloids and their influence on contaminant behavior are presented, along with problems associated with sampling and analyses of environmental colloids.

INTRODUCTION

The occurrence and distribution of natural and anthropogenic chemicals in natural waters is often strongly influenced by their partitioning between solid and aqueous phases (Allan, 1986; Chapter 6, this volume). It is common therefore to think in terms of "two-phase" solution–solid interactions where contaminants are present either in the stationary solid phases or as mobile, aqueous-phase solutes. For ground water the solid phase is considered to be the stationary vadose zone or aquifer material. In surface water the solid phases generally considered are the bottom sediments.

The reality is, however, that most water systems are three-phase systems, the third phase being a mobile solid phase in the form of suspended matter. Turbidity is the macroscopic, visual evidence of suspended matter in a water system. A certain portion of suspended matter, the largest particles first, will settle out over time and therefore be removed from the mobile phase. However, by some estimates (Moran and Moore, 1989), as much as one-half of the suspended matter in natural waters is less than 1 μm in diameter and does not settle out, even over long periods of time. By definition these particles are "colloids" and represent a "permanent" phase which, of all the solid phases, can potentially exert the most significant control over the fate and transport of waterborne chemicals. This importance arises largely from the small size of colloidal particles, which gives them a very high surface area per unit mass. This facilitates all surface-area-dependent interactions between solution and solid components. Extensive laboratory studies of contaminant sorption to colloidal phases (i.e., metal oxides, clays, etc.) has resulted in a reliable but evolving fundamental physicochemical understanding of these processes (Morel and Gschwend, 1987). However, a similar understanding of the abundance, chemical nature, and reactivity of colloids in natural waters is yet to be developed.

In surface waters, the available evidence shows that significant amounts of certain trace elements and organic contaminants occur in the colloidal component of suspended solid phases. Interest in the environmental importance of colloids has been further heightened in recent years as a result of evidence which suggests that colloids can be mobile in porous media and may therefore act as transport phases for contaminants in soils and ground waters (McCarthy and Zachara, 1989). This complicates the traditional view of contaminant transport in ground water as consisting of a two-phase chromatographic process where solutes are partitioned between solution and an immobile aquifer matrix. Numerous two-phase models for contaminant transport have been devised which include equilibrium adsorption and ion exchange with respect to the stationary matrix (see IGWMC, 1992 for descriptions of specific computer codes). Where mobile colloids are present, contaminants having an affinity for solid surfaces may not be retarded as significantly as predicted by current physicochemical models because part of the solid phase is mobile.

It is the purpose of this chapter to facilitate an understanding of the nature of environmental colloids, their chemical and physical stability, their transport in natural waters, and their roles in contaminant transport and transformations. All of these are dependent on the physical and surface-chemical properties of the colloids. Aspects of colloid chemistry, physics, and geochemistry will be addressed in varying detail in this chapter, but emphasis will be given on the basic properties of colloidal systems and how they relate to colloid transport. Available data on the occurrence and environmental significance of colloidal particles will also be presented. The focus will primarily be on metal, organic, and radionuclide contaminants. Biocolloids, bacteria, and virus, while extremely important, are considered to be largely outside the scope of this chapter.

Major processes affecting pollutant transport in surface waters are illustrated in Figure 2.1, adapted from Salomons and Forstner (1984). "Dissolved Processes," shown on the right, include metal complexation by inorganic and organic ligands that are quantitatively described by thermodynamic constants, which are well known for a variety of metals and simple ligands. Also included are metal, metal–ligand, and ligand sorption reactions with well-defined solid components as described by empirical relationships and various physicochemical models. Among these models addressing sorption of ionic contaminants to solid surfaces are those employing conditional surface sorption constants which account for electrostatic and specific chemical interactions. Such constants are known for many metals and solid phases such as iron, manganese, and aluminum oxides and hydroxides and clays (Chapter 1, this volume). Likewise, for hydrophobic partitioning, rigorous, though empirical, relationships have been developed and formalized in various algorithms to relate partitioning to the properties of the contaminant and the solid phase (Chapter 5, this volume). Generally these are described by the fundamental properties of the contaminant through octanol–water partition co-

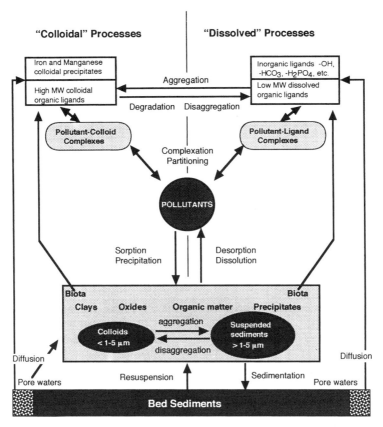

Figure 2.1. Major processes affecting pollutant speciation and transport in surface waters. (Adapted from Salomons and Forstner, 1984.)

efficients and by the properties of the solid components via the fraction organic carbon.

In contrast, "Colloidal Processes," shown on the left side of Figure 2.1, include metal complexation and sorption with less well-defined components such as complex organic and inorganic colloids. These processes are poorly understood. The limited amount of thermodynamic or even empirical data for such reactions is a reflection of the complexity and heterogeneity of environmental colloids and the difficulties associated with their isolation.

Uncertainty in partitioning constants also arises from the practical difficulties of separating colloids from dissolved phases in sorption experiments. Colloids behave similarly to solutes in that they are not effectively removed by centrifugation or filtration. The commonly observed apparent decrease in partitioning coefficients of organic and inorganic contaminants with increasing total solids concentration has been explained by the presence of colloids which are not completely separated from the aqueous phase (Morel and Gschwend, 1987). Contaminants associated with colloids are mistakenly measured as part of the solution phase, therefore reducing the apparent partitioning coefficient. Since the amount of colloidal material present increases with increasing total solids, taking colloids into account will generally produce partitioning coefficients that are invariant with suspended solids concentration.

Filtration through a 0.45-μm pore-size membrane has been widely, if arbitrarily, used as a means of separating particulate from "dissolved" species. Such an operational definition entirely ignores the continuum of suspended particle sizes and the issue of colloids. For environmental investigations this separation may be insufficient to fully describe the processes involved in contaminant transport. Ignoring the presence of colloids may cause errors in aqueous phase modeling, lead to misinterpretation of solid-solution partitioning data, and create serious errors in the prediction of contaminant mobility. Unfortunately, relatively few studies have attempted to rigorously differentiate between solutes, colloids, and suspended particles in natural waters.

DEFINITIONS OF ENVIRONMENTAL COLLOIDS

Chemical species present in natural waters occur in a variety of physical modes. These modes define an essentially continuous series which we will refer to as the solute–colloid–suspended-sediment continuum. While the pure end member modes are easily conceptualized as dissolved ions or molecules and macroscopic particles, the intermediate modes require further explanation. The word "colloid" as used here collectively identifies the intermediate continuum of modes. Chemical species which may be referred to as "colloidal" include (a) discrete molecules with sufficient size to behave as colloids (macromolecules), (b) amorphous or crystalline chemical compounds which exist in the solid phase as colloidal particles, and (c) chemical species associated with (usually by sorption) compositionally distinct, colloid-size particles.

A colloidal system is described as a two-phase system in which one phase is uniformly and permanently distributed or dispersed in the second phase. This is in contrast with both (a) a true solute–solvent system, which comprises a single phase, and (b) a suspended-sediment–solvent system, which is a two-phase system but is typically not uniform and never permanent. Although natural colloidal systems exist for every binary combination (except gas-in-gas) of solid, liquid, and gas phases (e.g., fog is a liquid-in-gas colloidal system), our sole focus here is on (a) systems with an aqueous dispersing media and (b) dispersed solid particles, known as *sols*. In a physicochemical sense, colloids behave like solids requiring consideration of surface area, charge, and other aspects of heterogeneous reaction theory. Hydrodynamically however, colloids behave somewhat like solutes, principally because of their small size.

Long-term kinetic stability as a suspension fundamentally defines a colloid and a colloidal system. An exceedingly wide variety of natural colloidal material has been identified in surface and ground waters. Examples are listed in Table 2.1. The list may be subdivided into the two main groups of colloids. The first four listings represent sparingly soluble minerals, for which lack of solubility suggests an inherent resistance to interaction with water. They are called lyophobic, solvophobic, or hydrophobic colloids. On the other hand, macromolecular organics (humic and fulvic acids, polysaccharides, proteins, peptides, amino acids, etc.) and polymeric precipitates (silica gel) tend to be polar and form direct hydrogen bonds with water. These materials are lyophilic, solvophilic, or hydrophilic colloids.

For hydrophobic solids, permanent dispersion or suspension is maintained by the random thermal activity of water molecules (Brownian movement). In order for this mechanism to be effective, the particle must be sufficiently small to allow spatially un-

TABLE 2.1.
Typical Aquatic Colloids

Hydrophobic Colloids
Phyllosilicates (clays)
Iron, manganese, and aluminum (hydr)oxides
Framework silicates
Precipitates (carbonates, sulfides, and phosphates)
Biocolloids (virus, bacteria, picoplankton)

Hydrophilic Colloids
Macromolecular organic matter (humics, micelles, etc.)
Polymeric precipitates (silica gels, etc.)

even bombardment by water molecules. If not for surface ions of like charge, the hydrophobic nature of these particles would tend to promote aggregation which would lead to larger particle sizes and eventual destruction of the colloidal system. Surface ions promote coulombic repulsion between particles and allow interactions with water molecules, thus making the particles effectively more hydrophilic.

Naturally hydrophilic macromolecules, which more closely approach true solute behavior, maintain their colloidal stability via interactions between charged functional groups and water molecule dipoles. Large molecules which are inherently hydrophobic (e.g., oils and greases) may form stable colloidal systems (an emulsion or liquid-in-liquid colloidal system) through interaction with surfactants, which are large molecules with both hydrophobic and hydrophilic portions. The hydrophobic molecule interacts with the hydrophobic portion of the surfactant while the surfactant's hydrophilic end interacts with water. A micelle is formed and the hydrophobic molecule is stabilized in a colloidal suspension.

The other fundamental property of colloidal systems, in addition to their physical stability, is their ability to scatter light, known as the *Tyndall effect*. Interactions between a beam of incident light and colloidal particles include refraction, polarization, reflection, and adsorption. The collective effect is that some of the incident light is randomly scattered and the colloidal suspension appears turbid. The light-scattering characteristics of a given suspension are related to concentration, size, shapes, and molecular weight of the colloidal particles, albeit in very complex ways (Moore, 1972). These relationships have led to the development of important tools for the investigation of colloids (Rees, 1987). Turbidity in itself, as unrelated to the properties of the suspended solid, is an important water quality issue (Greenberg et al., 1992).

Although particle size is not part of the technical definition of a colloid, in order for particles to be dispersed by Brownian motion and to produce the Tyndall effect, they must be very small, yet not small enough to be truly dissolved and part of the aqueous phase. Generally, "colloid" is only operationally defined by environmental scientists, typically in terms of the ability of a particle to pass through a certain filter or molecular sieve. Similarly, it is conceptually convenient to describe the differences between solutes, colloids, and suspended solids in terms of particle diameters or molecular weights. However, operational factors such as filter efficiency, density, and surface charge generally complicate separations based on size.

Colloids clearly lie between solutes and suspended solids in the size continuum of water-dispersed particulate matter (Figure 2.2), but the size cutoffs are a matter of debate. Generally the size of colloids is quoted as simply being submicron or, more specifically, ranging from 0.001 to 1.0 μm (1 to 1000 nm). A slightly larger lower size limit is used by some, while others extend the upper range to 2–5 μm to coincide with the classical clay–silt boundary, while still others further extend this range to 10 μm (Table 2.2). As is generally the case when describing a continuum, intermediate limits and bounds are necessarily arbitrary. It is suggested here that the size distinction between solute (i.e., truly dissolved) and colloidal constituents is best drawn at the upper size limit for thermodynamically well-defined phases (i.e., simple inorganic and organic aquo-ions). Assuming this to be the largest hydrated-ion size parameter used in Debye–Hückel equations (cf. Nordstrom and Munoz, 1986), this is 11 Å or 0.0011 μm which corresponds well to the 0.001 μm lower size limit for colloids accepted by many authors. Note that particle "sizes" are referenced either to linear particle dimensions (units of micrometers or nanometers) or to mass (usually in terms of molecular weights or daltons). The choice depends largely on the method used to study the particles. There is no universal conversion between the size and mass, but a few thousand daltons may be taken to be equivalent to a diameter of about 1 nm. The colloid-suspended particle boundary can be based on the widespread use of 0.45-μm filtration, but larger particles, up to a few micrometers, may in fact behave as colloids. An upper size limit in the range 2–5 μm better describes the hydrodynamic behavior of larger colloids.

Colloidal particles in natural waters may be composed of a wide variety of natural and synthetic materials ranging from macromolecules (molecular weight >1000) to oxides and phyllosilicates (Figure 2.2 and Table 2.1). The relative importance of various colloids will vary with the aquatic environment. For example, it is likely that in biologically productive surface waters, living microorganisms and nonliving organic materials will dominate the colloid population. In waters affected by mining operations, iron and aluminum oxides and clays may be the most abundant colloids.

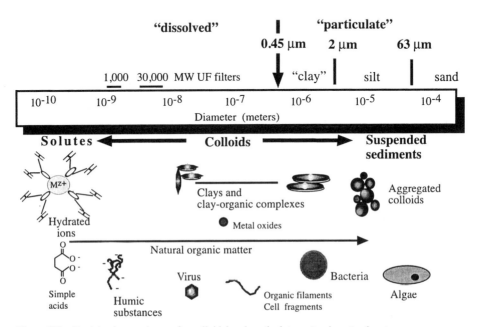

Figure 2.2.   Particle-size continuum for colloidal and particulate matter in natural waters.

TABLE 2.2.
Some Generally Accepted Colloid Size Ranges

Minimum Diameter (μm)	Maximum Diameter (μm)	Literature Source
0.001	1	Domenico and Schwartz (1990)
0.0005	5	Driscoll (1986)
0.001	1	Freeze and Cherry (1979)
0.003	1	Pankow (1991)
0.005	10	Stumm and Morgan (1981)

Colloids originate in at least three ways:

1. From the surrounding environment by disaggregation/dissolution and entrainment (e.g., suspension of primary or authigenic clays, acquisition of atmospheric dust by rain, weathering of framework silicates, suspension of oxides and oxyhydroxides, dissolution of fulvic acids in soil profiles, erosion of mineral and rock fragments from tailing piles or soil profiles)

2. As a result of *in situ* aqueous phase precipitation due to changing chemical or physical conditions (e.g., precipitation of calcite due to pH changes, precipitation of iron oxides and/or sulfides due to redox changes, formation of complex phosphate mineral suspensions at chemical interfaces in lakes, polymerization of dissolved silica due to pH changes)

3. From biological activity (e.g., growth and death of microbes, production of fibrils, organic skeletons, and protein-rich cell fragments (Leppard, 1992), or biologically mediated precipitation or dissolution)

SIGNIFICANCE OF ENVIRONMENTAL COLLOIDS

The environmental significance of colloids arises primarily from their very large specific surface areas which facilitate reactions between truly dissolved solutes and the particle surface. These heterogeneous reactions include simple electrostatic sorption (ion exchange), chemisorption, coadsorption, catalysis of complexation, redox and hydrolysis reactions among solute species, and dissolution, precipitation, and leaching. Increased surface area is an expressed or implied factor in increasing the equilibrium capacity and/or the rates of these reactions. The significance of small particles such as colloids in influencing the available surface area can be illustrated in several ways.

As particle size decreases, the total surface area per unit mass of suspended material increases dramatically. Since adsorption is, at least in part, a surface-area-extensive process, the total adsorption capacity of a volume of water containing suspended matter will be proportional to the particle size of the suspended matter. An estimate of the number of particles (N) and hence the associated surface area (A) is easily calculated for a known concentration of suspended matter (C_s) with a particular density (ρ), uniform particle size (d, sphere diameter or cube side length), and regular shape. Gregg and Sing (1982) and Parks (1990) provide formulae for different shapes and nonuniform size distributions. According to these formulae, 1-μm spheres of quartz would have a specific surface area of 2.26 m^2/g. Simple estimates such as these tend to underestimate surface areas of natural materials due to complex shapes, porosity, and surface defects (White and Peterson, 1990; Davis and Kent, 1990; and others). For example, Parks (1990) shows a plot of measured specific surface area as a function of grain size for natural quartz sand and crushed quartz which indicates that approximately 10 m^2/g is reasonable for 1-μm-diameter particles. Figure 2.3 illustrates the effect of dividing a 1-cm cube of hematite into smaller and smaller cubes. Not only does the number of cubes increase tremendously, the total surface area approaches 1000 m^2/g at the lower size limit for colloids. Taking this one step further, consider the reactivity of the hematite itself. Obviously any reaction must take place with the surficial atoms of iron and oxygen—for example, those within 2 Ås of the surface. As the particle size decreases below 1 μm, the percentage of atoms near the surface rises dramatically and approaches 100% at 1 nm. Thus almost the entire mass of hematite is immediately available for reaction.

Experimental adsorption data are typically reported in the form of isotherms which compare the dissolved concentration of some solute (mass/volume) with the sorbed concentration (mass/mass). The same experimental data may be normalized to the measured surface area of sorbent and reported as adsorption density (G, sites/nm^2) or G_{max} for conditions of maximum adsorption. As an example, site densities for α-FeOOH, as derived from adsorption experiments involving OH$^-$, H$^+$, F$^-$, SeO$_3^{2-}$, PO$_4^{3-}$, oxalate, and Pb^{2+}, vary from 0.8 to 7.3 sites/nm^2. The large range is in part due to in-

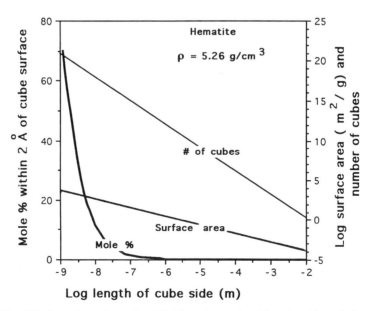

Figure 2.3. This figure shows the results of dividing a 1-cm cube of the mineral hematite into smaller and smaller cubes. Note that all scales are log$_{10}$ and that for colloids of 0.01-μm (1 × 10^{-6} cm) diameter the percentage of the moles of hematite within 2 Å of the particle surface is approaching 100%. Since Fe and O radii are on the order of 1 Å, this means that a significant number of atoms are subject to reactions which occur at the particle surface.

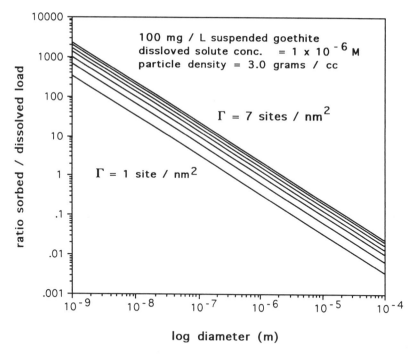

Figure 2.4. Theoretical comparison of dissolved and suspended loads of a hypothetical chemical constituent. Each line represents a different surface sorption site density, ranging from 1 to 7 sites/nm². Assuming complete occupancy of the surface sites and a dissolved concentration of 1×10^{-6} M, the suspended load can carry significantly greater amounts of the constituent relative to dissolved load.

trinsic differences in adsorption mechanisms for different sorbates and in part due to differences in experimental conditions.

Figure 2.4 illustrates the combined effects of particle size and site density on the suspended matter-sorbed solute load assuming full saturation of available sites with sorbate ions and 1:1 ratio of sites:ions. The plot suggests that a water sample with 100 mg/L of some suspended matter (e.g., α-FeOOH with sites = 1 to 7 sites/nm²) and 10^{-6} mol/L of a truly dissolved solute (e.g., 0.2 mg/L dissolved Pb^{2+}) could contain significantly greater concentrations of the solute as a suspended matter-sorbed phase. In this simplified scenario, as the particle size decreases, the moles of metal associated with colloids may equal or exceed the moles of dissolved metal. The picture is complicated in natural waters by (1) competition for the available sites, (2) aqueous complexation of the metal, and (3) variable site densities and variable aqueous concentrations.

Figure 2.5 illustrates the importance of colloid surface area relative to the surface area of solid aquifer material in a ground water. A cubic meter of hypothetical aquifer composed of 1-mm-diameter spherical sand grains ($\rho = 2.5$ g/cm³), and having 30% porosity, has a computed total surface area of 4200 m² and a specific surface area of 2.4×10^{-3} m²/g. An approximately equivalent total surface area would be available

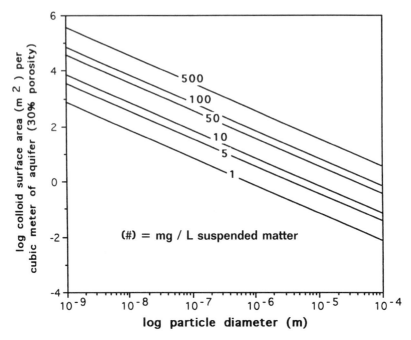

Figure 2.5. Particle-size dependence of total particulate surface area (m²) available in the liquid phase of 1 m³ of saturated 30% porosity aquifer. Lines represent different concentrations of suspended matter (mg/L).

in the liquid phase if the pore water contained 580 mg/L of similar-density, spherical, 0.1-μm-diameter particles or 5.8 mg/L of 0.001-μm particles. These suspended loads would have calculated specific surface areas of 24 and 2400 m²/g, respectively. Assuming the larger colloid size (0.1 μm) and the lower concentration (5.8 mg/L), a ground-water velocity of 1 m/day would in 100 days provide a mobile surface area in the cubic-meter volume equivalent to the aquifer matrix. These simple examples illustrate the potential importance of colloids to the chemistry of contaminants in ground-water and other natural aqueous systems.

SURFACE CHARGE ASPECTS OF COLLOIDS

A colloid, like any other particle in a aqueous media, is bounded by an interface which separates the two phases. Electrical charges are an integral and ubiquitous characteristic of solid–aqueous-phase interfaces and strongly influence the thickness and nature of the transition zone between the two bulk phases (i.e., the electrical double layer). The total amount of electrical charge at the interface is a surface area-intensive function. In the case of colloidal particles, or any suspended particle for that matter, this electrified nature of the interface has three very important consequences.

1. Electrical forces of attraction and repulsion control how closely two particles may approach one another and whether or not there will be sufficient tendency for them to remain together if the approach is close enough or forceful enough. This is the

basis for coagulation or agglomeration of suspended matter and controls the stability and transportability of the particles in the aqueous media.

2. The electrical charges associated with particle interfaces influences the processes of adsorption by which metal ions, anions, and organic molecules become associated with the particles. Therefore, electrical forces influence the identity and abundance of sorbed ions and molecules associated with colloids. These ions and molecules may be contaminants of environmental interest.

3. By a combination of the above factors, electrical interfaces influence the transportability of colloid-associated contaminants.

Origin of Colloid Surface Charge

All surface electrical charges have their ultimate origin in unsatisfied chemical bonds associated with the solid phase. The physical location of the unsatisfied bonds determines the characteristics of the electrified interface (Sposito, 1992). "Permanent" (i.e., not affected by the aqueous environment) surface charges have their origin within the bulk solid phase where substitutions of similar-sized but different-charged atoms lead to a net lattice charge imbalance. These charges are typically negative in sign (e.g., substitution of Al^{3+} for Si^{4+} in the tetrahedral layer or Mg^{2+} for Al^{3+} in the octahedral layer of smectite clays). The result of the lattice charge deficiencies is a non-point-source electrostatic surface charge which gives rise to such phenomena as cation exchange. Nonstoichiometric substitutions in other minerals such as carbonates can generate similar charges.

Another source of surface charge is found on developing or broken surfaces which abound in yet-unsatisfied or "dangling" bonds, respectively. These bonds are exposed to the mobile aqueous phase and tend to react strongly with water. Lasaga (1990) presents detailed theory and evidence for the nature of the water-surface reactions. In general, positive surface charges (e.g., due to an incompletely bonded Si atom) attract the negative charge of an OH^- ion or the negative dipole of a water molecule. Water bound in this way may dissociate to form a surface OH^- ($X-OH$). The $X-OH$ may deprotonate to form $X-O^-$ or may protonate to form $X-OH_2^+$. In short, surface-bound water behaves amphoterically (as an acid or base), and therefore the character of the surface and its charge is related to the hydrogen ion activity as well as to other solution characteristics such as ionic strength. Any ions which function like H^+ and OH^- in this case are referred to as potential-determining ions (PDIs). When water (OH^- and H_3O^+) is involved, the surface charge is a function of the pH of the aqueous medium (positively charged at lower pHs and negatively charged at higher pHs). The pH at which the surface is neutral is defined as the *zero point of charge* (ZPC).

Sorption of organic matter on suspended particles has been shown to control particle surface charge (Tipping and Cooke, 1982; Beckett, 1990) in most aqueous systems. Hunter and Liss (1979) concluded that adsorption of organic matter is responsible for the observation that particles have a net negative charge in almost all natural waters, regardless of the particle mineralogy. Natural organic material is a complex mixture of polyelectrolytes, whose charge primarily results from carboxylic acid functional groups ($-COOH$) with pK_a values in the range of 3–5. At typical environmental pHs these groups tend to be negatively charged ($-COO^-$). Adsorption of organic matter to mineral surfaces is primarily through a ligand exchange mechanism where

$-COO^-$ groups bond directly to the metalations (i.e., Al, Fe, Mn) and displace the surface OH group. Not all COO^- groups of the organic matter are involved in sorption to the particle surfaces, and those remaining provide the net negative surface charge on the particles. The sorption of organic matter in turn affects colloid stability (Tipping and Higgins, 1982; Liang and Morgan, 1990). Removal of organic coatings has been shown to reduce colloid stability significantly (Gibbs, 1983).

Electrical double-layer theory deals with the distribution of electrical forces and charged particles in the immediate vicinity of the surface (Chapter 1, this volume). For our purposes we will only consider that each particle is surrounded by a diffuse layer of charge which is opposite to the charge of the surface itself.

Surface Charge and Sorption on Colloids

In addition to the chemical load which may be transported as colloidal particles per se (e.g., iron in the form of suspended Fe-oxyhydroxides or organic carbon in the form of organic macromolecules), there is a potentially more significant chemical load present as sorbed species associated with colloids (Figure 2.5). Ionic sorption processes are generally interpreted with respect to surface electrical charges.

Models for the interpretation and prediction of adsorption of aqueous species onto solids, including colloids, are of two basic kinds: empirical and surface complexation (Davis and Kent, 1990). Empirical adsorption models include Langmuir, Freundlich, and other, more complex isotherms (Detenbeck and Brezonik, 1991) and the simple distribution coefficient (K_d) models, which are a special case of the more complex models. These useful models employ various conceptual statements of mass action to generate theoretical curves which relate the dissolved [S] and the sorbed [XS] concentrations of a constituent.

Although electrostatic attraction may be envisioned to be the driving force for ionic sorption, no specific mechanism is called upon by these empirical models. More sophisticated isotherm models contain more adjustable parameters to allow matching of theory to observation, but in general the models are highly specific to the experimental or observational conditions of pH and solution composition. In other words, experimentally determined constants cannot be used to accurately predict field behavior unless conditions are similar. Nevertheless, numerous examples of the application of distribution coefficient models to aqueous suspension systems are presented by Davis and Kent (1990).

Surface complexation models are physicochemical models developed to relate the pH, ionic strength, and aqueous phase composition to surface charge variations, the thickness and structure of the interface transition zone, and the position and bonding of ions and molecules in the interface zone. These models are collectively known as electric double-layer (EDL) models but include constant capacitance, diffuse-, triple-, and quadruple-layer models. Very briefly, the total surface charge density, which is equal to the sum of permanent and PDI densities, is rarely equal to zero and therefore results in a gradient in the electrical potential between the two bulk phases. Counterions, responding to this gradient, variously position themselves (depending on the specific model) in a compact (Stern) layer bound to the surface, or in more or less diffuse atmospheres which are dissociated from the surface (Davis and Kent, 1990; Bockris and Reddy, 1970). Electrical and chemical properties of colloid surfaces govern the adsorption of species from solution and can be modeled successfully using surface complexation models (Tessier, 1992). Modeling of metal sorption to amorphous iron oxyhydrite (ferrihydrite) has been carefully worked out by Dzomback and Morel (1990). Laboratory-derived constants for this system have been shown to accurately describe metal sorption in acid-mine streams (Smith et al., 1992).

Colloids, especially colloidal organic matter, also play an important role in the nonelectrostatic, hydrophobic partitioning of organic contaminants. Colloidal organic matter can provide regions of relatively nonpolar character into which hydrophobic contaminants can partition (dissolve). The general aspects of hydrophobic sorption are presented in Chapters 5 and 6 (this volume).

Surface Charge and Colloid Stability

The interactions of colloids with one another and with macroscopic solid phases are governed by hydrodynamic and chemical forces (O'Melia, 1990). The stability of colloids (not to be confused with chemical stability with respect to dissolution) refers to their ability to remain suspended. Many destabilizing mechanisms are operational in moving-water bodies. Physical interactions between particles and an aquifer matrix or stream bed may remove colloids from suspension. Hydrodynamic forces tend to bring suspended particles into contact with other particles and with bounding surfaces. Gravitational aggregation is caused by differential rates of settling due to size and mass differences, while random collisions of small particles due to Brownian motion results in perikinetic flocculation. Collisions also occur when particles are being transported in zones of steep fluid, and consequently particle, velocity profiles (orthokinetic flocculation). Colloidal particles in lakes are subject to aggregation that is dependent on both particle surface charge and initial particle size (Ali et al., 1985). Aggregated particles are then subject to sedimentation which transports adsorbed metals from the water column to the sediments (Sigg, 1985). The bulk-chemical properties of settling colloidal aggregates and nonsettling particles can be very different, especially with respect to their organic matter and trace metal content (Ranville et al., 1991a).

Collision of two particles can have two outcomes: Either the particles adhere to each other or they do not. The stability of a colloidal suspension is described by its stability ratio, alpha, which is simply the ratio of collisions which result in particles sticking to the total number of collisions (O'Melia, 1987). The value of alpha decreases from unity as stability increases. Interfacial electrostatic forces, as previously described, govern the magnitude of alpha and the stability of particle associations.

As two similar, like-charged hydrophobic particles approach each other they experience increasing electrostatic repulsion as the gap (D) between them closes. As D decreases, the repulsive forces ($\Psi_{repulsion}$) increase and more energy is required to continue the approach. If the particles approach each other with energy enough to make D very small, van der Waals attractive forces ($\Psi_{attraction}$) begin to come into play and can ultimately overwhelm the repulsive forces and aggregation can take place. The thickness of the diffuse charge layer surrounding the particle controls the magnitude of the repulsive force, and this thickness is decreased by the presence of ions in the aqueous media (i.e., ionic strength). As ionic strength increases, the diffuse charge cloud shrinks and the repulsive force preventing aggregation decreases, and therefore aggregation is more likely. Similarly, increases in temperature increase aggregation.

THE MOBILITY OF COLLOIDS IN POROUS MEDIA

The abundance of colloids in ground water is a function of the source area and/or *in situ* processes. Weathering and disaggregation results in the generation or release of colloids from the grain surface. Movement of colloids in porous media involves all the mechanisms of surface water transport with the added complications of finite pore sizes, very complex flow paths, and abundant opportunity to interact with the solid aquifer matrix. This is illustrated in Figure 2.6, is described in detail by McDowell-Boyer et al. (1986), and is summarized here. Assuming spherical suspended particles and matrix grains, a particle diameter/matrix diameter ratio (d_p/d_m) greater than 0.1 implies that particles are normally excluded from a porous medium and form external filter cakes. Particles entering the porous medium are subject to gravitational settling, depending on the flow velocity, and straining. Straining appears to be most important when $0.05 < d_p/d_m < 0.1$ and relatively ineffective below 0.05. Significant losses of porosity may occur due to straining, and the mechanism itself may become important with increasing d_p due to particle agglomeration and/or with decreasing d_m due to changes in lithology.

The combination of these mechanisms probably result in the upper size limit for suspended and/or mobile particles in porous media. According to these parameters, even the largest colloids (~ 1-μm diameter) could easily enter a fine sand aquifer with a uniform spherical grain size of 0.1 mm. Assuming clay-sized (0.001 mm) aquifer grains, colloids up to 50 nm could potentially move through the available pore spaces.

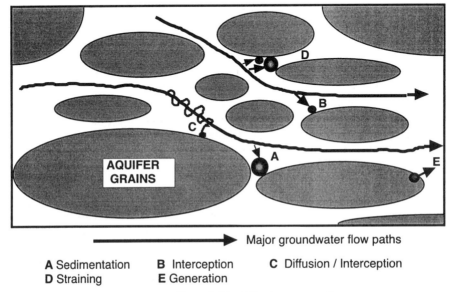

Figure 2.6.   Important processes controlling colloid mobility in porous media.

These are very conservative estimates that do not take into account irregular grain shapes and packing which would reduce the maximum size of transportable particles.

For particles small enough to penetrate a porous medium, the probability of remaining suspended or entrained in the prevailing flow largely depends on particle-media interactions. McDowell-Boyer et al. (1986) regard the process as physicochemical filtration comprised of the initial collision and subsequent attachment mechanisms. Larger suspended particles may collide with a matrix surface as a result of gravitational forces "pulling" it from the flow path. Alternately, the trajectory of the flow lines may result in direct interception of the particle by the surface.

For smaller particles and perhaps all colloids, Brownian-motion-based diffusion is the predominant collision mechanism. Brownian motion allows colloids to diffuse from the principal flow region between grains and closely approach the aquifer grains where attachment may occur. Attachment is determined by the complex nature of the electrified interface and is a balance between attractive and repulsive forces as previously described. Attractive forces are dependent upon the nature of the particles whereas repulsive forces, due to charge on the surface, are also dependent upon water chemistry. Water chemistry parameters of importance include pH (Kita et al., 1987) and ionic strength (Cerda, 1987; McDowell-Boyer, 1992).

Since there are processes which remove both "large" and "small" particles in flow through porous media, each medium has an optimum particle size for transport (Yao et al., 1971). In many aquifers, the greatest mobility occurs for particles near 1 μm. The processes controlling colloid flow in an aquifer have been clearly demonstrated by the work of Harvey et al. (1989). In these studies, tracer tests were conducted between monitoring wells in a glacial aquifer in Cape Cod, Massachusetts under conditions of natural and induced flow gradients. Chloride and bromide were used as conservative tracers, and fluorescent-labeled bacteria and submicrometer latex spheres were used to study the mobility of particles. The tests demonstrated the following.

1. Micrometer and submicrometer particulates can easily migrate in porous media (as shown by the similarity of the arrival times at the second well).

2. Average velocities of particulates may exceed conservative solutes.

3. As predicted, submicrometer particles travel more slowly than particles around 1 μm in size.

4. The chemical functionality of the spheres (carboxyl, carbonyl, uncharged) affects retention and retardation. The transport of these spheres was successfully modeled by colloid filtration theory (Harvey and Garabedian, 1991).

Another study reveals the potential impact of colloids on adsorbed contaminant transport in a porous medium. Puls and Powell (1992) compared the mobility of arsenic associated with iron oxide colloidal particles to truly dissolved arsenic in laboratory columns of natural aquifer material. Tritium was used as conservative tracer. The following was reported.

1. FeOx colloids moved very easily through the columns (shown by the similarity of the tritium and colloid arrival times).

2. Colloid-associated As was transported through the columns 21 times faster than the dissolved As.

3. Colloid velocity was a function of colloid size and the ionic environment of the solutions.

METHODS OF ANALYSIS OF AQUATIC COLLOIDS

Aspects of Sample Collection

A full understanding of the distribution of dissolved and particulate components in any natural water requires that special care be taken to obtain representative samples and to avoid the creation of sampling artifacts. While considerable attention has been given to the impact of sampling artifacts on dissolved or presumed-dissolved components (Barcelona, 1990; Kent and Payne, 1988; Smith et al., 1988; Herzog et al., 1991; Palmer et al., 1987; Barcelona and Helfrich, 1986; US EPA, 1986; and many others), less effort has been focused on particulates, and especially colloids. It is now clear that the act of sampling, especially of ground waters, can alter suspended particulate populations *in situ* and therefore bias information, not only about the particulates themselves, but also about dissolved constituents (Ryan and Gschwend, 1990; Backhus et al., 1986).

The problem of artifacts is especially acute in ground-water sampling because of the following:

1. Physical and chemical conditions in deep or confined aquifers and even some relatively shallow, unconfined aquifers may be significantly different than those at the surface.
2. The act of sampling can impose dramatic hydrodymanic changes in the immediate vicinity of the sampling point.

Regarding the first point, differences in total pressure (e.g., 2.94 ATM/100 feet of water column), partial pressures of reactive gases (e.g., high CO_2 or H_2S pressures in certain aquifers), temperature (as much as 37°C in extreme conditions), and light can affect dissolved gas compositions and other volatile components, heterogeneous equilibria involving gas phases, and reaction rates. These in turn can alter the redox state, carbonate equilibria, and pH and initiate precipitation-dissolution reactions, shifts in aqueous phase speciation, and changes in sorption–desorption equilibria. Such reactions may affect the identity, stability, size, and sorbed chemical composition of colloid phases. Secondly, well development activities (which often induce considerable turbulence), purging, and sample collection (especially when done with high-volume pumps) can mobilize particles from the aquifer matrix because these actions can impose higher-than-normal piezometric gradients that cause accelerated flow velocities. This results in the mobilization of particles that might otherwise remain attached to the aquifer media and/or disturbing the size distribution of existing suspended matter. Either of these factors could alter the distribution of analytes in the sample.

The effects of sampling on ground-water colloids have been studied only recently (Backhus et al., 1986; Puls et al., 1990). Puls et al. (1991) found that bailed samples contained significantly higher concentrations of particulate chromium and arsenic than pumped samples. This presumably was due to surging causing resuspension of particles in the bottom of the well or from the surrounding aquifer. Furthermore, Puls et al.

(1991) found differences between pump types. A bladder pump (flow rate = 0.6–1.1 L/min) produced 13 times fewer particles than a low-speed submersible pump (2.8–3.8 L/min) and 20 times fewer than a high-speed submersible pump (12–92 L/min). Light-scattering intensity, a measure of particle concentration, has been shown to decrease as pumping rates decreased (Ryan and Gschwend, 1990), indicating unnatural entrainment of particles by pumping. Light-scattering and turbidity measurements have also shown that in some cases many more well casing volumes must be purged to reach a stable particulate concentration than for stabilization of dissolved ground-water chemical parameters such as pH, dissolved oxygen, temperature, and Pt-electrode potential (Eh) (Ryan and Gschwend, 1990; Puls et al., 1991). Gschwend et al. (1990) employed a modified gear pump to obtain ground-water samples at very low rates in order to study colloids. These studies collectively suggest that ground-water samples should be taken at low pumping rates ($\sim$100 mL min^{-1}) and that particle-sensitive parameters such as turbidity should be monitored to determine purging volumes.

Logic would suggest that water should not be withdrawn from a well bore faster than it can recharge under normal hydrostatic conditions so as to not exceed normal interstitial flow velocities. This pertains to purging as well as sampling. To achieve this there should be no draw down during purging or sampling where a single aquifer is present. Where multiple aquifers are present the individual water-producing intervals should be isolated if possible and the *in situ* pressure monitored during withdrawal. Such practices raise the issue of practicality, particularly in the case of a slowly recharging monitoring well.

An additional concern for ground-water sampling where colloid-sensitive parameters are concerned is contact with the atmosphere. Formation of ferric oxyhydroxide precipitates and their subsequent flocculation are common occurrences when reducing waters are exposed to air. A study the colloid-facilitated ground-water transport of radionuclides in the Gorleben aquifer (Germany) avoided this problem by transferring samples directly into resin-coated aluminum drums equipped with an N_2 atmosphere (Lieser and Hill, 1992; Deerlove et al., 1991; Kim et al., 1992).

In sampling surface waters, precautions similar to those used in ground-water sampling to exclude exposure to the atmosphere must be taken when sampling lakes and reservoirs, especially where anaerobic zones exist. Redox-sensitive species such as iron exist as very different phases in oxic and anoxic waters and have been shown to exist in significant amounts as colloids at the oxic–anoxic boundary (Buffle et al., 1989). For flowing surface waters (rivers, canals, etc.) the nonhomogeneous distribution of particles vertically and horizontally must be taken into account. Techniques to obtain discharge-weighted and vertically integrated samples should be used (Lenheer et al., 1989). These procedures require a series of samples to be taken at points across the channel to provide horizontal integration. Pressure-compensating, isokinetic bag-samplers allow collection of sample volumes at each point which represent the proportion of flow at that point. From these data the mass flux of colloidal particles can be computed. These methods have recently been used to investigate colloidal size distributions in the Mississippi River (Rees and Ranville, 1990).

Current Standard Filtration Practices

Sample filtration is a critical aspect of natural water sampling, both to the environmental chemist interested in details of solute speciation, modeling, and transport mechanisms and to those concerned with regulatory compliance and/or water facilities operations. Filtration is also a point of considerable disagreement concerning suspended particles and colloids (Nielsen, 1991). Standard practice for filtration of those natural waters intended for analyses of "dissolved" (cf. "suspended," "total," or "acid-extractable") species is to filter through a submicrometer, typically 0.45 μm, membrane filter. The selection of the 0.45-μm pore size appears to have its origin in bacteriology and the smallest dimensions of some bacteria. It is not apparent that such a size was ever intended to address solutes.

Greenberg et al. (1992) specify use of either vacuum or pressure filtration through a 0.4- or 0.45-μm polycarbonate or cellulose acetate filter for dissolved metals, certain inorganic nonmetals like sulfate and phosphate, and waters to be analyzed by ion chromatography, but make no recommendations for other constituents. The US EPA (1986) specifies 0.45-μm filtration for metals analysis on one split of the sample and no filtration of the other split. The US Geological Survey (1982) recommends filtration for all dissolved inorganic constituents through a nonmetallic 0.45-μm filter. Barcelona (1990) cautions that dissolved gases and volatile organic constituents are notable exceptions to the above recommendations and should never be determined in filtered samples. "Dissolved" is thus operationally defined as anything passing a 0.45-μm filter, but Hem (1985) alludes to the inadequacy of this definition.

Several studies (*vide infra*) reveal the particulate-associated nature of significant portions of ions and molecular populations in the <0.45-μm fraction of natural waters. Such observations imply a compromise in the 0.45-μm filter pore size-based operational term "dissolved," which ignores colloids.

Particle Size Analysis

Many techniques exist for the size analysis of colloidal particles. These include methods based on light-scattering, microscopy, sedimentation, filtration, and dialysis. Coupling of some size-analysis methods with a wide variety of chemical analyses gives additional information of the composition of various-sized particles. Since the focus of this discussion is contaminant transport, only those methods providing chemical information will be presented.

Scanning electron microscopy (SEM) and transmission electron microscopy (TEM) are extremely useful techniques for examining colloids (Nomizu et al., 1988). Microscopy provides the most direct way of sizing particles but is very time-consuming and care must be taken to avoid creation of artifacts. Limited chemical information is provided by energy-dispersive X-ray analysis (EDAX) commonly available with SEM instruments. Electron microscope techniques for examining colloids in natural waters have been reviewed by Leppard (1992).

Unlike the suspended sediment-colloid distinction (easily, if arbitrarily, identified with some filter pore size such as 1 μm), considerable difficulty exists in differentiating between colloidal and dissolved species. For this task, filtration through micro-porous membranes, dialysis, and centrifugation at high gravities have been employed (Batley, 1990). The distinction between "colloidal" and "dissolved" is also operational; furthermore, many factors affect the actual physical size or molecular weight of particles passing through a given filter, especially when membrane filtration is employed. Filtration through successively smaller pore sizes has most commonly been used to give size distributions of colloidal-sized particles and to ultimately obtain the "dissolved" fraction of natural waters (Kennedy et al., 1974; Wagemann and Brunskill, 1975; Danielsson, 1982; Laxen and Chandler, 1982). Filtration has been performed in series (cascade filtration), in parallel, or in combination. In the case of cascade filtration, each filtrate is passed through a successively smaller pore size (Hoffman et al., 1981). In parallel filtration, splits of a sample are each filtered through a different-pore-size filter. The size distribution for a given analyte is calculated by subtracting the results for each filtrate from the unfiltered concentration (Tanizaki et al., 1992).

Factors affecting the results of filtration were carefully examined by Buffle et al. (1992), including such variables as filter type, filter loading, and flow rates. Two major types of commercial filters exist and were examined. The two types are depth filters, composed of a complex mat of fibers which retain particles within the matrix of the filter, and screen filters which consist of nonporous materials containing discrete holes which retain particles on the surface of the filter. The most important physicochemical processes occurring during filtration are (a) concentration polarization (i.e., buildup of particles) at the membrane surface which promotes coagulation and alteration of the size distribution and (b) membrane exclusion effects which inhibit the passage of very small particles through pores with similar diameters. Honeyman and Santschi (1991) documented the failure of 0.25-μm ^{59}Fe-hematite particles to pass a 0.40-μm membrane filter, apparently as a result of aggregation at the filter surface. The study of Buffle et al. (1992) illustrates that the results of filtration are dependent on procedural technique and suggests that any interpretation of particle-size measurements based on filtration must be carefully examined within the context of how the filtration was performed. Often filtration is routinely performed in the field with little control over the methods used. This poses serious questions as to the validity of our current database on the "true" distribution of solutes, colloids, and suspended particles in natural waters.

Cross-flow filtration is a very promising filtration method which may overcome some of the aforementioned problems associated with filtration (Gutman, 1987). In cross-flow or tangential filtration, a high flow rate is maintained parallel to the filter membrane while a much lower flow rate of filtrate passes through the membrane. This method induces large shear forces near the membrane surface and thus reduces both the concentration polarization region and membrane fouling. Filtration membranes are generally stacked in order to provide large areas, on the order of a few square meters, to further limit filter fouling. The method allows processing of large volumes of water to isolate colloids in significant quantities without reduced filtration rates. Cross-flow filtration uses systems based on either hollow-fibers (4–8 L min^{-1} @ 0.2 μm; Kuwabara and Harvey, 1990; Marley et al., 1991) or flat membranes (10–12 L hr^{-1} @ 10,000 molecular weight (MW) or $\approx$10 nm, Whitehouse et al., 1986, 1990). A number of studies have utilized cross-flow filtration techniques with success in producing gram quantities of material in the colloid size range. The technique of cascade tangential-flow filtration using different sized membranes has been used successfully to isolate particles in various colloidal size classes. The accuracy of the size cutoff for tangential-flow filtration has not as yet been investigated

Continuous-flow centrifugation is a useful technique for the large-scale isolation of suspended particles and larger colloids. In this method a suspension is continuously passed through a spinning centrifuge. Larger particles are retained on the walls of the centrifuge bowl, whereas solutes and smaller particles are carried through (Horowitz et al., 1989). Continuous-flow centrifugation has been used to separate river water particles down to about 1.0 to 0.3 μm (Leenheer et al., 1989; Rees et al., 1991). It has been noted that suspended particles retained by continuous flow filtration, when dissaggregated, show similar particle-size distributions to the colloids not retained by the centrifuge (Rees et al., 1991). This suggests that either the particles occur as aggregates in the sample or aggregation occurs in the centrifuge.

A new and effective method of colloid size analysis is field-flow fractionation (FFF; Beckett et al., 1988). This method, somewhat analogous to chromatography, provides high resolution and a continuous size distribution (unlike sequential filtration, which has a low resolution defined by the number of pore sizes used). Various subtechniques of FFF exist which allow analysis of particles ranging in size from 0.001 to 50 μm. Because separated particle-size fractions are eluted, FFF fractions can be further analyzed for their composition. This technique has been shown to be useful to study colloid-pollutant interactions (Beckett et al., 1990), especially with the recent direct coupling of FFF to inductively coupled plasma–mass spectrometry (ICP-MS; Taylor et al., 1992; Murphy et al., 1993).

COLLOIDS IN NATURAL WATERS

Surface Water Colloids

Most data on the importance of colloids in aquatic systems come from studies of surface waters. Fluvial suspended sediments are generally composed of aggregates and contain variable percentages of particles less than 2 μm. Walling and Moorehead (1989) reported that the proportion <2-μm material in the suspended matter of fluvial systems was about 10–30%. After disaggregation the <2-μm material increased to 40–80%, with 5–30% <0.2 μm.

Some researchers have found little or no differences between 0.45-μm filtrates and ultrafiltrates (10–100 kilodaltons) for most elemental concentrations, suggesting little colloidal material less than 0.45 μm (Taylor et al., 1990; Waber et al., 1990). In contrast, Hoffman et al. (1981) used a cascade ultrafiltration scheme to investigate trace metal concentrations in sizes ranging from less than 10,000 to greater than 100,000 molecular weight. Their results show that significant amounts of trace metals were present as colloidal particles in the upper Mississippi River. Salbu et al. (1985) studied colloidal metals in surface and ground waters using filtration, dialysis, and centrifugation. This study found, for example, that the majority of Cd and Cr were in the 0.1 to 0.005-μm fraction. Moran and Moore (1989), using cross-flow filtration, found 15% of the <0.45-μm aluminum to be present in the $<10,000$ MW (1–10 nm) fraction of open sea water. Honeyman and Santschi (1991) showed that up to 30% of the total tin in a 10^{-8} M solution could be partitioned onto colloidal hematite.

The size distribution and concentrations of 46 elements in eight Japanese river waters were determined by Tanizaki et al. (1992) using filtration and ultrafiltration performed in parallel. They were able to divide elements into groups based on their affinities for various size classes. Size classes were: suspended sediments (>0.45 μm); colloids $>10,000$ molecular weight, colloids 10,000–500 molecular weight; and solutes <500 molecular weight ($<\sim1$ nm). Transition elements were largely present in the >0.45, 10,000–500, and <500 MW fractions. The most dramatic effects are evident in the trace element determinations, particularly rare earth elements (REEs), where typically less than 20% of the <0.45-μm fraction content is present in the <500 MW fraction. These results compare favorably with those of Kim et al. (1992), who studied REE elements in ground waters (*vide infra*).

Natural organic matter (NOM) is ubiquitous in all surface and ground waters. Humic and fulvic acid components of NOM have molecular sizes which place them at the boundary between dissolved and colloidal materials (700–2000 MW). NOM components are excellent complexers of metals and radionuclides (Buffle, 1988) and are directly involved in the macromolecular transport and transformation of a variety of metals (Dunnivant et al., 1992) and organic contaminants (Enfield and Bengtsson, 1988, Perdue and Wolfe, 1982; Chapter 5 and references therein, this volume). As discussed above, NOM components dominate particle surfaces in natural waters and generally affect metal sorption by increasing sorption at low pH (Davis, 1984; Tipping et al., 1983). In acid-mine drainage, however, a lesser role for particle-associated NOM in metal transport has recently been suggested (Smith, 1992). This is presumably due to competition from abundant dissolved iron for surface binding sites.

Higher-molecular-weight organic colloids have been found to occur in estuarine waters (Sigleo and Helz, 1981) and pore waters at high concentration (Chin and Gschwend, 1991). These colloids have been shown to affect the partitioning of hydrophobic organic chemicals (Means and Wijayaratne, 1982; Chin and Gschwend, 1992) and are expected to also bind trace metals. Pore water organic colloids can diffuse into overlying waters and may act to enhance the flux of contaminants from these sediments (Thoma et al., 1991).

Colloidal iron and manganese phases are common in surface waters (Laxen, 1983) and are often present in regions of redox boundaries such as the oxycline in lakes, where dissolved oxygen disappears due to biological respiration. These colloidal Fe and Mn phases are generally a few tenths of a micrometer in size and are associated with organic matter and phosphorous (Buffle et al., 1989).

In rivers and streams receiving acid-mine drainage, ferrous iron is oxidized to ferric iron which then hydrolyzes and subsequently precipitates as colloidal hydrous iron oxides and hydrous iron hydroxysulfates. Colloidal iron can make up a significant portion of total suspended iron in these systems and can also adsorb and transport trace metals. Roughly an order of magnitude difference between 0.45-μm and 0.01-μm filtered iron has been shown to persist in a river for hundreds of kilometers downstream of acid-mine drainage inputs (Kimball et al., 1995). Solubility calculations for iron hydroxides will be affected not only by the enhanced solubility of small particles (Langmuir and Whittemore, 1971) but, in a more practical sense, by the choice of filter pore size used to determine "dissolved" iron (Kimball et al., 1992). Acid-mine drainage systems are the only environment where positively charges particles have been clearly demonstrated to exist (Newton and Liss, 1987; Ranville et al., 1991b). This occurs primarily as a result of insufficient organic matter to fully coat the extensive iron hydroxide surfaces formed in these systems. This may affect both metal interaction with iron oxyhydroxide colloids and particle stability, and hence transport.

Soil and Ground-Water Colloids

Many of the mineral and organic components of soil are colloidal in size, usually bound up in aggregates which make up the texture of soils. The properties of some soil colloids were studied by Bremner and Genrich (1990). In a study of 10 Iowa Mollisols they found that, on average, 30% of the soil mass was less than 2 μm. Even more importantly, the <2-μm fraction contained 68%, 73%, and 80% of the total organic matter, iron, and cation exchange capacity, respectively. Mobilization of soil colloids can carry colloid-associated contaminants vertically into the underlying aquifer. Vertical transport of DDT adsorbed on sewage particles and Paraquat adsorbed to montmorillonite was observed in soil columns by Vinten et al. (1983). The addition of calcium chloride significantly reduced the amount of pesticide transported, presumably either by aggregation and straining in the pores or by sorption of the particles to the soil matrix. The physical structure of soils will greatly influence colloid transport. Macropores have been shown to facilitate the transport of DOC and mineral colloids in soils (Chittleborough et al., 1992). The presence of these macropores will complicate transport modeling which is based on flow through porous media.

Natural and anthropogenic changes to the chemistry of soil and ground waters can result in the formation or liberation of colloids. This is particularly true for changes which may remove cementing agents which bind together soil and aquifer matrix colloids. Ryan and Gschwend (1990) found that dissolution of iron hydroxides by infiltration waters with low E_h released colloidal kaolinite and other aluminosilicates. Dissolution of iron hydroxides might also be expected to release organic matter which was associated with the iron hydroxides (Tipping and Cooke, 1982), thus mobilizing contaminants associated with the released organic matter. Gschwend et al. (1990) found that perturbation of the carbonate equilibrium of a ground water near a coal ash disposal site resulted in dissolution of calcite and decementation, which released colloids into the ground water. SEM-EDAX analysis suggested that the colloids are composed of aluminosilicates and residual carbonate cements. Changes in ground-water conductivity or pH can mobilize otherwise immobile colloids. For example, significant amounts of colloidal particles were shown to be mobilized by artificial recharge of an aquifer by low-conductivity surface water (Nightingale and Bianchi, 1977). A groundwater particle concentration of 9.93 mg/L was determined from turbidity data. From this value it was calculated that approximately 148 metric tons of particles were liberated from the overlying sandy soils during one year of recharge. The liberated colloids appeared to be effectively transported since no decrease in recharge rate was observed. Application of gypsum prevented further mobilization of colloids by destabilizing the colloids.

Precipitation of colloids can occur in aquifers resulting from anthropogenic perturbations in water chemistry. In a sand-and-gravel aquifer in Cape Cod, Massachusetts, Gschwend and Reynolds (1987) confirmed the reaction of phosphorous-rich sewage effluent with aquifer iron, resulting in the precipitation of ferrous phosphate colloids which were ~0.1 μm in size and fairly monodisperse. Particle concentration was estimated at 10^6 particles/liter.

Very few studies have investigated either the nature or the amount of colloids naturally present in ground water (Reynolds, 1985). Salbu et al. (1985) used centrifugation, filtration, hollow-fiber ultrafiltration, and dialysis to determine colloid concentrations in ground water and lake water. The concentrations of 20 elements were determined in four size fractions: >0.45 μm, 0.45–0.1 μm, 0.1–0.005 μm, <0.005 μm. In the ground water, iron, zinc, and copper were primarily present as particles >0.45 μm, whereas chromium, cadmium, and manganese showed significant amounts in each of the size fractions. Somwhat different results were observed by Puls and Barcelona (1989), working with ground waters from the Globe, Arizona area. They found a steady decrease in major, minor, and transition elements in the filtrates from 10-, 0.4-, and 0.1-μm filters respectively, indicating that these elements were concentrated in the smaller particle-size fractions.

Colloidal particles have been found in deep fractures in granite at a concentration of 10^{10} particles/L (Degueldre et al., 1989). These colloids consist primarily of aluminosilicates. Waber et al. (1990) studied the relationship between colloid content of the river Glatt and a shallow aquifer which is continuously recharged by the river. Colloid (<0.45 μm) concentrations in the river were computed to be approximately 2.3 mg/L, of which >90% was composed of organic matter. In contrast, the colloid concentration in the ground water was an order of magnitude less, thus leading the authors to conclude that transport of colloids from the river to the ground water was insignificant.

Biocolloids

Much of the information on colloid transport in ground waters comes from work with bacteria and viruses, which have been shown to be transported over significant distances (Gerba and Bitton, 1984). Laboratory experiments have shown pH to be a major factor in interaction between bacteria and the aquifer matrix (Scholl and Harvey, 1992). The deposition of bacteria in soil columns can be modeled by using DLVO theory to explain bacterial stability and clean-bead filtration theory to describe the physical processes involved (Martin et al., 1992).

Radionuclides

One of the most important areas for concern over colloid-facilitated transport of contaminants is of that of radionuclide transport. Most radio nuclides are considered extremely insoluble or highly sorbing, and hence immobile in the aquatic environment. A number of studies have indicated the radionuclides are indeed mobile in ground water, and this may be attributed to colloid transport. Bates et al. (1992) demonstrated the formation of colloids containing radionuclides by the interaction of waste glass with water. These colloids form near the surface of the glass and can be released into solution. Colloid-associated radionuclides were found at the Nevada test site, where after a nuclear test, colloid concentrations in the cavity formed were on the order of 10 mg/L (Buddemeier and Hunt, 1988). Hollow-fiber ultrafiltration was used by Lowson and Short (1990) to investigate colloids in ground-water downgradient from a uranium ore body. They found colloids of 1.0- to 0.018-μm sizes that were composed primarily of iron and silica but contained uranium and thorium. Although only a minor proportion of the total uranium was colloid-associated, a significant amount of ^{230}Th was associated with colloids. In another study, enhanced transport of americium and plutonium in an aquifer in an arid climate was demonstrated by Penrose et al. (1990). Predictions based on K_d values from soil materials indicated transport of a few meters, but observations showed transport of several kilometers.

Kim et al. (1992), using parallel membrane filtration, determined the abundance of a wide variety of elements in >0.4-, 0.4- to 0.1-, 0.1- to 0.002 , and 0.002- to 0.001-μm filter fractions of deep (150–300 m) ground waters from a sedimentary sequence overlying a salt dome intended for disposal of radionuclides. Their results showed that

70–87% of REE, selected transition metals, uranium and thorium, 99% of americium (Am[III]) and curium (Cm[III]), and 90+% of DOC found in the <0.4-μm filtrates were removed by filtration through 0.001-μm filters. Thus the 0.4-μm filtrate provided a significant overestimate of the dissolved content of these constituents. Acidification, however, allowed effectively all Am and Cm to pass the 0.001-μm filter, and time-resolved laser fluorescence spectroscopy (TRLFS) indicated that these elements were bound to fulvic acid colloids.

CONCLUSION

Colloids are recognized as a distinct mobile solid phase in the majority of natural water systems. Such particles are capable of hosting heterogeneous reactions typically ascribed to immobile solid phases but exhibit mobilities similar to those of true solutes and the aqueous media itself. An important role for colloidal particles in the facilitated transport of contaminants in surface and ground waters is theoretically reasonable and supported by a relatively small but rapidly growing number of field and laboratory studies.

The mobility of particles in the micrometer size range, as shown by laboratory studies and the limited field data, suggests that standard filtration practices (0.45 μm) may overestimate truly dissolved concentrations (including contaminant loads). On the other hand, the use of ultrafilters may result in seriously underestimating the total transported contaminant load which includes that portion carried by colloids. Furthermore, filtration results are greatly affected by factors such as flow rate through the filter, total volume filtered through a given filter apparatus, concentration of particles, filter type, and sample storage time. Often these variables are not considered during the design of field sampling schemes, and/or filtering practices are not reproducible from one sampling event to the next. Even if considered, these variables are often not closely controlled in the field due to the pressure to obtain samples in a rapid, cost-effective manner. For this reason the U.S. EPA has suggested the collection of both unfiltered and filtered water samples (Puls et al., 1990). If significant differences are found, further examination is recommended to evaluate the possibility of colloidal transport. It should be pointed out that comparisons between filtered and unfiltered samples yield little direct evidence about colloid-associated loads if relatively large pore-size filters are employed.

The extent and general importance of colloids and colloid-facilitated transport is yet to be determined, especially in ground-water systems. For the case of ground waters it appears that the most important factor determining the mobility of natural colloids is ground-water chemistry. In particular, changes to ground-water chemistry (i.e., pH, ionic strength, redox, and chemical composition) due to human activities can influence colloid stability. Colloids may be formed by *in situ* precipitation reactions or liberated from aquifer material. Similarly, alterations in the ambient chemistry may result in destabilization of existing colloids through flocculation or adhesion to the matrix. In the course of investigations of ground-water contamination sites, determination of the spatial distributions of colloids may give important clues to understanding the migration of contaminants and the most effective means of remediation.

Various aspects of ground-water sampling protocols, such as pump type, purging schedules, sampling rate, and control of redox and dissolved gases, can influence colloid populations and hence the results of chemical analyses. This is particularly true of unfiltered, filtered, or inadequately filtered samples. Improper sampling techniques which disturb the ground water in the vicinity of the well bore may result in artificially high levels of particulates. Observed temporal variations in dissolved and particulate concentrations may therefore be as much due to differences in sampling and filtration techniques as due to real physicochemical speciation. It cannot be strongly enough stressed that our understanding of the nature of contaminants in surface and ground water will be incomplete until uniform sampling protocols, which accurately determine the presence of micrometer- and submicrometer-sized particles, are developed.

Despite the current lack of a clear understanding of the role of colloids in contaminant transport, interest in colloids will continue to grow. It is now necessary for careful work to be done to develop the techniques and databases which will, in time, provide a clearer picture of the importance of the role that colloids play in the overall environmental fate and transport of contaminants in surface and ground waters.

References

Ali, W., O'Melia, C. R., and Edzwald, J. K., 1985, Colloidal stability of particles in lakes: measurements and significance. *Water Sci. Technol.*, 17, 701–712.

Allan, R. J., 1986, The role of particulate matter in the fate of contaminants in aquatic ecosystems. In: *Water Quality Management: The Role of Particulate Matter in the Transport and Fate of Pollutants*, Water Studies Centre, Melbourne, Australia, pp. 1–56.

Backhus, D. A., Gschwend, P. M., and Reynolds, M. D., 1986, Sampling colloids in ground water. *EOS*, 67, 954, abstract H41D-03 0900H.

Barcelona, M. J., 1990, Uncertainties in ground water chemistry and sampling problems. In: D. Melchior and R. L. Bassett, eds. *Chemical Modeling in Aqueous Systems, II*, ACS Symposium Series, No. 416, American Chemical Society, Washington, D.C., pp. 310–320.

Barcelona, M. J., and Helfrich, J. A., 1986, Well construction and purging effects on groundwater samples. *Environ. Sci. Technol.*, 20, 1179–1184.

Bates, J. K., Bradley, J. P., Teetsov, A., Bradley, C. R., and Buchholtz ten Brink, M., 1992, Colloid formation during waste-form reaction: implications for nuclear waste disposal. *Science*, 256, 649–651.

Batley, G. E., 1990, Physiochemical separation methods for trace element speciation in aquatic samples. In: Batley, G. E., ed., *Trace Element Speciation: Analytical Methods and Problems*, CRC Press, Boca Raton, FL, pp. 43–76.

Beckett, R., 1990, The surface chemistry of humic substances. In: Beckett, R., ed., *Surface and Colloid Chemistry in Natural Waters and Water Treatment*, Plenum Press, New York, pp. 3–20.

Beckett, R., Nicholson, G., Hart, B. T., Hansen, M., and Giddings, J. C., 1988, Separation and size characterization of colloidal particles in river water by sedimentation field-flow fractionation. *Water Res.*, 22, 1535–1545.

Beckett, R., Hotchin, D. M., and Hart, B. T., 1990, The use of field-flow fractionation to study pollutant-colloid interactions. *J. Chromatogr.*, 517, 435–477.

Bockris, J. O., and Reddy, A. K. N., 1970, *Modern Electrochemistry*, Plenum Press, New York, 1432 pp.

Bremner, J. M., and Genrich, D. A., 1990, Characterization of the sand, silt and clay fractions of some Mollisols. In: DeBoodt, M. F., Hayes, M. H. B., and Herbillon, A., eds.,

Soil Colloids and Their Association in Aggregates, Plenum Press, New York, pp. 423–483.

Buddemeier, R. W., and Hunt, J. R., 1988, Transport of colloidal contaminants in ground water: radio nuclide migration at the Nevada Test Site, *Applied Geochem.*, 3, 535–548.

Buffle, J., 1988, *Complexation Reactions in Aquatic Systems: An Analytical Approach*, Ellis Horwood, Chichester, UK, 692 pp.

Buffle, J., DeVitre, R. R., Perret, D., and Leppard, G. G., 1989, Physicochemical characteristics of a colloidal iron phosphate species formed at the oxic–anoxic interface of a eutrophic lake, *Geochim. Cosmochim. Acta*, 53, 399–408.

Buffle, J., Perret, D., and Newman, M., 1992, The use of filtration and ultrafiltration for size fractionation of aquatic particles, colloids, and macromolecules. In: J. Buffle, and H. van Leeuwen, eds., *Environmental Particles*, Vol. 1, Lewis Publishers, Boca Raton, FL, pp. 171–230.

Cerda, C. M., 1987, Mobilization of kaolinite fines in porous media, *Colloids Surf.*, 27, 219–241.

Chin, Y. P., and Gschwend, P. M., 1991, The abundance distribution, and configuration of porewater organic colloids in recent sediments. *Geochim. Cosmochim. Acta*, 55, 1309–1318.

Chin, Y. P., and Gschwend, P. M., 1992, Partitioning of polycyclic aromatic hydrocarbons to marine porewater colloids. *Environ. Sci. Technol.*, 26, 1621–1626.

Chittleborough, D. J., Smettem, K. R., Costsaris, E., and Leaney, F. W., 1992, Seasonal changes in pathways of dissolved organic carbon through a hillslope soil (Xeralf) with contrasting texture. *Aust. J. Soil Res.*, 30, 465–476.

Danielsson, L. G., 1982, On the use of filters for distinguishing between dissolved and particulate fractions in natural waters, *Water Res.*, 16, 179–182.

Davis, J. A., 1984, Complexation of trace metals by adsorbed organic matter, *Geochim. Cosmochim. Acta*, 48, 679–691.

Davis, J. A., and Kent, D. B., 1990, Surface complexation modeling in aqueous geochemistry. In: M. E. Hochella, Jr., and A. F. White, eds., *Mineral–Water Interface Geochemistry*, Reviews in Mineralogy, Vol. 23, Mineralogical Society of America, Washington, D.C., pp. 177–248.

Deerlove, J. P. L., Longworth, G., Ivanovich, M., J. I., Delakowitz, B., and Zeh, P., 1991, A study of ground water-colloids and their geochemical interactions with natural radio nuclides in Gorleben aquifer systems. *Radiochim. Acta*, 52/53, 83–89.

Degueldre, C., Baeyans, B., Goerlich, W., Riga, J., Verbist, J., and Stadelmann, P., 1989, Colloids in water from a subsurface fracture in granitic rock, Grimsel Test Site, Switzerland. *Geochim. Cosmochim. Acta*, 53, 603–610.

Detenbeck, N. E., and Brezonik, P. L., 1991, Phosphorus sorption by sediments from a softwater seepage lake. 1. An evaluation of kinetic and equilibrium models, *Environ. Sci. Technol.*, 25, 395–402.

Domenico, P. A., and Schwartz, F. W., 1990, *Physical and Chemical Hydrogeology*, John Wiley & Sons, New York, 824 pp.

Driscoll, F. G., 1986, *Groundwater and Wells*, 2nd ed., Johnson Division, St. Paul, MN, 1089 pp.

Dunnivant, F. M., Jardine, P. M., and Taylor, D. L., 1992, Transport of naturally occurring dissolved organic carbon in laboratory columns containing aquifer material, *Soil Sci. Soc. Am. J.*, 56, 437.

Dzomback, D. A., and Morel, F. M. M., 1990, *Surface Complexation Modeling: Hydrous Ferric Oxide*, John Wiley & Sons, New York, 393 pp.

Enfield, C. G., and Bengtsson, G., 1988, Macromolecular transport of hydrophobic contaminants in aqueous environments. *Ground Water*, 26, 64–70.

Freeze, R. A., and Cherry, J. A., 1979, *Groundwater*, Prentice-Hall, Englewood Cliffs, NJ, 604 pp.

Gerba, C. P. and Bitton, G., 1984, Microbial pollutants: their survival and transport pattern to ground water. In: G. Bitton and C. P. Gerba, eds., *Ground water Pollution Microbiology*, Wiley-Interscience, New York.

Gibbs, R. J., 1983, Effect of natural organic coatings on the coagulation of particles, *Environ. Sci Technol.*, 17, 237–240.

Greenberg, A. E., Clesceri, L. S., and Eaton, A. D., eds., 1992, *Standard Methods for the Examination of Water and Wastewater*, 18th edition, American Public Health Association, American Water Works Association, Water Environment Federation, Washington, D.C.

Gregg, S. J., and Sing, K. S. W., 1982, *Adsorption, Surface Area and Porosity*, 2nd edition, Academic Press, London, 303 pp.

Gschwend, P. M., and Reynolds, M. D., 1987, Monodisperse ferrous phosphate colloids in an anoxic ground water plume, *J. Contam. Hydrol.*,1, 309–327.

Gschwend, P. M., Backhus, D. A., MacFarlane, J. K., and Page, A. L., 1990, Mobilization of colloids in ground water due to infiltration of water at a coal ash disposal site, *J. Contam. Hydrol.*, 6, 307–320.

Gutman, R. G., 1987, *Membrane Filtration: The Technology of Pressure Driven Crossflow Processes*, Adam Hilger, Bristol, UK, 210 pp.

Harvey, R. W., and Garabedian, S., 1991, Use of colloid filtration theory in modeling movement of bacteria through a contaminated sandy aquifer, *Environ. Sci Technol.*, 25, 178–185.

Harvey, R. W., George, L. H., Smith, R. L., and Leblanc, D. R., 1989, Transport of microspheres and indigenous bacteria through a sandy aquifer: results of natural and forced-gradient tracer experiments, *Environ. Sci. Technol.*, 23, 51–56.

Hem, J. D., 1985, USGS. Water-Supply Paper, v. 2254. *Study and Interpretation of the Chemical Characteristics of Natural Water*, Washington, D.C., 263 pp.

Herzog, B., Pennino, J., and Nielsen, G., 1991, Ground-water sampling. In: Nielsen, D. M., ed., *Practical Handbook of Ground-Water Monitoring*, Lewis Publishers, Chelsea, MI, pp. 449–499.

Hoffman, M. R., Yost, E. C., Eisenreich, S. J., and Maier, W., 1981, Characterization of soluble and colloidal-phase metal complexes in river water by ultrafiltration. A mass balance approach. *Environ. Sci. Technol.*, 15, 655–661.

Honeyman, B. D., and Santschi, P. H., 1991, Coupling adsorption and particle aggregation: laboratory studies of "colloid pumping" using ^{59}Fe-labeled hematite. *Environ. Sci. Technol.*, 25, 1739–1746.

Horowitz, A. J., Elrick, K. A., and Hopper, R. C., 1989, A comparison of dewatering methods for the separation and concentration of suspended sediment for sequential trace element analysis. *Hydrol. Process.*, 2, 163–184.

Hunter, R. J., 1981, *Zeta Potential in Colloid Science: Principles and Applications*, Academic Press, Orlando, FL, 386 pp.

Hunter, K. A., and Liss, P. S., 1979, The surface charge of suspended particles in estuarine and coastal waters, *Nature*, 282, 823–825.

IGWMC, 1992, *International Ground Water Modeling Center, Software Catalog*, Institute for Ground-Water Research & Education, Golden, CO 80401-1887, 34 pp.

Kennedy, V. C., Zellweger, G. W., and Jones, B. F., 1974, Filter pore-size effects on the analysis of Al, Fe, Mn and Ti in water, *Water Resources Res.*, 10, 785–790.

Kent, R. T., and Payne, K. E., 1988, Sampling ground water monitoring wells—special quality assurance and quality control considerations. In: L. Keith, ed., *Principles of Environmental Sampling*, American Chemical Society, Washington, D.C., Chapter 15, pp. 231–246.

Kim, J. I., Zeh, P., and Delakowitz, B., 1992, Chemical interactions of actinide ions with ground water colloids in Gorleben aquifer systems, *Radiochim. Acta*, 58/59, 147–154.

Kimball, B. A., McKnight, D. M., Wetherbee, G. A., and Harnish, R. A., 1992, Mechanisms of iron photoreduction in a metal-rich, acidic stream (St. Kevin Gulch, Colorado), *Chem. Geol.*, 96, 227–239.

Kimball, B. A., Callendar, E. C., and Axtmann, E. V., 1995, Effects of colloids on metal transport in a river receiving acid mine drainage, upper Arkansas River, Colorado: *Applied Geochem.*, 10, 285–306.

Kita, S. F., Folger, H. S., and Reed, M. G., 1987, Effect of pH on colloidally induced fines migration, *J. Colloid Interface Sci.*, 118, 158–168.

Kuwabara, J. S., and Harvey, R. W., 1990, Application of hollow fiber, tangential flow device for sampling suspended bacteria and particles from natural waters, *J. Environ. Qual.*, 19, 625–629.

Langmuir, D., and Whittemore, D. O., 1971, Variations in the stability of precipitated ferric oxyhydroxides. In: Hem, J., ed., *Non-equilibrium Systems in Natural Water Chemistry*, Advances in Chemistry Series, No. 106, American Chemical Society, Washington, D.C., pp. 209–234.

Lasaga, A. C., 1990, Atomic treatment of mineral-water surface reactions. In: M. E. Hochella, Jr., and A. F. White, eds., *Mineral–Water Interface Geochemistry*, *Reviews in Mineralogy*, Vol. 23, Mineralogical Society of America, Washington, D.C., pp. 17–86.

Laxen, D. P. H., 1983, Size distribution of iron and manganese in freshwaters. *Geochim. Cosmochim. Acta*, 47, 731–741.

Laxen, D. P. H., and Chandler, I. M., 1982, Comparison of filtration techniques for size distribution in freshwaters. *Anal. Chem.*, 54, 1350–1355.

Leenheer, J. A., Meade, R. H., Taylor, H. E., and Pereira, W. E., 1989, Sampling, fractionation, and dewatering of suspended sediment from the Mississippi River for geochemical and trace-contaminant analysis. In: G. E. Mallard and S. E. Ragone, eds., *US Geological Survey Toxic Substances Hydrology Program—Proceedings of the Technical Meeting, Phoenix, Arizona, September, 26–30, 1988*, Water-Resources Investigations Report 88-4 220.

Leppard, G. G., 1992, Evaluation of electron microscope techniques for the description of aquatic colloids. In: J. Buffle and H. P. van Leeuwen, eds., *Environmental Particles*, Vol. 1, Lewis, Boca Raton, FL, pp. 231–290.

Liang, L., and Morgan, J. J., 1990, Coagulation of iron oxide particles in the presence of organic materials. In: D. Melchior and R. L. Bassett, eds., *Chemical Modeling in Aqueous Systems II*, *ACS Symposium Series*, No. 416, American Chemical Society, Washington, D.C., pp. 293–309.

Lieser, K. H., and Hill, R., 1992, Chemistry of thorium in the hydrosphere and in the geosphere. *Radiochim. Acta*, 56, 141–151.

Lowson, R. T., and Short, S. A., 1990, Application of the uranium decay series to the study of ground water colloids. In: R. Beckett, ed., *Surface and Colloid Chemistry in Natural Waters and Water Treatment*, Plenum Press, New York, pp. 71–86.

Marley, N. A., Gaffney, J. S., Orlandini, K. A., and Dugue, C. P., 1991, An evaluation of an automated hollow-fiber ultrafiltration apparatus for the isolation of colloidal materials in natural waters. *Hydrol. Process*, 5, 291–299.

Martin, R. E., Bouwer, E. J., and Hanna, L. M., 1992, Application of clean-bed filtration theory to bacteria deposition in porous media. *Environ. Sci. Technol.*, 26, 1053–1058.

McCarthy, J. F., and Zachara, J. M., 1989, Subsurface transport of contaminants. *Environ. Sci. Technol.*, 23, 496–502.

McDowell-Boyer, L. M., Hunt, J. R., and Sitar, N., 1986, Particle transport through porous media. *Water Resources Res.*, 22, 1901–1921.

McDowell-Boyer, L. M., 1992, Chemical mobilization of micron-sized particles in saturated porous media under steady flow conditions: *Environ. Sci. Technol.*, 26, 586–593.

Means, J. C., and Wijayaratne, R., 1982, Role of natural colloids in the transport of hydrophobic pollutants, *Science*, 215, 968–970.

Moore, W. J., 1972, *Physical Chemistry, 4th edition*, Prentice-Hall, Englewood Cliffs, NJ, 977 pp.

Moran, S. B., and Moore, R. M., 1989, The distribution of colloidal aluminum and organic carbon in coastal and open waters off Nova Scotia. *Geochim. Cosmochim. Acta*, 53, 2519–2527.

Morel, F. M. M., and Gschwend, P. M., 1987, The role of colloids in the partitioning of solutes in natural waters, in: Stumm W., ed., *Aquatic Surface Chemistry*, John Wiley & Sons, New York, pp. 405–522.

Murphy, D. M., Garbarino, J. R., Taylor, H. E., Hart, B. T., and Beckett, R., 1993, Determination of size and element composition distributions of complex colloids using sedimentation field-flow fractionation/inductively coupled plasma-mass spectroscopy, *J. Chromatogr.*, 642, 457–467

Newton, P. P., and Liss, P. S., 1987, Positively charged suspended particles: Studies in an iron-rich river and its estuary. *Limnol. Oceanogr.*, 32, 1267–1276.

Nielsen, D. M., ed., 1991, *Practical Handbook of Ground-Water Monitoring*, Lewis, Chelsea, MI, 717 pp.

Nightingale, H. I., and Bianchi, W. C., 1977, Ground water turbidity resulting from artificial recharge. *Ground Water*, 15, 146–152.

Nomizu, T., Goto, K., and Mizuike, A., 1988, Electron microscopy of nanometer particles in freshwater. *Anal. Chem.*, 60, 2653–2656.

Nordstrom, D. K., and Munoz, J. L., 1986, *Geochemical Thermodynamics*, Blackwell Scientific Publications, Palo Alto, CA, 477 pp.

O'Melia, C. R., 1987, Particle-particle interactions. In: Stumm W., ed., *Aquatic Surface Chemistry*, John Wiley & Sons, New York, pp. 385–403

O'Melia, C. R., 1990, Kinetics of colloid chemical processes in aquatic systems. In: Stumm, W., ed., *Aquatic Chemical Kinetics*, John Wiley & Sons, New York, pp. 447–474.

Palmer, C. D., Keely, J. F., and Fish, W., 1987, Potential for solute retardation on monitoring well sand packs and its effect on purging requirements for ground water sampling. *Ground Water Monit. Rev.*, Spring 1987, pp. 40–47.

Pankow, J. F., 1991, *Aquatic Chemical Concepts*, Lewis, Chelsea, MI, 673 pp.

Parks, G. A., 1990, Surface energy and adsorption at mineral/water interfaces: an introduction. In: M. E. Hochella, Jr., and A. F. White, eds., *Mineral–Water Interface Geochemistry, Reviews in Mineralogy*, Vol. 23, Mineralogical Society of America, Washington, D.C., pp. 133–175.

Penrose, W. R., Polzer, W. L., Essington, E. H., Nelson, D. M., and Orlandini, K. A., 1990, Mobility of plutonium and americium through a shallow aquifer in a semiarid region. *Environ. Sci. Technol.*, 24, 228–234.

Perdue, E. M., and Wolfe, N. L., 1982, Modification of pollutant hydrolysis kinetics in the presence of humic substances. *Environ. Sci. Technol.*, 16, 847–852.

Puls, R., and Powell, R., 1992, Transport of inorganic colloids through natural aquifer material: implications for contaminant transport, *Environ. Sci. Technol.*, 26, 614–621.

Puls, R. W., and Barcelona, M. J., 1989, *Filtration of Ground Water Samples for Metals Analysis*, EPA/600/J-89/278, R. S. Kerr Environmental Research Laboratory, US Environmental Protection Agency, Ada, OK, and Illinois State Water Survey, 10 pp.

Puls, R. W., Eychaner, J. H., and Powell, R. M., 1990, Colloid-facilitated transport of inorganic contaminants in ground water, Part I. Sampling considerations, Environmental Research Brief, EPA/600/M-90/023, US Environmental Protection Agency, Washington, D.C.

Puls, R. W., Powell, R. M., Clark, D. A., and Paul, C. J., 1991, Facilitated transport of inorganic contaminants in ground water: Part II. Colloid transport, Environmental Research Brief, EPA/600/M-91/040, US Environmental Protection Agency, Washington, DC.

Ranville, J. F., Harnish, R. A., and McKnight, D. M., 1991a, Particulate and colloidal organic material in Pueblo Reservoir, Colorado: Influence of autochthonous source on chemical composition. In: R. Baker, ed., *Organic Substances and Sediments in Water*, Vol. 1, Lewis, Chelsea, MI, pp. 47–74.

Ranville, J. F., Smith, K. S., McKnight, D. L., Macalady, D. L., and Rees, T. F., 1991b, Effect of organic matter co-precipitation and sorption with hydrous iron oxides on electrophoretic mobility of particles in acid mine drainage. In G. E. Mallard and D. A. Aronson, eds., *US Geological Survey Toxic Substances Hydrology Program: Proceedings of the Technical Meeting*, Monterey, California, March 11–15, 1991; US Geological Survey, Water Resources Investigations Report 91-4034, Washington, DC, pp. 422–427.

Rees, T. F., 1987, A review of light-scattering techniques for the study of colloids in natural water. *J. Contam. Hydrol.*, 1, 425–439.

Rees, T. F., and Ranville, J. F., 1990, Collection and analysis of colloidal particles transported in the Mississippi River, USA. *J. Contam. Hydrol.*, 6, 214–250.

Rees, T. F., Leenheer, J. A., and Ranville, J. F., 1991, Use of a single-bowl continuous-flow centrifuge for dewatering suspended sediments: effect on the sediment physical and chemical characteristics. *Hydrol. Processes*, 5, 201–214.

Reynolds, M. D., 1985, *Colloids in Ground water*, M.S. Thesis, Massachusetts Institute of Technology, Cambridge, MA, 92 pp.

Ryan, J. N., and Gschwend, P. M., 1990, Colloid mobilization in two Atlantic coastal plain aquifers: field studies. *Water Resources Res.*, 26, 307–322.

Salbu, B., Bjørnstad, H. E., Lindstrøm, N. S., Lydersen, E., Brevik, E. M., Rambaek, J. P., and Paus, P. E., 1985, Size fractionation techniques in the determination of elements associated with particulate or colloidal material in natural fresh waters. *Talanta*, 32, 907–913.

Scholl, M. A., and Harvey, R. W., 1992, Laboratory investigations on the role of sediment surface and ground water chemistry in the transport of bacteria through a contaminated sandy aquifer. *Environ. Sci. Technol.*, 26, 1410–1417.

Sigg, L., 1985, Metal transfer mechanisms in lakes: the role of settling particles. In: W. Stumm, ed., *Chemical Processes in Lakes*, John Wiley & Sons, New York, pp. 283–310.

Sigleo, A. C., and Helz, G. R., 1981, Composition of estuarine colloidal material: major and trace elements, *Geochim. Cosmochim. Acta*, 45, 283–310.

Smith, J. S., Steele, D. P., Malley, M. J., and Bryant, M. A., 1988, Ground water sampling. In: Keith, L., ed., *Principles of Environmental Sampling*, American Chemical Society, Washington, D.C., pp. 255–260, Chapter 17.

Smith, K. S., 1992, Sorption of trace elements by earth minerals: an oveview with examples relating to mineral deposits. In: Plumlee, G. S. and M. J. Logsdon, eds., *The Environmental Geochemistry of Mineral Deposits*, *Reviews in Economic Geology*, Vol. 6A, Society of Economic Geologists, Chapter 7.

Smith, K. S., Ficklin, W. H., Plumlee, G. S., and Meier, A. L., 1992, Metal and arsenic partitioning between water and suspended sediment at mine-drainage sites in diverse geologic settings. In: Y. K. Kharaka, and A. S. Maest, eds., *Water-Rock Interactions*, *Proceedings of the 7th International Symposium on Water–Rock Interactions*, WRI-7, Balkeme, Rotterdam, pp. 443–447.

Salomons, W., and Forstner, U., 1984, *Metals in the Hydrocycle*, Springer-Verlag, New York, p. 349.

Sposito, G., 1992, Characterization of particle surface charge. In: Buffle, J., and van Leeuwen, H. P., eds., *Environmental Particles*, Vol. 1, Lewis, Boca Raton, FL, pp. 291–314.

Stumm, W., and Morgan J. J., 1981, *Aquatic Chemistry*, John Wiley & Sons, New York, 780 pp.

Tanizaki, Y., Shimokawa, T., and Nakamura, M., 1992, Physiochemical separation of trace elements in river waters by size fractionation. *Environ. Sci. Technol.*, 26, 1433–1444.

Taylor, H. E., Garbarino, J. R., and Brinton, T. I., 1990, The occurrence and distribution of trace metals in the Mississippi River and its tributaries. *Sci. Total Environ.*, 97/98, 369–384.

Taylor, H. E., Garbarino, J. R., Murphy, D. M., and Beckett, R., 1992, Inductively coupled plasma-mass spectrometry as an element-specific detector for field flow fractionation separation. *Anal. Chem.*, 64, 2036–2041.

Tessier, A., 1992, Sorption of trace elements on natural particles in oxic environments, in J. Buffle and van Leeuwen, H., eds., *Environmental Particles*, Vol. 1, Lewis Publishers, Boca Raton, FL, pp. 425–454.

Thoma, G. J., Koulermos, A. C., Valsaraj, K. T., Reible, D. D., and Thibodeaux, L. J., 1991, The effects of pore-water colloids on the transport of hydrophobic organic compounds from bed sediments. In: Baker, R., ed., *Organic Substances and Sediments in Water*, Vol. 1, Lewis, Chelsea, MI.

Tipping, E., and Higgins, D. C., 1982, The effect of adsorbed humic substances on the colloid stability of hematite particles. *Colloids Surf.*, 5, 85–92.

Tipping, E., and Cooke, D. C., 1982, The effects of adsorbed humic substances on the surface charge of goethite (α FeOOH) in freshwaters. *Geochim. Cosmochim. Acta* 46, 75 80.

Tipping, E., Griffith, J. R., and Hilton, J., 1983, The effect of adsorbed humic substances on the uptake of copper(II) by goethite, *Croatica. Chem. Acta*, 56, 613–621.

US Geological Survey, 1982, *National Handbook of Recommended Methods for Water-Data Acquisition*, U.S. Department of the Interior, Reston, VA.

US EPA,1986, *RCRA Ground-Water Monitoring Technical Enforcement Guidance Document (TEGD)*, OSWER-9950.1, U.S. Government Printing Office Washington, D.C., 208 pp. plus appendices.

Vinten, A. J. A., Yaron, B., and Nye, P. H., 1983, Vertical transport of pesticides into soil when adsorbed on suspended particles. *J. Agric. Food Chem.*, 31, 662–664.

Waber, U. E., Lienert, C., and Von Gunten, H. R., 1990, Colloid-related infiltration of trace metals from a river to shallow ground water. *J. Contam. Hydrol.*, 6, 251–265.

Wagemann, R., and Brunskill, G. J., 1975, The effect of filter pore-size on analytical concentrations of some elements in filtrates of natural water. *Intern. J. Environ. Anal. Chem.*, 4, 75–84.

Walling, D. E., and Moorehead, P. W., 1989, The particle size characteristics of fluvial suspended sediment: an overview. *Hydrobiologia*, 176/177, 125–149.

White, A. F., and Peterson, M. L., 1990, Role of reactive-surface-area characterization in geochemical kinetic models. In: D. C. Melchior and R. L. Bassett, eds., *Chemical Modeling in Aqueous Systems II*, ACS Symposium Series, No. 416, American Chemical Society, Washington, D.C., pp. 461–477.

Whitehouse, B. G., Petrick, G., and Ehrhardt, M., 1986, Crossflow filtration of colloids from Baltic Sea water. *Water Res.*, 20, 1599–1601.

Whitehouse, B. G., Yeats, P. A., and Strain, P. M., 1990, Cross-flow filtration of colloids from aquatic environments. *Limnol. Oceanogr.*, 35, 1368–1375.

Yao, K., Habibian, M. T., and O'Melia, C. R., 1971, Water and wastewater filtration: concepts and applications. *Environ. Sci. Technol.*, 5, 1105–1112.

CHAPTER 3

Environmental Chemistry of Trace Metals

JANET G. HERING AND STEPHAN KRAEMER

ABSTRACT. A common set of geochemical and biogeochemical processes controls the cycling of trace metals in various aquatic systems, ranging from ground waters to open ocean waters. Knowledge of these processes is critical for understanding the impact of anthropogenic metal pollution on ecosystems and human health. The thermodynamics as well as the mechanisms and kinetics of biogeochemical reactions involved in these processes are discussed in this chapter.

Speciation of metals is given special consideration because it influences not only the mobility of metals, but also their bioavailability and toxicity. For several metals, significant complexation by organic ligands has been observed in sea water. It has been proposed that microorganisms may be a source of such organic ligands, which in turn may facilitate biological uptake of metals or lower their toxicity. Various organic ligands have been identified in freshwater, including strong anthropogenic ligands. Trace metal speciation is an area of active research, and many questions regarding the structure of metal–organic complexes and their fate in the environment have yet to be answered.

Speciation and partitioning of metals in the environment have been modeled successfully using hydrochemical equilibrium and transport models. However, physical constraints in heterogeneous systems and slow reaction kinetics can lead to slow attainment of equilibrium in natural waters. While the evaluation of the kinetics of hydrochemical processes introduces an additional level of complexity, a critical assessment of the equilibrium approach is required for various hydrochemical processes. Dissolution of silicate and oxide minerals, slow ligand exchange reactions, and heterogeneous reactions of metal organic complexes with solid phases are among the examples for possible kinetic limitations in aquatic systems.

"...the strongest argument of detractors is that the fields are devastated by mining operations.... Further, when the ores are washed, the water which has been used poisons the brooks and streams."

DE RE METALLICA, 1556
GEORGIUS AGRICOLA

INTRODUCTION

The occurrence of artifacts crafted from native copper and dating to the Chalcolithic Age (ca. 5000 B.C.E.) is evidence of the importance of metals and metal-working throughout human history (Tylecote, 1992). Environmental pollution from mining of metal ores and smelter emissions reached significant levels long before the onset of the industrial revolution. Greenland ice cores document air pollution with lead on a hemispheric scale since the rise of the Greek civilization, two millennia ago (Hong et al., 1994). The adverse impact of such metal pollution to the environment has been known for centuries (Agricola, 1950). The hazard to human health associated with occupational exposure to mercury, which inspired the common expression "mad as a hatter," was documented in 1860 (Follmann, 1987).

The exploitation of mineral reserves and metal production and use have increased exponentially since the industrial revolution. The magnitude of the environmental impacts of these activities have increased consequently. Recent world production figures for metals ranging from ca. 10^4 to 10^7 metric tons/year are given in Table 3.1, and estimated global discharges of metals to the environment are listed in Table 3.2 (Moore, 1991). Lead (Pb) concentrations in Greenland ice indicate a gradual increase in hemispheric lead pollution from ca. 1750 to 1950, a dramatic increase in the 1950s and 1960s, and a significant decrease from the mid-1960s to the present day, attributable to the phase-out of leaded gasoline in the United States (Boutron et al., 1991). Even "local" pollution problems associated with metals can be quite widespread. The largest sites regulated under the Comprehensive Environmental Response, Compensation and Liability Act (Superfund), the Clark Fork sites, comprise approximately 50,000 acres along 140 miles of the Clark Fork River (Montana). These sites have been contaminated by decades of mining, milling, and smelting operations; principal contaminants include cadmium, copper, lead, and zinc (Miller, 1992; Kemble et al., 1994).

Contemporary concern over the environmental chemistry of trace metals arises from the potential deleterious effects of metal pollution on ecosystems and human health. Unlike xenobiotic organic compounds, metals are natural, not synthetic, materials. Human activities may, however, profoundly alter the distribution of metals in the

TABLE 3.1
Current World Metal Production[a]

Metal	Quantity (10^3 metric tons/year)	Year
Aluminum	15,300	1986
Cadmium	19,000	1986
Chromium	11,000	1980
Cobalt	31	1980
Copper	7,660	1980
Lead	3,100	1980
Mercury	7,100	1980
Nickel	760	1980
Silver	11	1980
Tin	250	1980
Zinc	5,230	1980

[a]*Data from Moore (1991).*

TABLE 3.2
Estimated Global Discharges of Metals (in 10^3 metric tons/year)[a]

Metal	Water	Air	Soil
Cadmium	9.4	7.6	22
Chromium	142	30	896
Copper	112	35	954
Lead	138	332	796
Mercury	4.6	3.6	8.3
Nickel	113	56	325
Tin	—	6.4	—
Zinc	226	132	1,370

[a]*Data from Moore (1991).*

environment, introducing metals into sensitive ecosystems and increasing human exposure to them. The extent to which the natural geochemical cycling of metals is perturbed by anthropogenic inputs—or, conversely, the extent to which metallic contaminants are incorporated into natural cycles—is critical for determination of the ecosystem and human health effects of metal pollution.

This chapter will briefly review the chemistry of metals in natural systems and the processes involved in the biogeochemical cycling of metals. The extent and significance of metal–organic interactions will be discussed. Current approaches to the study of the environmental chemistry of trace metals will be presented; the limitations of these approaches and possible avenues for further research will be addressed.

GEOCHEMICAL AND BIOGEOCHEMICAL PROCESSES IN TRACE METAL CYCLING

Even in widely varying aquatic environments (ranging, for example, from contaminated ground waters to pristine open ocean waters), the cycling of trace metals involves a common set of biogeochemical processes, which are illustrated schematically in Figure 3.1. Processes that remove metals to solid phases are particularly important in mitigating the effects of contaminant metals introduced into aquatic ecosystems. Ultimately, the immobilization of such metal inputs (e.g., by permanent burial in sediments) reduces the exposure of the biota to potentially toxic metal species. A critical distinction between the ecotoxicology of trace metals and toxic organic compounds should be noted. Toxic organic compounds can be broken down into nontoxic constituents; barring transmutation, the potential toxicity of metals can never be completely eliminated.

The biogeochemical processes outlined in Figure 3.1 determine the distributions of metals among the various environmental compartments. Obviously, in different aquatic systems (i.e., lakes, aquifers, etc.), the predominance of specific processes will vary. Thus, for example, adsorption of metals onto the mineral surfaces is an important mechanism for metal removal from ground water because of the high mineral surface area and low biomass that are characteristic of aquifers (Freeze and Cherry, 1979). In contrast, in oceanic surface waters where the particulate load is primarily composed

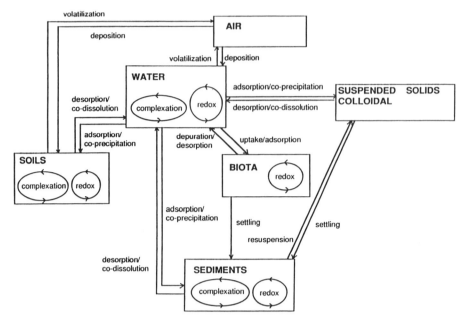

Figure 3.1. Schematic diagram of processes controlling the biogeochemical cycling of metals in aquatic environments.

of phytoplankton, biological uptake, and/or adsorption onto algal surfaces dominate metal removal processes. Consequently, the depth profiles of many metals bear a marked similarity to nutrient profiles (Bruland, 1980; Landing and Bruland, 1980; Gordon et al., 1982).

Although not easily included in Figure 3.1, oxic–anoxic boundaries, which may occur in the water column, sediments, or soils, have a dramatic influence on trace metal chemistry. Metal solubilities often vary significantly with the oxidation state of the metal; thus redox reactions of metals, such as iron, manganese, and chromium, in natural waters result in the precipitation or dissolution of solid phases (Davison, 1985). The formation or elimination of surfaces resulting from such redox reactions can, in turn, control the dissolved concentrations of potential adsorbates, including other trace metals, metal oxyanions, inorganic anions, and organic micropollutants (Armstrong et al., 1987; Chambers and Odum, 1990). Furthermore, the precipitates can serve either as reactant or as catalyst in redox reactions, influencing the cycling of redox active metals (Stumm and Sulzberger, 1992; Deng and Stumm, 1994).

Redox cycling of metals can also occur in oxic surface waters as a result of photochemical processes. Diel cycles (Collienne, 1983; McKnight et al., 1988; Behra and Sigg, 1990; Sulzberger et al., 1990) in the concentration of iron(II), depth profiles (Moffett and Zika, 1987; Sunda and Huntsman, 1988) of copper(I) and manganese(II), and irradiation experiments with natural water samples (Sunda et al., 1983; Wells et al., 1991; Waite and Szymczak, 1993) indicate that low concentrations of metals in their thermodynamically unstable reduced forms are maintained by the coupled processes of photoreduction and reoxidation.

Importance of Trace Metal Speciation

The partitioning of metals between dissolved and particulate phases critically influences metal transport, reactivity, and bioavailability. It is widely recognized, however, that metal bioavailability and toxicity are not determined merely by the total, dissolved metal concentrations. The distribution of the metal among various inorganic and organic metal species (i.e., metal speciation) governs bioavailability, which is related to the free metal concentration or activity (as a thermodynamic measure of metal reactivity) rather than to the total, dissolved metal concentration (Anderson and Morel, 1982; Morel, 1986). Thus the deleterious effects of metal toxicity can be mitigated by complexation of the metal with organic or inorganic ligands (Sunda and Guillard, 1976; Anderson and Morel, 1978). Competitive interactions among metals may also influence metal bioavailability. For example, copper toxicity to algae becomes more pronounced if the organisms are stressed by manganese limitation (Sunda et al., 1981; Sunda and Huntsman, 1983). Such effects can usually be explained with an equilibrium model of metal complexation, although kinetic effects, where complexation reactions are slow compared to biological uptake, have been observed in culture studies (Anderson and Morel, 1978). Complexation kinetics may be even more important at the low metal concentrations typical of natural waters (Hudson and Morel, 1990, 1993). An exception to the dependence of biological effects on free metal ion concentration is observed when an intact complex is taken up by the organism, as may occur by passive diffusion of lipid-soluble species through cell membranes (Luoma, 1983; Morel, 1986; Stauber and Florence, 1987; Phinney and Bruland, 1994).

Metal speciation may also affect the mobility of metals in aquatic environments and thus the exposure of organisms to metals. The effects of complexation on processes by which metals are removed from the aqueous phase, particularly adsorption onto mineral surfaces, may influence the residence time of dissolved metals in aquatic environments. Laboratory studies have demonstrated that adsorption of metals to oxide surfaces can be markedly influenced by synthetic organic ligands and by natural dissolved organic matter (DOM). Organic ligands can both increase metal adsorption, through coadsorption of metals and organics (usually at low pH), and decrease metal adsorption, through competition between the dissolved organic ligand and surface binding sites for the metal (usually at high pH) (Davis and Leckie, 1978; Davis, 1984; Girvin et al., 1996; Zachara et al., 1995a,b).

Questions concerning the effects of metal speciation on the biogeochemical cycling of metals become increasingly important as evidence for metal complexation in natural waters accumulates. Strong metal complexation by humic and fulvic substances (McKnight et al., 1983; Cabaniss and Shuman, 1988a,b) and root exudates (Ochs et al., 1993) have been demonstrated in laboratory experiments. Anthropogenic ligands with high affinities for metals such as EDTA and NTA have been found in surface waters receiving industrial and domestic wastewater (Gardiner, 1975; Kuehn and Brauch, 1988) Field measurements indicate that the extent of metal complexation in seawater is close to 100% for reactive metals such as copper. Significant complexation has been reported even for less reactive metals like zinc and cadmium (Table 3.3). Few, if any, analytical techniques can provide a direct measurement of metal–organic complexes, and the ligands responsible for metal complexation under ambient conditions have not been identified. However, marine phytoplankton have been shown to synthesize metal

TABLE 3.3
Extent of Organic Complexation of Metals in Surface Waters[a]

Metal	Percent Complexed	Water Type	Method	Reference
Cu	>99.4	Oceanic	Electrochemical	1
	>99	Coastal	Electrochemical	2
	>99.9		Ligand competition	3
	>99		Bioassay	4
	>99.99	Estuarine	Electrochemical	5
	80–92	Coastal/inland	Electrochemical, chromatography	6
Fe	~99	Oceanic/coastal	Electrochemical	7
Zn	>95	Oceanic	Electrochemical	8
	60–95	Estuarine	Electrochemical	5
Pb	50–70	Oceanic	Electrochemical	9
	~55		Electrochemical	10
Cd	0	Oceanic	Radiotracer addition, chromatography	11
	0–70		Electrochemical	12
Mn	0	Oceanic	Radiotracer addition, chromatography	11
Ni	30–50	Coastal	Electrochemical	13
	30–50	Coastal/inland	Electrochemical, chromatography	6
Co	45–100	Oceanic	Electrochemical	14

[a]*Adapted from Morel and Hering (1993).*
References: (1) Coale and Bruland (1990), (2) Hering et al. (1987), (3) Moffett and Zika (1987), (4) Sunda and Ferguson (1983), (5) van den Berg et al. (1987), (6) Donat et al. (1994), (7) Gledhill and van den Berg (1994), (8) Donat and Bruland (1990), (9) Capodaglio et al. (1990), (10) Scarponi et al. (1995), (11) Sunda (1984), (12) Bruland (1992), (13) van den Berg and Nimmo (1987), (14) Zhang et al. (1990).

coordinating ligands for intracellular trace metal regulation (Ahner et al., 1994). Strong organic ligands have also been found in culture media of marine and freshwater algae (McKnight and Morel, 1979; Imber and Robinson, 1983; Trick et al., 1983; Zhou and Wangersky, 1989; Kozarac et al., 1989; Moffett et al., 1990). Release of such ligands might serve to facilitate metal uptake by microorganisms (Wilhelm and Trick, 1994) or lower metal toxicity. These regulatory mechanisms have received particular attention since it has been suggested that availability of iron (Martin et al., 1991; Wells et al., 1995) or zinc (Morel et al., 1994) limits phytoplankton productivity in nutrient-rich areas of the oceans. While most studies on the inhibition of plankton growth have focused on single metals, it has been recognized that complex synergistic and antagonistic effects among multiple trace metals may have an important influence on biological communities (Bruland et al., 1991).

Current Approaches

The current understanding of the biogeochemical cycling of metals in aquatic systems derives from both empirical observations (laboratory and/or field studies) and modeling efforts. Field studies in surface waters have generally focused on the evaluation of the residence time of metals and the mechanisms of metal removal from the water column (Santschi, 1984). To this end, concentrations of total or total dissolved metals, the depth variations in concentrations, and the concentrations of metals in particulate material captured in sediment traps have been extensively studied (Sigg, 1985; Murray, 1987). Both field and laboratory studies (Santschi, 1984; Sigg, 1985; Murray,

1987; Honeyman et al., 1988; Evans, 1989; Muller and Sigg, 1990; Morel and Hering, 1993, Benoit et al., 1994) have identified adsorption as a critical mechanism for removal of metals from the dissolved phase and the consequent maintenance of many metal concentrations below the solubility limit of pure solid phases of the metal (e.g., oxides, carbonates, etc.).

Although metal speciation in natural waters has also been widely studied, it has only infrequently been considered as a possibly important factor influencing the flux of metals from or to the water column. Under conditions of estuarine mixing, remobilization of cadmium from riverine suspended particles has been shown to result from the increasing stability of dissolved cadmium chloro complexes with increasing chloride concentrations (Comans and van Dijk, 1988). A recent study of metal cycling in lakes has suggested that the lack of correspondence between the ratios of zinc and copper in settling particles and the ratios of the total, dissolved concentrations of the metals in the water column may be attributable to the greater complexation of copper by naturally occurring organic ligands (Sigg et al., 1995).

Generally, field studies of metal speciation have been related to the bioavailability of metals to aquatic organisms, usually phytoplankton. A goal of such studies is to determine the ambient free metal ion concentration (or activity) in natural waters, which may then be compared with the optimal growth conditions for the phytoplankton species present. This comparison allows an evaluation of the nutritional stress on the organisms under ambient conditions arising from either metal limitation or toxicity (Sunda and Guillard, 1976; Coale and Bruland, 1988, 1990).

Equilibrium models of metal complexation are intrinsic to the interpretation of studies of metal bioavailability to phytoplankton in laboratory cultures, to the extrapolation of these results to the ambient conditions of free metal ion concentrations in field studies of metal speciation, and, to some extent, to the estimation of such ambient conditions. Often, ambient free metal ion concentrations are not directly measurable but are instead calculated from titrations of natural water samples, in which the total, dissolved metal concentration is increased to titrate the naturally occurring ligands. One of several models (Dzombak et al., 1986; Bartschat et al., 1992; Westall et al., 1995) for metal complexation by naturally occurring ligands (e.g., as a set of ligands with discrete or continuous distributions of metal affinity which may or may not explicitly consider electrostatic contributions) may be applied to the resulting titration curve and the extracted model parameters used to extrapolate to ambient conditions. Thus, field studies of metal speciation tend to be more closely coupled with laboratory studies and models of metal complexation than are field studies of metal distributions and fluxes in surface waters.

In field studies of metal transport in ground waters, concentrations of total, dissolved metals and of conservative tracers in ground water may be compared to evaluate the mobility of metals in the subsurface and identify mechanisms for their immobilization. Hydrological transport models have been expanded to incorporate the chemistry of contaminants, including metal complexation and adsorption (Dzombak and Ali, 1993). Hydrochemical ground water models have been developed assuming local equilibrium chemistry (Jennings et al., 1982; Cederberg et al., 1985; Furrer et al., 1989, 1990) or incorporating reaction kinetics (Kirkner et al., 1985; Liu and Narasimhan, 1989). Various models are also available for metal transport and fate in surface waters (Chapman, 1982; Felmy et al., 1984).

Kinetic Considerations in Metal Speciation

As the importance of metal speciation in the biogeochemical cycling of metals is increasingly recognized, models of metal complexation become an integral part of the interpretation of field and laboratory studies of metals in aquatic environments. Use of equilibrium models of metal complexation in the interpretation of laboratory studies of metal bioavailability has been, for the most part, enormously successful, but has also raised the question of whether the pseudoequilibrium assumption (i.e., the assumption that complexation reactions are fast compared with other processes, such as metal uptake by biota) is necessarily valid in natural waters. Even though a complete kinetic description of metal complexation may not be currently attainable, it is worth considering a few cases that illustrate some of the limitations of the equilibrium approach to modeling metal speciation.

Physical constraints in heterogeneous systems

Reactions in heterogeneous systems, such as soils or sediments, may be limited by physical constraints that cannot be adequately addressed by equilibrium models. Reactions of solids necessarily occur at the interface between the solid and solution; altered interfacial layers can present a physical barrier to reaction of the bulk solid. For example, retardation of the rates of oxidative dissolution of pyrite and iron silicates has been attributed to the formation of oxidized surface layers composed of iron oxide or iron(III) in a silicate matrix (Schott and Berner, 1983; Nicholson et al., 1990; Casey et al., 1993).

In soils and aquifers, amorphous coatings are often present on mineral surfaces. For example, element distribution maps of sediment samples from an acidic lake show a uniform distribution of aluminum throughout the sample in contrast to the localized distributions of iron and silicon (Giovanoli et al., 1988; Sulzberger et al., 1990).

Some possible effects of such coatings on the weathering of the bulk minerals can be illustrated by laboratory experiments (Hering, 1995). Dissolution of oxide minerals can be enhanced by organic ligands; this effect is attributed to formation of a surface complex between the organic ligand and surface metal centers followed by a slow detachment step in which the organically complexed metal is released into solution (Stumm and Furrer, 1987). As shown in Figure 3.2, the reaction of the bulk oxide hematite with the organic ligand 8-hydroxyquinoline-5-sulfonic acid (HQS) is markedly affected by surface coatings of adsorbed aluminum and/or amorphous aluminum hydroxide. In these experiments, hematite was preequilibrated with aluminum before the addition of HQS. At the highest aluminum-to-hematite ratio (83 μmol Al/g hematite), the bulk oxide is initially protected against the dissolving action of HQS (Figure 3.2a). With decreasing aluminum-to-hematite ratios, ligand-promoted dissolution of the bulk oxide is observed concomitant with desorption/dissolution of the aluminum surface coating (Figure 3.2b,c). At low aluminum-to-hematite ratios, the steady-state dissolution of the bulk oxide is unaffected although the rapid, initial dissolution observed in the absence of aluminum is inhibited.

These experiments suggest that amorphous coatings on soil and aquifer minerals may similarly protect bulk minerals against changes in soil-water or ground-water composition that would increase weathering (e.g., decreased pH or increased concentrations of organic ligands). Prediction of the composition of soil water or ground water

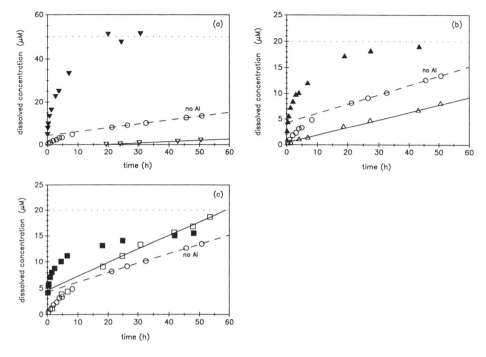

Figure 3.2. Concentrations of dissolved metals over time during ligand-promoted desorption/dissolution of hematite preequilibrated with aluminum. Conditions: 0.01 M NaClO$_4$, pH 7.5, total ligand (HQS) concentration 500 μM, hematite concentration 0.6 g/L (a, b), 1.2 g/L (c), total aluminum concentration 50 μM (a), 20 μM (b, c) indicated by dotted lines. Symbols: closed symbols, dissolved aluminum concentrations; open symbols: dissolved iron concentrations. For comparison, results from hematite dissolution (without aluminum coating) are given for 0.6 g/L hematite (open circles, dashed lines indicate steady-state dissolution rate). (Adapted from Hering, 1995.)

based on equilibrium with all the various solid phases present could not account for the effects of sequential reaction of surface and bulk layers. Although the equilibrium solution composition could eventually be reached by subsequent reequilibration of the soil solution with mineral surfaces, limited contact time between the solids and soil solution or other physical constraints, such as restricted mixing, might result in a very slow progress toward equilibrium.

Slow complexation kinetics in solution

Even in homogeneous solution, equilibrium distributions of metal species are not necessarily rapidly attained after some perturbation of the system. Slow complexation kinetics are most likely to be observed when competing metals and ligands are present at low concentrations, conditions directly relevant to natural waters. When such systems are perturbed, the distribution of metal and ligand species that is attained most rapidly is not necessarily the most stable (i.e., equilibrium) distribution.

An example of the types of conditions that lead to slow complexation kinetics is provided by an experiment in which a mixture of two ligands, ethylenediaminetetraacetic acid (EDTA) and humic acid, is perturbed by the addition of copper (Hering and Morel, 1989, 1990). The kinetics of the reaction, to form the thermodynamically favored copper–EDTA complex, is markedly affected by the presence of a competing metal, calcium. In the absence of calcium, the copper reacts rapidly with EDTA. The absence of any reaction of copper with the humic acid is evidenced by the maintenance of the native fluorescence of the humic acid, which is quenched upon formation of copper–humate complexes. In the presence of calcium, addition of copper results in immediate quenching of the humate fluorescence; slow recovery of the fluorescence signal indicates the progress of an exchange reaction in which the copper–EDTA complex is formed at the expense of the initial copper–humate complex (Figure 3.3). Depending on the time scale of interest, predictions based on an equilibrium model of metal complexation may diverge significantly from the actual composition of the system. Such kinetic effects may have implications for metal toxicity to aquatic organisms, which may be exposed to transient conditions of elevated free metal ion concentrations.

Effects of speciation on metal transport in ground water: possible nonequilibrium effects

Complexation by organic ligands has been proposed as a mechanism for facilitated transport of metals in ground water, particularly in cases of radionuclide migration from waste disposal sites (Toste et al., 1984; Means et al., 1978; Cleveland and Rees, 1981; Killey et al., 1984). For metals introduced in an organically complexed form into the subsurface environment, immobilization of the metal by adsorption onto soil or aquifer minerals requires either adsorption of the intact complex or dissociation of the complex. The mobility of a contaminant metal in the subsurface will then reflect both the equilibrium distribution of the metal between the surface of aquifer solids

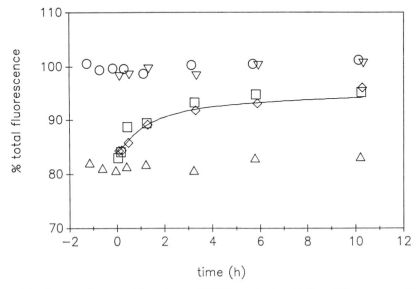

Figure 3.3.   Reaction of copper with a mixture of EDTA and humic acid. Percent fluorescence of the humic acid is shown over time after addition of copper, at $t = 0$, to the ligand mixture without calcium (∇), or with 0.01 M calcium ($\square$, $\diamond$). For reference, the percent fluorescence in the absence of copper ($\bigcirc$) and of humic acid and copper in the absence of EDTA ($\triangle$) are shown. (Adapted from Hering and Morel, 1989.)

and the mobile phase and the kinetics of the processes by which the metal is removed from the mobile phase.

Recent field studies, in which zinc and EDTA were injected into a sewage-contaminated aquifer, have illustrated the effects of metal speciation on metal mobility in groundwater (Kent et al., 1991; Rea et al., 1991; Davis et al., 1993). Conservative transport of zinc in the sewage-contaminated zone (pH ~6.0) was attributed to the lack of adsorption and consequent mobility of zinc–EDTA complexes. In the recharge zone (with slightly lower pH ~5.0), transport of zinc was considerably retarded. Displacement of zinc from the EDTA complex, possibly by iron, and adsorption of inorganic zinc (i.e., Zn^{2+} and inorganic zinc complexes) onto aquifer material was proposed as a mechanism for retardation of zinc. This mechanism was supported by the observed breakthrough curves for "excess metals" (i.e., in excess of their background concentrations), which was taken as a qualitative measure of EDTA transport.

Both the (thermodynamic) stability and (kinetic) lability of the zinc–EDTA complexes govern the mobility of zinc in this system. Even if the displacement of zinc from the zinc–EDTA complex, through equilibration of the complex with the aquifer minerals, is thermodynamically favorable, the rate of the reaction is not necessarily fast under natural conditions. Recent research on the transport of the cobalt(II)–EDTA complex indicated an even higher level of complexity for the transport of redox active metals. In the presence of iron oxides and aluminum oxides, cobalt was displaced from the cobalt(II)–EDTA complex by iron and aluminum (Girvin et al., 1993; Jardine et al., 1993; Szecsody et al., 1994). However, in the presence of manganese oxides, cobalt(II)–EDTA was oxidized to cobalt(III)–EDTA, which greatly enhanced the mobility of the complex (Jardine et al., 1993, 1995; Zachara et al., 1995a,b). The transport of cobalt is then influenced by the kinetics of a redox reaction at the manganese oxide surface.

Evaluation of the kinetics of such reactions require consideration of the possible reaction pathways; unfortunately, this is a more complicated undertaking than mere equilibrium calculations. Kinetic limitations in such systems may be indicated by discrepancies between observations of the mobility of contaminant metals and predictions of equilibrium calculations.

FURTHER RESEARCH NEEDS

The investigation of metal speciation in natural waters and the influence of speciation on the biogeochemical cycling of metals is an area of active research. Much of the previous research on these subjects has been inspired by and focused on the dependence of metal bioavailability (i.e., uptake, assimilation, and toxicity of metals) to aquatic biota on metal speciation. More recently, attention has been focused on the influence of metal speciation on removal of metals to solid materials in both surface waters and ground waters. In general, such investigations have been predicated on an equilibrium model of metal complexation. In some cases, however, this (pseudo)equilibrium assumption is of questionable validity. Some critical research needs must be addressed for better understanding of metal speciation and its consequences in natural waters.

Because of analytical limitations, ambient metal–organic complexes are rarely characterized or even directly measured in studies of metal complexation in aquatic environments. Modeling of metal speciation requires information on ligand concentrations and metal affinities, yet field measurements of metal complexation provide these parameters only indirectly. Direct information on distribution and structures of naturally occurring metal–organic complexes is a crucial link between measuring and modeling metal speciation in natural waters.

Further research is also needed into the influence of metal speciation on biogeochemical cycling of metals, particularly on fluxes of metals from the water column in lakes and oceans and on transport of metals in ground water. These studies should include a critical assessment of (pseudo)equilibrium assumptions concerning metal complexation reactions and, if warranted, incorporation of complexation kinetics.

ACKNOWLEDGMENTS

This work was supported in part by the National Science Foundation (BCS9258431) and by a gift from the Fluor Foundation.

References

Agricola, G., 1950, *De Re Metallica*, H. C. Hoover, and L. H. Hoover, translators, Dover, New York, p. 8.

Ahner, B. A., Price, N. M., and Morel, F. M. M., 1994, Phytochelatin production by marine phytoplankton at low free metal ion concentrations: laboratory studies and field data from Massachusetts Bay. *Proc. Natl. Acad. Sci. USA*, 91, 8433.

Anderson, D. M., and Morel, F. M. M., 1978, Copper sensitivity of *Gonyaulax tamarensis*. *Limnol. Oceanogr.*, 23, 283.

Anderson, M. A. and Morel, F. M. M., 1982, The influence of aqueous iron chemistry on the uptake of iron by the coastal diatom *Thalassiosira weissflogii*. *Limnol. Oceanogr.*, 27, 789.

Armstrong, D. E., Hurley, J. P., Swackhamer, D. L., and Shafer, M. M., 1987, Cycles of nutrient elements, hydrophobic organic compounds, and metals in Crystal Lake: role of particle-mediated processes in regulation. In: R. A. Hites and S. J. Eisenreich, eds., *Sources and Fates of Aquatic Pollutants, Advances in Chemistry Series*, No. 216, American Chemical Society, Washington, D.C., p. 491.

Bartschat, B. M., Cabaniss, S. E., and Morel, F. M. M., 1992, Oligoelectrolyte model for cation binding by humic substances. *Environ. Sci. Technol.*, 26, 284.

Behra, P., and Sigg, L., 1990, Evidence for redox cycling of iron in atmospheric water droplets, *Nature*, 344, 419.

Benoit, G., Oktay-Marshall, S. D., Cantu, A., Hood, E. M., Coleman, C. H., Coapcioglu, M. O., and Santschi, P. H., 1994, Partitioning of Cu, Pb, Ag, Zn, Fe, Al, and Mn between filter retained particles, colloids, and solution in 6 Texas estuaries. *Mar. Chem.*, 45, 307.

Boutron, C. F., Gorlach, U., Candelone, J. P., Bolshov, M. A., Delmas, R. J., 1991, Decrease in anthropogenic lead, cadmium, and zinc in Greenland snows since the late 1960's. *Nature*, 353, 153.

Bruland, K. W., 1980, Oceanographic distributions of cadmium, zinc, nickel, and copper in the North Pacific. *Earth Planet. Sci. Lett.*, 47, 176.

Bruland, K. W., 1992, Complexation of cadmium by natural organic ligands in the central North Pacific. *Limnol. Oceanogr.*, 37, 1008.

Bruland, K. W., Donat, J. R., and Hutchins, D. A., 1991, Interactive influences of bioactive tract metals on biological production in ocean waters. *Limnol. Oceanogr.*, 36, 1555.

Cabaniss, S. E., and Shuman, M. S., 1988a, Copper binding by dissolved organic matter: I. Suwannee River fulvic acid equilibria. *Geochim. Cosmochim. Acta*, 52, 185.

Cabaniss, S. E., and Shuman, M. S., 1988b, Copper binding by dissolved organic matter. II. Variation in type and source of organic matter. *Geochim. Cosmochim. Acta*, 52. 195.

Capodaglio, G., Coale, K. H., and Bruland, K. W., 1990, Lead speciation in surface waters of the eastern North Pacific. *Mar. Chem.*, 29, 221.

Casey, W. H., Banfield, J. F., Westrich, H. R., and McLaughlin, L., 1993, What do dissolution experiments tell us about natural weathering? *Chem. Geol.*, 105, 1.

Cederberg, G. A., Street, R. L., and Leckie, J. O., 1985, A groundwater mass transport and equilibrium chemistry model for multicomponent systems. *Water Resources Res.*, 21, 1095.

Chambers, R. M., and Odum, W. E., 1990, Porewater oxidation, dissolved phosphate and the iron curtain. *Biogeochemistry*, 10, 37.

Chapman, B. M., 1982, Numerical simulation of the transport and speciation of nonconservative chemical reactants in rivers. *Water Resources Res.*, 18, 155.

Cleveland, J. M., and Rees, T. F., 1981, Characterization of plutonium in Maxey Flats radioactive trench leachates. *Science*, 212, 1506.

Coale, K. H., and Bruland, K. W., 1988, Copper complexation in the northeast Pacific. *Limnol. Oceanogr.*, 33, 1084.

Coale, K. H., and Bruland, K. W., 1990, Spatial and temporal variability in copper complexation in the North Pacific. *Deep-Sea Res.* 37, 317.

Collienne, R. H., 1983, Photoreduction of iron in the epilimnion of acidic lakes. *Limnol. Oceanogr.*, 28, 83.

Comans, R. N. J., and van Dijk, C. P. J., 1988, Role of complexation processes in cadmium mobilization during estuarine mixing. *Nature*, 336, 151.

Davis, J. A., 1984, Complexation of trace metals by adsorbed natural organic matter. *Geochim. Cosmochim. Acta*, 48, 679.

Davis, J. A., and Leckie, J. O., 1978, Effect of adsorbed complexing ligands on trace metal uptake by hydrous oxides. *Environ. Sci. Technol.*, 12, 1309.

Davis, J. A., Kent, D. B., Rea, B. A., Maest, A. S., and Garabedian, S. P., 1993, Influence of redox environment and aqueous speciation on metal transport in groundwater: preliminary results of tracer injection studies. In: H. E. Allen and E. M. Perdue, eds., *Metals in Groundwater*, Lewis, Boca Raton, FL, p. 223.

Davison, W., 1985, Conceptual models for transport at a redox boundary. In: Stumm, W., ed., *Chemical Processes in Lakes*, Wiley-Interscience, New York, p. 31.

Deng, Y. W., and Stumm, W., 1994, Reactivity of aquatic iron(III)oxyhydroxides: Implications for redox cycling of iron in natural waters. *Appl. Geochem.*, 9(1), 23–36.

Donat, J. R., and Bruland, K. W., 1990, A comparison of 2 voltammetric techniques for determining zinc speciation in northeast Pacific ocean waters, *Mar. Chem.* 28, 301.

Donat, J. R., Lao, K. A., and Bruland, K. W., 1994, Speciation of dissolved copper and nickel in South San Francisco Bay—a multi-method approach. *Anal. Chim. Acta*, 284, 547.

Dzombak, D. A., and Ali, M. A., 1993, Hydrochemical modeling of metal fate and transport in freshwater environments. *Water Poll. Res. J. Canada*, 28, 7.

Dzombak, D. A., Fish, W. and Morel, F. M. M., 1986, Metal–humate interactions. 1. Discrete ligand and continuous distribution models. *Environ. Sci. Technol.*, 20, 669.

Evans, L. J., 1989, Chemistry of metal retention by soils. *Environ. Sci. Technol.*, 23, 1046.

Felmy, A. R., Brown, S. M., Onishi, Y., Yabusaki, S. B., and Argo, R. S., 1984, *MEXAMS: The Metal Exposure Analysis Modeling System*, U.S. Environmental Protection Agency, Environmental Research Laboratories, Athens, GA.

Follmann, J. F., 1987, *The Economics of Industrial Health*, AMACOM, New York, p. 14.

Freeze, R. A. and Cherry, J. A., 1979, *Groundwater*, Prentice-Hall, Englewood Cliffs, NJ.

Furrer, G., Westall, J. C., and Sollins, P., 1989, The study of soil chemistry through quasi-steady-state models: 1. Mathematical definition of model. *Geochim. Cosmochim. Acta*, 53, 595.

Furrer, G., Sollins, P., and Westall, J. C., 1990, The study of soil chemistry through quasi-steady-state models. II. Acidity of soil solution. *Geochim. Cosmochim. Acta*, 54, 2363.

Gardiner, J., 1976, Complexation of trace metals by ethylenediaminetetraacetic acid (EDTA) in natural waters, *Water Res.*, 10, 507.

Giovanoli, G., Schnoor, J. L., Sigg, L., Stumm, W., and Zobrist, J., 1988, Chemical weathering of crystalline rocks in the catchment area of acidic Ticino lakes, Switzerland. *Clays Clay Min.*, 36, 521.

Girvin, D. C., Gassman, P. L., and Bolton, H., Jr., 1993, Adsorption of aqueous cobalt ethylenediaminetetraacetate by δ-Al_2O_3. *Soil Sci. Soc. Am. J.*, 57, 47.

Girvin, D. C., Gassman, P. L., and Bolton, H., Jr., 1996, Adsorption of nitrilotriacetate (NTA), Co, and CoNTA by gibbsite. *Clays Clay Minerals*, in press.

Gledhill, M., and van den Berg, C. M. G., 1994, Determination of complexation of iron(III) with natural organic complexing ligands in seawater using cathodic stripping voltammetry. *Mar. Chem.*, 47, 41.

Gordon, R. M., Martin, J. H., and Knauer, G. A., 1982, Iron in North-East Pacific waters. *Nature*, 299, 611.

Hering, J. G., 1995, Implications of complexation, sorption, and dissolution kinetics for metal transport in soils. In: H. E. Allen and C. P. Huang, eds., *Metal Speciation and Contamination of Soils*, Lewis, Boca Raton, FL, p. 59.

Hering, J. G., and Morel, F. M. M., 1989, Slow coordination reactions in seawater. *Geochim. Cosmochim. Acta*, 53, 611.

Hering, J. G. and Morel, 1990, F. M. M., The kinetics of trace metal complexation: implications for metal reactivity in natural waters. In: W. Stumm, ed., *Aquatic Chemical Kinetics*, Wiley-Interscience, New York, p. 145.

Hering, J. G., Sunda, W. G., Ferguson, R. F., and Morel, F. M. M., 1987, A field comparison of two methods for the determination of copper complexation: bacterial bioassay and fixed-potential amperometry. *Mar. Chem.*, 20, 299.

Honeyman, B. D., Balistrieri, L. S., and Murray, J. W., 1988, Oceanic trace metal scavenging: the importance of particle concentration. *Deep-Sea Res.*, 35, 227.

Hong, S., Candelone, J.-P., Patterson, C. C., and Boutron, C. F., 1994, Greenland ice evidence of hemispheric lead pollution two millennia ago by Greek and Roman civilizations. *Science*, 265, 1841.

Hudson, R. J. M., and Morel, F. M. M., 1990, Iron transport in marine phytoplankton—kinetics of cellular and medium coordination reactions. *Limnol. Oceanogr.*, 35, 1002.

Hudson, R. J. M., and Morel, F. M. M., 1993, Trace metal transport by marine microorganisms—implications of metal coordination kinetics. *Deep-Sea Res.*, 40, 129.

Imber, B. E., and Robinson, M. G., 1983, Complexation of zinc by exudates of *Thalassiosira fluviatilis* grown in culture. *Mar. Chem.*, 14, 31.

Jardine, P. M. and Taylor, D. L., 1995, Fate and transport of ethylenediaminetetraacetate-chelated contaminants in subsurface environments. *Geoderma*, 67(1–2), 125–140.

Jardine, P. M., Jacobs, G. K., and O'Dell, J. D., 1993, Unsaturated transport processes in undisturbed heterogeneous porous media. II. co-contaminants. *Soil Sci. Soc. Am. J.*, 57,954.

Jennings, A. A., Kirkner, D. J. and Theis, T. L., 1982, Multicomponent equilibrium chemistry in groundwater quality models. *Water Resources Res.*, 18, 1089.

Kemble, N. E., Brumbaugh, W. G., Brunson, E. L., Dwyer, F. J., Ingersoll, C. G., Monda, D. P., and Woodward, D. F., 1994, Toxicity of metal-contaminated sediments from the Upper Clark Fork River, Montana, to Aquatic Invertebrates and Fish in Laboratory Exposures. *Environ. Toxicol. Chem.*, 13, 1985.

Kent, D. B., Davis, J. A., Anderson, L. D., and Rea, B. A., 1991, Transport of zinc in the presence of a strong complexing agent in a shallow aquifer on Cape Cod, MA. In *U.S. Geological Survey Water Resources Investigations Report 91-4034*.

Killey, R. W. D., McHugh, J. O., Champ, D. R., Cooper, E. L., and Young, J. L., 1984, Subsurface cobalt-60 migration from a low-level waste disposal site. *Environ. Sci. Technol.*, 18, 148.

Kirkner, D. J., Jennings, A. A., and Theis, T. L., 1985, Multisolute mass transport with chemical interaction kinetics. *J. Hydrol.*, 76, 107.

Kozarac, Z., Plavsic, M., Cosovic, B., and Vilicic, D., 1989, Interaction of cadmium and copper with surface-active organic matter and complexing ligands released by marine phytoplankton. *Mar. Chem.*, 26, 313.

Kuehn, W., and Brauch, H.-J., 1988, Organische Mikroverunreinigungen im Rhein. Bestimmung und Aussagekraft fuer die Gewaesserbeurteilung. *Fresenius Z. Anal. Chem.*, 330, 324.

Landing, W. M., and Bruland, K. W., 1980, Manganese in the North Pacific. *Earth Planet. Sci. Lett.*, 49, 45.

Liu, C. W., and Narasimhan, T. N., 1989, Redox-controlled multiple-species reactive chemical transport, 1. Model development. *Water Resources Res.*, 25, 869.

Luoma, S. M., 1983, Bioavailability of trace metals to aquatic organisms—a review. *Sci. Tot. Environ.*, 28, 1.

Martin, J. H., Gordon, R. M., and Fitzwater, S. E., 1991, The case for iron. *Limnol. Oceanogr.*, 36, 1793.

McKnight, D. M., and Morel, F. M. M., 1979, Release of weak and strong copper-complexing agents by algae. *Limnol. Oceanogr.*, 24, 823.

McKnight, D. M., Feder, G. L., Thurman, E. M., Wershaw, R. L., and Westall, J. C., 1983, Complexation of copper by aquatic humic substances from different environments. *Sci. Tot. Environ.* 28, 65.

McKnight, D. M., Kimball, B. A., and Bencala, K.E., 1988, Iron photoreduction and oxidation in an acidic mountain stream, *Science*, 240, 637.

Means, J. L., Crerar, D. A., and Duguid, J. O., 1978, Migration of radioactive wastes: radionuclide mobilization by complexing agents. *Science*, 200, 1477.

Miller, S., 1992, Cleanup delays at the largest Superfund sites. *Environ. Sci. Technol.*, 26, 658.

Moffett, J. W., and Zika, R. G., 1987, Photochemistry of copper complexes in sea water. In: R. G. Zika and W. J. Cooper, eds., *Photochemistry of Environmental Aquatic Systems*, ACS Symposium Series No. 327, American Chemical Society, Washington, D.C., p. 116.

Moffett, J. W., Zika, R. G., and Brand, L. E., 1990, Distribution and potential sources and sinks of copper chelators in the Sargasso Sea. *Deep-Sea Res.*, 37, 27.

Moore, J. W., 1991, *Inorganic Contaminants in Surface Water*, Springer-Verlag, New York.

Morel, F. M. M., 1986, Trace metals-phytoplankton interactions: an overview. In: P. Lasserre and J. M. Martin, eds., *Biogeochemical Processes at the Land-Sea Boundary*, Elsevier, Amsterdam, 177.

Morel, F. M. M, and Hering, J. G., 1993, *Principles and Applications of Aquatic Chemistry*, Wiley-Interscience, New York.

Morel, F. M. M., Reinfelder, J. R., Roberts, S. B., Chamberlain, C. P., Lee, J. G., and Yee, D., 1994, Zinc and carbon co-limitation of marine phytoplankton. *Nature*, 369, 740.

Muller, B., and Sigg, L., 1990, Interaction of trace metals with natural particle surfaces: comparison between adsorption experiments and field measurements. *Aquatic Sci.*, 52, 1015.

Murray, J. W., 1987, Mechanisms controlling the distribution of trace elements in oceans and lakes. In: R. A. Hites and S. J. Eisenreich, eds., *Sources and Fates of Aquatic Pollutants*, Advances in Chemistry Series, No. 216, American Chemical Society, Washington, D.C., p. 153.

Nicholson, R. V., Gilham, R. W., and Reardon, E. J., 1990, Pyrite oxidation in carbonate-buffered solution. 1. Rate control by oxide coatings, *Geochim. Cosmochim. Acta*, 54, 395.

Ochs, M., Brunner, I., Stumm, W., and Cosovic, B., 1993, Effects of root exudates and humic substances on weathering kinetics. *Water Air Soil Pollut.*, 68, 213.

Phinney, J. T., and Bruland, K. W., 1994, Uptake of lipophilic organic Cu, Cd, and Pb complexes in the coastal diatom *Thalassiosira weissflogii*. *Environ. Sci. Technol.*, 28, 1781.

Rea, B. A., Kent, D. B., LeBlanc, D. R., and Davis, J. A., 1991, Mobility of zinc in a sewage-contaminated aquifer. In: *U.S. Geological Survey Water Resources Investigations Report 91-4034*.

Santschi, P. H., 1984, Particle flux and trace metal residence time in natural waters. *Limnol. Oceanogr.* 29, 1100.

Scarponi, G., Capodaglio, G., Toscano, G., Barbante, C., and Cescon, P., 1995, Speciation of lead and cadmium in antarctic seawater: comparison with areas subject to different anthropic influence. *Microchem. J.*, 51, 214.

Schott, J., and Berner, R. A., 1983, X-ray photoelectron studies of the mechanism of iron silicate dissolution during weathering. *Geochim. Cosmochim. Acta*, 47, 2233.

Sigg, L., 1985, Metal transfer mechanisms in lakes: the role of settling particles. In: Stumm, W., ed., *Chemical Processes in Lakes*, Wiley-Interscience, New York, p. 283.

Sigg, L., Kuhn, A., Xue, H., Kiefer, E., and Kistler, D., 1995, Cycles of trace elements (copper and zinc) in an eutrophic lake: role of speciation and sedimentation. In: C. P. Huang, C. R. O'Melia and J. J. Morgan, eds., *Aquatic Chemistry*, ACS Symposium Series, No. 244, American Chemical Society, Washington, D.C.

Stauber, J. L., and Florence, T. M., 1987, Mechanism of toxicity of ionic copper and copper complexes to algae. *Mar. Biol.* 94, 511.

Stumm, W., and Furrer, G., 1987, The dissolution of oxides and aluminum silicates: examples of surface-coordination-controlled kinetics. In: W. Stumm, ed., *Aquatic Surface Chemistry*, Wiley-Interscience, New York, p. 197.

Stumm, W., and Sulzberger, B., 1992, The cycling of iron in natural environments: considerations based on laboratory studies of heterogeneous redox processes. *Geochim. Cosmochim. Acta*, 56, 3233.

Sulzberger, B., Schnoor, J. L., Giovanoli, R., Hering, J. G., and Zobrist, J., 1990, Biogeochemistry of iron in an acidic lake. *Aquatic Sci.*, 52, 1015.

Sunda, W. G., 1984, Measurement of manganese, zinc and cadmium complexation in seawater using chelex ion exchange equilibria. *Mar. Chem.*, 14, 365.

Sunda, W. G., and Ferguson, R. L., 1983, Sensitivity of natural bacterial communities to additions of copper and to cupric ion activity: a bioassay of copper complexation in sea water. In: C. S. Wong, E. Boyle, K. W. Bruland, J. D. Burton, and E. D. Goldberg, eds., *Trace Metals in Sea Water*, Plenum, New York.

Sunda, W. G., and Huntsman, S. A., 1983, Effect of competitive interactions between manganese and copper on cellular manganese and growth in estuarine and oceanic species of the diatom *Thalassiosira*. *Limnol. Oceanogr.*, 28, 924.

Sunda, W. G., and Huntsman, S. A., 1988, Effect of sunlight on redox cycles of manganese in the southwestern Sargasso Sea. *Deep-Sea Res.*, 35, 1297.

Sunda, W. G., Barber, R. T., and Huntsman, S. A., 1981, Phytoplankton growth in nutrient rich seawater: importance of copper–manganese cellular interactions. *J. Mar. Res.*, 39, 567.

Sunda, W. G., Huntsman, S. A., and Harvey, G. R., 1983, Photoreduction of manganese oxides in seawater and its geochemical and biological implications. *Nature*, 301, 234.

Sunda, W. G., and Guillard, P. A., 1976, The relationship between cupric ion activity and the toxicity of copper to phytoplankton. *J. Mar. Res.* 34, 511.

Szecsody, J. E., Zachara, J. M., and Bruckhart, P. L., 1994, Adsorption–dissolution reactions affecting the distribution and stability of co(ii)edta in iron oxide-coated sand. *Environ. Sci. Technol.*, 28, 1706.

Toste, A. P., Kirby, L. J., and Pahl, T. R., 1984, Role of organics in the subsurface migration of radionuclides in groundwater. In: G. S. Barney, J. D. Navratil, and W. W. Schulz, eds., *Geochemical Behavior of Disposed Radioactive Wastes*, ACS Symposium Series, No. 246, American Chemical Society, Washington, D.C., p. 251.

Trick, C. G., Andersen, R. J., Gillam, A. and Harrison, P. J., 1983, Prorocentrin: an extracellular siderophore produced by the marine dinoflagellate *Prorocentrum minimum*. *Science*, 219, 306.

Tylecote, R. F., 1992, *A History of Metallurgy*, 2nd edition, Institute of Materials, London.

van den Berg, C. M. G., and Nimmo, M., 1987, Determinations of interactions of nickel with dissolved organic material in seawater using cathodic stripping voltammetry. *Sci. Total Environ.*, 60, 185.

van den Berg, C. M. G., Merks, A. G. A., and Duursma, E. K., 1987, Organic complexation and its control of the dissolved concentrations of copper and zinc in the Scheldt Estuary. *Est. Coast. Shelf Sci.*, 24, 785.

Waite, T. D., and Szymczak, R., 1993, Manganese dynamics in surface waters of the eastern Caribbean. *J. Geophys. Res.*, 98, 2361.

Wells, M. L., Mayer, L. M., Donard, O. F. X., de Souza Sierra, M. M., and Ackelson, S. G., 1991, The photolysis of colloidal iron in the oceans. *Nature*, 353, 248.

Wells, M. L., Price, N. M., and Bruland, K. W., 1995, Iron chemistry in seawater and its relationship to phytoplankton: a workshop report. *Mar. Chem.*, 48, 157.

Westall, J. C., Jones, J. D., Turner, G. D., and Zachara, J. M., 1995, Models for association of metal ions with heterogeneous environmental sorbents. I. complexation of co(II) by leonardite humic acid as a function of pH and $NaClO_4$ concentration. *Environ. Sci. Technol.*, 29, 951.

Wilhelm, S. W., and Trick, C. G., 1994, Iron-limited growth of cyanobacteria: multiple siderophore production is a common response. *Limnol. Oceanogr.*, 39, 1979.

Zachara, J. M., Gassman, P. L., Smith, S. C., and Taylor, D., 1995a, Oxidation and adsorption of $Co(II)EDTA^{2-}$ complexes in subsurface materials with Fe and Mn oxide grain coatings. *Geochim. Cosmochim. Acta*, 59(21), 4449–4463.

Zachara, J. M., Smith, S. C., and Kuzel, L. S., 1995b, Adsorption and dissociation of Co–EDTA complexes in Fe oxide-containing subsurface sands. *Geochim. Cosmochim. Acta*, 59(23), 4825–4844.

Zhang, H., van den Berg, C. M. G., and Wollast, R., 1990, The determination of interactions of cobalt(II) with organic compounds in seawater using cathodic stripping voltammetry. *Mar. Chem.*, 28, 285.

Zhou, X., and Wangersky, P. J., 1989, Production of copper-complexing organic ligands by the marine diatom *Phaeodactylum tricornutum* in a cage culture turbidostat. *Mar. Chem.*, 26, 239.

Metal-Catalyzed Hydrolysis of Organic Compounds in Aquatic Environments

ALAN T. STONE

ABSTRACT. Dissolved metal ions and metal-containing mineral surfaces found in aquatic environments are capable of forming complexes with a number of synthetic organic chemicals. Complex formation, like any other change in chemical speciation, can have a profound effect on pathways and rates of chemical transformation. Using carboxylate ester and phosphorothioate ester pesticides as illustrative examples, the nature of the complex formation process will be explored, and effects on pesticide hydrolysis will be examined. Metals differ substantially in their natural abundance and distribution among free, complexed, adsorbed, and precipitated forms. These differences, in turn, lead to substantial differences in the ability of naturally occurring metals to catalyze carboxylate ester and phosphorothioate ester hydrolysis reactions.

INTRODUCTION

Upon release into aquatic environments, organic pollutants encounter a wide variety of ambient chemical constituents. Inorganic anions and cations, oxides, and aluminosilicate clays are ultimately derived from rock weathering. Natural organic matter consisting of various functional groups and structural moieties are generated by biological activity. These and other ambient chemical constituents may react directly with organic pollutants as stoichiometric reagents and may affect pollutant chemistry through a number of indirect mechanisms. For this reason, the abundance and reactivity of ambient chemical constituents must be accounted for when assessing pollutant transformation and ultimate fate.

The objective of this chapter is to examine how metal-containing species affect the transformations of organic pollutants in aquatic environments. Using hydrolysis reactions as illustrative examples, effects of metal–pollutant complex formation will be explored. The abundance and speciation of metals in the environment will be discussed along with details concerning their reactions with hydrolyzable pollutants. Using this approach, susceptibilities of important hydrolyzable pollutants toward metal catalysis can be evaluated under conditions typical of aquatic environments. Interest in this subject was inspired by the earlier work of Hoffmann (1980), Plastourgou and Hoffmann (1984), and Mill and Mabey (1988).

THE EFFECT OF DISSOLVED METAL IONS ON CARBOXYLIC ACID ESTER HYDROLYSIS

The inherent reactivity of an organic pollutant ultimately depends upon the distribution of electron density within the molecule. Electron-poor (Lewis acid) atoms within pollutant molecules become sites of nucleophilic attack, while electron-rich (Lewis base) atoms serve as ligand donor substituents toward protons and metal ions. The chemical literature contains several illustrative examples of the interconnections between ester properties and susceptibility toward metal complexation and hydrolysis (Stone and Torrents, 1995). In this and the following section, we will review some of these examples and ideas.

In the case of carboxylic acid esters, the higher electronegativity of oxygen relative to carbon shifts electron density away from the carbonyl carbon, making it susceptible toward nucleophilic attack. "Hydrolysis" denotes attack by water molecules ($[H_2O(aq)] \approx 55.6$ M) or by hydroxide ion ($[OH^-] \approx 10^{-7}$ M at pH 7). The reaction stoichiometry is given below:

$$RC(O)OR' \xrightarrow{+H_2O} RCOO^- + R'OH + H^+ \qquad (4.1)$$

In the absence of other participating species, the rate equation for ester hydrolysis is composed of three terms:

$$\frac{-d[\text{Ester}]}{dt} = k_a[H^+][\text{Ester}] + k_n[\text{Ester}] + k_b[OH^-][\text{Ester}] \qquad (4.2)$$

Acid-Catalyzed Reaction	Neutral Reaction	Base-Catalyzed Reaction

The second and third terms of the rate law correspond to nucleophilic attack onto the parent ester. Complete reaction mechanisms, involving the formation of a carbon-centered tetrahedral intermediate and eventual loss of the leaving group, are presented in Mabey and Mill (1978) and Schwarzenbach et al. (1993). At constant pH, the three terms can be combined into a single pseudo-first-order rate constant, k_h, which has units of sec^{-1}:

$$k_h = k_a[H^+] + k_n + k_b[OH^-] \qquad (4.3)$$

The first, or "acid-catalyzed," term in the rate equation deserves special attention. Protonation of the carbonyl oxygen draws electron density away from the carbonyl carbon atom, activating it toward nucleophilic attack. Even if the proportion of ester molecules in the protonated form is small (e.g., one molecule in one thousand), the catalytic effect of protons is important if $k_a[H^+]$ is comparable in magnitude to the quantity $(k_n + k_b[OH^-])$.

Reaction stoichiometries for the two-step hydrolysis of dimethyl oxalate are given below:

$$H_3COC(O)C(O)OCH_3 \xrightarrow{+H_2O} H_3COC(O)C(O)O^- + CH_3OH$$

$$\begin{aligned} k_a &\quad 3.3 \times 10^{-4} \text{ M}^{-1}\text{s}^{-1} \\ k_n &\quad 4.0 \times 10^{-5} \text{ s}^{-1} \\ k_b &\quad 3.0 \times 10^{+4} \text{ M}^{-1}\text{s}^{-1} \end{aligned}$$

(4.4)

$$H_3COC(O)C(O)O^- \xrightarrow{+H_2O} {}^-OC(O)C(O)O^- + CH_3OH$$

$$\begin{aligned} k_a &\quad - \\ k_n &\quad 2.7 \times 10^{-7} \text{ s}^{-1} \\ k_b &\quad 1.5 \quad \text{M}^{-1}\text{s}^{-1} \end{aligned}$$

(4.5)

Rate constants from the literature (Guthrie and Cullimore, 1980) used in Figure 4.1 indicate that in the absence of catalysts, hydrolysis of the monoester anion is substantially slower than hydrolysis of the parent diester. The half-life for hydrolysis at pH 6, for example, is 0.56 hours for the diester and 28 days for the monoester. Thus, the monoester intermediate persists for a considerable period of time. This is a common occurrence; unfavorable electrostatic interactions between the monoester anions and incoming nucleophiles retard hydrolysis in comparison to the diesters (Westheimer, 1992). The picture presented in Figure 4.1 is not complete, since the neutral, protonated form of the monoester will be the predominant species below the pK_a of the carboxylate group (pH ~4.5). This protonated form should hydrolyze more quickly than the monoester anion, because unfavorable electrostatic interactions have been eliminated.

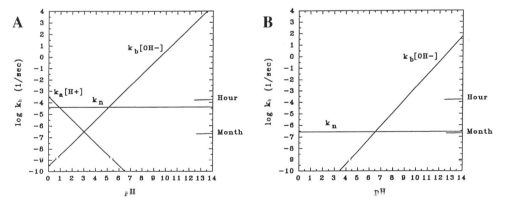

Figure 4.1. Rate constants for hydrolysis of (A) dimethyl oxalate and (B) monomethyl oxalate as a function of pH. (Based upon constants from Guthrie and Cullimore, 1980.)

Metal ions are Lewis acids and should therefore influence ester hydrolysis in a manner analogous to that of protons. Complex formation with metal cations lowers ester charge, thereby lessening unfavorable electrostatic interactions with incoming nucleophiles. Because protons have an extremely high charge density and a tremendous capacity to shift electron density within coordinated ligands, acid catalysis is typically more important than metal catalysis under acidic conditions (Hay, 1987). Metal ions have the advantage, however, in that they can readily coordinate two or more ligand donor sites on a hydrolyzable molecule and can vastly outnumber protons under neutral and alkaline conditions (Hoffmann, 1980; Plastourgou and Hoffmann, 1984).

In order for complex formation with metal ions to have a significant effect on ester hydrolysis, two conditions must be met: (i) The metal–ester complex formation constant must be sufficiently large, and (ii) the metal–ester complex must hydrolyze more rapidly than the free ester. Examination of Reactions (4.4) and (4.5) indicates that catalysis should have a greater effect on monoester hydrolysis than on diester hydrolysis. First, the carboxylate plus carbonyl ligand-donor groups of the monoester form a more stable metal–substrate complex than the carbonyl plus carbonyl ligand–donor groups of the diester. Second, formation of the metal–substrate complex removes unfavorable electrostatic interaction between the anionic monoester and OH^- nucleophile.

Information regarding metal complexation and catalysis is not available for the methyl esters discussed above. A careful study of metal-catalyzed hydrolysis of the ethyl oxalate monoester is, however, available (Johnson and Angelici, 1971). Formation of a metal–monoester complex has been postulated, followed by OH^- nucleophilic attack:

Preequilibrium Step:

$$K_f = \frac{[EtOxMe^{2+}]}{[EtOx^{2+}][Me^{2+}]} \qquad (4.6)$$

Hydrolysis Step:

$$EtOx^- + OH^- \xrightarrow{k_{OH}} Products \qquad k_{OH} = 6.9 \times 10^{-5} \ M^{-1}s^{-1} \qquad (4.7)$$

$$EtOxMe^+ + OH^- \xrightarrow{k_M} Products \qquad (4.8)$$

Rate Law:

$$\frac{d[Products]}{dt} = k_{OH}[OH^-][EtOx^-] + k_M K_f [OH^-][Me^{2+}][EtOx^-] \qquad (4.9)$$

The rate law illustrates that the magnitude of the metal–ester complex formation constant (K_f) and the rate constant for nucleophilic attack (k_M) are both important in establishing the role of metal catalysis relative to uncatalyzed hydrolysis.

Values of K_f and k_M are listed in Table 4.1. Focusing on the first-row divalent transition metal ions, the complex formation constant K_f increases from left to right, reaching a maximum value for Cu^{2+}. This trend is observed for most oxygen- and nitrogen-donor ligands and is known as the *Irving–Williams series*. Differences in ionic radii are primarily responsible for this trend in K_f, although increases in covalent bond character (polarizability) with increasing atomic number are also important (Houghton, 1979). Increases in polarizability enable the metal ion to have a greater effect on the electronic structure of the coordinated organic compound, reflected in increases in the hydrolysis rate constant k_M. Mg^{2+} does not possess the polarizable d orbitals found on the other metals, and as a consequence the Mg^{2+}–ester complex hydrolyzes $10^{3.7}$ times more slowly than the Cu^{2+}–ester complex.

The product $k_M K_f$ provides a first indication of the relative catalytic effectiveness of the metals listed in Table 4.1. Our analysis is not complete, however, until a full description of metal and ester speciation has been developed. Aqueous speciation arising from the addition of 10 μM Cu^{II} shown in Figure 4.2A was calculated using the equilibrium computer program MINEQL (Westall et al., 1976) as modified by Papelis et al. (1988). Although the hexaaquo complex (Cu^{2+}(aq)) is predominant at low pH, Cu^{II}-hydroxo species grow in importance as the pH is increased. At sufficiently high pH (near 7.5 in the present example), the solid phase $Cu(OH)_2$(s) forms, drastically limiting Cu^{II} solubility.

The observed rate constant for ethyl oxalate hydrolysis in the presence of 10 μM Cu^{II}, Zn^{II}, and Mg^{II} is shown in Figure 4.2B. Metal-free solutions and Mg^{II}-containing solutions yield the same results; log k_h increases by one log unit for every unit increase in pH, reflecting the predominance of OH^- nucleophilic attack on the uncomplexed ester. The catalytic effect is greatest for Cu^{II} throughout most of the pH range shown, reflecting the high values of both K_f and k_M. The extent of Cu–ester complex formation (and therefore the rate of the metal-catalyzed pathway) is directly proportional to [Cu^{2+}(aq)]. As the pH is increased beyond values where $Cu(OH)_2$(s) precipitates, [Cu^{2+}(aq)] decreases substantially, causing a drop in k_h toward values calculated for the uncatalyzed reaction. Because $Zn(OH)_2$(s) is more soluble than $Cu(OH)_2$(aq), catalysis by Zn^{II} actually exceeds catalysis by Cu^{II} within a narrow range of conditions (7.7 < pH < 8.7).

Within natural waters, complexation of Cu^{II} by inorganic and organic ligands lowers concentrations of both the free metal ion and any metal–ester complex. To illustrate, a system will be considered that contains Cu^{II} and the $EtOx^-$ monoester, as well

TABLE 4.1

Complex Formation Constants (K_f) and Metal-Catalyzed Hydrolysis Rate Constants (k_M) for Ethyl Oxalate

	Metal Ion:				
Constant	Mg^{2+}	Co^{2+}	Ni^{2+}	Cu^{2+}	Zn^{2+}
log K_f (M^{-1})	0.63	1.2	1.3	2.1	1.3
k_M (M^{-1} s^{-1})	0.0030	0.29	0.37	14.1	0.63
$k_M K_f$	1.3×10^{-2}	4.6	7.4	1800	13.

[a]*From Johnson and Angelici (1971).*

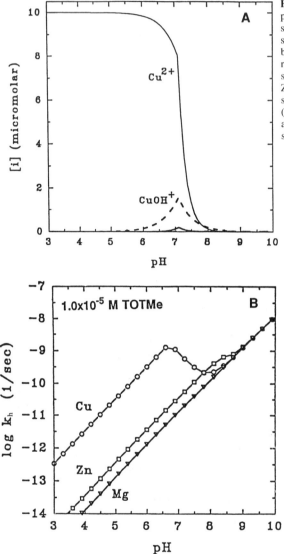

Figure 4.2. (A) Concentrations of important species as a function of pH in a solution containing 10.0 μM Cu^{II}; the solid phase Cu(OH)$_2$(s) limits Cu solubility above pH 7.5. (B) k_h, the observed rate constant for ethyl oxalate hydrolysis in the presence of 10.0 μM Cu^{II} (○), Zn^{II} (◇), and Mg^{II} (△), based upon constants from Johnson and Angelici (1971). (k_h values in metal-free solutions and in Mg^{II}-containing solutions are the same.)

as the ligand glycine. The following mass balance equations and equilibrium relationships apply:

$$\text{Cu}_T = [\text{Cu}^{2+}] + [\text{CuOH}^+] + [\text{Cu(OH)}_2{}^0(\text{aq})]$$

$$+ [\text{EtOxCu}^+] + [\text{CuGly}^+] + [\text{Cu(Gly)}_2{}^0(\text{aq})] \qquad (4.10)$$

$$\text{GLY}_T = [\text{H}_2\text{Gly}^+] + [\text{HGly}^0] + [\text{Gly}^-] + [\text{CuGly}^+] + [\text{Cu(Gly)}_2{}^0(\text{aq})] \qquad (4.11)$$

$$\text{H}_2\text{Gly}^+ = \text{HGly}^0 + \text{H}^+ \qquad K_{a1} = 10^{-2.35} \qquad (4.12)$$

$$\text{HGly}^0 = \text{Gly}^- + \text{H}^+ \qquad K_{a2} = 10^{-9.78} \qquad (4.13)$$

$$Cu^{2+} + Gly^- = CuGly^+ \qquad\qquad K_1 = 10^{8.6} \tag{4.14}$$

$$Cu^{2+} + Gly^- = Cu(Gly)_2^0(aq) \qquad \beta_2 = 10^{15.6} \tag{4.15}$$

Under acidic conditions, the protonated glycine species H_2Gly^+ and $HGly^0$ predominate. As a consequence, glycine has little effect on Cu^{II} speciation and metal-catalyzed hydrolysis. Under neutral and alkaline conditions, less protonated glycine species predominate, and the percentage of Cu^{II} bound to glycine becomes significant. Increasing total dissolved glycine (GLY_T) extends the pH range where Cu^{II}-glycine complexes form, thereby causing catalysis to diminish (see Figure 4.3).

Although natural aquatic environments contain a much more complex mixture of ligands, the procedure for evaluating their effect on metal speciation and hydrolysis is the same. The extent of metal–ester complex formation and hence the metal-catalyzed hydrolysis rate are a function of the free metal ion concentration.

THE EFFECT OF DISSOLVED METAL IONS ON PHOSPHORO(THIO)ATE ESTER HYDROLYSIS

Phosphate and phosphorothioate esters are intentionally released into the environment to control insects. The enzyme acetylcholineesterase serves as the target within the or-

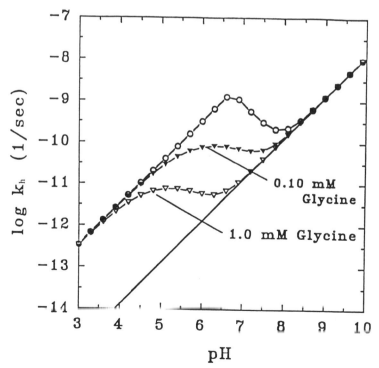

Figure 4.3. Cu^{II} complexation by 0.10 and 1.0 mM glycine lessens the extent of Cu^{II}–ester complex formation and metal-catalyzed hydrolysis, thereby lowering k_h.

ganism; it forms an adduct with the insecticide via nucleophilic attack at the electrophilic phosphorus. According to Schrader's Rule for pesticidal activity (Fest and Schmidt, 1983), one of the three leaving groups of the insecticide must have a pK_a below pH 8.0 in order for adduct formation to occur at sufficient rates. In addition, the pK_a must not be less than pH 6.0, since hydrolysis would then occur before the insecticide had a chance to form an adduct with acetylcholineesterase.

Methyl chlorpyrifos serves as an illustrative example; hydrolysis pathways are shown in Figure 4.4. When nucleophilic attack occurs at the phosphorus atom, the trichloropyridinol moiety is the preferred leaving group because of its lower pK_a relative to methanol. The alkyl carbon represents an alternative "soft" electrophile; nucleophilic attack at one of these carbon sites yields methanol as the leaving group. It should also be noted that dethiolation can occur during each step in the hydrolysis process (the P=S moiety is replaced by the P=O moiety). Hydrolysis of the chlorpyrifos triester yields a complex mixture of diester and monoester intermediates on the way to the final hydrolysis products.

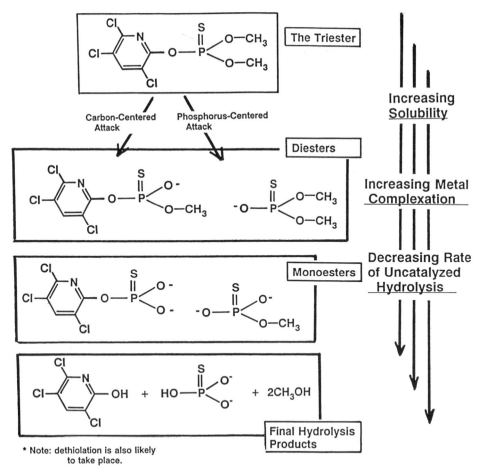

Figure 4.4. Overall scheme for the hydrolysis of chlorpyrifos-methyl via carbon-centered attack and via phosphorus-centered attack.

The chlorpyrifos triester is sufficiently hydrophobic that significant bioaccumulation and sorption into particulate natural organic matter is expected to occur. Hydrolysis generates diesters, monoesters, and inorganic phosphoro(thio)ates that are all anionic within the pH range of natural waters, substantially increasing solubility. With each step in the sequence of hydrolysis reactions, nucleophilic attack occurs slowly, in part because of the development of unfavorable electrostatic interactions between the anionic esters and the attacking nucleophile (Westheimer, 1992). As we shall see, the extent of metal binding increases substantially as progressive phosphoro(thio)ate groups become accessible toward complex formation.

Chlorpyrifos susceptibility toward Cu^{II}-catalyzed hydrolysis is firmly established (Mortland and Raman, 1967; Blanchet and St. George, 1982). It is believed that Cu^{II} coordinates the phosphorothioate sulfur, thereby increasing the electrophilicity of the phosphorothioate phosphorus and activating it toward nucleophilic attack. The pyridyl nitrogen may facilitate Cu^{II} coordination through the formation of a five-membered chelate ring (Figure 4.5). Unfortunately, many details concerning this metal-catalyzed reaction are unavailable. Cu^{II}–chlorpyrifos complex formation constants are not known, so it is difficult to evaluate how complex formation relates to Cu^{II} catalysis. Separate information about the effect of Cu^{II} on diester and monoester hydrolysis is not available. Finally, complex formation constants (and hydrolysis rate data) for other metal ions more widespread in the environment are not known. For these reasons, it is difficult to evaluate whether metal-catalyzed chlorpyrifos hydrolysis is an important process in the environment.

Some progress can be made by examining metal complexation and metal-catalyzed hydrolysis of some analogous phosphate esters. Monomethyl 8-quinolyl phosphate and 8-quinolyl phosphate, shown in Figure 4.6, bear some resemblance to diesters and monoesters produced during chlorpyrifos hydrolysis. The pyridyl nitrogen is suitably placed for chelate formation in a manner comparable to chlorpyrifos and its hydrolysis intermediates. It should be noted, however, that the 8-hydroxyquinoline leaving group has a pK_a of 9.81, which is substantially higher than the pK_a of 5.45 for the trichloropyridinol group of chlorpyrifos (Dilling et al., 1984). As a consequence, hydrolysis in the absence of metal catalysts should be slower for the compounds shown in Figure 4.6 than for analogous chlorpyrifos intermediates.

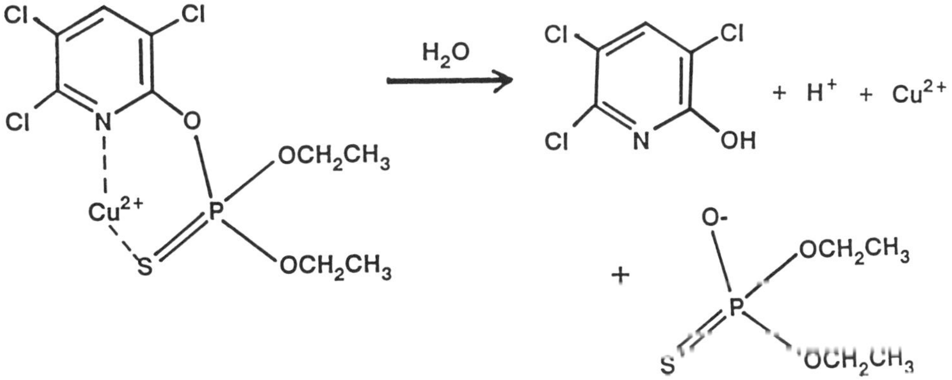

Figure 4.5. Cu^{II}-catalyzed hydrolysis of chlorpyrifos-methyl. [After Mortland and Raman (1967) and Blanchet and St. George (1982).]

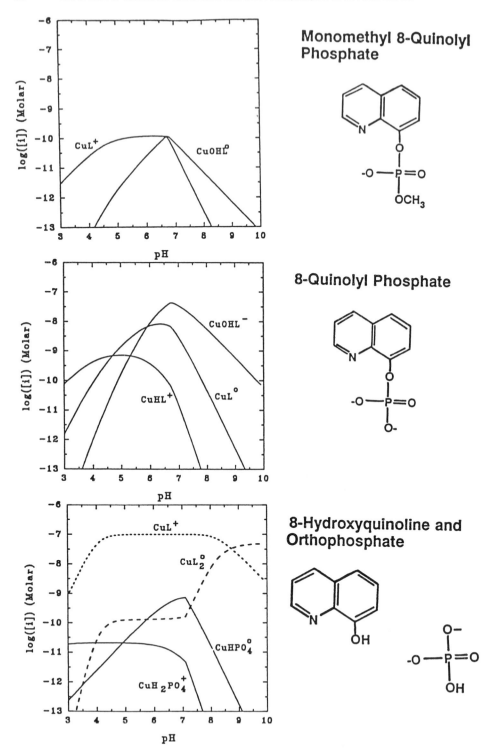

Figure 4.6.   Results from complex equilibrium calculations involving 10.0 μM Cu^{II} and 0.10 μM (A) monomethyl 8-quinolyl phosphate, (B) 8-quinolyl phosphate, and (C) 8-hydroxyquinoline and orthophosphate.

Acidity constants and Cu^{II}–ester complex formation constants for monomethyl 8-quinolyl phosphate and 8-quinolyl phosphate have recently become available in the literature (Armstrong et al., 1989), making equilibrium calculations possible. As shown in Figure 4.6, Cu^{II}–diester and Cu^{II}–monoester complex formation drops off under strongly acidic conditions, because Cu^{II} must compete with protons for binding the ester. Complex formation also drops off under strongly alkaline conditions, because di- and monoesters must compete with hydroxide ions for binding Cu^{II}. Metal–ester complex formation reaches a maximum near neutral pH. Since complex formation is a prerequisite for metal catalysis, metal ions will exert their strongest effect under these same conditions. There is one other reason why metal catalysis is most important within this pH range: Rates of uncatalyzed hydrolysis are the lowest under neutral and weakly acidic pH conditions, because contributions arising from acid-catalyzed and base-catalyzed hydrolysis pathways are relatively small.

The calculations presented in Figure 4.6 involve two different stoichiometries for Cu^{II}–diester complexes (denoted CuL^+ and $CuOHL^0$) and three different stoichiometries for Cu–monoester complexes (denoted CuL^0, $CuHL^+$, and $CuOHL^-$). In the final analysis, it is possible that one stoichiometry in particular may dominate metal-catalyzed hydrolysis. Particular attention should be paid to ternary complexes where OH^- and ester are simultaneously bound to the central Cu^{II} atom. Binding of the nucleophile in this manner may increase its encounter frequency with the ester, promoting attack at the electrophilic site (see Sutton and Buckingham, 1987).

Complex formation constants between first-row divalent transition metal ions and 8-quinolyl phosphate esters are presented in Figure 4.7. Trends in log K_M obey the Irving–Williams ordering: Cu^{II} binding to both the monoester and diester exceeds that of the other divalent metals. It is interesting to note that the complex formation constant between Mn^{II} and the monoester is still quite significant. Thus, the potential exists for more environmentally relevant metals to complex phosphoro(thio)ate esters and hence affects hydrolysis rates.

Now that the equilibrium aspects of ester–metal binding have been discussed, preliminary reports of the susceptibility of the 8-quinolyl monoester toward hydrolysis should be mentioned. Murakami and Sunamoto (1971) reported that 2.0 mM Ni^{II} and Cu^{II} exerted only a slight catalytic effect at pH 2.0. Our calculations shown in Figure 4.6 indicate that this choice of pH was perhaps not an appropriate one. A much greater catalytic effect is expected near neutral pH, where the extent of metal–ester complex formation is two or more orders of magnitude higher and where the contribution from acid-catalyzed hydrolysis is minimal.

In recent years, attempts have been made to quantitatively explain trends in both the acid–base and metal complex formation behavior of simple phosphoro(thio)ate species (Massoud and Sigel, 1988; Martin and Sigel, 1988). Much of this work emphasizes the importance of solvation forces on complex formation (Massoud and Sigel, 1988). Geometric requirements for binding some phosphoro(thio)ate compounds can raise complex formation constants for Mn^{II} above those for Ni^{II} and Co^{II}, contrary to expectations based upon the Irving Williams Series (Massoud and Sigel, 1988). For the alkaline earth metals Mg^{II} and Ca^{II} (and possibly some divalent transition metals), outer-sphere complexes may surpass inner sphere complexes in importance (Brützinger, 1965). An important task for environmental chemists is to extend existing knowledge of this kind toward understanding the coordination chemistry and reactivity of complex phosphoro(thio)ate esters such as methyl chlorpyrifos under realistic environmental conditions.

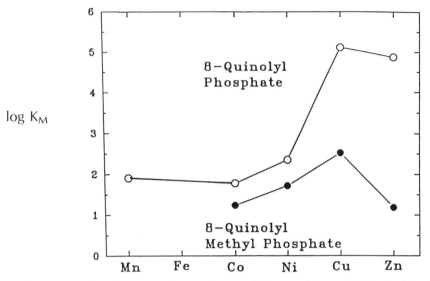

Figure 4.7. Complex formation constants between first-row, divalent metal ions and (A) 8-quinolyl methyl phosphate (a diester) and (B) 8-quinolyl phosphate (a monoester). (Adapted from Armstrong et al., 1989.)

METAL SPECIATION AND ABUNDANCE

Metals reside within a number of different "compartments" within aquatic environments and are associated with a variety of other constituent chemicals. Dissolved metal ions may occur as aquo complexes, complexes with inorganic anions (e.g., Cl^-, SO_4^{2-}), and complexes with natural organic matter. Metal ions may be found sorbed onto the surfaces of minerals, or bound within interlayer regions and cavities. (Partial loss of coordinated water upon entry into interlayer regions and cavities may have a profound effect on catalytic activity). Metals and ligands can be associated with one another on mineral surfaces through ternary surface complexes (Schindler and Stumm, 1987). Lattice-bound metals may fill vacancies within tetrahedrally or octahedrally coordinated layers of clays, or may be incorporated into oxides, carbonates, sulfides, and other minerals through isomorphic substitution. Because adsorbed and lattice-bound metals on the surfaces of minerals are still in contact with the surrounding aqueous medium, they are potentially available as catalysts. Details concerning the nature and catalytic activity of various metal species are largely unknown, thus providing impetus for new research.

Much of the literature concerning metal–ester complex formation and metal catalysis is devoted to metals exhibiting relatively simple aqueous speciation and metals which have been identified as cofactors for important hydrolytic enzymes (e.g., Ni^{II}, Cu^{II}, and Zn^{II}). As indicated in Table 4.2, however, metals to the right of iron in the periodic table exist only as trace constituents in the earth's crust; their naturally occurring aqueous concentrations seldom exceed 1.0 micromolar. It is likely that Ni^{II}, Cu^{II}, and Zn^{II} only play a catalytic role under isolated natural conditions or with hydrolyzable pollutants that exhibit pronounced selectivity toward these metals.

Catalysis by the more abundant metals in the earth's crust deserves serious attention. We can use Pearson's (1966, 1988) classification of "Hard," "Borderline," and

"Soft" metal ions and ligands as a basis for exploring trends in complex formation behavior. Most aquatic environments possess an abundance of classic "Hard" metal ions such as Mg^{2+} and Ca^{2+} in solution, along with Al^{III} and Ti^{IV} bound to mineral surfaces. Under oxidizing conditions, two more metals can be added to the list of surface-bound hard metals: Mn^{IV} and Fe^{III}. Bonds to ligands involving "Hard" metals are primarily ionic in character; stability constants generally increase as the charge of the metal is increased and the ionic radius is decreased. Although oxygen and nitrogen are both classified as "Hard" ligand donor atoms, nitrogen is the more polarizable (i.e., "Softer") of the two, and therefore it experiences a greater convalent contribution toward bond formation. This imparts a selectivity toward binding transition metal ions. Fe^{III}, for example, is coordinated by nitrogen-donor groups in preference to Al^{III} (Evers et al., 1989). Solubilities of $Al_2O_3(s)$ and $Fe_2O_3(s)$ at pH 4.0 are calculated below:

$Al^{3+}(aq)$	3.1×10^{-4} M	$Fe^{3+}(aq)$	1.0×10^{-9} M
$AlOH^{2+}(aq)$	3.2×10^{-5}	$FeOH^{2+}(aq)$	6.5×10^{-8}
$Al(OH)_3^0(aq)$	3.2×10^{-7}	$Fe(OH)_2^+(aq)$	2.1×10^{-7}
$Al(OH)_4^-(aq)$	3.2×10^{-11}	$Fe(OH)_4^-$	2.6×10^{-15}
TOTAL:	3.6×10^{-4} M	TOTAL:	2.7×10^{-7} M

Although somewhat higher solubilities can be calculated for meta stable phases (e.g., $Fe(OH)_3(s)$), the overall conclusion is the same: Al^{III} and Fe^{III} dissolved species are present at catalytically significant levels only within the most acidic range of natural aquatic environments. For Ti^{IV} and Mn^{IV}, the situation is even more pronounced; reactions involving these metals must take place at the mineral–water interface because of solubility limitations.

Reducing conditions can yield significant amounts of the "borderline" metals Mn^{II} and Fe^{II}. Although their radii are comparable to that of Mg^{2+}, increased polarizability (and crystal field stabilization energy in the case of Fe^{II}) yield higher values for complex formation constants. Within most low-sulfur environments, carbonates limit Mn^{II}

TABLE 4.2
The Average Crustal Abundance of Selected Metals[a]

Element	Average Crustal Abundance (%)
Aluminum	8.3
Silicon	27.
Magnesium	2.8
Calcium	4.7
Titanium	0.63
Manganese	0.11
Iron	6.2
Cobalt	0.0029
Nickel	0.0099
Copper	0.0068
Zinc	0.0072

[a]From Emsley (1989).

and Fe^{II} solubility. Production of bisulfide (HS^-) under reduced, high-sulfur conditions limits Fe^{II} solubility through formation of mackinawite ($FeS(s)$) and pyrite ($FeS_2(s)$).

CATALYSIS INVOLVING SURFACE-BOUND METAL SPECIES

Mineral surface areas in contact with soil water and ground water can range from a few tens to perhaps a thousand square meters per liter of water. When the density of adsorbed and lattice-bound metals is taken into account, it is apparent that this reservior of potentially catalytic metals can easily surpass that of dissolved metal species. Much less is known about the chemistry of surface-bound metal atoms, however, because many of the spectroscopic and other analytical techniques developed for studying aqueous-phase speciation can only be applied to the mineral–water interface with considerable difficulty. [For a comprehensive review of these methods, see Hochella and White (1990) and references therein.]

Principles governing metal speciation and reactivity in solution also apply to the mineral–water interface, although some modifications must be made. The following areas are pertinent to this discussion: (i) metal–ligand coordination chemistry, (ii) solvation of both dissolved species and surface sites, and (iii) long-range electrostatic interactions near charged surfaces.

We can begin with a statement by Jorgensen (1965): if most of the ligands surrounding a metal atom are "Hard," then the remaining coordinative positions show preference to "Hard" ligands over "Soft" ones. Conversely, if most of the ligands are "Soft," then the remaining coordinative positions show preference for "Soft" ligands. Fe^{III} can serve as an illustrative example. It possesses a coordination number of six, which is satisfied by placing donor ligands in an octahedral arrangement around the central iron atom. As we move from a neutral ligand (e.g., H_2O) to a mono-anionic ligand (e.g. OH^-) and finally to a di-anionic ligand (e.g., O^{2-}), the relative "Hardness" is increased. Since the number of OH^- and O^{2-} anions within the coordination shell of surface-bound Fe^{III} is greater than for dissolved Fe^{III} species, Jorgensen's statement implies that surface-bound Fe^{III} will show a pronounced preference for adsorbing "Hard" ligand species. There is, however, one additional factor to take into account. Unlike dissolved monomers, Fe^{III} atoms within an oxide are electronically coupled to one another (Sherman, 1985a,b) which may affect their coordination chemistry and reactivity.

As metal ions and ligands form complexes in solution, both reactants must lose or modify their solvation shells. Pearson's (1966, 1988) classification of "Hard" versus "Soft" metal ions and ligands as well as the well-known "Chelate Effect" both arise, in part, from the effects of solvation on complex formation constants. It is likely that surface-bound metal atoms and adsorbed organic ligands are less strongly solvated than their counterparts in aqueous solution. The role played by solvation in surface complex formation and surface site reactivity are poorly understood at this point and require further investigation.

The acquisition of charge through the adsorption of H^+, OH^-, and other ions onto oxides and other metal-containing surfaces can have a strong influence on organic substrate adsorption. Electrostatic interactions lead to the accumulation of counterions and the depletion of co-ions in the vicinity of the oxide–water interface, and the development of an electrical double layer (see Stumm, 1992). These effects typically diminish as the electrolyte concentration in solution is increased. The Poisson–Boltzmann equation provides a quantitative basis for evaluating electrical double-layer effects on organic ligand adsorption (Stone et al., 1993).

Illustrative Reactions at Mineral Surfaces

Aqueous solutions in contact with mineral surfaces are inherently complex systems; several phenomena involving the mineral/water interface can promote hydrolysis. To illustrate this point, surface-catalyzed reactions of four hydrolyzable organic compounds will now be discussed.

Phenyl picolinate (Figure 4.8) is a likely candidate for surface catalysis because of its well-documented susceptibility toward catalysis by dissolved metal ions (Fife and Przystas, 1985). In the absence of metal catalysts, the half-life for phenyl picolinate hydrolysis at pH 7 is 86 hours; addition of 500 m^2/L TiO_2(s) or FeOOH(s) lowers the half-life to less than 5 hours (Torrents and Stone, 1991). When oxide particles are removed by filtration, hydrolysis slows to the rate observed in catalyst-free solution, confirming that surface catalysis is taking place. The meta-isomer, phenyl isonicotinate, is not susceptible toward catalysis by dissolved or surface-bound metals, indicating that chelate formation with the pyridyl nitrogen must occur in order for catalysis to take place (Torrents and Stone, 1991).

As indicated in Figure 4.8, methyl chlorpyrifos is also capable of forming a surface chelate via a pyridyl nitrogen. Although catalysis has been observed in TiO_2(s), FeOOH(s), and Al_2O_3(s) suspensions (Torrents and Stone, 1994), the role of the sur-

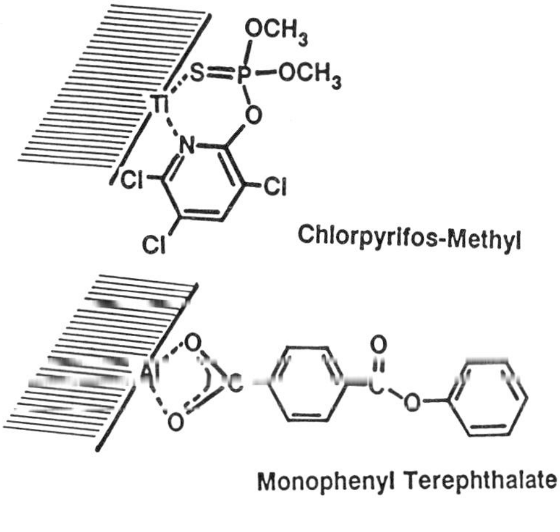

Figure 4.8. Three organic esters subject to surface-catalyzed hydrolysis: phenyl picolinate, chlorpyrifos-methyl, and monophenyl terephthalate.

Phenyl Picolinate

Chlorpyrifos-Methyl

Monophenyl Terephthalate

face chelate has not been confirmed. Establishment of a catalytic mechanism is further complicated by the observation that methyl parathion and ronnel are also susceptible toward surface-catalyzed hydrolysis. Although both pesticides are capable of forming weak monodentate complexes with surface-bound metals, neither has the capacity to form a surface chelate, suggesting that other catalytic phenomena are involved (Torrents and Stone, 1994).

Hydrolysis of monophenyl terephthalate (Figure 4.8) is catalyzed by the addition of aluminum oxide, which possesses a net positive charge throughout the pH range of the experiments. In this case, surface complex formation anchors the ester onto the oxide surface, where it is subject to attack by hydroxide ions accumulated near the surface through electrostatic interactions or specifically adsorbed to surface-bound Al atoms (Stone, 1989). Increases in electrolyte concentration cause rates of surface-catalyzed hydrolysis to drop to nearly negligible levels, confirming that electrostatic interactions are important (Stone, 1989).

Surface complex formation is not necessarily required in order for catalysis to occur. Silica-catalyzed hydrolysis of the fungicide oxycarboxin (Figure 4.9) provides an illustrative example of this. Catalysis arises from the ability of the silica surface to act as a general base catalyst, accepting a proton from oxycarboxin (Stanton, 1987). The ability of weak acids and bases to act as general acid and base catalysts depends upon their concentration and acidity constants, as well as the magnitude of the catalysis rate constant (Perdue and Wolfe, 1983). In saturated soils and aquifer sediments, proton and hydroxide ion-binding sites on mineral surface sites can represent a substantial fraction of the available pool of weak acids and bases; general acid and base catalysis under such conditions remains to be explored.

Most literature reports of surface-catalyzed hydrolysis deal with clays, zeolites, and complex minerals and often involve dehydrated and partially hydrated surfaces (Theng, 1982; Voudrais and Reinhard, 1986). Interpreting results of these experiments on a mechanistic level is considerably more difficult than for the simple oxides discussed above. Spatial constraints within interlayers or cavities may impede molecular

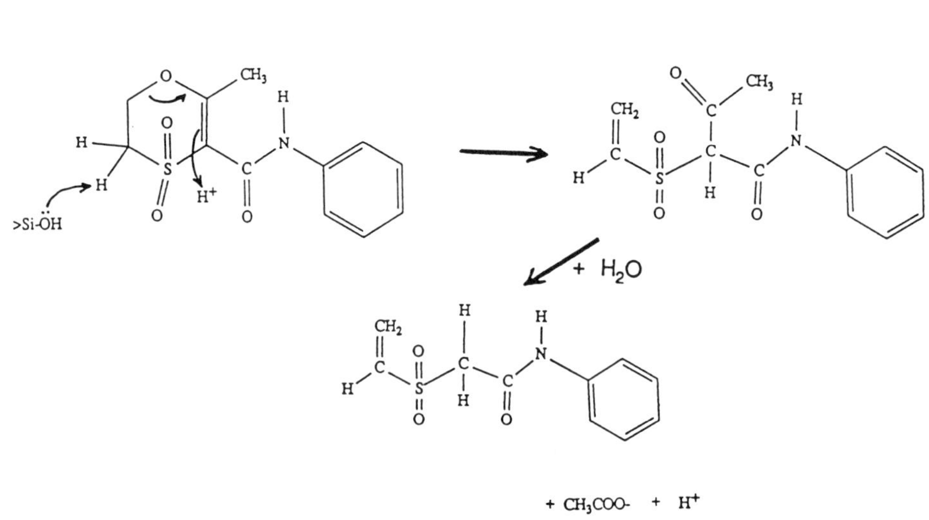

Figure 4.9. Hydrolytic decomposition of the fungicide oxycarboxin in the presence of silica surfaces. (Adapted from Stanton, 1987.)

motions and disturb the hydration shells of metal ions and organic substrates in ways facilitating reaction. Dehydration increases the Lewis acidity of metal ions, substantially increasing their ability to coordinate and polarize organic substrates. Despite these complexities, information concerning reactions of this kind is crucial to understanding and predicting transformations of hydrolyzable organic compounds under realistic field conditions.

CONCLUSION

In most instances, metal ion catalysis in solution involves the formation of metal–organic substrate complexes. For this reason, metal speciation is critically important in determining the magnitude of catalytic effects observed. The susceptibility of various organic compounds toward hydrolysis, in turn, is determined by the nature and placement of ligand donor groups within the molecule.

Within soils, sediments, and aquifers, natural waters come into contact with immense areas of mineral surface. Under such conditions, catalysis by adsorbed and lattice-bound metals may completely overshadow catalysis involving dissolved metal species. The ability to predict pathways and rates of hydrolysis therefore requires a thorough understanding of adsorption phenomena and the nature of the mineral–water interface.

ACKNOWLEDGMENTS

Support of this work by the Office of Exploratory Research of the US Environmental Protection Agency (Grant R-818894) is gratefully acknowledged.

References

Armstrong, M. M., Kramer, U., Linder, P. W., and Torrington, R. G., 1989, Equilibrium studies of the protonation of 8-quinolyl phosphate, 1-naphthyl phosphate and 8-quinolyl methyl phosphate and their complexation with copper(II), zinc(II), nickel(II), cobalt(II) and manganese(II) ions in aqueous solutions. *J. Coord. Chem.*, 20, 81.

Blanchet, P.-F., and St. George, A., 1982, Kinetics of chemical degradation of organophosphorus pesticides; hydrolysis of chlorpyrifos and chlorpyrifos-methyl in the presence of Cu(II). *Pesticide Sci.*, 13, 85.

Britzinger, H., 1965, Intraspharishe und extraspharische komplexe mit phosphatliganden infrarotspektren von phosphatkomplexen in wasseriger losung. *Helv. Chim. Acta*, 48, 47.

Dilling, W. L., Lickly, L. C., Lickly, T. D., Murphy, P. G., and McKellar, R. L., 1984, Organic photochemistry. 19. Quantum yields for O,O-diethyl-O-(3,5,6-trichloro-2-pyridinyl) phosphorothioate (chlorpyrifos) and 3,5,6-trichloro-2-pyridinol in dilute aqueous solutions and their environmental phototransformation rates. *Environ. Sci. Technol.*, 18, 540.

Emsley, J., 1989, *The Elements*, Clarendon, Oxford, England.

Evers, A., Hancock, R. D., Martell, A. E., and Motekaitis, R. J., 1989, Metal ion recognition in ligands with negatively charged oxygen donor groups. Complexation of Fe(III), Ga(III), In(III), Al(III), and other highly charged metal ions. *Inorg. Chem.*, 28, 2189.

Fest, C., and Schmidt, K.J., 1983, Organophosphorus insecticides. In: K. H. Buchel, ed., *Chemistry of Pesticides*, Wiley-Interscience, New York, Chapter 2.3.

Fife, T. H., and Przystas, T. J., 1985, Divalent metal ion catalysis in the hydrolysis of esters of picolinic acid. Metal ion promoted hydroxide ion and water catalyzed reactions. *J. Am. Chem. Soc.*, 107, 1041.

Guthrie, J. P., and Cullimore, P. A., 1980, Effect of the acyl substituent on the equilibrium constant for hydration of esters, *Canad. J. Chem.*, 58, 1281.

Hay, R. W., 1987, Lewis acid catalysis and the reactions of coordinated ligands. In: G. Wilkinson, R. D. Gillard, and J. A. McCleverty, eds., *Comprehensive Coordination Chemistry*, Vol. 6, Pergamon Press, Oxford, England, Chapter 61.4.

Hochella, M. F., and White, A. F., eds.,1990, *Mineral–Water Interface Geochemistry, Reviews in Mineralogy*, Vol. 23, Mineralogical Society of America, Washington, D.C.

Hoffmann, M. R., 1980, Trace metal catalysis in aquatic environments. *Environ. Sci. Technol.*,14, 1061.

Houghton, R. P., 1979, *Metal Complexes in Organic Chemistry*. Cambridge University Press, Cambridge, England.

Johnson, G. L., and Angelici, R. J., 1971, Metal-ion catalysis of ethyl oxalate hydrolysis. *J. Am. Chem. Soc.*, 93, 1106.

Jorgensen, C. K., 1965, Symbiotic ligands, hard and soft central atoms. *Inorg. Chem.*, 3, 1201.

Mabey, W., and Mill, T., 1978, Critical review of hydrolysis of organic compounds in water under environmental conditions. *J. Phys. Chem. Ref. Data*, 7, 383.

Martin, R. B., and Sigel, H., 1988, Quantification of intramolecular ligand equilibria in metal–ion complexes. *Comments Inorg. Chem.*, 6, 285.

Massoud, S. S., and Sigel, H., 1988, Metal ion coordinating properties of pyrimidine-nucleoside 5'-monophosphates (CMP, UMP, TMP) and of simple phosphate monoesters, including D-ribose 5'-monophosphate. Establishment of relations between complex stability and phosphate basicity. *Inorg. Chem.,* 27, 1447.

Mill, T., and Mabey, W., 1988, Hydrolysis of organic chemicals. In: O. Hutzinger, ed., *The Handbook of Environmental Chemistry: Reactions and Processes,* Vol. 2, Part D, Springer-Verlag, Berlin.

Mortland, M. M., and Raman, K. V., 1967, Catalytic hydrolysis of some organic phosphate pesticides by copper(II). *J. Agric. Food Chem.*, 15, 163.

Murakami, Y., and Sunamoto, J., 1971, Solvolysis of organic phosphates. IV. 3-Pyridyl and 8-quinolyl phosphates as effected by the presence of metal ions. *Bull. Chem. Soc. Jpn.*, 44, 1827.

Papelis, C., Hayes, K. F., and Leckie, J. O., 1988, *HYDRAQL: A Computer Program for the Calculation of Chemical Equilibrium Composition of Aqueous Batch Systems Including Surface-Complexation Modeling of Ion Adsorption at the Oxide/Solution Interface.* Technical Report No. 306, Department of Civil Engineering, Stanford University, Stanford, CA.

Pearson, R. G., 1966, Acids and bases. *Science*, 151, 172.

Pearson, R. G., 1988, Absolute electronegativity and hardness: application to inorganic chemistry. *Inorg. Chem.*, 27, 734.

Perdue, E. M., and Wolfe, N. L., 1983, Prediction of buffer catalysis in field and laboratory studies of pollutant hydrolysis reactions. *Environ. Sci. Technol.*, 17, 635.

Plastourgou, M., and Hoffmann, M. R., 1984, Transformation and fate of organic esters in layered-flow systems: the role of trace metal catalysis. *Environ. Sci. Technol.*, 18, 756.

Schindler, P. W., and Stumm, W., 1987, The surface chemistry of oxides, hydroxides, and oxide minerals. In: W. Stumm, ed., *Aquatic Surface Chemistry*, Wiley-Interscience, New York, Chapter 4.

Schwarzenbach, R. P, Gschwend, P. M., and Imboden, D. M., 1993, *Environmental Organic Chemistry*. Wiley-Interscience, New York.

Sherman, D. M., 1985a, The electronic structures of Fe^{3+} coordination sites in iron oxides; applications to spectra, bonding, and magnetism. *Phys. Chem. Minerals*, 12, 161.

Sherman, D.M., 1985b, SCF-XaSW MO study of Fe-O and Fe-OH chemical bonds; applications to the Mossbauer spectra and magnetochemistry of hydroxyl-bearing Fe^{3+} oxides and silicates. *Phys. Chem. Minerals*, 12, 311.

Stanton, D. T., 1987, Glass-catalyzed decomposition of oxycarboxin in aqueous solution. *J. Agric. Food Chem.*, 35, 856.

Stone, A. T., 1989, Enhanced rates of monophenyl terephthalate hydrolysis in aluminum oxide suspensions, *J. Colloid Interface Sci.*, 127, 429.

Stone, A. T., and Torrents, A., 1995, The role of dissolved metals and metal-containing surfaces in catalyzing the hydrolysis of organic pollutants. In: P. M. Huang, ed., *Environmental Impact of Soil Component Interactions*, Lewis, Chelsea, MI.

Stone, A. T., Torrents, A., Smolen, J., Vasudevan, D., and Hadley, J., 1993, Adsorption of organic compounds possessing ligand donor groups at the oxide/water interface, *Environ. Sci. Technol.*, 27, 895.

Stumm, W., 1992, *Chemistry of the Solid–Water Interface*, Wiley-Interscience, New York.

Sutton, P. A., and Buckingham, D. A., 1987, Cobalt(III)-promoted hydrolysis of amino acid esters and peptides and the synthesis of small peptides, *Acc. Chem. Res.*, 20, 364.

Theng, B. K. G., 1982, Clay-activated organic reactions. In: H. van Olphen, and F. Veniale, eds., *International Clay Conference 1981*, Developments in Sedimentology 35 Elsevier, Amsterdam, The Netherlands, pp 197–238.

Torrents, A., and Stone, A. T., 1991, Hydrolysis of phenyl picolinate at the mineral/water interface. *Environ. Sci. Technol.*, 25, 143.

Torrents, A., and Stone, A. T., 1994, Oxide surface-catalyzed hydrolysis of carboxylate esters and phosphorothioate esters. *J. Soil Sci. Soc. Am.*, 58, 738.

Voudrias, E. A., and Reinhard, M., 1986, Abiotic organic reactions at mineral surfaces. In: J. A. Davis and K. F. Hayes, eds., *Geochemical Processes at Mineral Surfaces*, ACS Symposium Series, No. 323, American Chemical Society, Washington, DC, Chapter 22.

Westall, J. C., Zachary, J., and Morel, F., 1976, *MINEQL: A Computer Program for the Calculation of Chemical Equilibrium Composition of Aqueous Systems*, Technical Note No. 18, Ralph M. Parsons Laboratory, MIT, Cambridge, MA.

Westheimer, F. H., 1992, The role of phosphorus in chemistry and biochemistry. In: E. N. Walsh, E. J. Griffith, R. W. Parry, and L. D. Quin, eds., *Phosphorus Chemistry; Developments in American Science*, ACS Symposium Series, No. 486, American Chemical Society, Washington, D.C., Chapter 1

The Chemistry and Geochemistry of Natural Organic Matter (NOM)

DONALD L. MACALADY AND JAMES F. RANVILLE

ABSTRACT. Natural organic substances in aquatic and terrestrial ecosystems are described in terms of their origins, compositions, chemical and physical characteristics, and effects on geochemical processes. Discussion of the roles of natural organic matter (NOM) in the determination of the rates and pathways of the transport and transformation of anthropogenic chemicals provides an additional focus. The particular importance of NOM as a mediator of oxidation/reduction reactions of both natural and anthropogenic substances receives special attention in this chapter.

INTRODUCTION

Natural organic matter (NOM), in its broadest sense, refers to the complex and chemically and physically diverse group of substances that result directly or indirectly from the photosynthetic activity of plants. In environmental science, however, one typically limits the use of this term to substances resulting from the partial decay of detrital materials from terrestrial and aqueous plants. This definition includes everything along the diagenic pathway from living plants to fossil fuels such as coal and petroleum.

For the purposes of this chapter, a further limitation is evoked. In this discussion, nothing further along the diagenic pathway than materials roughly resembling peat will be considered. In other words, the NOM discussed here has retained a considerable portion of its original oxygen and hydrogen. It may have picked up a bit of additional sulfur or metal content along the diagenic pathway or in other geologic processes not accurately described as diagenic. It's the kind of stuff you're likely to find as you dig in your garden, walk through a bog or fen, or examine the colored substances found naturally in lakes, rivers, and ground waters.

This chapter is not about the detailed geochemistry of the diagenesis of organic carbon compounds. Rather, it concerns (a) the origin and nature of NOM as defined above and (b) the role of NOM in the transport and transformations of substances, natural and anthropogenic, in aquatic systems. Thus, the goal of this discussion is to provide the background and technical understanding necessary for the evaluation of the role of NOM in the wide variety of systems one encounters in the consideration of environmental problems. Hopefully, it will also stimulate the reader to pursue additional study into the nature of the involvements of NOM in chemical and physical processes in aquatic systems.

THE NATURE AND ORIGIN OF NATURAL ORGANIC MATTER

Natural organic matter has been classified according to a wide variety of schemes, one of which involves the origin of the plants which serve as a starting material for the NOM (Figure 5.1). *Aquogenic* refers to NOM originating from the decomposition of aquatic organisms (macrophytes, algae, bacteria); *pedogenic* refers to NOM from terrestrial plants and microorganisms, including material leached from soils into an aquatic system. For most catchments which do not include a large lake, river, or reservoir, as well as for all soils, pedogenic materials can be expected to dominate. For large lakes and, of course, the open oceans, aquogenic materials are the overwhelming majority. Another pair of terms is often used to refer to a similar distinction among types of organic materials is *autochtonous* and *allochtonous*. Some authors are careful to distin-

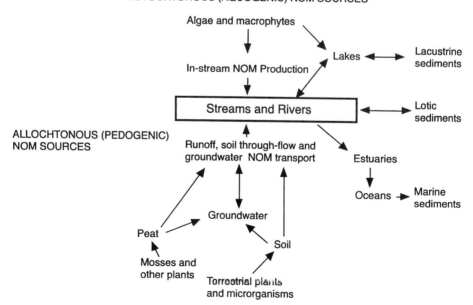

Figure 5.1. Some of the many possible flow paths for NOM. Adapted from Aiken, G. R., et al., An introduction to humic substances in soil, sediment and water, in Aiken, G. R., McKnight, D. M., Wershaw, R. L., and MacCarthy, P. (eds.), *Humic Substances in Soil, Sediment and Water: Geochemistry, Isolation, and Characterization.* Copyright © 1985 John Wiley & Sons, New York. Reprinted by permission of John Wiley & Sons, Inc.

guish between these pairs of terms (Buffle, 1984), particularly with reference to dissolved or colloidal organic materials.

Terrestrial Detritus and NOM

Detritus is a general term which refers to dead plant parts which are in various stages of decomposition. The production of plant litter fall has well-understood geographical patterns which roughly follow global patterns in net primary production. Generally speaking, production of terrestrial detritus varies inversely with latitude from tropical to boreal forests (Lonsdale, 1988).

Most terrestrial detritus is pedogenic and results from either the above-ground portions or the root systems of plants and is thus located in the upper layers of the soil. Here it is subject to a variety of decomposition processes, mostly of which involve the heterotrophic metabolism of bacteria, fungi, or other microfauna (Swift et al., 1979). One result of such processes is the release of CO_2, H_2O, and nutrients such as nitrogen and phosphorus compounds. Another important result is the production of a group of substances which are highly resistant to further aerobic or anaerobic decomposition, collectively known as *humus*. These early diagenic processes can be viewed in terms of the rapid turnover of the majority of the detrital material near the soil surface and the much slower production, accumulation, and turnover of humus in the deeper soil layers (Schlesinger, 1977).

A simple, first-order model for the rate of decomposition of forest litter fall provides a basis for comparing a variety of ecosystems, allowing the estimation of a mean residence time for plant debris as $1/k$, where k is the first-order decay rate constant (yr^{-1}). Values for many tropical rain forest ecosystems, for example, are greater that 1.0, indicating little surface accumulation and rapid turnover of organic matter (Cuevas and Medina, 1988). In contrast, some peat-lands have k values of ~ 0.001 (Olson, 1963). Esser et al. (1982) estimated a global mean value of $k = 0.33$ for soil surface litter. In various geographical and ecological regions, the rate of decomposition may be limited by a variety of environmental factors, including temperature, moisture, and the origin of the litter material.

Humus and Humic Materials

Humus results from the incomplete turnover of plant litter in most of the world's ecosystems. Soil NOM-accumulation rates vary from about 0.2 g $C/m^2/yr$ in polar desert environments up to 15 in cool, wet locations such as boreal forests (Bockheim, 1979; Ugolini, 1968). Most of this accumulated material is soil humus, which is extremely resistant to further degradation. Radiocarbon dating techniques indicate ages of several thousand years for soil organic matter in lower soil profiles (O'Brien and Stout, 1978; Schlesinger, 1977).

Humus is composed of a complex group of organic molecules that defy complete chemical characterization. Classification of humus is consequently usually based on empirical separation schemes such as the solubility separation originally developed by soil scientists (Stevenson, 1982, p. 43, see also Stevenson, 1986). Soil or sediment is first extracted with alkali. The insoluble organic matter remaining after alkaline extraction is termed *humin*. The portion of the soil solubilized by the alkali is subsequently treated with acid to near-neutral pH. The material which precipitates in this step is termed *humic acid*, while the organic material remaining in solution is called *fulvic acid*. The precipitated humic acid is sometimes further differentiated into *hymatomelanic acid* (alcohol-soluble portion of humic acid), gray humic acid (precipitated from an alkaline humic acid solution with added electrolyte), and brown humic acid (soluble in alkaline solution with added electrolyte), but these latter categories are not widely discussed in environmental chemistry.

More recently, NOM fractions in aquatic systems have been redefined in terms of a separation scheme based on retention characteristics on macroreticular nonionic resin (Figure 5.2, Thurman and Malcolm, 1981). In the most commonly applied method, the hydrophobic characteristics of different humic fractions are exploited to enable separation after adsorption (at pH = 2) on a column of a nonionic acrylic resin. Application of a pH gradient to the material adsorbed onto the columns provides a separation of the aquatic NOM into fractions identifiable as fulvic and humic acids (MacCarthy et al., 1979), which typically accounts for about 50% of the total dissolved organic carbon (DOC, Figure 5.3).

Regardless of the isolation/definition scheme, humic and fulvic acids constitute a major fraction of any NOM sample. Soil NOM typically has a larger humic/fulvic ratio and smaller contents of identifiable nonhumic materials. Humic and fulvic acids account for widely varying proportions of aquatic NOM, from around 25% for seawater to almost 90% for some wetlands and highly colored rivers. The remaining NOM in

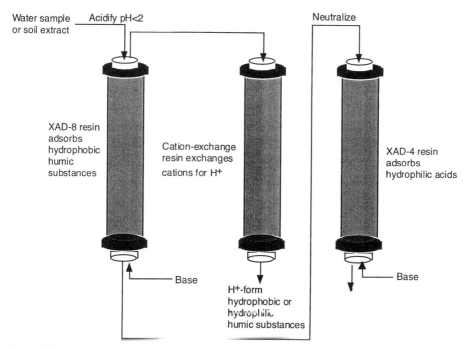

Figure 5.2. Separation of humic substances and hydrophilic acids by adsorption and ion exchange chromatography. Adapted from Thurman, E. M., *Organic Geochemistry of Natural Waters*. Martinus Nijhoff/Dr. W. Junk Pubs., Dordrecht, 1985. With kind permission from Kluwer Academic Publishers.

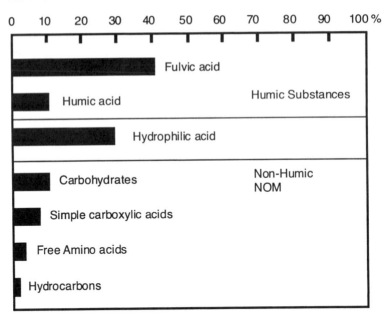

Figure 5.3. Dissolved organic carbon histogram for an average river water with a DOC of 5 mg/L. Adapted from Thurman, E. M., *Organic Geochemistry of Natural Waters*. Martinus Nijhoff/Dr. W. Junk Pubs., Dordrecht, 1985. With kind permission from Kluwer Academic Publishers.

aquatic samples consists of more hydrophilic organic acids and a variety of identifiable compounds including amino acids, some simple sugars, and several small organic acids (Thurman, 1985).

Much of the effort to chemically and physically characterize NOM has focused on the humic and fulvic acid fractions. In general, aquatic NOM can be described as a polymolecular array of organic molecules which contain about 50–60% carbon, 4–6% hydrogen, 23–40% oxygen, and 0.3–6% nitrogen (by weight), with traces of phosphorus, sulfur, and "ash" (residue after combustion). Soil NOM is similar in elemental composition, with naturally larger proportions of highly insoluble and/or hydrophobic constituents such as lipids, proteins, and polysaccharides. Isolates of soil NOM generally have considerably higher ash contents (Figure 5.4).

In addition to elemental composition, NOM fractions have been characterized according to molecular weight distributions, the presence of certain functional groups, acid–base and redox characteristics, aromatic/aliphatic character, chelating/complexation abilities, and trace element composition. An excellent summary of many characterizations of aquatic NOM fractions is given by Thurman (1985). A similarly useful source for soil NOM is given by Stevenson (1982).

The utility of such analytical efforts, for the purposes of this chapter, is limited by the fact that the NOM has been fractionated into fulvic acid, humic acid, and so on, prior to analysis. It is a given that, for the purposes of studies into processes which occur in nature, one is interested in the behavior of NOM as it exists *in situ*, not in the behavior of the humic or fulvic acid fraction of that NOM. Thus, the large amount of scientific endeavor which has been expended to isolate and characterize NOM fractions, though extremely useful in providing a basis for the understanding of the physical and chemical characteristics of NOM, is not directly applicable to the considerations of this chapter. The following brief look at some of these characteristics is intended to provide a general picture of the types of interactions we might expect between NOM and the surrounding environmental matrices.

The fractions of NOM from soil or aquatic environments called fulvic and humic acids differ primarily in their hydrophobicity and solubility characteristics, with concomitant differences in average molecular size and mass. In general, fulvic acids are more soluble, smaller in average molecular weight, less aromatic, and more highly charged than humic acids. Fulvic acids also typically have higher oxygen content, with higher carboxylic acid (COOH) and lower phenolic (ArOH) content than humic acids (Hayes et al., 1989).

Fulvic acids are therefore probably more representative of aquatic NOM, especially the "dissolved" fraction (Malcolm, 1985). A typical "molecule" of fulvic acid might contain one COOH per 6 carbon atoms, with about 65% aliphatic (cf. aromatic) carbons (Thurman, 1985, p. 356). Fulvic acid and NOM in general are known from electron spin resonance (ESR) studies to produce free radicals in aqueous solutions (Senesi and Steelink, 1989). The most likely groups to contribute this free radical character are quinones, and quinone functionalities have recently been shown by ^{13}C nuclear magnetic resonance (NMR) evidence to be present (Thorn et al., 1992). The emerging picture of aquatic NOM is one of polyfunctional organic acids containing significant metal-complexing and redox capabilities. Furthermore, various portions of individual fulvic and humic acid "molecules" are expected to exhibit significant hydrophobic character, more so for humic acids and particulate NOM (PNOM) than for the "dissolved" fraction of aquatic NOM (DNOM). Finally, the dominance of car-

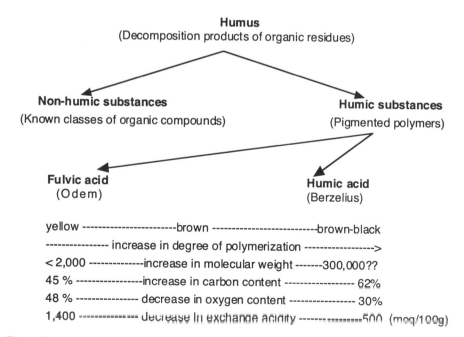

Figure 5.4. Classification and chemical properties of humic substances. Adapted from Stevenson, F. J., *Humus Chemistry: Genesis, Composition, Reactions.* Copyright © 1982 John Wiley & Sons, New York. Reprinted by permission of John Wiley & Sons, Inc.

boxylic acid and phenolic functional groups in NOM means that the net "surface" charge of aqueous NOM and the overwhelming majority of soil or sediment particles, which contain NOM-dominated surfaces, can be expected to be negative at essentially all environmental pH values (Tipping and Cooke, 1982)

Aquatic Organic Matter

Dissolved and colloidal forms

Organic matter present in raw natural waters may be considered to physically occur as an essentially continuous series which we will refer to as the *dissolved/ colloidal/particulate organic matter continuum* (Figure 5.5). While the pure end members are easily conceptualized as (a) dissolved (solvated) organic molecules and (b) macroscopic solid particles which may occur with or without sorbed ions or molecules, the intermediate colloidal range requires further explanation. The word "colloid" as used here collectively identifies this intermediate part of the size continuum. Organic matter species that can be referred to as 'colloidal' include:

a. discrete chemical species with sufficient size or mass to behave as colloids (macromolecules),

b. aggregates of smaller organic molecules, and

c. organic matter associated with (usually by sorption) compositionally distinct, colloid-size particles, such as clays or oxides.

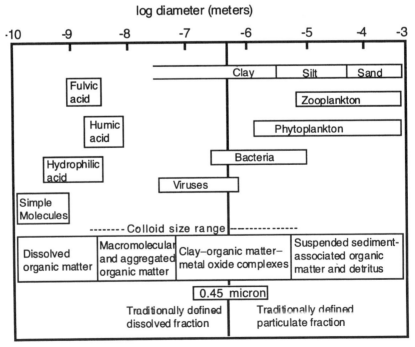

Figure 5.5.   Continuum of particulate and dissolved organic carbon in natural waters. (Redrawn from Thurman, 1985, p. 3.)

Traditionally, filtration through a 0.45-μm pore-size membrane, although quite arbitrary, has been widely used as a means of separating particulate from "dissolved" species. Such an operational definition entirely ignores the continuum of suspended particle sizes and the issue of colloids. For environmental investigations, this separation may be insufficient to fully describe the processes involved in both organic matter and contaminant transport. Unfortunately, it has only been recently that studies have attempted to rigorously differentiate among dissolved, colloidal, and particulate organic matter in natural waters.

Dissolved natural organic matter (DNOM)

Dissolved NOM, in its most rigorous sense, refers to those compounds which are fully solvated by water. This consists of a wide range of compounds which generally share the properties of relatively low molecular weight (<1000 daltons) and significant numbers of polar and/or ionizable functional groups. These compounds include some members of the complex mixtures of polyelectrolytes called *humic substances* which are described above. Humic and fulvic acid are the major classes of humic substances and are defined by their pH-dependent solubility and/or their adsorption onto nonionic acrylic resins. Another class of complex mixtures is known as *hydrophilic acids*. Hydrophilic acids are similar to fulvic acid but are more soluble, higher in oxygen content, and less readily isolated on acrylic resins. DNOM also includes simpler organic compounds, the major classes being carbohydrates, carboxylic acids, amino acids, and hydrocarbons (Thurman, 1985, Chapter 10).

Experimentally, DNOM is often determined by measurements of dissolved organic carbon (DOC). In general DOC can be converted to DNOM by multiplying by a factor of ~2. Most modern DOC determinations rely on either chemical or thermal oxidation of the organic matter to CO_2 followed by detection using spectrophotometric or coulometric techniques (see, for example, Lee and Macalady, 1989). DOC measurements are complicated by the low values often encountered, differing oxidation efficiencies for components of DNOM, and the presence of interfering species such as chloride (Hedges and Lee, 1992). Distinction between particulate organic carbon (POC) and DOC is made by analyzing raw and filtered samples. As previously pointed out, the result of this-two phase classification scheme is that most DOC and POC values will contain variable amounts of colloidal organic carbon (COC). The amount of COC will vary depending on the source of NOM and the aqueous environment.

Colloidal natural organic matter (CNOM)

The major components classified as colloidal organic matter are: high-molecular-weight compounds, aggregates of smaller molecules, organic coatings on other colloidal-sized particles (i.e., clays and oxides), and viral and bacterial cells (both viable and senescent). These colloidal materials can be classified as either hydrophilic or hydrophobic colloids. Macromolecular organics such as humic and fulvic acids, polysaccharides, proteins, peptides, and amino acids tend to be polar and maintain their colloidal stability via interactions between charged functional groups and water molecule dipoles. They are best classified as hydrophilic colloids. For hydrophobic solids, permanent dispersion or suspension is maintained by the random thermal activity of water molecules (Brownian movement). In order for this mechanism to be effective, the particle must be sufficiently small to allow spatially uneven bombardment by water molecules and must have at least some electrostatic charge. If not for electrostatic repulsions, the hydrophobic nature of these particles would tend to force them together during particle collisions and promote aggregation which would lead to larger and larger particle sizes and eventual destruction of the colloidal system. These electrostatic repulsions promote coulombic repulsion between particles and also allow interactions with water molecules, thus making the colloids effectively more hydrophilic. The importance of organic coatings to establish a substantial negative charge on mineral colloids and thereby impart significant stability has been widely demonstrated (Tiller and O'Melia, 1993).

The "molecular weight" ranges of aquatic fulvic and humic acids are generally considered to be approximately 500–1000 and 1000–5000 daltons, respectively (Thurman, 1985, Chapter 10). These values, along with the significant number of ionizable functional groups, suggest that fulvic acid may best be considered a dissolved species, while humic acid molecules probably span the dissolved–colloidal interface. It should be noted that these molecular weights are generally determined on de-salted, isolated fractions of NOM. Fulvic and humic acids may occur in natural waters as aggregates with much higher apparent molecular weights and exist in colloidal forms. Aggregation is enhanced by the presence of multivalent cations which reduce the electrical double layer. Ultrafiltration studies of NOM have suggested that a significant portion of NOM exists as colloids with "molecular weights" exceeding 10,000 daltons. In general, it is not known whether this CNOM exists as discrete, high-molecular-weight molecules or represents some of the other forms of CNOM mentioned above such as aggregates or NOM coatings (see below). Ultrafiltration results must be interpreted carefully because artifacts can result from processes such as charge repulsion by the ultrafilter membrane, clogging of the membrane, and aggregation at the membrane surface (Buffle et al., 1992). Some of the experimental difficulties of using ultrafiltration, particularly membrane fowling, have been obviated by the introduction of tangential-flow filtration methods (Figure 5.6; Ranville et al., 1991).

Some of the CNOM is present as adsorbed coatings on mineral particles. Particles with very similar negative surface charge, regardless of underlying mineralogy, dominate in most aquatic environments, a reflection of the universal presence of organic coatings (Hunter, 1980). These NOM coatings significantly modify the ability of mineral colloids to partition metal and organic contaminants in natural waters. The negative charge imparted by these coatings also facilitates the transport of colloids and associated contaminants in surface and ground waters by preventing aggregation in the surface waters and capture by the aquifer matrix in ground waters. In addition to altering the properties of the mineral colloid, the adsorption process can also significantly alter the distributions of NOM fractions within an aquatic system. McKnight et al. (1992) documented preferential sorption of certain components of NOM to iron and aluminum oxides in an acidic stream. The NOM remaining in solution was shown to be less aromatic than that adsorbed to the oxides. Changes also occurred in the elemental composition and molecular weight.

Pedogenic and aquogenic materials

As outlined above, dissolved and colloidal organic matter in natural waters may be composed of an exceedingly wide variety of natural and synthetic materials, in-

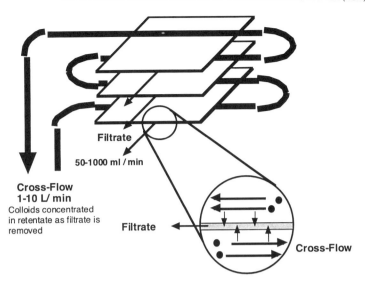

Cross-Flow
1-10 L/ min
Colloids concentrated
in retentate as filtrate is
removed

Figure 5.6. Diagram showing the operation of cross-flow ultrafiltration for collection of a colloid sample.

cluding simple compounds, macromolecules (molecular weight >1000), biological particles, NOM-coated oxides, and phyllosilicates (clays). The relative importance of these various types of organic colloids will vary with the nature of the aqueous environment. For example, it is likely that in productive surface waters, living microorganisms and detrital organic materials will dominate the colloid population. In waters affected by mining wastes or agricultural runoff, iron and aluminum oxides and clays with adsorbed organic matter may be the most abundant colloids. Aquatic colloids originate in at least three ways:

1. Pedogenic NOM includes both DNOM leached out of the soil of the drainage basin (Leppard et al., 1986) and particulate NOM carried in runoff. This latter form, at least in streams and rivers, is comprised principally of plant fragments and woody debris and may be chemically degraded during transport to the aquatic system (Ranville et al., 1991).

2. Aquogenic NOM, on the other hand, forms from two distinct types of processes:

   a. as a result of biological activity [e.g., growth and death of microbes, production of fibrils, organic skeletons, and protein-rich cell fragments (Leppard, 1992)] and

   b. as a result of *in situ* aqueous phase inorganic precipitation due to changing chemical or physical conditions, followed by NOM adsorption [e.g., precipitation of iron oxides and/or sulfides due to redox changes, formation of complex phosphate mineral suspensions at chemical interfaces in lakes (Buffle et al., 1989)].

GEOCHEMICAL REACTIONS OF NATURAL ORGANIC MATTER

The geochemistry of NOM is as complex and varied as the diverse origins and nature of the material suggest. In this section, some of the more important geochemical interactions of NOM with unaltered geological systems are briefly reviewed. For a more complete discussion of any of the topics in this section, refer to the reference section.

Weathering and NOM

The occurrence of specific interactions involving NOM in the chemical weathering of soils and minerals is difficult to determine because such interactions are universally accompanied by a complex array of biological and biochemical reactions as well as simple hydrolysis (Figure 5.7). The role of microbial processes in weathering has been discussed in detail by Robert and Berthelin (1986). It is the opinion of these and other (e.g., Loughman, 1969) authors that the direct involvement of NOM in chemical weathering is extremely limited, with carbonic and other mineral acids produced by biological activity being the primary agents in chemical weathering.

Nevertheless, both low-molecular-weight biochemical compounds and humic and fulvic acids have been implicated in the degradation of mineral matter in nature. The ability of many microorganisms, particularly lichen fungi, to bring about the weathering of rocks and minerals due to the synthesis of biochemical chelating agents is well known (Stevenson, 1982). Specific weathering effects of low-molecular-weight organic acids of natural origin are well documented (Bennett and Siegel, 1987; Mast and Drever, 1987; Schalscha et al., 1967) with both the acidic and chelating characteristics of the acids implicated in weathering processes. Antweiler and Drever (1983) correlated the weathering of volcanic ash with the DOC of percolating waters. The DOC

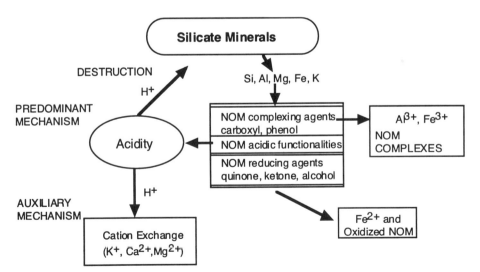

Figure 5.7. Mechanisms in the alteration of silicates by organic acids. (Redrawn from Robert and Berthelin, 1986, p 479.)

consisted predominately of humic materials but contained traces of oxalate, acetate, and formate with pH values between 4.3 and 5.2.

Data obtained by Baker (1973) show that humic acids exhibit an activity of the same order as that of several simple organic acid chelating agents in the weathering of a number of minerals and metals. The ability of NOM, more specifically humic substances, to decompose common soil minerals such as biotite, muscovite, illite, and kaolinite has been demonstrated by a number of authors, including Huang and Keller (1971), Schnitzer and Kodama (1976), and Tan (1980). Because of their low molecular weights and abundance in soil solutions, fulvic acids may be particularly effective in dissolving silicate minerals (Stevenson, 1985). Zunino and Martin (1977) have advanced a detailed concept of the involvement of humic substances in weathering and subsequent translocation of trace elements to biological systems.

An additional role ascribed to humic substances in weathering processes depends upon their redox properties. The reduction of iron minerals, causing the release of ferrous iron into solution, is an obvious possibility. However, little direct evidence has been reported for direct involvement of NOM in redox transformations of minerals. The role of NOM in redox transformations of metals, however, is well established (see below).

NOM Profiles in Soils and Associated Aqueous Systems

The general pattern of NOM content in a soil profile includes a gradation from the highly organic layers containing litterfall from the surface vegetation to the lower highly mineral layers containing relatively little organic content. The precise pattern for a particular catchment depends upon a variety of geographical, ecological, and mineralogical factors, as well as the extent of agricultural and other anthropogenic disturbances (Figure 5.8). For example, in typical tropical forest soils, the organic matter is largely recycled in the surface soil zones with little or no transport of soluble NOM fractions to underlying layers. In contrast, in regions between the Arctic and cooler temperate climates, much of the forested area is dominated by coniferous forests, which produce litterfall rich in phenolic compounds and organic acids (Cronan and Aiken, 1985). In these ecosystems, decomposition is also slower and correspondingly less complete, resulting in a soil solution rich in NOM percolating into the lower soil horizons (Dethier et al., 1988). A detailed discussion of the variations in soil development and concomitant NOM profiles is beyond the scope of this review, but a brief description of soil NOM profiles in a typical Northern forest ecosystem will be presented as an example. For further discussion the reader is referred to Schlesinger's treatise on biogeochemistry (Schlesinger, 1991). Much of the discussion below is based on Schlesinger's treatment of temperate forest soils.

Soil profiles are typically separated into layers, or horizons, which define the characteristics of the soil as a function of depth. Forest soils generally contain an organic layer which is clearly separated from the underlying mineral layers. The organic, or O, layer can be further divided into zones of increasingly decomposed organic materials on the forest floor, from the largely cellular surface material to a humus layer consisting of amorphous, degradation-resistant organic substances and often containing significant quantities of minerals. Differentiation of sublayers within this surface zone is often difficult, with large regional and seasonal variations. In temperate and boreal

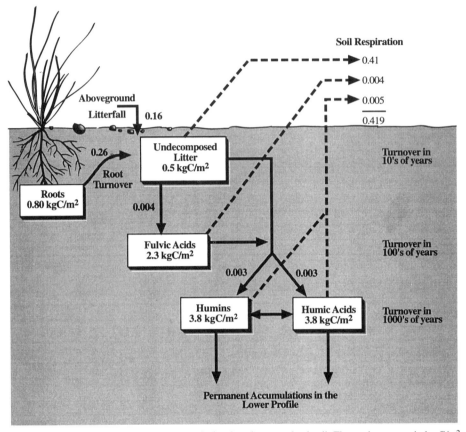

Figure 5.8.   Turnover of litter and soil organic fractions in a grassland soil. Flux estimates are in kg C/m²
yr. (Redrawn from Schlesinger, 1991, p. 136.)

coniferous forests, this organic layer often accumulates into a thick organic mat, known
as a *mor*, that is sharply differentiated from the underlying soil (Romell, 1935). Many
Arctic soils are waterlogged and contain primarily organic material in the entire root-
ing zone (peatland soils, or Histosols).

Beneath this organic layer lies the upper mineral soil, designated the A horizon,
which has a significant NOM fraction and varies from several centimeters to over 1 m
in thickness. This is a zone of eluvial processes, characterized by significant weathering by organic and mineral acids contained in the water percolating through the forest
floor. Iron and aluminum are typically removed by chelation with NOM. Such downward movement of Fe and Al in conjunction with NOM is known as *podzolization*
(Antweiler and Drever, 1983; Chesworth and Macias-Vasquez, 1985). Known throughout the world, podzolization is particularly intense in temperate to boreal coniferous
forest ecosystems (Figure 5.9). In such soils, the pH of the soil solution is often as low
as 4.0 (Dethier et al., 1988). When podzolization is particularly pronounced, a whitish
layer of nearly pure quartz (the E horizon) can form at the base of the A horizon (Pedro et al., 1978). There is not universal agreement that organic acids pay a significant
role in podzolization (Drever, personal communication).

Substances leached from the A and E horizons are deposited during soil development in the underlying, or B, horizon, defined as a zone of deposition or illuvial processes, where secondary clay minerals accumulate. These clay minerals retard the downward movement of soluble NOM fractions that are carrying Fe and Al (Greenland, 1971). Soils of varying degrees of podzolization are characterized by B horizons which contain varying amounts of clay and organic matter. Spodosols are highly podzolized soils which have an upper B horizon (B_s) which is dark and rich in Fe and NOM. Less highly podzolized soils may have B horizons which vary from nearly pure clays to zones which are orange-red to yellow from varying iron contents. In forests in New England, the accumulation of NOM in the B_s horizon appears to control the loss of solution-phase NOM to streams (McDowell and Wood, 1984).

Beneath the B horizon, the C horizon is characterized by soil material with little NOM content. Depending upon whether or not the soil has developed from local materials, the C horizon may or may not resemble the underlying bedrock. In any case, carbonation weathering is generally the dominant process in the C horizon (Ugolini et al., 1977). In many temperate regions, anthropogenic activities such as agriculture and subsequent erosion substantially alter these forest profiles. For example, in the Piedmont region of the southeastern United States, the forest floor often resides directly atop the B horizon due to such erosion. Acid rain can also substantially alter this picture through the introduction of strong acids such as sulfuric into the soil solution, particularly with respect to the weathering of Al.

The NOM content of aqueous systems associated with a given catchment is largely controlled by the soil processes outlined above. The export of NOM from a catchment is largely through DNOM and CNOM in surface and ground waters. The NOM con-

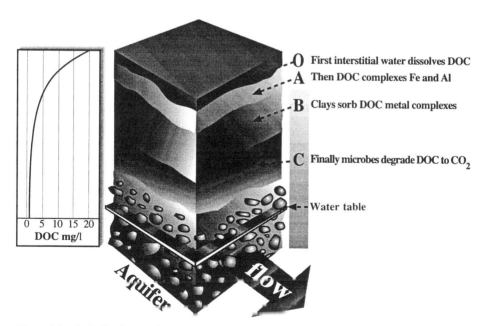

Figure 5.9. Podzolization and decrease of organic carbon in interstitial water of soils. Adapted from Thurman, E. M., *Organic Geochemistry of Natural Waters*. Martinus Nijhoff/Dr. W. Junk Pubs., Dordrecht, 1985. With kind permission from Kluwer Academic Publishers.

tent of ground water is generally a direct result of the nature and extent of the soil processes, though some NOM is imported into a given catchment through rainfall, which typically contains about 1 mg C/L above the forest canopy and 2–3 mg C/L below the canopy. [Rainfall that drips from leaf and plant matter may contain 5–25 mg C/L (Thurman, 1985, p. 21).] Because of variations in soil development processes and hydrological considerations, the measured NOM content of ground water and streams draining various catchments shows dramatic regional variations. For example, streams and ponds in very sandy regions, which show little retention of NOM in soils, may have up to several hundred mg/L of organic carbon in the DNOM and CNOM forms. Shallow ground waters in peat lands or bogs may have similarly high NOM contents (see below).

However, NOM concentrations in both surface and ground waters are generally much lower than these extremes (Figure 5.10). In a broad study of ground water NOM concentrations, Leenheer et al. (1974) analyzed 100 ground-water samples from a variety of aquifer types in the United States, finding a median DOC concentration of 0.7 mg C/L, with a range of 0.2 to 15 mg C/L. Their study and others show that the majority of ground waters have a DOC concentration of 2 mg C/l or less, provided that areas dominated by kerogen or petroleum and oil-field brines are excluded. Surface waters in rivers and streams show generally higher NOM levels, with significant fractions often present as DNOM, CNOM, and larger particulate forms. Thurman (1985, Chapter 1) lists DNOM concentrations, expressed as DOC, ranging from 2 mg C/L in Arctic and alpine streams to above 25 mg C/L in streams draining swamps or wetlands. Inclusion of particulate NOM forms can increase these organic stream loadings

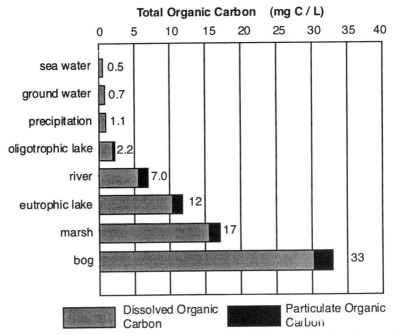

Figure 5.10. Approximate concentrations of dissolved and particulate organic carbon in natural waters. Adapted from Thurman, E. M., *Organic Geochemistry of Natural Waters*. Martinus Nijhoff/Dr. W. Junk Pubs., Dordrecht, 1985. With kind permission from Kluwer Academic Publishers.

considerably. Particulate organic carbon loadings have been estimated to show a range of 1 to 30 mg C/L for 99% of all rivers (Meybeck, 1981, 1982).

Thus, the export of NOM by ground and surface water flow is typically a relatively minor process in terms of the overall effect on mass balance in a given drainage system. The fraction of pedogenic NOM generated that is exported via ground and surface water flow is difficult to estimate in a general sense, but these fractions can be estimated for a given catchment given adequate data on NOM production, ground- and surface-water compositions, and hydrology (see, for example: Hornberger et al., 1994; Chittleborough et al., 1992; Jardine et al., 1989, 1990; Wilson et al., 1990). Concomitant roles of such exported NOM on the mass balances of other natural and anthropogenic materials may, as discussed in subsequent sections of this review, be a more significant consideration than the fractional export of NOM may indicate.

Redox Chemistry of Metal–Organic Complexes

One of the most important considerations in the geochemistry and environmental chemistry of NOM is its role in the transport and transformations of metals. The interactions of NOM with base and soil minerals has already been discussed briefly in the preceding sections on weathering and soil profile development. The transport of NOM—and, necessarily, associated organic and inorganic materials—has also been discussed. This and the following section will more fully explore the role of NOM in determining the ultimate fate of the metals released in weathering and soil development processes (see also Chapter 3, this volume).

NOM plays an important role in the cycling of iron in surface waters. Reduction of Fe(III) to Fe(II) in oxygenated waters by NOM can occur by both photolytic and nonphotolytic reactions. Nonphotolytic reactions involving organic compounds such as oxalate can promote the dissolution of iron oxides by forming soluble iron complexes. These may in turn be reduced in solution by species such as sulfide. Catalyzed reductive dissolution of Fe(III) oxides by Fe(II) oxalate complexes is another important mechanism. Organic compounds such as ascorbate can form inner-sphere surface complexes with Fe(III) oxides and directly transfer electrons. Photolytic reduction of Fe(III) oxides is greatly enhanced by formation of surface complexes with electron-donating ligands such as oxalate. The cycling of iron in surface waters is shown schematically in Figure 5.11, taken from Stumm (1992).

In ground waters, NOM (2 mg C/L) has been found to increase the rate of Fe(II) oxidation at low partial pressures of oxygen ($P_{O2} = 0.005$ ATM.) by a factor of five (Figure 5.12, Liang et al., 1993). In contrast, NOM showed little effect on oxidation rates of Fe(II) at high partial pressures of oxygen ($P_{O2} = 0.2$ atm). In addition, total concentrations of Fe(III) species in solution were enhanced by Fe(III)–NOM complex formation and stabilization of colloidal Fe(III) hydroxides by adsorption of NOM.

Finally, recent observations in our laboratories (Peiffer and Macalady, unpublished data) indicate that NOM may play an additional role in metal redox reactions in the presence of varying amounts of sulfide and oxygen. Reduction of ferric iron, present as nonfilterable (0.2 μm) ferric NOM complex, in an iron- and NOM-rich ground water from a shallow aquifer beneath a wetland (pH ~6.5), proceeded rapidly upon the introduction of millimolar quantities of sulfide. Upon removal of the sulfide (by N_2 sparging) and the introduction of air, the iron was quickly reoxidized. However, the

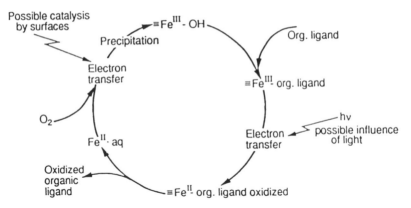

Figure 5.11. Schematic representation of the aquatic redox cycling of iron. Triple lines represent the lattice surface of an iron(III)(hydr)oxide. (Adapted from Stumm, 1992, p. 362.)

NOM was apparently reoxidized much more slowly. After exposure to air for periods of up to several days, the NOM retained the ability to reduce the iron back to the ferrous state after the solution was sparged with N_2 to remove oxygen. This suggests that NOM reduced by sulfide is reoxidized very slowly by molecular oxygen, providing a kind of redox-buffering of kinetic origin. The extent to which these observations are related to sulfide–NOM interactions as a unique form of redox reactions remains to be demonstrated.

NOM and Metal Ion Transport

The binding of a metal ion to an organic complexing agent is generally expressed in terms of a *binding constant*. Binding constants can take a variety of forms depending upon the protonation state of the complexing agent used in the expressions, but they generally represent (for monodentate complexes) the equilibrium ratio of bound metal (metal–organic complex, ML) concentration (or activity) to the product of the concentrations (activities) of unbound ligand (L) and free aqueous metal ion (M); for example,

$$HL + M^{n+} \longleftrightarrow ML^{(n-1)+} + H^+$$

$$K_b = [ML^{(n-1)+}]\,[H^+]/[M^{n+}]\,[HL]$$

where K_b is the binding constant and the square brackets represent concentrations or, more accurately, activities.

Because of the multiplicity of sites for metal complex formation on NOM "molecules," precisely defined binding constants for NOM or its defined fractions are conceptually difficult. Nevertheless, the binding constant concept can be used in a general way to discuss the relative strengths of metal-binding interactions of NOM in aqueous systems. Binding constants can be used as a surrogate for binding strength.

As previously discussed, the presence of functional groups which may form bonds with metals suggests NOM is likely to play a major role in metal transport. The strength of the binding of most metals to these groups generally follows the series

Figure 5.12. The effect of NOM on the rate of oxidation of Fe(II) at ~20°C. under high (upper frame, P_{O2} = 0.2 atm), intermediate (middle frame, P_{O2} = 0.02 atm), and low (lower frame, P_{O2} = 0.005 atm) dissolved oxygen concentrations. Reprinted with permission from Liang, L., et al., Kinetics of Fe(II) oxygenation at low partial pressure of oxygen in the presence of natural organic matter. *Environ. Sci. Technol.*, 27: 1864–1870. Copyright 1993 American Chemical Society.

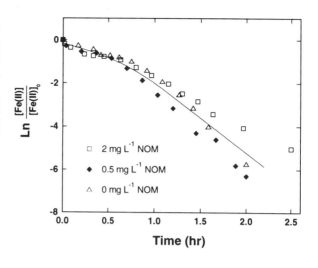

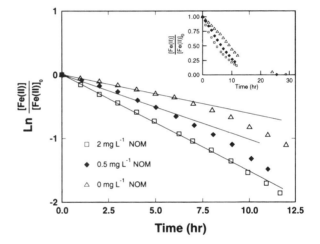

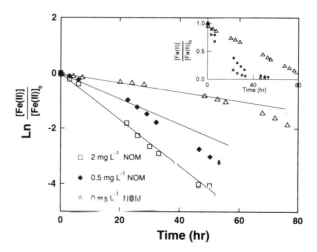

$$-O^- > -NH_2 > -N{=}N{-} > -COO^- > -O{-} > C{=}O$$

The order of metal binding affinity is then enolate followed by amines, azo compounds, ring nitrogen, carboxyl, ether, and ketones (Charbeck and Martell, 1959). In some cases the adjacent placement of two groups on the NOM "molecule" is favorable for bidentate complexes, (M_2L, Figure 5.13), creating sites with very large binding constants. Some examples of the more important chelation sites, according to Thurman (1985), are as follows: (a) salicylic acid type: aromatic carboxyl and a phenol; (b) phthalic acid type: two aromatic carboxyls; (c) picolinic acid type: aromatic carboxyl and a ring nitrogen; and (d) malonic acid type: two aliphatic carboxyls (especially with a nearby ether group).

The components of NOM likely to be responsible for metal mobility include humic and fulvic acids, hydrophilic acids, pigments, and amino acids. However, humic and fulvic acids account for roughly 50–90% of DNOM, so they are often considered to be the major complexers of metals in most natural waters. Certain other components of NOM, however, such as pigments (e.g.. chlorophyll), other porphyrins, and low-molecular-weight polyfunctional organic acids such as oxalate and citrate are known to have very strong binding constants, perhaps several orders of magnitude greater than humic substances. Though these components are generally present at very low concentrations in natural waters, their extremely high binding constants may impart a measurable impact on metal mobility in certain systems. An example is the upper leaf litter zone of forest floor soils, where pigments may influence metal transport. Amino acids constitute about 3–5% of DNOM. Their binding constants are similar to that of humic substances (Tuschall and Brezonik, 1980). Simple acids such as acetate, formate, and so on, have been identified in water, and their metal binding constants are

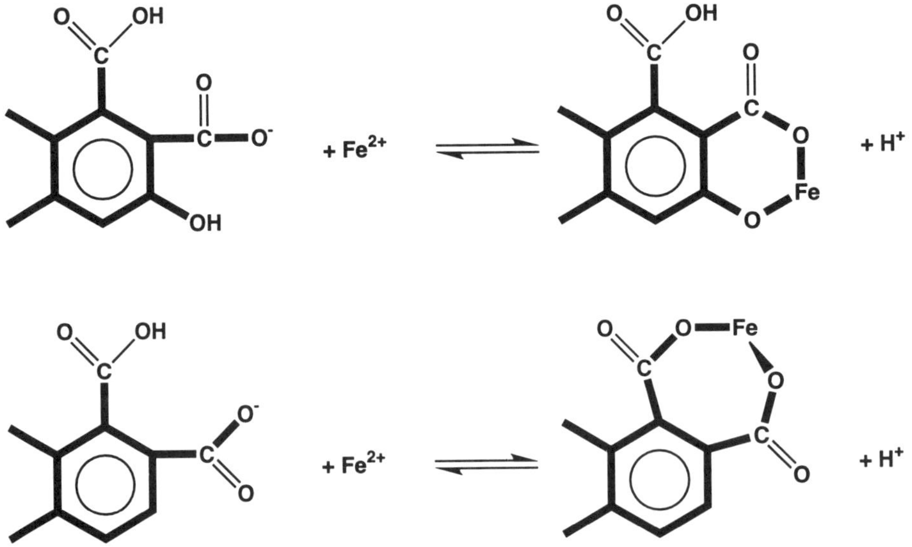

Figure 5.13.   An example of the formation of a bidentate complex with a metal ion by portions of an NOM molecule. Reprinted with permission from Liang, L., et al., Kinetics of Fe(II) oxygenation at low partial pressure of oxygen in the presence of natural organic matter. *Environ. Sci. Technol.*, 27: 1864–1870. Copyright 1993 American Chemical Society.

generally well known (Martel and Smith, 1977, 1982). Little is known, however, about the more complex components of the hydrophilic acid class, which can account for up to 50% of DNOM. McKnight et al. (1983) determined copper binding constants for hydrophilic acids and found them similar to those of fulvic acid.

Schnitzer and Kahn (1978) reviewed the early work on metal binding by pedogenic humic substances. Most workers found that in general the order of metal binding constants follow the Irving–Williams series (1948, as quoted by Thurman, 1985, p. 415). Part of the series found for pedogenic humic substances is (neutral pH)

$$Pb^{2+} > Cu^{2+} > Ni^{2+} > Co^{2+} > Zn^{2+} > Cd^{2+} > Fe^{2+} > Mn^{2+} > Mg^{2+}$$

and at pH = 3

$$Fe^{3+} > Al^{3+} > Cu^{2+} > Ni^{2+} > Co^{2+} > Pb^{2+} > Ca^{2+} > Zn^{2+} > Mn^{2+} > Mg^{2+}$$

For aquogenic humic substances, Mantoura et al. (1978) found

$$Hg^{2+} > Cu^{2+} > Ni^{2+} > Co^{2+} > Mn^{2+} > Cd^{2+} > Ca^{2+} > Mg^{2+}$$

In addition to the magnitude of the binding constants, the number of sites present on DNOM is important to the mobility of metals. The estimates of the number and strength of sites varies considerably based on the fraction of NOM studied, the environment of isolation, and the metal of interest. As an example, McKnight and Wershaw (1989) found that the Cu^{2+} binding to aquatic fulvic acid could be modeled using two types of sites: a strong binding site with a log $K_b = 8.1$, present at a level of 2.7×10^{-7} moles per milligram of carbon; and a weak site with log $K_b = 5.9$, present at 1.6×10^{-6} moles per milligram of carbon. This result suggests that a few micromoles/liter of copper can be complexed by fulvic acids in most freshwaters. This will also be influenced by competition with major divalent cations and other trace metals. The exact role of DNOM in metal transport is still poorly understood.

INTERACTIONS BETWEEN NATURAL ORGANIC MATTER AND ANTHROPOGENIC CHEMICALS

Transport of Pollutant Metals as NOM Complexes

NOM in an aquifer exists as DNOM or CNOM in the aqueous phase and as adsorbed organic coatings on the aquifer matrix. Consequently, the role of NOM as either a sink for metal contaminants or a facilitator of transport is complex and difficult to ascertain. It is likely that NOM may act in both roles with different NOM components contributing in different ways. Consider, for example, the role of NOM in radionuclide transport. Righetto et al. (1991) studied the sorption of actinides to mineral oxides in the presence of humic acid (Figure 5.14). The humic acid was strongly sorbed to the oxides and significantly enhanced the actinide sorption. Thus the humic acid component of NOM can act as a sink for actinides through adsorption. In addition, D. M. McKnight (personal communication, 1995) observed that 70–85% of the Pu present in surface and ground water at a nuclear weapons facility was associated with fulvic acid. Their results further suggested that fulvic acid contains very strong binding sites for Pu. Marley et al. (1993) performed a forced-gradient tracer injection using

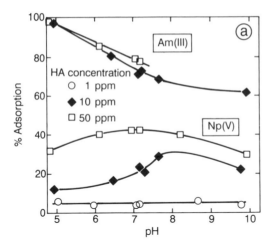

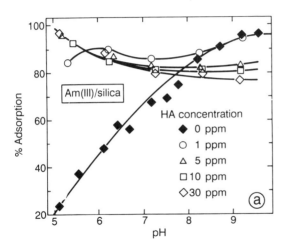

Figure 5.14. **Upper:** pH dependence of Am(III) and Np(V) association with particulate humic acids. **Lower:** Adsorption behavior of Am(III) in the amorphous silica (1200 ppm)–humic acid system as a function of pH and humic acid concentration. Reprinted with permission from Righetto, L., et al., Competitive actinide interactions in colloidal humic acid-mineral oxide systems. *Environ. Sci. Technol.*, 25: 1913–1919. Copyright 1991 American Chemical Society.

surface water NOM and monitored the breakthrough of NOM and actinides. Their NOM consisted of 81–98% fulvic acid and only 2–19% humic acid. Their results indicated that Am was transported by lower-molecular-weight components which were less sorbed than the more hydrophobic components. These studies and others (Figure 5.15) suggest that the fulvic acid component of NOM may provide a means of facilitated transport of actinides.

Another example of metal transport by NOM complexes was the work of Dunnivant et al. (1992a). They performed column experiments using aquifer sediments and NOM collected from a stream near a peat deposit to study cadmium transport. At an NOM concentration of 5.2 mg C/L the breakthrough volume was reduced by roughly 25% over that when NOM was absent. Increasing the NOM concentration further enhanced the breakthrough, although the relative effect was not as great per mg C/L (Figure 5.16)

These two examples, along with numerous other literature sources and the large body of evidence on metal NOM complexes in surface waters, suggest that metal NOM complexes are a very important, if not predominant, transport mechanism.

Figure 5.15. Variation of K_d of reduced plutonium as a function of total DOC. Reprinted with permission from Nelson, D. M., et al., Effects of dissolved organic carbon on the adsorption properties of plutonium in natural waters. *Environ. Sci. Technol.*, 19: 127–131. Copyright 1985 American Chemical Society.

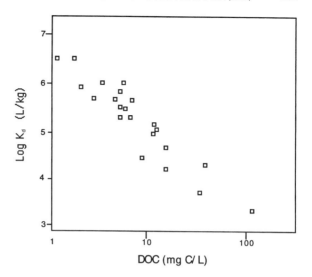

Effects of Sorption/Partitioning to NOM on the Transport of Organic Contaminants

Hydrophobic interactions between soil or sediment organic matter and sparingly soluble, nonionic organic chemicals has been characterized in terms of a partitioning (or sorption) of the organic chemical into the NOM fraction of the soil or sediment. In

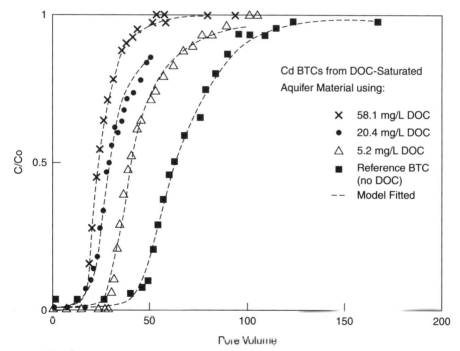

Figure 5.16. Breakthrough of cadmium from DOC-saturated aquifer material (0.058% by weight) in the presence of four different DOC concentrations. Dashed lines represent model-fitted curves. Reprinted with permission from Dunnivant, F. M., et al., Co-transport of cadmium and hexachlorobiphenyl by dissolved organic carbon through columns containing aquifer material. *Environ. Sci. Technol.*, 26: 360–368. Copyright 1992 American Chemical Society.

a series of papers (Karickhoff et al., 1979; Karickhoff, 1980, 1981, 1984; Chiou et al., 1983; Voice and Weber, 1983), this partitioning was shown to be adequately described in terms of an equilibrium between the dissolved and particulate phases. The partition coefficient, K_p, defines the ratio of the concentration of the organic contaminant in the solid soil or sediment phase, C_s, expressed on a weight or mole of contaminant per kilogram of solid basis, to the concentration in the aqueous phase, C_w, also expressed on a weight or mole per kilogram basis, or

$$K_p = C_s/C_w$$

(In some literature, K_p is called K_d, the solid–water distribution coefficient.) This relationship is valid over a wide range of conditions for dilute (C_s less than 10^{-5} M or one-half the aqueous solubility) solutions of hydrophobic contaminants. The value of K_p for a particular soil or sediment and a particular contaminant is of course related to both the characteristics of the solid phase and the nature of the contaminant organic chemical. For most soils or sediment samples [fraction organic carbon, f_{oc}, greater than about 0.002 (Schwarzenbach et al., 1993; Abdul et al., 1987)], the only relevant soil characteristic is its NOM fraction, and, for a given contaminant and a series of soils or sediments,

$$K_p = K_{oc} (f_{oc})$$

where f_{oc} represents the NOM fraction of the soil or sediment expressed as a fraction of the soil mass which is organic carbon (Figure 5.17). K_{oc} is a characteristic of the contaminant chemical and represents its hypothetical partitioning to the organic car-

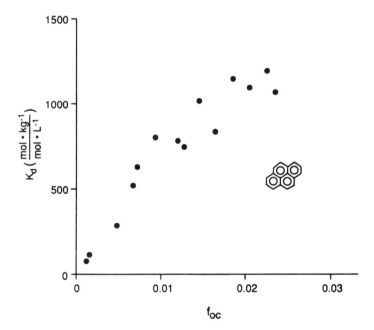

Figure 5.17. Observed increase in solid–water distribution ratios for a hydrophobic compound, pyrene, as a function of the NOM content of the solid (measured as organic carbon) in a variety of soils and sediments. Data from Means, et al., 1980. Adapted from Schwarzenbach, R. P., et al., *Environmental Organic Chemistry.* Copyright © 1993 Wiley-Interscience, New York. Reprinted by permission of John Wiley & Sons, Inc.

bon fraction of the soil or sediment. For soils with a dominant sand size fraction, the NOM content of the finer size fractions (<50 μm) has been shown to be most useful for predictions of solid–water partitioning. For most soils and sediments, whole-soil f_{oc} values can be used without substantial error (Karickhoff, 1981).

Values of K_{oc} for contaminant organic chemicals were in turn shown to be related to a variety of molecular parameters which attempt to quantify the hydrophobicity of the organic substance. The most widely discussed of such parameters is the octanol–water partitioning coefficient, K_{ow}, which has been measured for a large number of chemicals of interest in studies of contaminant behavior (Pomona College, 1984). Measured values of K_{ow} for pollutant organic chemicals range over at least seven orders of magnitude. Many empirical and semiempirical relationships between K_{ow} values and other properties such as water solubility and molecular size (Karickhoff, 1981; and others), as well as molecular size and topology indices such as molecular connectivity (Bahnick and Doucette, 1988), have been developed. The most widely used relationships for the estimation of K_{oc} values involve empirical relationships of the form

$$\log K_{oc} = a \log K_{ow} + d$$

A large number of such relationships (Figure 5.18) have been reported in the literature (see, for example, Brown and Flagg, 1981; Schwarzenbach and Westall, 1981; Karickhoff, 1981; Hassett et al., 1983; Kenaga and Goring, 1978; Chiou et al., 1983). Baker and Mihelcic (1996) recently developed a relationship valid for log K_{ow} values between 1.59 and 7.32 (spanning a wide variety of chemical classes) which gives a value of

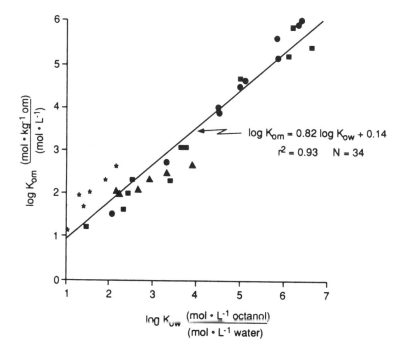

Figure 5.18. Relationship of log K_{om} (K_{oc}) and log K_{ow} for a series of neutral organic compounds: ●, aromatic hydrocarbons; ■, chlorinated hydrocarbons; ▲, chloro-S-triazines; and ★, phenyl ureas. Data compiled by Karickhoff, 1981. Adapted from Schwarzenbach, R. P., et al., *Environmental Organic Chemistry.* Copyright © 1993 Wiley-Interscience, New York. Reprinted by permission of John Wiley & Sons, Inc.

1.015 for a and -0.404 for d and which accounts for 90% of the variability in the data. Such correlations can provide an adequate approximation for the prediction of the equilibrium partitioning of a wide variety of hydrophobic organic chemicals to soil or sediment NOM, but more precise predictions of K_{oc} are not possible using existing data and models.

Nevertheless, the role of NOM in solid phases in the transport of hydrophobic organic chemicals in soil and aqueous systems is well established and widely accepted as a primary means of the retardation of pollutant migration in soils, streams, and ground waters. Knowledge of the organic carbon content of a soil or sediment can be used to predict the aqueous phase concentration of a given hydrophobic chemical in an associated aqueous phase and provide an accurate indication of the retardation of pollutant migration in a given system.

Numerous problems with such predictions have been discussed in the literature, including nonequilibrium between solid and aqueous phases caused by slow kinetics in the partitioning process (Karickhoff, 1984), the role of mobile colloidal phases in considerations of pollutant transport (see below), and the difficulties associated with predictions for marginally hydrophilic substances and ionizable organic compounds (Grundl and Small, 1993) and for solid phases of very low organic carbon contents (Southworth and Keller, 1986). All of these illustrate that the application of such partitioning estimates to a given system must be made with attention to the limitations of the formalism used. In the best of situations, the partition coefficient can be predicted for a given system to within 10–20% based on f_{oc} and K_{ow} values. Values accurate to within a factor of two are possible for most chemicals in most soils or sediments (Karickhoff, 1981).

NOM Effects on Hydrolytic Reactions

The degradation chemistry of many anthropogenic organic compounds is dominated by abiotic or biologically mediated hydrolysis reactions. Hydrolysis refers to nucleophilic attack on a molecule by water, resulting in the breakdown of the molecule into smaller, more hydrophilic products. Examples of hydrolysis reactions relevant to considerations of ground and surface water pollution are shown in Figure 5.19.

Abiotic hydrolysis reactions, which are commonly fast compared to biologically mediated reactions for many contaminant molecules in most aquatic systems (Macalady et al., 1989), are formally second-order reactions and involve pH-independent and base- or acid-catalyzed components (also see Harris, 1983). At constant pH, hydrolysis reactions assume a pseudo-first-order character, with the observed rate constant expressed formally as

$$k_{obs} \text{ (time}^{-1}) = k_A[H^+] + k_W + k_B[OH^-]$$

where k_A, k_W, and k_B are the acid-catalyzed, pH-independent, and base-catalyzed rate constants, respectively, and $[H^+]$ and $[OH^-]$ represent acid and base concentrations. Observation of hydrolysis kinetics at constant temperature and a series of fixed pH values enables determination of all three rate constants. Studies of the organophosphorothioate insecticide chlorpyrifos, for example, reveal measurable pH-independent and base-catalyzed contributions only, with values in distilled water at 298 K of $(6.2 \pm 0.9) \times 10^{-6}$ min^{-1} and 0.5 ± 0.2 M^{-1}min^{-1}, respectively (Macalady and Wolfe,

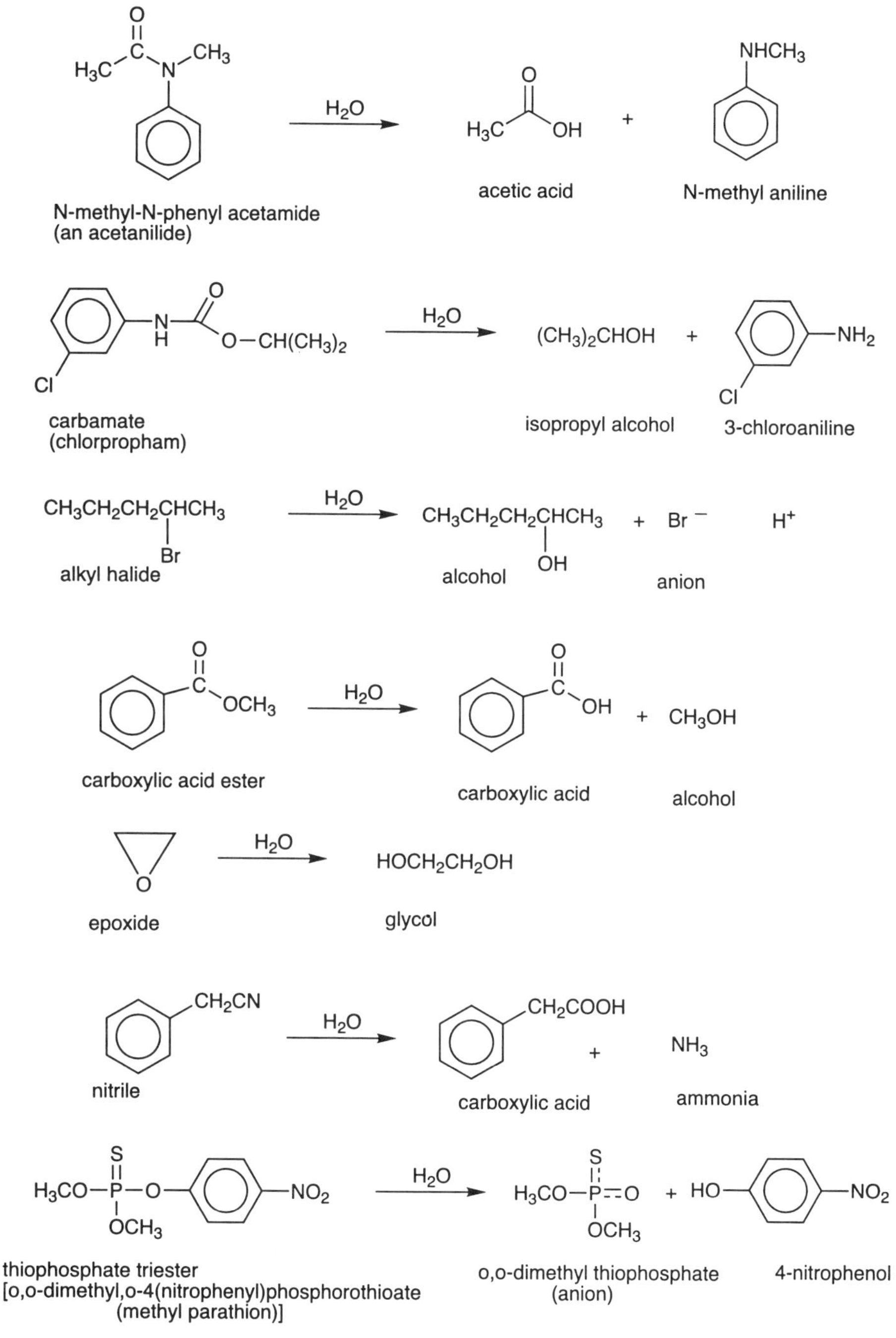

Figure 5.19. Some examples of hydrolysis reactions and reaction products.

1983). Figure 5.20 shows acid- and base-catalyzed hydrolysis data for the herbicide atrazine.

Since most pesticides and many other organic contaminants are quite hydrophobic and would therefore be expected to partition to NOM fractions of soils, sediments, and colloids, it is important to assess the effect of NOM–pollutant interactions on hydrolysis reactions. Beginning with the work of Macalady and Wolfe (1984, 1985), a substantial body of evidence has established the effects of NOM in both solid and aqueous phases on hydrolysis reactions. Partitioning to sediment or colloidal aqueous phase NOM has no measured effect on values of pH-independent hydrolysis rate constants (cf. distilled water). Base-catalyzed hydrolysis reactions, on the other hand, are generally considerably retarded (by factors of 10–1000) for the fraction of a given substance associated with particulate, colloidal, or dissolved NOM as compared to the rate for the molecules in the associated dissolved phase (Figure 5.21). The effects of NOM on acid-catalyzed hydrolysis reactions are less well characterized, partly due to the fact that observations of acid-catalyzed hydrolysis are relatively uncommon for organic contaminants, especially at pH values at which NOM retains its net negative surface charge. In addition, other pH effects, such as inactivation of soil enzymes, may be more important in the acid regime (Macalady et al., 1989; see also Chapter 4, this volume).

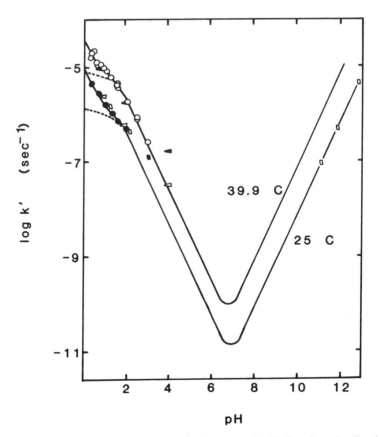

Figure 5.20. The pH–rate profile for the hydrolysis of the aromatic chlorine of atrazine. Reprinted with permission from Plust, S.J., et al. Kinetic and mechanism of hydrolysis of chloro-1,3,5-triazines. Atrazine. *J. Organic Chem.*, 26: 2133–2141. Copyright 1981 American Chemical Society.

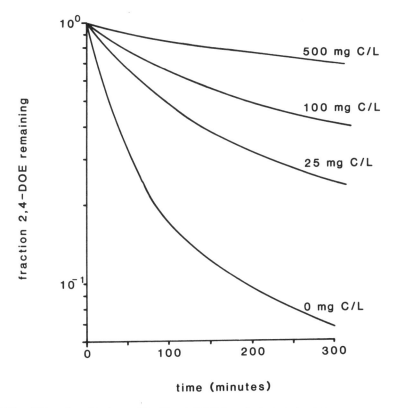

Figure 5.21. Effect of partitioning to DNOM and CNOM on the hydrolysis of the *n*-octyl ester of 2,4-dichlorophenoxyacetic acid at pH 10 with varying amounts of NOM. Reprinted with permission from Perdus, E.M. and Wolfe, N.L. Modification of pollutant hydrolysis kinetics in the presence of humic substances. *Environ. Sci. Technol.*, 16: 847–852. Copyright 1982 American Chemical Society.

Thus, the presence of NOM may have a negligible or substantial effect on the rates of hydrolyses of organic contaminants depending upon the relative importance of acid-, base-, and pH-independent contributions to the hydrolysis reactions at the ambient pH of the aqueous system and, of course, on the hydrophobicity of the contaminant molecule.

Oxidation/Reduction Reactions Facilitated by NOM

Oxidation–reduction (redox) reactions represent another important mode of transformation of anthropogenic organic chemicals in aqueous systems. These reactions can be defined as electron-transfer reactions, in which electrons are exchanged between a contaminant molecule and some component of the aqueous system. As a consequence of a redox reaction, both dramatic and relatively subtle changes in the characteristics of organic contaminants can be realized. For example, biologically mediated oxidations of a wide variety of organic substrates, both natural and anthropogenic, result in the conversion of the carbon–hydrogen framework of organic molecules to carbon dioxide and water, a process often referred to as "mineralization."

Less complete oxidations also occur, as in conversions of carbohydrates to simple organic alcohols or acids, thioethers to sulfoxides and sulfones, or amines to the corresponding nitroso or nitrate derivatives. Reduction reactions generally produce less dramatic changes, including the important class of reductive dehalogenation reactions and the inverse reactions of the partial oxidation processes listed above. Comprehensive reviews of redox reactions of anthropogenic organic chemicals in aquatic systems have been provided by Macalady et al. (1986) and by Wolfe and Macalady (1992).

A clear definition of redox reactions is complicated by the wide variety of such reactions which have been characterized for organic molecules. Simple definitions of redox reactions as involving the addition or removal of hydrogen or oxygen to an organic substrate are helpful but not sufficiently comprehensive or unambiguous. A more useful definition involves the separation if the reaction into oxidation and reduction *half-reactions*. In aqueous systems, this procedure involves separating related pairs of reactants and products and adding water and/or protons to provide an elemental balance for the reactant–product pair. Electrons are then added to provide charge balance. Half-reactions with electrons as reactants are reductions, and those with electrons as products are oxidations. For example, the oxidation of ethanol to acetic acid by molecular oxygen is written

$$CH_3CH_2OH + O_2 \rightarrow CH_3COOH + H_2O$$

The relevant half-reactions are

$$CH_3CH_2OH + H_2O \rightarrow CH_3COOH + 2H^+ + 2e^- \text{ (oxidation)}$$

$$O_2 + 2H^+ + 2e^- \rightarrow 2H_2O \text{ (reduction)}$$

Another example is the reduction of 1,2-dichloroethane to ethene by iodide ions:

$$CH_2Cl-CH_2Cl + 2I^- \rightarrow H_2C{=}CH_2 + I_2 + 2Cl^-$$

which has the following half-reactions:

$$CH_2Cl-CH_2Cl + 2e^- \rightarrow H_2C{=}CH_2 + 2Cl^- \text{ (reduction)}$$

$$2I^- \rightarrow I_2 + 2e^- \text{ (oxidation)}$$

Since the actual oxidation or reducing agents for redox transformations of anthropogenic chemicals in environmental aqueous systems are often unknown, it is often necessary to express the pollutant transformation half-reaction without being able to write the corresponding half-reactions for the environmental oxidant or reductant.

Because oxidation of organic molecules is the primary activity of a diverse and widespread group of biological systems and because many organic molecules of concern in the pollution of aqueous systems are intentionally designed to be resistant to oxidation, reductive processes are of considerably more interest in considerations of the redox transformations of anthropogenic organic chemicals. A wide variety of reductive transformations of anthropogenic organic chemicals have been observed in natural systems. A few examples are illustrated in Figure 5.22.

Considerations of the role of NOM in reductive transformations have only been studied in detail relatively recently. In some senses the present understanding of the role of NOM in reductive transformations began with the observation of the extremely rapid reduction of the nitroaromatic group on the pesticide methyl parathion to the corresponding aromatic amine. Experiments in our laboratories at the Colorado School of Mines with sediment–water slurries sampled in a variety of local ponds and reservoirs

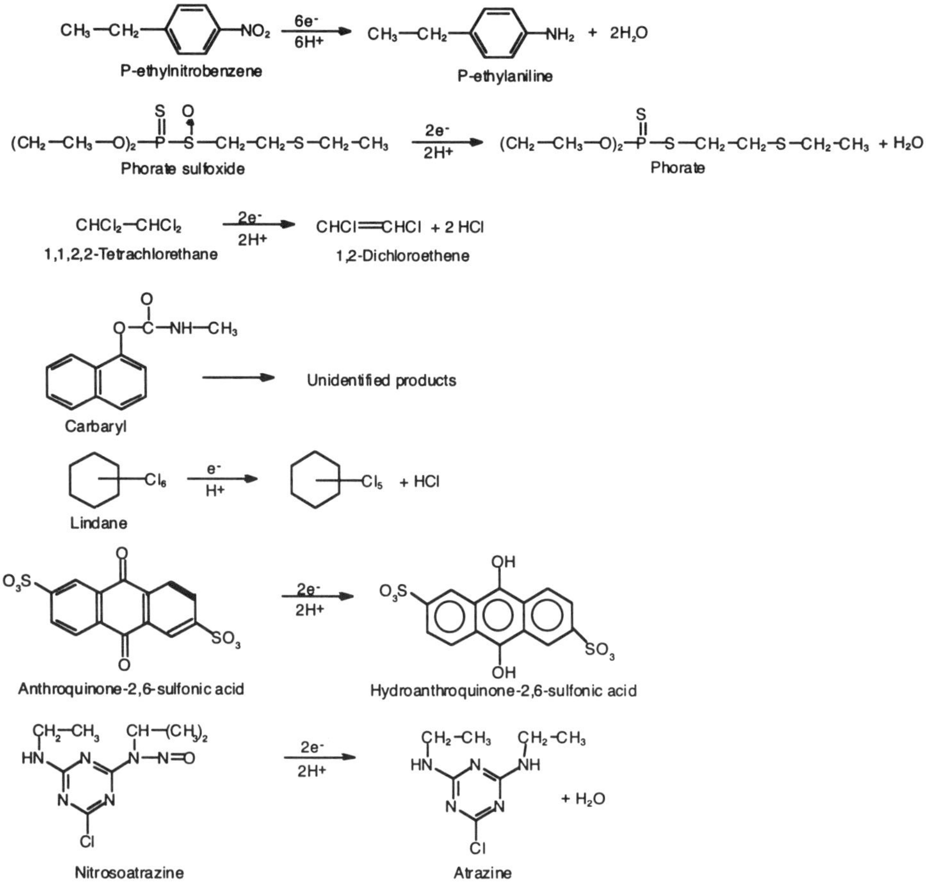

Figure 5.22. Examples of reductions which have been shown to occur in environmental samples under laboratory conditions. Reprinted from *J. Contam. Hydrol.*, 9, Wolfe, N. L., and Macalady, D. L., New perspectives in aquatic redox chemistry: Abiotic transformations of pollutants in ground water and sediments, pp. 17–34, 1993, with kind permission of Elsevier Science—NL, Sara Burgerhartstraat 25, 1055 KV Amsterdam, The Netherlands.

demonstrated first-order reductive transformations with half-lives from several hours to as fast as 15 seconds. Reactions proceeded in open beakers of unmodified sediment slurries, with no precautions to exclude atmospheric oxygen. Related studies at the U.S. Environmental Protection Agency (USEPA) laboratories subsequently demonstrated rapid reduction of methyl parathion in additional sediment–water samples.

Efforts to correlate reduction rates with sediment parameters revealed a direct relationship of the rate of these reductions to the NOM content of the sediments, with reaction rates also directly proportional to the sediment-to-water ratio in the slurries (Wolfe et al., 1987).

Similar results were obtained for reductions of azo compounds (Weber and Wolfe, 1987) and halogenated organic compounds (Peijnenburg et al., 1992). Direct involvement of microbial systems in these reactions seemed improbable due to the extremely fast reaction rates observed in some systems. The emerging hypothesis was that NOM played a key role in the observed transformations.

Additional efforts were clearly necessary in order to more precisely define the nature of the role of NOM in the reduction of nitroaromatic compounds. Nitro-reduction can be viewed as a series of two-electron, two-proton reductions to intermediate reduction products:

$$-NO_2 \xrightarrow{2e^-, 2H^+} -NO \ (+H_2O) \xrightarrow{2e^-, 2H^+} -NHOH \xrightarrow{2e^-, 2H^+} -NH_2$$

However, there is little evidence that these are the actual mechanistic steps in the reaction.

One series of efforts to characterize the role of NOM in these reductive processes began with the hypothesis that functional groups in NOM similar to quinone–hydroquinone redox pairs (Figure 5.23) are responsible for the observed NOM-related redox activity. Reduction of methyl parathion and a series of substituted nitrobenzenes and nitrophenols in homogeneous aqueous systems containing reduced model quinones such as anthroquione disulfonate and the common redox indicator dye indigo carmine (Tratnyek and Macalady, 1989) and natural napthoquinone compounds such as juglone and lawsone (Schwarzenbach et al., 1990) demonstrated the ability of quinones to reduce nitroaromatic compounds.

In the studies using model, highly water-soluble quinones, an important aspect of the reduction of nitroaromatics by hydroquinones was revealed. Studies of the reaction rate as a function of pH demonstrated increasing reaction rates as the pH increased from 5 to 9. This is precisely the opposite pH trend expected based on the involvement of six protons in the overall reduction sequence. This suggests two important features of quinone-mediated reductions. First, the rate-limiting step in the reduction reaction probably does not involve the addition of a proton to the molecule. Second, the reductive reactivity of hydroquinones must increase dramatically as the pH is increased. Since hydroquinones are diphenols, this further suggests that the anionic forms of hydroquinones are more effective reductants than the neutral molecule. These possibilities were supported in the first study quoted above by the fact that the rate of the reduction of methyl parathion by the model quinone compounds was shown to be directly proportional over the pH range 5–9 to the concentration of the hydroquinone anion.

In experiments using natural quinones, an important additional feature of this NOM-mediated reduction was elucidated. Since juglone, lawsone, and many other nat-

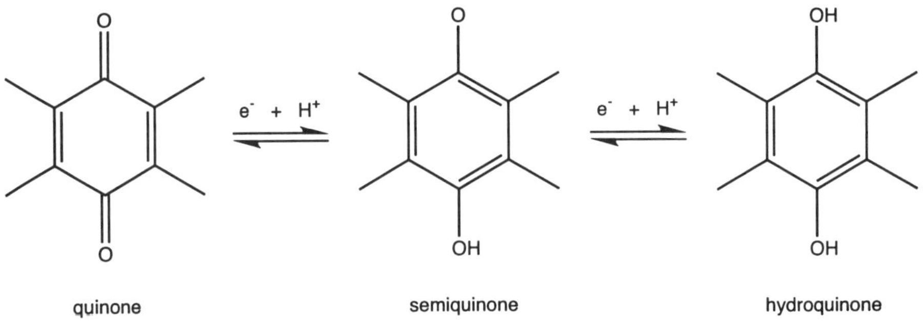

Figure 5.23. Redox half-reactions for the conversions among quinones, semiquinones and hydroquinones. Reprinted from *J. Contam. Hydrol.*, 1, Macalady, et al., Abiotic reduction reactions of anthropogenic chemicals in anaerobic systems: A critical review, pp. 1–28, 1986, with kind permission of Elsevier Science—NL, Sara Burgerhartstraat 25, 1055 KV Amsterdam, The Netherlands.

ural quinones have limited water solubility, and thus very limited capacity to reduce nitroaromatic compounds, a more abundant bulk reducing agent must be involved if such compounds and similar functional groups in the larger array of NOM molecules are to be postulated as responsible for the reduction of nitrobenzenes in homogeneous systems. Addition of a large excess of sulfide, which itself reduces nitrobenzenes only very slowly, to solutions of juglone or lawsone facilitated rapid reduction of nitroben- zenes to the corresponding anilines. A model of the role of quinones in these reduc- tive processes emerged, illustrated for a more general case in Figure 5.24, in which a bulk electron donor, in this case sulfide, rapidly reacts with the mediator, which in this case would be represented as quinones, to reduce them (to the corresponding hydro- quinones), which in turn react more slowly to reduce the nitroaromatic compounds. Sulfide reacts much more slowly with the nitroaromatic compounds, producing little or no observed reductions in the time scale of these experiments (several days). Fur- thermore, pH variations in these reaction rates followed a pattern similar to that ob- served in the sulfide-free solutions of the more water-soluble model quinones (Trat- nyek and Macalady, 1989) . Studies with a series of substituted nitrobenzenes further revealed a clear relationship between the reduction rate constants in a solution of fixed pH and sulfide and napthoquinone concentrations and the one-electron redox potential of the nitrobenzenes (Schwarzenbach et al., 1990). This confirms the notion that the rate-limiting step for these reactions is the addition of a single electron to the nitro group.

Subsequent studies by Macalady and Dunnivant, conducted at the Swiss Federal Institute for Water Resources and Water Research (EAWAG) with homogeneous (0.2 μm filtered) solutions of natural organic matter demonstrated the ability of NOM to act as an effective reductant for aromatic compounds in the presence of a bulk reduc- ing agent such as sulfide (Figure 5.25). Furthermore, a wide variety of NOM sources, including simple water leachates of leaf-litter and forest by-products, ground waters with high NOM concentrations (10–30 mg C/L), surface waters from NOM-rich rivers and ponds, and landfill leachates, all exhibited carbon-normalized reduction rate con- stants of very similar magnitude. At fixed pH and fixed sulfide concentration, the sec- ond-order reduction rate constants agreed within a factor of about 3, suggesting the ubiquitous presence of similar reductive ability in NOM. The pH dependencies and structure–reactivity relationships (Figure 5.26) observed for the natural napthoquinones were paralleled for the NOM samples (Dunnivant et al., 1992b).

Several additional features of the experiments with NOM are worth highlighting. Analyses of the metals contents of the NOM solutions revealed wide variations in the

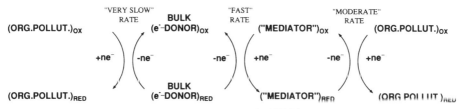

Figure 5.24. Model for the mediation of redox reactions by NOM in natural waters. Reprinted with per- mission from Dunnivant, F. M., et al., Reduction of substituted nitrobenzenes in aqueous solutions contain- ing natural organic matter. *Environ. Sci. Technol.*, 26: 2133–2141. Copyright 1992 American Chemical So- ciety.

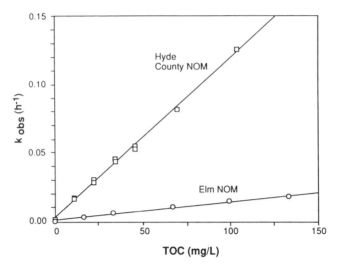

Figure 5.25.   Data plot showing the pseudo-first-order rate constants for the reduction of 3-chloronitrobenzene at 25°C as a function of NOM concentration for two water sources. Reprinted with permission from Dunnivant, F. M., et al., Reduction of substituted nitrobenzenes in aqueous solutions containing natural organic matter. *Environ. Sci. Technol.*, 26: 2133–2141. Copyright 1992 American Chemical Society.

contents of redox-active metals such as iron and manganese in these solutions, with no discernible relationships between metal contents and reductive reactivity. This strongly suggests that, at least over the pH ranges investigated in these studies (6–9), metal–organic complexes do not play a primary role in the observed reduction reactions. Work by Schwarzenbach et al. (1990), among others, has demonstrated that metal organic complexes such as iron porphyrins can also act as effective reductants for nitroaromatic compounds in the presence of a bulk reducing agent such as sulfide. Deviations

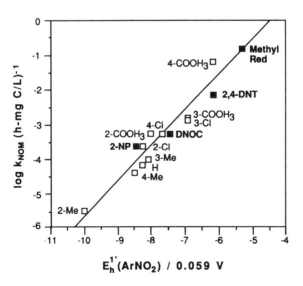

Figure 5.26.   Plot of log k_{NOM} vs. the one-electron redox potential for a series of aromatic nitro compounds in Hyde County NOM solutions at pH 7.5 and 5 mM sulfide. The regression line has a slope of 1.0 and an R^2 value of 0.91. Reprinted with permission from Dunnivant, F. M., et al., Reduction of substituted nitrobenzenes in aqueous solutions containing natural organic matter. *Environ. Sci. Technol.*, 26: 2133–2141. Copyright 1992 American Chemical Society.

from expected behavior of NOM-mediated reductions at the extremes of the pH range studied, however, suggest the possibility of the involvement of other, non-quinone-like reactants, especially at lower pH values.

Thus a considerable body of evidence has been accumulated supporting the hypothesis that quinone-like functionalities in NOM can act as effective mediators of reductive transformations in natural systems containing a bulk reducing agent such as sulfide. However, recent research by Judith Perlinger at EAWAG which attempted to further verify this hypothesis strongly suggests that this interpretation may be an oversimplified view of the actual redox pathways in NOM-mediated reductions of anthropogenic organic compounds such as nitroaromatics and halogenated organics. Electron-spin-resonance studies, which can detect the free radical semiquinone intermediates known to be present in hydroquinone/quinone redox reactions, indicate that other types of free radicals may be involved, perhaps more characteristic of the bulk reductant used (sulfide) than the naturally present quinones (Perlinger, 1994; Perlinger et al., 1996). Further research is clearly necessary in order to fully understand the mechanisms by which aqueous-phase NOM participates in the reductions of anthropogenic organic compounds.

Other efforts to characterize the nature of the involvement of NOM in reductions of nitroaromatic and halogenated organic compounds have focused directly on reactions mediated by PNOM in sediments. Continuing work in the laboratories of N. Lee Wolfe at the USEPA Laboratories in Athens, Georgia and others has resulted in isolation and characterization of a several robust sediment enzymes which facilitate reductions of anthropogenic chemicals. Immunospecific assay techniques have been developed which provide a method to identify the sources of biological redox proteins in soils, sediments, and aquifer materials. Proteins isolated from plant materials around soils and ponds have been shown to exhibit reductive activity which can account for a substantial fraction of the redox activities observed in sediment slurries (Schnoor et al., 1995; Vanbeelen and Burris, 1995). Thus, the reductive activity of sediment NOM may be largely due to biological components such as proteins.

Clearly, NOM can facilitate the rapid reduction of a variety of anthropogenic chemicals in sediment–water and homogeneous aqueous systems. It is equally clear that we have an incomplete understanding of the nature and scope of such redox activities.

Colloidal NOM and Facilitated Transport

The importance of colloids, particularly colloidal NOM, in the transport of contaminants in the environment has recently received a great deal of attention. Part of this interest arose from "anomalous" observations obtained in laboratory investigations of hydrophobic organic contaminant partitioning. It was observed that as the total solids content increased, the observed partitioning coefficient, which should be invariant, decreased. This variation was termed the "solids effect" and has since been explained by the presence of colloids which were not efficiently separated from the aqueous phase by standard techniques (i.e., filtration, centrifugation) due to their small size and/or low density (Morel and Gschwend, 1991). These results highlighted the potential for colloidal influences on contaminant distributions.

Further evidence of the role of colloids in contaminant transport came from investigations of the distribution of radionuclides in the environment. For example, Penrose et al. (1990) found elevated levels of Pu and Am in ground water more than 3 km from a radioactive liquid waste outfall. Laboratory partitioning studies with soils from the site indicated the plutonium should be highly retarded by sorption to the soil and should not have traveled more than a few meters. Ultrafiltration studies showed that Pu was associated with material in the size range of 0.025–0.45 μm. This process of enhanced transport (over model predictions) of strongly sorbing contaminants has been termed "facilitated" transport.

In the traditional view of partitioning in ground water, a solute exists as partitioned to an immobile solid phase and dissolved in the mobile aqueous phase. It has become clear that a third phase must be considered. Facilitated transport occurs as the result of the presence of this third mobile sorbing phase. This phase shares the property of the immobile solid phase in that it binds the solute. However, due to its small particle size (generally submicron) and/or hydrophilic character, it is mobile and transported in the aqueous phase. The transport of the contaminant in the environment thereby becomes linked to the transport of the colloidal phases. Colloid transport in surface waters is mainly influenced by aggregation followed by settling. In ground waters, colloid transport is influenced by straining of larger particles by pores in the porous media and by interception and capture of smaller colloids by the surfaces of the aquifer particles. These processes are all dependent on colloid size and surface charge.

Colloidal NOM is particularly well-suited to provide a means of facilitated transport. CNOM, although not well-characterized, has ionizable functional groups of the same type as DNOM (primarily carboxyl) which may bind metals and other positively charged species. Some of these groups are in favorable positions to allow bidentate chelate formation and therefore have large binding constants for metals (Schnitzer and Kahn, 1972). Other minor components of CNOM, nitrogen in particular, may provide additional sites for metal interactions. Electrostatic repulsion between the negatively charged CNOM and soil minerals, which are predominantly negatively charged aluminosilicates (clays) and quartz, promotes the transport of the CNOM. Adsorption of NOM to iron(III) oxides will inhibit transport, but under anoxic, reducing conditions common in contaminated ground waters, NOM has been shown to stabilize iron oxide colloids and thereby promote transport (Liang et al., 1993). Also present on CNOM are regions of relatively hydrophobic character (i.e., aromatic rings, aliphatic chains) which can bind hydrophobic contaminants. It is suspected that CNOM, in part due to its higher apparent molecular weight, should provide a greater amount of hydrophobic character than DNOM and should therefore exert a greater influence on the transport of sorbed hydrophobic contaminants.

Considerable evidence demonstrates the importance of the association of organic pollutants with CNOM. Less is known about the specific role of CNOM in enhancing the transport of contaminants in aqueous systems, especially ground water. A number of studies have demonstrated that CNOM may adsorb nonpolar organic pollutants in a manner similar to soil and sediment organic matter. Chiou et al. (1986) demonstrated an increase in the "apparent solubility" of DDT and PCBs in the presence of humic colloids (Figure 5.27). Estuarine colloids have been shown to be 10–35 times better than soil or sediment organic matter at adsorbing the herbicides atrazine and linuron (Means and Wijayaratne, 1982) and 10 times greater for PAHs (Wijayaratne and Means, 1984). Accounting for the presence of CNOM improved the modeling of the partitioning of PCBs in Lake Superior (Baker et al., 1986). This study indicated that a three-phase model, including nonfilterable CNOM, best explained the data and also suggested that colloid-associated PCBs may be the dominant species in surface waters.

In addition to hydrophobic partitioning, which is related to the organic carbon content of the colloid and the hydrophobicity of the pollutant (Karickhoff et al., 1979), specific interactions between certain types of pollutants are also important. Interaction of pollutants with NOM can occur by various mechanisms including hydrogen bonding, charge transfer complexes through aromatic pi electrons, cation exchange, and conjugate formation through biochemical processes (Leenheer, 1991). As an example, the sorption of benzidine and toluidine onto estuarine colloids was enhanced over that explained by hydrophobic partitioning (Means and Wijayaratne, 1989). Under pH conditions favoring the cationic form of the amine, sorption was enhanced, presumably by cation exchange.

Despite overwhelming evidence showing interaction of NOM with organic and inorganic pollutants, the role of NOM in enhancing transport, especially in ground waters, is poorly known.

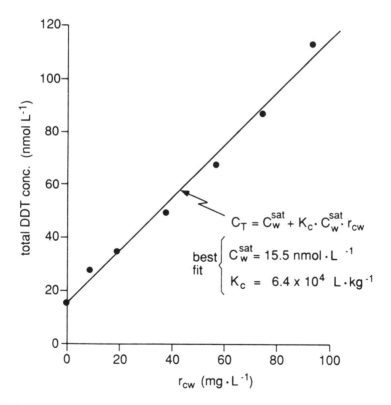

Figure 5.27. Plot of apparent DDT concentration in the aqueous phase as a function of the concentration of humic colloids, r_{cw}. Data from Schwarzenbach, et al., 1993, p. 201. Reprinted with permission from Chiou, C. T., Water solubility enhancement of some organic pollutants and pesticides by dissolved humic and fulvic acids. *Environ. Sci. Technol.*, 20: 502–508. Copyright 1986 American Chemical Society.

NOM.... SINK FOR POLLUTANTS OR FACILITATOR OF TRANSPORT? MYSTERIES AND RESEARCH QUESTIONS

NOM is a complex mixture of components, with a wide range of molecular weight and functional group contents. As scientists, we seek to understand how this complex mixture of chemical and physical properties determines NOM's distribution among various environmental compartments (i.e. water, soil, sediment), its role in soil development and weathering processes, its ability to bind metals and moderate hydrolysis reaction rates and its role in inorganic and organic redox reactions. It should be clear from the discourse above that our present understanding of relationships between the properties of NOM, the time, source and spatial variations in these properties, and the role of NOM in geochemical processes is superficial and incomplete. Consequently, research to deepen and broaden the role of NOM in these processes must proceed along a wide variety of pathways.

The research outlined above provides a confusing array of conclusions which indicate that, in various soil and aqueous environments, NOM can act either as an agent for retardation of contaminant transport and transformation or as facilitator of such processes. If we have as our goal the ability to predict the roles of NOM in geochemical processes, both natural and those induced by anthropogenic disturbances, a considerable effort must continue to define in a fundamental way the nature of the chemistry and physics of NOM. There is no lack of mysteries, and the attempt to obtain a coherent picture of the interplay of NOM with its surrounding chemical and geological environments will continue to provide exciting and important research challenges.

References

Abdul, A. S., Gibson, T. L., and Rai, D. N.,1987, Statistical correlations for predicting the partition coefficients for non-polar organic contaminants between aquifer organic carbon and water. *Hazardous Waste and Hazardous Materials*, 4, 211.

Aiken, G. R., McKnight, D. M., Wershaw, R. L., and MacCarthy, P., 1985, An introduction to humic substances in soil, sediment and water. In: Aiken, G. R., McKnight, D. M., Wershaw, R. L., and MacCarthy, P., eds., *Humic Substances in Soil, Sediment and Water: Geochemistry, Isolation, and Characterization*, John Wiley & Sons, New York, p. 6.

Antweiler, R. C., and Drever, J. I., 1983, The weathering of a late Tertiary volcanic ash: importance of organic solutes. *Geochim. Cosmochim. Acta*, 47, 623–629.

Bahnick, D. A., and Doucette, W. J., 1988, Use of molecular connectivity indices to estimate soil sorption coefficients for organic chemicals. *Chemosphere*, 17, 1703.

Baker, J. R., and Mihelcic, J. R., 1996, Development and evaluation of models for predicting soil/water partition coefficients. (submitted for publication). See also Mihelcic, J. R., M.S. thesis, Michigan Technological University, 1995.

Baker, J .E., Capel, P. D., and Eisenreich, S. J., 1986, Influence of colloids on sediment–water partition coefficients of polychlorobiphenyl congeners in natural waters. *Environ. Sci. Technol.*, 20, 1136–1143.

Baker, W. E., 1973, The role of humic acids from Tasmanian podzolic soils in mineral degradation and metal mobilization. *Geochim. Cosmochim. Acta*, 37, 269–281.

Bennett, P., and Siegel, D. I., 1987, Increased solubility of quartz in water due to complexing by organic compounds. *Nature*, 326, 684–686.

Bockheim J. G., 1979, Properties and relative ages of soils of southwestern Cumberland Peninsula, Baffin Island, N.W.T., Canada. *Arctic and Alpine Res.*, 11, 289–306.

Brown, D. S., and Flagg, E. W., 1981, Empirical prediction of pollutant sorption in natural sediments. *J. Environ. Qual.*, 10, 382–386.

Buffle, J., 1984, Natural organic matter and metal–organic interactions in aquatic systems. In: H. Siegel, ed., *Metal Ions in Biological Systems*, Marcel Dekker, New York, pp. 154–221.

Buffle, J., DeVitre, R. R., Perret, D., and Leppard, G. G., 1989, Physicochemical characteristics of a colloidal iron phosphate species formed at the oxic–anoxic interface of a eutrophic lake. *Geochim. Cosmochim. Acta*, 53, 399–408.

Buffle, J., Perret, D., and Newman, M., 1992, The use of filtration and ultrafiltration for size fractionation of aquatic particles, colloids, and macromolecules. In: J. Buffle and H. P. van Leeuwen, eds., *Environment Particles*, Vol. 1, Lewis, Boca Raton, FL, pp. 171–230.

Charbeck, S., and Martell, A. E., 1959, *Organic Sequestering Agents*, John Wiley & Sons, New York.

Chesworth, W., and Macias-Vasquez, F., 1985, Pe, pH, and podzolization. *Am. J. Sci.*, 285, 128–146.

Chiou, C. T., Porter, L. J., and Schmedding, D. W., 1983, Partition equilibria of nonionic organic compounds between soil organic matter and water. *Environ. Sci. Technol.*, 17, 227–231.

Chiou, C. T., Malcolm, R. L., Brinton, T. I., and Kile, D. E., 1986, Water solubility enhancement of some organic pollutants and pesticides by dissolved humic and fulvic acids. *Environ. Sci. Technol.*, 20, 502–508.

Chittleborough, D. J., Smettem, K. R. J., Cotsaris, E., and Leaney, F. W., 1992, Seasonal changes in the pathways of dissolved organic carbon through a hillslope soil (Xeralf) with contrasting texture. *Aust. J. Soil. Res.*, 30, 465–76.

Cronan, C. S., and Aiken, G. R., 1985, Chemistry and transport of soluble humic substances in forested watersheds of the Adirondack Park, New York. *Geochim. Cosmochim. Acta*, 49, 1697–1705.

Cuevas, E., and Medina, E., 1988, Nutrient dynamics within Amazonia forests. II. Fine root growth, nutrient availability and leaf litter decomposition, *Oecologia*, 76, 222–235.

Dethier, D. P., Jones, S. B., Fiest, T. P., and Ricker, J. E., 1988, Relations among sulfate, aluminum, iron, dissolved organic carbon, and pH in upland forest soils of northwestern Massachusetts. *Soil Sci. Soc. Am. J.*, 52, 506–512.

Dunnivant, F. M., Jardine, P. M., Taylor, D. L., and McCarthy, J. F., 1992a, Cotransport of cadmium and hexachlorobiphenyl by dissolved organic carbon through columns containing aquifer material. *Environ. Sci. Technol.*, 26, 360–368.

Dunnivant, F. M., Schwarzenbach, R. P., and Macalady, D. L., 1992b, Reduction of substituted nitrobenzenes in aqueous solutions containing natural organic matter. *Environ. Sci. Technol.*, 26, 2133–2141.

Esser, G., Aselmann, I., and Lieth, H., 1982, Modeling the carbon reservoir in the system compartment 'litter'. In: *Mitteilungen aus dem Geologien Palantologia der Institut der Universität Hamburg*, Vol. 52, University of Hamburg, Germany, pp 39–58.

Greenland, D. J., 1971, Interactions between humic and fulvic acids and clays. *Soil Sci.*, 111, 34–41.

Grundl, T. J., and Small, G., 1993, Mineral contribution to atrazine and alachlor sorption in soil mixtures of variable organic carbon. *J. Contam. Hydrol.*, 14, 117–128.

Harris, J. C., 1983, Rate of hydrolysis. In: W. Lyman, W. Reehl, and D. Rosenblatt, eds., *Handbook of Chemical Property Estimation Methods*, McGraw-Hill, New York, Chapter 7.

Hassett, J. J., Banwart, W. L., and Griffen, R. A., 1983, Correlation of compound properties with sorption characteristics of non-polar compounds by soils and sediments: Concepts and limitations. In: C. W. Francis and S. I. Aurebach, eds., *Environmental and Solid Waste Characterization, Treatment and Disposal*, Butterworth, Woburn, MA.

Hayes, M. H. B., MacCarthy, P., Malcolm, R. L., and Swift, R. S., 1989, Structures of humic substances: the emergence of 'forms'. In: M. H. B. Hayes, P. MacCarthy, R. L. Malcolm, and R. S. Swift, eds., *Humic Substances II. In Search of Structure*, John Wiley & Sons. New York. pp. 690–733.

Hedges, J. I., and Lee, C., 1992, Measurement of dissolved organic carbon in natural waters. *Marine Chem.*, 41, 290.

Hornberger, G. M., Bencala, K. E., and McKnight, D. M., 1994, Hydrological controls on dissolved organic carbon during snowmelt in the Snake River near Montezuma, Colorado. *Biogeochemistry*, 25, 147–165.

Huang, W. H., and Keller, W. D., 1971, Dissolution of clay minerals in dilute organic acids at room temperature. *Am. Mineral.*, 56, 1082–1095.

Hunter, K. A., 1980, Microelectrophoretic properties of natural surface-active organic matter in coastal seawater. *Limnol.. Oceanogr.*, 25, 807–822.

Jardine, P. M., Wilson, G. V., Luxmoore, R. J., and McCarthy, J. F., 1989, Transport of inorganic and natural organic tracers through an isolated pedon in a forest watershed. *Soil Sci. Soc. Am. J.*, 53, 317–323.

Jardine, P. M., Wilson, G. V., McCarthy, J. F., Luxmoore, R. J., Taylor, D. L., and Zelazny, L. W., 1990, Hydrogeochemical processes controlling the transport of dissolved organic carbon through a forested hillslope. *J. Contam. Hydrol.*, 6, 3–19.

Karickhoff, S. W., 1980, Sorption kinetics of hydrophobic pollutants in natural sediments. In: R. A. Baker, ed., *Contaminants and Sediments*, Vol. 2: Analysis, Chemistry, and Biology. Ann Arbor Science, Ann Arbor, pp. 193–205.

Karickhoff, S. W., 1981, Semi-empirical estimation of sorption of hydrophobic pollutants on natural sediments and soils. *Chemosphere*, 10, 833–846.

Karickhoff, S. W., 1984, Organic pollutant sorption in aquatic systems. *J. Hydraulic Eng.*, 110, 707–735.

Karickhoff, S. W., Brown, D. S., and Scott, T. A., 1979, Sorption of hydrophobic pollutants on natural sediments. *Water Res.*, 13, 241–248.

Kenaga, E. E., and Goring, C. A. I., 1978, Relationship between water solubility, soil sorption, octanol–water partitioning and bioconcentration of chemicals in biota. *Proceedings of the American Society for Testing and Materials, 3rd Aquatic Toxicology Symposium*, No. STP 707, pp. 78–115.

Lee, C. M. and Macalady, D. L., 1989, Towards a standard method for the measurement of organic carbon in sediments. *Intern. J. Environ. Anal. Chem.*, 35, 219–225.

Leenheer, J. A., 1991, Organic substance structures that facilitate contaminant transport and transformations in aquatic sediments. in R. A. Baker, ed., *Organic Substances and Sediments in Water*, Vol. 1: *Humics and Soils*, Lewis, Chelsea, MI, pp. 3–22.

Leenheer, J. A., Malcolm, R. L., McKinley, P. W., and Eccles, L. A., 1974, Occurrence of dissolved organic carbon in selected ground water samples in the United States. *U.S. Geol. Surv. J. Res.*, 2, 361–369.

Leppard, G. G., 1992, Evaluation of electron microscope techniques for the description of aquatic colloids. In: J. Buffle and H. P. van Leeuwen, eds., *Environmental Particles*, Vol. 1, Boca Raton, FL, Lewis, pp. 231–290.

Leppard, G. G., Buffle, J., and Baudat, R., 1986, Description of the aggregation properties of aquatic pedogenic fulvic acids—combining physico-chemical data and microscopical observations. *Water Res.*, 20, 185–196.

Liang, L., McNabb, J. A., Paulk, J. M., Gu, B., and McCarthy, J. F., 1993, Kinetics of Fe(II) oxygenation at low partial pressure of oxygen in the presence of natural organic matter. *Environ. Sci. Technol.*, 27, 1864–1870.

Lonsdale, W. M., 1988, Predicting the amount of litterfall in forests of the world. *Ann. of Botany*, 61, 319–324.

Loughman, F. C., 1969, *Chemical Weathering of Silicate Minerals*. Elsevier, New York, 154 pp.

Macalady, D. L., and Wolfe, N. L., 1983, New perspectives on the hydrolytic degradation of the organophosphorothioate insecticide chlorpyrifos. *J. Agric. Food Chem.*, 31, 1139–1147.

Macalady, D. L., and Wolfe, N. L., 1984, Abiotic hydrolysis of sorbed pesticides. In: R. F. Krueger and J. N. Seiber, eds., *Treatment and Disposal of Pesticide Wastes*, ACS Symposium Series 259. American Chemical Society, Washington, D.C., pp. 221–244.

Macalady, D. L., and Wolfe, N. L, 1985, Effects of sediment sorption and abiotic hydrolyses. 1. Organophosphorothioate esters. *J. Agric. Food Chem.*, 33, 167–173.

Macalady, D. L., Tratnyek, P. G., and Grundl, T. J., 1986, Abiotic reduction reactions of anthropogenic chemicals in anaerobic systems: A critical review. *J. Contam. Hydrol.*, 1, 1–28.

Macalady, D. L., Tratnyek, P. G., and Wolfe, N. L., 1989, Influence of natural organic matter on the abiotic hydrolysis of organic contaminants in aqueous systems. In: I. H. Suffet and P. MacCarthy, eds., *Aquatic Humic Substances: Influence on Fate and Treatment of Pollutants*. Advances in Chemistry Series 219, American Chemical Society, Washington, D.C., pp. 323–332.

MacCarthy, P., Peterson, M. J., Malcolm, R. L., and Thurman, E. M., 1979, Separation of humic substances by pH gradient desorption from a hydrophobic resin. *Anal. Chem.*, 51, 2041–2043.

Malcolm, R. L., 1985, Humic substances in rivers and streams. In: G. A. Aiken, D. M. McKnight, and R. L. Wershaw, eds., *Humic Substances in Soil, Sediment, and Water: Geochemistry, Isolation, and Characterization*, Wiley-Interscience, New York, 692 pp.

Mantoura, R. F. C., Dickson, A., and Riley, J. P., 1978, The complexation of metals with humic materials in natural waters. *Estuarine and Coastal Marine Sci.*, 6, 387–408.

Marley, N. A., Gaffney, J. S., Orlandini, K. A., and Cunningham, M. M., 1993, Evidence for radionuclide transport and mobilization in a shallow, sandy aquifer. *Environ. Sci. Technol.*, 27, 2456–2461.

Martell, A. E., and Smith, R. M., 1977, *Critical Stability Constants.* Vol. 3, Plenum Press, New York.

Martell, A. E., and Smith, R. M., 1982, *Critical Stability Constants.* Vol. 5, Plenum Press, New York.

Mast, M. A., and Drever, J. I., 1987, The effect of oxalate on the dissolution rates of oligoclase and tremolite. *Geochim. Cosmochim. Acta*, 51, 2559–2568.

McDowell, W. H., and Wood, T., 1984, Podzolization: Soil processes control dissolved organic carbon concentrations in stream water. *Soil Sci.*, 137, 23–32.

McKnight, D. M., and Wershaw, R. L., 1989, Complexation of copper by fulvic acid from the Suwannee River-Effect of counter-ion concentrations. In: R. C. Averett, J. A. Leenheer, D. M. McKnight, and K. A. Thorn, eds., *Humic Substances in the Suwannee River, Georgia: Interactions, Properties, and Proposed Structures.* U.S. Geol. Survey, Open-File Report No. 87-557, pp. 59–80.

McKnight, D. M., Feder, G. L., Thurman, E. M., Wershaw, R. L., and Westall, J. C., 1983, Complexation of copper by aquatic humic substances from different environments. In: R. E. Wildung and E. A. Jenne, eds., *Biological Availability of Trace Metals.* Elsevier, Amsterdam, pp. 65–76.

McKnight, D. M., Bencala, K. E., Zellweger, G. W., Aiken, G. R., Feder, G. L., and Thorn, K. A., 1992, Sorption of dissolved organic carbon by hydrous aluminum and iron oxides occurring at the confluence of Deer Creek with the Snake River, Summit County, Colorado. *Environ. Sci. Technol.*, 26, 1388–1396.

Means, J. C., and Wijayaratne, R., 1982, Role of natural colloids in the transport of hydrophobic pollutants. *Science*, 215, 968–970.

Means, J. C., and Wijayaratne, R., 1989, Sorption of benzidine, toluidine, and azobenzene on colloidal organic matter. In: I. H. Suffet and P. MacCarthy, eds., *Aquatic Humic Substances: Influence on Fate and Treatment of Pollutants*, Advances in Chemistry Series, No. 219, American Chemical Society, Washington, D.C., pp. 209–222.

Means, J. C., Wood, S. C., Hassett, J. J. and Banwart, W. L., 1980, Sorption of polynuclear aromatic hydrocarbons by sediments and soils. *Environ. Sci. Technol.*, 14, 1524–1528.

Meybeck, M., 1981, River transport of organic carbon to the ocean. In: G. E. Likens, ed., *Flux of Organic Carbon by Rivers to the Ocean.* U.S. Department of Energy. NTIS Report No. CONF-8009140, UC-11, Springfield, VA, pp. 219–269.

Meybeck, M., 1982, Carbon, nitrogen, and phosphorous transports by world rivers. *Am. J. Sci.*, 282, 401–450.

Morel, F. M. M., and Gschwend, P. M., 1991, The role of colloids in the partitioning of solutes in natural waters. In: W. Stumm, ed., *Aquatic Surface Chemistry*, John Wiley & Sons. New York, 508 pp.

Nelson, D.M., Penrose, W.R., Karttunen, J.O., and Melhoff, P., 1985, Effects of dissolved organic carbon on the adsorption properties of plutonium in natural waters. *Environ. Sci. Technol.*, 19, 127–131.

O'Brien, B. J., and Stout J. D., 1978, Movement and turnover of soil organic matter as indicated by carbon isotope measurements. *Soil Biol. Biochem.*, 10, 309–317.

Olson, J. S., 1963, Energy storage and the balance of producers and decomposers in ecological systems. *Ecology*, 44, 322–331.

O'Melia, C. R., and Tiller, C. L., 1993, Physicochemical aggregation and deposition in aquatic environments. In: J. Buffle and H. P. van Leeuwen, eds., *Environmental Particles. Vol. 2.* Lewis, Boca Raton, FL, pp. 353–386.

Pedro, G., Jamagne, M., and Begon, J. C., 1978, Two routes in the genesis of strongly differentiated acid soils under humid, cool-temperate conditions. *Geoderma*, 20, 173–189.

Peijnenburg, W. J. G. M., 't Hart, M. J., den Hollander, H. A., van de Meent, D., Verboom, H. H., and Wolfe, N. L., 1992, Reductive transformations of halogenated aromatic hydrocarbons in anaerobic water–sediment systems: kinetics, mechanisms, and products. *Environ. Toxicol. Chem.*, 11, 289–300.

Penrose, W. R., Polzer, W. L., Essington, E. H., Nelson, D. M., and Orlandini, K. A., 1990, Mobility of plutonium and americium through a shallow aquifer in a semiarid region. *Environ. Sci. Technol.*, 24, 228–234.

Perdue, E. M., and Wolfe, N. L., 1982, Modification of pollutant hydrolysis kinetics in the presence of humic substances. *Environ. Sci. Technol.*, 16, 847–852.

Perlinger, J., 1994, *Reduction of Polyhalogenated Alkanes by Electron-Transfer Mediators in Aqueous Solution*, Ph.D. thesis, Swiss Federal University (ETH), Zurich, Dissertation No. 10892, 142 pp.

Perlinger, J. Angst, W., and Schwarzenbach, R., 1996, Kinetics of the reduction of hexachloroethane by juglone in solutions containing hydrogen sulfide. *Environ. Sci. Technol.*, submitted for publication.

Plust, S. J., Loehe, J. R., Feher, F. J., Benedict, J. H., and Herbrandson, H. F., 1981, Kinetics and mechanism of hydrolysis of chloro-1,3,5-triazines. Atrazine. *J. Org. Chem.*, 46, 3661–3665.

Pomona College, 1984, *Log P and Parameter Database*. Pomona College, Claremont, CA. Seaver Chemistry Laboratory, Medicinal Chemistry Project. Technical Database Services, Inc., New York.

Ranville, J. F., Harnish, R. A., and McKnight, D., 1991, Particulate and colloidal organic material in Pueblo Reservoir, Colorado: influence of autochthonous source on chemical composition. In: R. A. Baker, ed., *Organic Substances and Sediments in Water*, Vol. 1. Lewis, Chelsea, pp. 47–74.

Righetto, L., Bidoglio, G., Azimonti, G., and Bellobono, I. R., 1991, Competitive actinide interactions in colloidal humic acid–mineral oxide systems. *Environ. Sci. Technol.*, 25, 1913–1919.

Robert, M., and Berthelin, J., 1986, Role of biological and biochemical factors in soil mineral weathering. In: P. M. Huang and M. Schnitzer, eds., *Interactions of Soil Minerals with Natural Organics and Microbes*, SSSA Special Publication No. 17, Madison, WI, pp. 453–496.

Romell, L. G., 1935, *Ecological Problems of the Humus Layers in the Forest*. Cornell University, Agricultural Experiment Station Memoir 170, Ithaca, New York.

Schalscha, E. B., Appelt, H., and Schatz, A., 1967, Chelation as a weathering mechanism. I. Effect of complexing agents on the solubilization of iron from minerals and granodiorite. *Geochim. Cosmochim. Acta*, 31, 587–596.

Schlesinger, W. H., 1977, Carbon balance in terrestrial detritus. *Annu. Rev. Ecology Syst.*, 8, 51–81.

Schlesinger, W. H., 1991, *Biogeochemistry: An Analysis of Global Change*. Academic Press, San Diego, 443 pp.

Schnitzer, M., and Kahn, S. U., 1972, *Humic Substances in the Environment*, Marcel Dekker, New York.

Schnitzer, M., and Kahn, S. U., eds., 1978, *Soil Organic Matter*, Elsevier, Amsterdam, 319 pp.

Schnitzer, M. and Kodama, H., 1976, The dissolution of micas by fulvic acid. *Geoderma*, 15, 381–391.

Schnoor, J, L,, Licht, J. A., McCutcheon, S. C., Wolfe, N. L. and Carreira, L. H., 1995, Phytoremediation of organic and nutrient contaminants. *Environ. Sci. Technol.*, 29(7), A318 A323.

Schwarzenbach, R. P., and Westall, J., 1981, Transport of non-polar organic compounds from surface water to ground water: laboratory sorption studies. *Environ. Sci. Technol.*, 15, 1360–1367.

Schwarzenbach, R. P., Stierli, R., Lanz, K., and Zeyer, J., 1990, Quinone and iron porphyrin mediated reduction of nitroaromatic components in homogeneous aqueous solution. *Environ. Sci. Technol.*, 24, 1566–1574.

Schwarzenbach, R. P., Gschwend, P. M., and Imboden, D. M., 1993, *Environmental Organic Chemistry*. Wiley-Interscience. New York, 681 pp.

Senesi, N., and Steelink C., 1989, Application of ESR spectroscopy to the study of humic substances. In: M. H. B. Hayes, P. MacCarthy, R. L. Malcolm, and R. S. Swift, eds., *Humic Substances. II. In Search of Structure*, John Wiley & Sons, New York, pp. 373–408.

Southworth, G. R., and Keller, J. L., 1986, Hydrophobic sorption of polar organics by low organic carbon soils. *Water Air Soil Pollut.*, 28, 239–248.

Stevenson, F. J., 1982, *Humus Chemistry*. John Wiley & Sons, New York, 443 pp.

Stevenson, F. J., 1985, Geochemistry of soil humic substances. In: G. A. Aiken, D. M. Mc-Knight, and R. L. Wershaw, eds., *Humic Substances in Soil, Sediment, and Water: Geochemistry, Isolation, and Characterization*, Wiley-Interscience, New York, pp. 13–52.

Stevenson, F. J., 1986, *Cycles of Soil*, John Wiley & Sons, New York.

Stumm, W., 1992, *Chemistry of the Solid–Water Interface*, John Wiley & Sons, New York, 428 pp.

Swift, M. J., Heal, O. W., and Anderson, J. M., 1979, *Decomposition in Terrestrial Ecosystems*, University of California Press, Berkeley, CA.

Tan, K. H., 1980, The release of silicon, aluminum, and potassium during decomposition of soil minerals by humic acid. *Soil Sci.*, 129, 5–11.

Thorn, K. A., Arterburn, J. B. and Mikita, M. A., 1992, [15] N and [13] C NMR investigation of hydroxylamine-derivatized humic substances. *Environ. Sci. Technol.*, 26, 107–116.

Thurman, E. M., 1985, *Organic Geochemistry of Natural Waters*, Martinus Nijhoff/Junk Publishers, Dordrecht, 497 pp.

Thurman, E. M., and Malcolm, R. L., 1981, Preparative isolation of aquatic humic substances. *Environ. Sci. Technol.*, 15, 463–466.

Tiller, C. L., and O'Melia, C. R., 1993, Natural organic matter and colloidal stability: models and measurements. *Colloids Surf. A: Physiochem. Eng. Aspects*, 73, 89–102.

Tipping, E., and Cooke, D., 1982, The effects of adsorbed humic substances on the surface charge of goethite (α-FeOOH) in freshwaters. *Geochim. Cosmochim. Acta*, 46, 75–80.

Tratnyek, P. G., and Macalady, D. L., 1989, Abiotic reduction of nitroaromatic pesticides in anaerobic laboratory systems. *J. Agric. Food. Chem.*, 37, 248–254.

Tuschall, J. R., and Brezonik, P. L., 1980, Characterization of organic nitrogen in natural waters: its molecular size, protein content, and interactions with heavy metals. *Limnol. Oceanog.*, 25, 495–504.

Ugolini, F. C., 1968, Soil development and alder invasion in a recently degalaciated area of Glacier Bay, Alaska. In: J. M. Trappe, J. F. Franklin, R. F. Tarrant, and G. M. Hansen, eds., *Biology of Alder*, US. Forest Service, Pacific Northwest Forest and Range Experiment Station, Portland, OR, pp. 115–140.

Ugolini, F. C., Minden, R., Dawson, H., and Zachara, J., 1977, An example of soil processes in the *Abies amabilis* zone of central Cascades, Washington, *Soil Sci.*, 124, 291–302.

Vanbeelen, P., and Burris, D. R., 1995, Reduction of the explosive 2,4,6-trinitrotoluene by enzymes from aqueous sediments, *Environ. Toxicol. Chem.*, 14, 2115–2123.

Voice, T. C., and Weber, W. J., 1983, Sorption of hydrophobic compounds by sediments, soils, and suspended solids. I. Theory and background. *Water Res.*, 17, 1433–1441.

Weber, E. J., and Wolfe, N. L., 1987, Kinetic studies of the reduction of aromatic azo compounds in anaerobic sediment/water systems. *Environ. Toxicol. Chem.*, 6, 911–919.

Wijayaratne, R. D., and Means, J. C., 1984, Sorption of polycylic aromatic hydrocarbons by natural estuarine colloids. *Mar. Environ. Res.*, 11, 77–89.

Wilson, G. V., Jardine, P. M., Luxmoore, R. J., and Jones, J. R., 1990, Hydrology of a forested watershed during storm events. *Geoderma*, 46, 119–138.

Wolfe, N. L., Macalady, D. L, Kitchens, B. E., and Grundl, T. J., 1987, Physical and chemical factors that influence the anaerobic degradation of methyl parathion in sediment systems. *Environ. Toxicol. Chem.*, 6, 827–837.

Wolfe, N. L. and Macalady, D. L., 1992, New perspectives in aquatic redox chemistry: abiotic transformations of pollutants in ground water and sediments. *J. Contam. Hydrol.*, 9, 17–34.

Zunino, H., and Martin, J. P., 1977, Metal-binding organic macromolecules in soils. *Soil Sci.*, 123, 65–76.

Suggested Readings

Aquatic Humic Substances: Influence on Fate and Treatment of Pollutants, 1989, I. H. Suffet and P. MacCarthy, eds., Advances in Chemistry Series, No. 219, American Chemical Society, Washington, D.C., 864 pp.

Biogeochemistry: An Analysis of Global Change, 1991, Schlesinger, W. H. Academic Press, San Diego, CA, 443 pp.

Environmental Organic Chemistry, 1993, Schwarzenbach, R. P., Gschwend, P. M., and Imboden, D. M., Wiley-Interscience. New York, 681 pp.

Geochemical Processes: Water and Sediment Environments, 1979, Lerman, A., John Wiley & Sons, New York, 481 pp.

Humic Substances in Soil, Sediment, and Water: Geochemistry, Isolation, and Characterization, 1985, G. A. Aiken, D. M. McKnight, and R. L. Wershaw, eds., Wiley-Interscience, New York, 692 pp.

Humic Substances II: In Search of Structure, 1989, M. H. B. Hayes, P. MacCarthy, R. L. Malcolm, and R. S. Swift, eds., Wiley-Interscience, Chichester, England, 764 pp.

Humus Chemistry: Genesis, Composition, Reactions, 2nd ed., 1994, Stevenson, F. J., Wiley-Interscience. New York, 496 pp.

Interactions of Soil Minerals with Natural Organics and Microbes, 1986, P. M. Huang and M. Schnitzer, eds., SSSA Special Publication Number 17, Soil Science Society of America, Inc., Madison, WI, 606 pp.

Organic Geochemistry of Natural Waters, 1985, Thurman, E. M., Martinus Nijhoff/Junk Publishers, Dordrecht, 497 pp.

Organic Substances and Sediments in Water. Volume 1: Humics and Soils, 1991, R. A. Baker, ed., Lewis Publishers Chelsea, MI, 392 pp.

The Geochemistry of Natural Waters, 1988, Drever, J. I., 2nd ed., Prentice-Hall, Englewood Cliffs, NJ, 437 pp.

Assessing the Dynamic Behavior of Organic Contaminants in Natural Waters

RENÉ P. SCHWARZENBACH, STEFAN B. HADERLEIN, STEPHAN R. MÜLLER, AND MARKUS M. ULRICH

ABSTRACT. This chapter provides an overview of the various scales of observations and approaches that must be considered in environmental (organic) chemistry. The major goal is a quantitative assessment of the distribution of an organic contaminant in space and time. Achieving this goal hinges on the ability to quantify all pertinent processes and their interplay in a given natural system. This requires both (a) knowledge of the molecular interactions that govern a given process and (b) detailed understanding of the nature and dynamics of the respective environmental compartment(s). The transfer of information and concepts derived from laboratory experiments and basic theory to the field is one of the key issues and major challenges in environmental chemistry. Research at the various scales covered by environmental organic chemistry is illustrated by a laboratory study on the mechanism of adsorption of nitroaromatic compounds to clay minerals and by a field study on the fate of organic contaminants in lakes in which a combination of field measurements and computer model calculations was applied.

INTRODUCTION

The contamination of natural waters (e.g., lakes, rivers, ground waters, marine waters) by anthropogenic organic chemicals has become a major issue in environmental protection. When considering that the global consumption of mineral oil exceeds 3 billion tons per year and that the production of man-made organic material in industrialized countries is within the same order of magnitude of that synthesized by nature (Stumm et al., 1983), it is obvious that a vast number of xenobiotic chemicals are introduced continuously into the environment. As has been documented by numerous studies and is illustrated by Figure 6.1, many synthetic organic compounds, although applied or introduced to confined locations, become widely dispersed even to the "ends of the earth" (see also Ballschmiter, 1992).

When addressing the issue of the contamination of natural waters by anthropogenic organic chemicals, one often tends to (over)emphasize the consequences of spectacular accidents or the problems connected with hazardous waste management (e.g., waste-

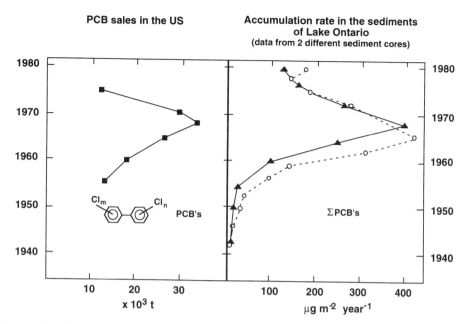

Figure 6.1. Historical record of the sales of polychlorinated biphenyls (PCBs) in the United States, and the similarity of this time-varying trend to the accumulation rates of these compounds in the sediments of Lake Ontario. (Adapted from Eisenreich et al., 1989.)

water treatment, waste incineration, dump sites). There is no question that these are significant problems, but of at least equivalent importance is the chronic contamination of the environment due to the daily use of chemicals (e.g., solvents, fossil fuels, components of detergents, dyes and varnishes, additives in plastics and textiles, chemicals used for construction, antifouling agents, herbicides, insecticides, fungicides, and many more). In this context, a major present and future task encompasses identification and possibly replacement of those widely used synthetic chemicals that may pose hazards to human health as well as to natural ecosystems. Furthermore, new chemicals must be designed to be "environmentally compatible." As is indicated in Figure 6.2, all of these tasks require knowledge of (1) the use patterns of a given chemical in human society (i.e., in the "anthroposphere"), (2) the processes that govern the transport, distribution, and transformations of the chemicals in the environment, and (3) the chemicals effects on organisms (including man), organism communities, and whole ecosystems. In this chapter, we will focus on the second topic, and we will emphasize the aquatic environment.

PROCESSES THAT DETERMINE THE DISTRIBUTION AND FATE OF ORGANIC CONTAMINANTS IN THE AQUATIC ENVIRONMENT

As is exemplified in Figure 6.3 for a lake system, an organic chemical that is introduced into a given natural water body is subjected to various transport, mixing, trans-

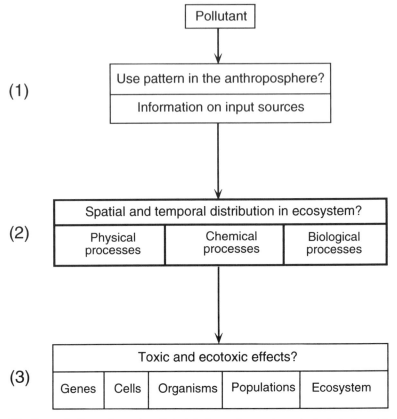

Figure 6.2. Fate and effects of organic pollutants in the environment—ecotoxicological considerations.

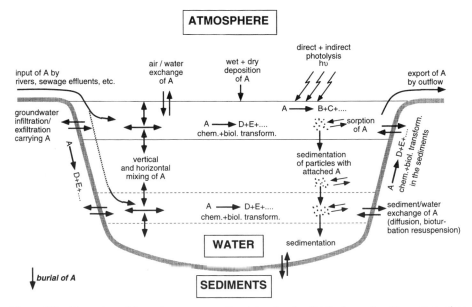

Figure 6.3. Illustration of the various processes that determine the distribution and residence time of an anthropogenic compound "A" in a lake.

fer, and transformation processes. These processes can be divided into two major categories: those that do not affect the structure of the chemical and those that transform the chemical into one or several products (of different environmental behavior and effect(s)!). The first category of processes includes advective, dispersive, and diffusive transport phenomena within a given environmental compartment (e.g., in the water column of the lake), as well as transfer processes between different phases and/or compartments (e.g., water–air exchange, sorption and sedimentation, sediment–water exchange, bioaccumulation). Alterations of the structure of a given compound may occur in each of the different compartments by chemical, photochemical, and/or biological (mostly microbial) reactions. As will become evident from the discussions in this and in other chapters of this book, each of these processes is determined by various and varying environmental factors which, themselves, are governed by numerous physical and biogeochemical processes. Furthermore, the different processes that determine the environmental behavior of a compound may occur simultaneously and, therefore, may strongly influence each other.

THE "PLAYGROUNDS" OF ENVIRONMENTAL ORGANIC CHEMISTRY

Figure 6.4 provides a general overview of the "playgrounds" on which environmental organic chemists can be found. As is indicated, a quantitative assessment of the dis-

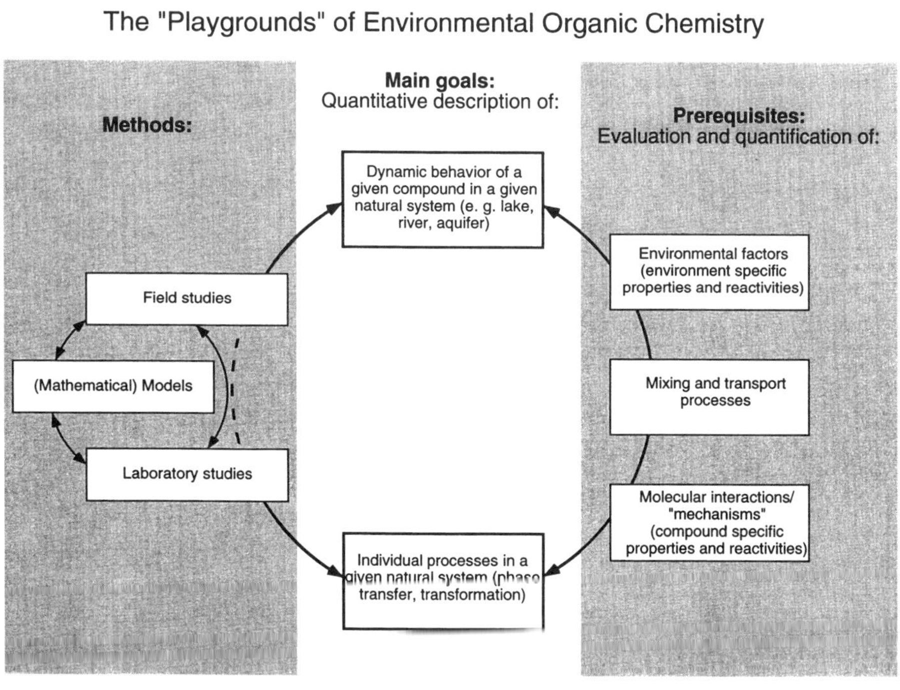

The "Playgrounds" of Environmental Organic Chemistry

Figure 6.4. The "playgrounds" of environmental organic chemistry; an attempt to depict the various tasks that one has to cope with when describing quantitatively the environmental behavior of organic contaminants.

tribution in space and time of an organic contaminant in the environment hinges on the ability to quantify all pertinent processes and their interplay in a given natural system. This requires, on the one hand, knowledge of the molecular interactions that govern a given process (i.e., knowledge of "mechanisms") and, on the other hand, insights into the nature and the dynamics of the (micro)environment(s) that an organic compound encounters in the natural system. Hence, research in environmental organic chemistry (as in general in environmental chemistry) takes place in the laboratory test tube as well as in the macroscopic world. The transfer of concepts and data derived from basic theory and from laboratory experiments to the field is one of the key issues and one of the major challenges of environmental organic chemistry. Since the nature and the dynamics of potential natural "interactants" and "reactants" may be extremely complex (e.g., natural organic matter, solid surfaces, microorganisms), it is often necessary to take somewhat unusual and rather pragmatic theoretical and experimental approaches to arrive at an answer to a given question. Although environmental organic chemistry lives, of course, on the theories and methods of fundamental organic, physical, inorganic, and biological chemistry, it cannot just be regarded as an application of basic chemistry to environmental problems. An extensive treatment of the current approaches used to assess the dynamic behavior of organic contaminants in the aquatic environment can be found in recently published textbooks (Schwarzenbach et al., 1993, 1995).

Considering all these aspects and considering its multidisciplinary character, environmental organic chemistry would seem to be a very attractive (and important) addendum to any university or college chemistry program. Unfortunately, to date, throughout the world, very few chemistry departments seem to have realized this fact. It is hoped that this and other chapters in this book may help to improve this situation. In the following, two studies that have been conducted at our institute will be used to illustrate some approaches and topics in environmental organic chemistry. First, an investigation of the sorption behavior of nitroaromatic compounds on clay minerals (Haderlein and Schwarzenbach, 1993, 1994; Haderlein et al., 1996; Weissmahr et al. 1997) will serve to address the advantages and the limitations of "test tube research" conducted in the laboratory. In the second example (Ulrich et al., 1994) dealing with the distribution and residence time of selected organic contaminants in a lake, the combined use of field measurements and model calculations for deriving information that cannot be obtained from laboratory studies will be demonstrated.

SORPTION OF NITROAROMATIC COMPOUNDS TO CLAY MINERALS—AN EXAMPLE OF A LABORATORY STUDY

Motivation and Goals of the Study

As is illustrated in Figure 6.5, sorption from aqueous solution to solid surfaces is a key process in determining the transport and the distribution of an organic contaminant in natural waters. Furthermore, the reactivity and bioavailability of a given compound depends strongly upon the degree to which it is associated with inorganic and organic nonaqueous phases (e.g., particles, colloids, soil and aquifer materials) in a given (micro)environment.

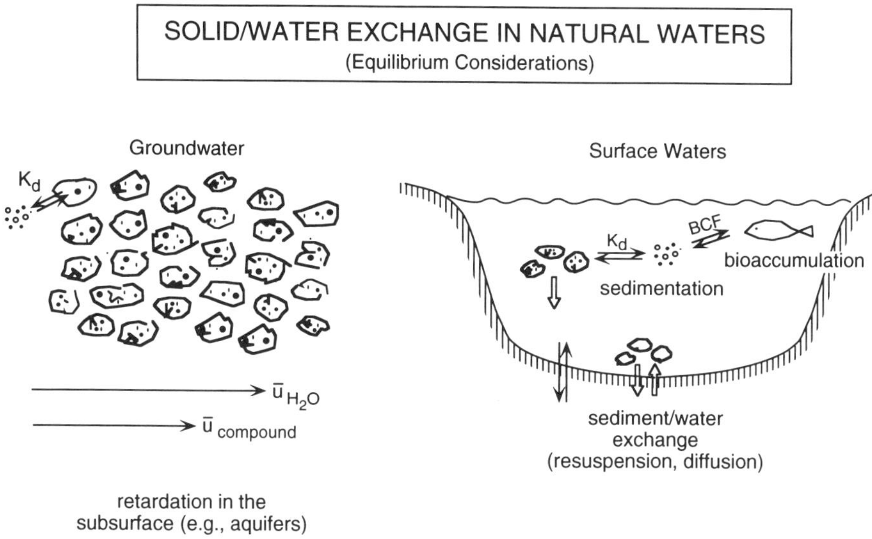

Figure 6.5. Illustration of solid–water exchange processes (sorption) in natural waters. In the subsurface, for example, sorption leads to retardation of a compound relative to the transport of water; in lakes, sorbed compounds are transported to the sediments from where they may reenter the water column again by re-suspension of particles. Sorption also determines the bioavailability (and thus the biological effect(s)) of a given compound.

For natural environments exhibiting an appreciable amount of "particulate" organic matter (i.e., mass fraction of organic carbon, f_{oc}, greater than 10^{-3}), partitioning of nonpolar organic contaminants into the organic phases is commonly assumed to be the major sorption mechanism, at least for those compounds that do not specifically interact with surface sites. In such cases, predictive quantitative models based on laboratory-derived data have been applied with reasonable success to the field scale (see, e.g., Schwarzenbach et al., 1993, Chapter 11). There are, however, situations in which adsorption to mineral surfaces may be equally important or may even dominate the overall sorption process. This may be the case for environments in which very little organic material is present, as is encountered, for example, in many aquifers or in clay mineral liners of landfills and hazardous waste sites. Furthermore, for compounds that interact specifically with surface sites of the solid matrix [e.g., carboxylic acids by surface complexation (Davis and Kent, 1990) or alkylammonium cations and N-heterocyclic aromatic compounds by ion exchange (Brownawell et al., 1990; Zachara et al., 1990)], adsorption to mineral surfaces may, in general, not be neglected. To date, predictive models for describing the sorption behavior of organic compounds by processes other than hydrophobic partitioning in "real" natural environments are scarce and have rarely been validated.

In the study discussed here, the adsorption of a series of nitrobenzenes, nitronaphthalenes, and nitrophenols to selected mineral surfaces has been systematically investigated. The reason for choosing nitroaromatic compounds ("NACs") as model sorbates is twofold. First, NACs are widely used (e.g., as explosives, as intermediates in the synthesis of pesticides and dyes, as solvents, and as herbicides and insecticides). They have been found to be ubiquitous environmental pollutants particularly in subsurface environments (Haderlein and Schwarzenbach, 1995). Second, as one might recall from organic chemistry textbooks, due to the strong electron withdrawing effects of nitro substituents on the π-electron distribution in the molecule, NACs may specifically interact with electron-donating species by forming so-called electron donor–acceptor (EDA) complexes (Foster, 1969). In fact, in a field study in which the transport of munition residues was investigated in a sandy aquifer (Spalding and Fulton, 1988), it was found that 2,4,6-trinitrotoluene (TNT) exhibited a much larger retardation factor than expected if sorption is assumed to be dominated by hydrophobic partitioning.

The purposes of this laboratory study were (1) to identify interactions of NACs with mineral surfaces that may determine the sorption of NACs in the environment, (2) to study the influence of water constituents (pH, major ion composition, presence of other organic species) on such sorption process(es), and (3) to evaluate the relationship between the structure (i.e., type of π-electron system, electronic and steric effects of substituents) and the sorption behavior of NACs. The results of this study form an important base for the quantification of sorption of NACs in natural systems (in particular, in the subsurface environment).

Experimental Approach and Summary of Results

The sorption behavior of over 50 different NACs was investigated in batch experiments using a variety of model sorbents that represent important naturally occurring mineral surfaces. The minerals used include aluminum (hydr)oxides, SiO_2, iron (hydr)oxides, and a series of clay minerals (i.e., kaolinite, illite, montmorillonite). The clay minerals were used in various homoionic forms because initial experiments showed a strong dependence of the sorption properties of the clays on the type of cations adsorbed to these minerals.

Batch experiments were typically performed by equilibrating a given NAC in a suspension of a given sorbent for 30–60 minutes. Kinetic experiments conducted with a series of NACs showed that sorption was fast (i.e., equilibrium was reached in less than 10 minutes) and reversible. Since mass-balance measurements showed excellent recoveries, concentrations in the sorbed phase were routinely calculated from the difference between initial and equilibrium solution-phase concentrations. The concentrations of NACs in solution were determined by reverse-phase high-performance liquid chromatography (HPLC). Experimental details can be found in Haderlein and Schwarzenbach (1993).

The following summary of some of the results of this study will illustrate how a series of simple batch experiments, in which the affinities of structurally related compounds to model sorbents are evaluated, can provide important information on the type of interactions and factors that govern the sorption of a given class of compounds to natural sorbents. Note that for our discussion we will use kaolinite as representative of the clay minerals, since very similar results were obtained with illites and montmorillonites (Haderlein et al., 1996).

Let us first consider the difference in affinity of a given NAC to the various mineral surfaces. As is illustrated by the surface normalized adsorption constants [K_d values, see Eq. (6.1)] of 1,4-dinitrobenzene given in Table 6.1, the majority of the NACs investigated sorbed much more strongly to the homoionic clays (Cs^+, K^+, NH_4^+-kaoli-

TABLE 6.1
Surface Area Normalized K_d Values of 1,4-Dinitrobenzene on Various Sorbents[a]

Sorbent	Formula	K_d (L m^{-2})
Cs$^+$-kaolinite		3×10^{-1}
K$^+$-kaolinite	$[Al_2 + x\ Si_2 - xO_5(OH)_4]^{x-}$	3×10^{-2}
NH$_4^+$-kaolinite		1×10^{-2}
Aerosil 380	SiO_2	5×10^{-4}
Kieselgur	SiO_2	2×10^{-4}
Alumina	δ-Al_2O_3	$<3 \times 10^{-6}$
Gibbsite	γ-$Al(OH)_3$	$<1 \times 10^{-4}$

[a] *Data from Haderlein and Schwarzenbach (1993).*

nite) as compared to the other mineral surfaces. Hence we will focus our interest primarily on the sorption of NACs to clay minerals.

As is illustrated by Figure 6.6 for two compounds in Cs$^+$-kaolinite suspensions, at higher NAC concentrations, sorption isotherms generally exhibited a saturation type curvature. For the linear part of this isotherm [i.e., at low concentration (see insert in Figure 6.6)], a (constant) distribution ratio K_d can be calculated:

$$K_d = \frac{[NAC_{sorb}]}{[NAC_{aq}]} \tag{6.1}$$

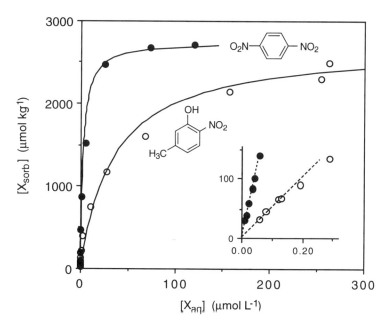

Figure 6.6. Adsorption isotherms for two representative NACs in suspensions of homoionic Cs$^+$-kaolinite ($I = 0.1$ M CsCl, 10–50 g kaolinite per liter, pH = 4.0). The insert shows the low concentration range where the isotherms can be considered to be linear. Reprinted with permission from Haderlein, S. B., and Schwarzenbach, R. P., Adsorption of substituted nitrobenzenes and nitrophenols to mineral surfaces. *Environ. Sci. Technol.*, 27: 316–326. Copyright 1993 American Chemical Society.

where [NAC$_{sorb}$] and [NAC$_{aq}$] are the total concentrations in the sorbed and in the solution phase, respectively.

The first question to be answered is how the K_d value of a given compound is influenced by the nature of the cations adsorbed to the clay mineral surface. As is indicated by Table 6.1 and shown for two other compounds in Figure 6.7, the K_d value of a given NAC depends on the nature of the cation adsorbed to the clay mineral surface. In the presence of strongly hydrated cations (e.g., Li$^+$, Na$^+$, Mg^{2+}, Ca^{2+}, Ba^{2+}), very small K_d values are observed as would be expected if solely nonspecific; that is, hydrophobic sorption would be important. For more weakly hydrated cations (e.g., NH$_4$$^+$, K$^+$, Rb$^+$, Cs$^+$), however, the log K_d value increases linearly with decreasing free energy of hydration of the cation, at least for compounds that do not contain bulky substituents (see Figure 6.8).

Before we turn to discuss the effect of structural moieties on K_d, we need to consider another important factor, the solution pH. Figure 6.9 shows that for neutral species (e.g., substituted nitrobenzenes), K_d does not change significantly over the ambient pH range. For weak acids such as the nitrophenols, however, a strong pH dependence of the apparent K_d value is observed in the pH region corresponding to the pK_a ($K_a =$ acidity constant) of the compound, indicating that the anionic species is only very poorly adsorbed as compared to the neutral (nondissociated) compound.

The effects of substituents on K_d are illustrated by Figure 6.10 and may be summarized as follows. One obvious feature is that substituents with strong electron-withdrawing and electron-delocalizing properties (e.g., CHO, COCH$_3$, CN, and, particularly, NO$_2$) strongly enhance sorption (compare, for example, the K_d values of the mono- versus the dinitrophenols in Figure 6.10). Furthermore, substituents other than OH in position *ortho* to the nitro group as well as bulky substituents in any position on the ring drastically diminish K_d. These latter findings suggest that only NACs ex-

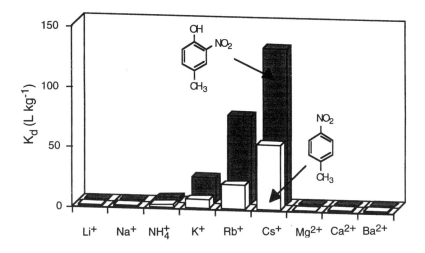

Figure 6.7. Comparison of the K_d values of 4-methyl-2-nitrophenol and 4-methylnitrobenzene for various homoionic kaolinites ($I = 0.1$ M, pH = 4.0). Reprinted with permission from Haderlein, S. B., and Schwarzenbach, R. P., Adsorption of substituted nitrobenzenes and nitrophenols to mineral surfaces. *Environ. Sci. Technol.*, 27: 316–326. Copyright 1993 American Chemical Society.

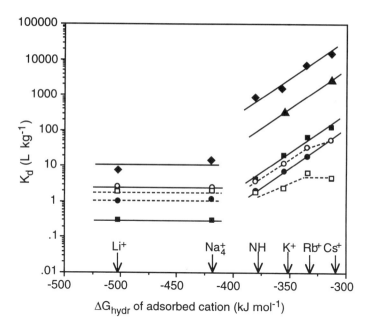

Figure 6.8. Adsorption of a series of NACs on various homoionic kaolinites: Plot of K_d (log scale) versus the free energy of hydration (in solution) of the adsorbed cation (◆, dinitro-*o*-cresol (DNOC); ▲, 1,5-dinitronaphthaline; ■, 4-methyl-2-nitrophenol; ●, 4-methylnitro benzene; ○, 2-*sec*-butyl-2,4-dinitrophenol (Dinoseb); □, 4-*sec*-butyl-2-nitrophenol). Reprinted with permission from Haderlein, S. B., and Schwarzenbach, R. P., Adsorption of substituted nitrobenzenes and nitrophenols to mineral surfaces. *Environ. Sci. Technol.*, 27: 316–326. Copyright 1993 American Chemical Society.

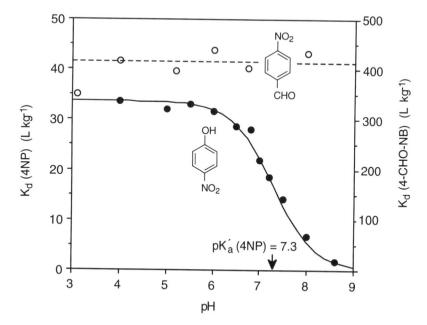

Figure 6.9. Effect of pH on the K_d value of 4-formyl-nitrobenzene and 4-nitrophenol (pK'_a = 7.3) in Cs$^+$-kaolinite suspensions (I = 0.1 M). The solid line represents K_d (pH) = $\alpha_a K_d^{HA}$, where K_d^{HA} is the K_d value of nondissociated 4-nitrophenol and α_a = 1/[1 + 10$^{(pH−pKa)}$] is the fraction of the nondissociated species. Reprinted with permission from Haderlein, S. B., and Schwarzenbach, R. P., Adsorption of substituted nitrobenzenes and nitrophenols to mineral surfaces. *Environ. Sci. Technol.*, 27: 316–326. Copyright 1993 American Chemical Society.

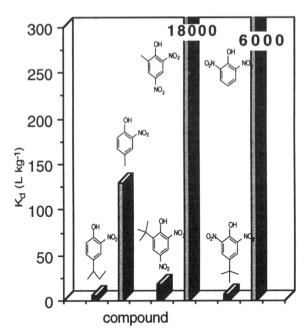

Figure 6.10. Effect of bulky substituents on K_d (sorbent: Cs^+-kaolinite, pH = 4.0, I = 0.1 M). Reprinted with permission from Haderlein, S. B., and Schwarzenbach, R. P., Adsorption of substituted nitrobenzenes and nitrophenols to mineral surfaces. *Environ. Sci. Technol.*, 27: 316–326. Copyright 1993 American Chemical Society.

hibiting a planar structure that allows the π-electron system to approach closely to the surface can significantly interact with the surface.

From all the evidence collected in this study (note that only a few results have been presented here), it can be postulated that (particularly planar) NACs may interact specifically with siloxane surface sites, provided that weakly hydrated cations are adsorbed at these sites. The hypothesis of a specific surface interaction mechanism is corroborated by the large negative adsorption enthalpies ($\Delta H_{ads} < -40$ kJ mol^{-1}) determined for a series of NACs. The most plausible explanation for the observed strong adsorption is the formation of an electron donor–acceptor (EDA) complex between electron donor functions at the siloxane surface (i.e., surface oxygen atoms) and the aromatic ring system of the NAC. Various *in situ* spectroscopic measurements (UV-visible, CIR-FTIR, ^{13}C-NMR, and XRD) fully agree with this proposed mechanism (Weissmahr et al., 1997).

Environmental Significance of the Results and Outlook

The results obtained to date from this laboratory study indicate that planar NACs exhibiting electron-withdrawing and electron-delocalizing substituents can interact specifically and reversibly with siloxane surfaces of clay minerals, provided that weakly hydrated cations (e.g., NH_4^+, K^+, Rb^+, Cs^+) are adsorbed to these surfaces. Although the K_d values derived from such model systems cannot be directly applied to field situations, the environmental significance of the process investigated can already be assessed with some simple "back-of-the-envelope" calculations.

In many soils and aquifers and, of course, in the clay liners of landfills, a significant part of the total mineral surface area available for sorption consists of clay mineral surfaces. The major cations adsorbed to these surfaces include Na^+, NH_4^+, K^+, Mg^{2+}, Ca^{2+}, and Al^{3+}. As shown above, among these cations only NH_4^+, and K^+ provide surfaces that allow EDA complexes with NACs to be formed. In aquifers, typically 1% and in landfills up to 10% of the cations adsorbed to the clay minerals is potassium (see refs. cited in Haderlein and Schwarzenbach, 1993). If one assumes that siloxane-type surfaces make up about 50% of the total clay mineral surfaces, a K_d value of about 2.5 L kg^{-1} is calculated for 1,4-dinitrobenzene (see K_d values in Table 6.1) for the aquifer material (assumed total surface area of ~5 m^2 g^{-1}) and up to 500 L kg^{-1} for a typical clay liner. Note that the K_d values expected if solely hydrophobic sorption to the mineral surfaces was important would be on the order of 0.05 L kg^{-1} and 1 L kg^{-1}, respectively. Consequently, for a typical aquifer material exhibiting very little organic material, a porosity, ϵ, of 0.2, and a density, ρ, of 2.5 kg dm^{-3}, a retardation factor

$$R_f = 1 + [\rho(1 - \epsilon)/\epsilon]K_d \tag{6.2}$$

of about 25 is calculated for 1,4-dinitrobenzene as compared to 1.5 when assuming only hydrophobic sorption. For the clay liners, the calculated R_f values are up to 5000 and only about 10, respectively.

For compounds that form strong EDA complexes to clay mineral surfaces (such as 1,4-dinitrobenzene), this sorption mechanism may even be important in organic-carbon-rich environments. Assuming a K_{oc} value of about 40 L kg$_{oc}$$^{-1}$ for 1,4-dinitrobenzene (estimated from its octanol–water partition constant), in the case of the aquifer, a fraction of natural organic carbon, f_{oc}, of 0.06 would be necessary to give a retardation factor similar to the one (i.e., 25) calculated above for EDA complex formation. Note that for the case of hydrophobic sorption to natural organic material, K_d is approximated by $f_{oc} \cdot K_{oc}$ (for details see Schwarzenbach et al., 1993, Chapter 11; or see Chapter 5, this volume).

These calculations illustrate that specific adsorption to clay mineral surfaces can, in principle, be a very important process in determining the transport and fate of NACs in the environment. Note that the simple retardation factor approach used above to compare the mobility of solutes is only applicable if the compound(s) exhibit linear adsorption isotherms, which is the case for the adsorption of NACs to clay minerals only at low concentrations (see Figure 6.6). Assessing the mobility of solutes that exhibit nonlinear isotherms is not so straightforward and requires more sophisticated approaches (Brusseau 1995; Fesch et al., 1997).

Further work was carried out to confirm the postulated sorption mechanism and to derive quantitative structure–adsorptivity relationships for sorption of NACs to clay mineral surfaces (Weissmahr, 1996). The major task to be carried out, however, will be the application of the results of this laboratory study to real field situations. This must include an evaluation and estimation of the distribution and of the surface chemistry (e.g., types of species adsorbed) of the relevant clay mineral surfaces present in a given environmental system.

As a first step in that direction, the sorption behavior of a series of selected model NACs on model and real soils, aquifer materials, and sediments is being investigated. These experiments are carried out in the laboratory using batch, flow-through reactor, and column systems. The results obtained so far show that an analysis of the relative sorption behavior of the selected model NACs under various solution conditions will yield important information on the availability and on the dynamics of sorptive sites in a more complex solid matrix such as a soil or sediment (Weissmahr, 1996).

Note that probing a complex system for a given process with a series of related compounds covering a wide range of electronic and steric properties is a common approach in environmental organic chemistry, particularly in cases for which the natural reactants are not known and/or are difficult to quantify. Another illustrative example in which natural reactants have been characterized by the relative reactivities of a series of model compounds is the reduction of NACs mediated by natural organic matter (see Dunnivant et al., 1992), and by reduced iron species (see Heijman et al., 1995; Klausen et al., 1995; Rügge et al., 1997).

FIELD STUDIES—THE CLUE TO HOW THINGS REALLY ARE (OR MIGHT BE!). EXAMPLE: ORGANIC CONTAMINANTS IN A SMALL LAKE

To date, the majority of the published field investigations of organic pollutants have been largely confined to reporting concentrations of specific contaminants in a given environmental compartment. Many of these monitoring-type studies have provided important insights into the occurrence and temporal and spatial variations in the concentrations of a great number of pollutants in natural waters. However, it is generally the case that only qualitative information on the processes that govern a given compound's behavior in a given environmental system can be obtained from such data sets. Field studies that allowed the derivation of quantitative and generalizable process information are still rather scarce, although a steady increase in number of such investigations can be observed in the literature. Some illustrative examples include an investigation of the dynamic behavior of (a) PCBs in the Great Lakes (Achman et al., 1993) and (b) some pesticides and volatile halogenated hydrocarbons in streams and rivers (Wanner et al., 1989; Cirpka et al., 1993).

In order to ensure an optimal collection and analysis of field data, field studies should preferentially be combined with model calculations. In many cases, a very simple model (e.g., a one- or two-box model for describing a lake) is sufficient for answering a variety of important questions. The combination of mathematical modeling and field measurements is particularly useful (1) for deriving information on processes that are difficult to quantify based on laboratory data (e.g., *in situ* microbial transformation rates), (2) for quantifying inputs to a given natural system, (3) for discovering unexpected pollutant behavior, and (4) for validating predictive models derived for describing pollutant behavior in a system. An extensive discussion on the role of mathematical models in environmental organic chemistry can be found elsewhere (Schwarzenbach et al., 1993, Chapter 15).

In the following, the spatial and temporal distributions of three organic contaminants, [i.e., tetrachloroethene (PER), atrazine, and NTA (for the structures see Figure 6.11)] in a small lake (Greifensee) in Switzerland are discussed. The results presented will serve primarily to illustrate some of the comments made above. A more detailed description of this study is given elsewhere (Ulrich, 1991; Ulrich et al., 1994).

Goals of the Field Study in Greifensee

The goals of this field investigation were severalfold. On the one hand, the study was designed to demonstrate various applications of a software package (MASAS, see

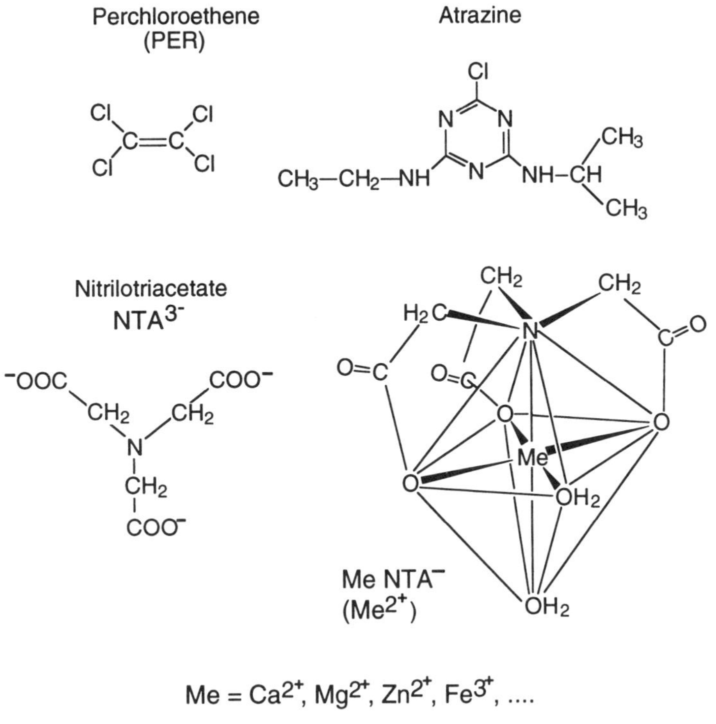

Perchloroethene
(PER)

Atrazine

Nitrilotriacetate
NTA^{3-}

Me NTA$^-$
(Me^{2+})

Me = Ca^{2+}, Mg^{2+}, Zn^{2+}, Fe^{3+}, ...

Figure 6.11. Structures of compounds investigated in the field study in Greifensee.

below) that has been developed for building computer models for evaluation and prediction of the dynamic behavior of organic contaminants in lakes. On the other hand, for each of the compounds investigated, some specific questions were addressed. Examples include the determination of the *in situ* (bio)degradation rate of the complexing agent NTA, the evaluation of a long-term (predictive) model for persistent volatile organic compounds (PER), and an assessment of the input dynamics and of the fate of triazines, particularly atrazine, in lakes.

Description of the Lake; Sampling Sites and Sampling Program

Figure 6.12 shows a map of Greifensee indicating the various sampling sites. Lake characteristics are summarized in Table 6.2. Greifensee is a eutrophic lake with regular overturn in winter (December through March). Regular successions of oxic (spring) and anoxic conditions (summer, fall) are observed in the hypolimnion of the lake.

The catchment area of Greifensee comprises about 160 km^2 with roughly 100,000 inhabitants. The high population density as well as some intensive agricultural activities cause a significant input of anthropogenic chemicals into the lake. The effluents of various treatment plants are discharged either directly or indirectly (via the two major tributaries, the Aa and Aabach Rivers) into the lake (see Figure 6.12). The only outflow of the lake is the River Glatt.

In the lake, samples were taken at the deepest point, commonly at monthly intervals (except for PER for which the intervals were usually longer) at 7–10 different

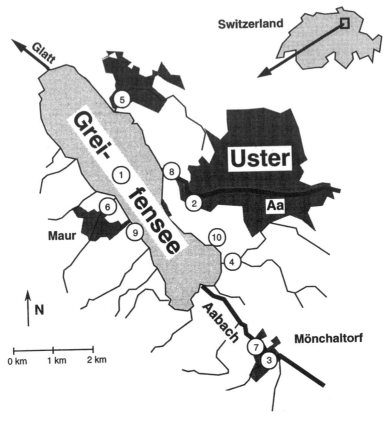

Figure 6.12. Map of Greifensee showing the major inflows, outflow, residential areas (■), and sampling sites. Site 1 is the sampling site in the lake at the deepest point; 2,3 and 4, 5, 6 are sampling locations at the major and minor tributaries, respectively; 7, 8, and 9 are effluents of wastewater treatment plants; 10 is the sampling site for rainwater.

depths. Concentration profiles were determined for NTA over a period of one year, for atrazine over a period of two years, and for PER over a period of nine years. Considering the relatively high horizontal mixing rates in lakes such as Greifensee, no significant horizontal concentration gradients can be expected (and this has been con-

TABLE 6.2
Some Characteristic Data of Greifensee[a]

Volume	1.51×10^8 m^3
Lake area at different depths	
Surface	8.49×10^6 m^2
10 m	6.55×10^6 m^2
20 m	3.51×10^6 m^2
30 m	1.02×10^6 m^2
Maximum depth	32 m
Mean depth	17.8 m
Mean residence time of water	440 days

[a]Data from Ulrich et al. (1994).

firmed for various water constituents), at least for compounds that are not highly re-active (Schwarzenbach et al., 1979; Imboden and Schwarzenbach, 1985; Schwarzenbach et al., 1993, Chapter 15). Thus, except for the regions close to a point source, the vertical concentration profiles taken at the deepest point of the lake can be considered to be representative of the whole lake.

An accurate determination of the input of a given compound to the lake is, in general, very difficult, particularly for compounds stemming primarily from diffuse sources such as the herbicide atrazine and, to a certain extent, the solvent PER. But even for the complexing agent NTA for which the effluents of the treatment plants represent the major input sources, a quantification of the average load and its range of variation is not that easy. In particular, the loads of NTA during rain events that lead to a direct input of untreated wastewater into the lake or its tributaries are difficult to determine. In the case of atrazine, a special effort was made to quantify the input, particularly during the main application period of this herbicide (May–June) and during the first big storm-water event after the application period. For PER, finally, only a few representative samples from the various input sources (i.e., treatment plants, tributaries) were analyzed during the first two years and in the last year of the investigation. A more detailed discussion of the sampling program, a description of the sampling procedure, and a description of the analytical methods used to determine the model compounds can be found elsewhere (Ulrich et al., 1994; Berg et al., 1995).

Short Description of the Computer Software

The general concept of the software package developed for *M*odeling of *A*nthropogenic *O*rganic *S*ubstances in *A*quatic *S*ystems (MASAS) is depicted in Figure 6.13.

In the present version of MASAS, a lake is described in terms of a one-dimensional vertical transport-reaction model including the water column, the sediment–water interface, and the sediment. Such a model, in which horizontal concentration gradients are neglected, is well suited for the description of moderately reactive compounds in deep lakes (as compared to very shallow lakes in which horizontal transport is significant). MASAS computes the concentration of the compound of interest in the water column (and in the sediments, if required) as a dynamic model variable. Transport, mixing, and transformation processes (i.e., loading, dilution, vertical mixing, air–water exchange, sorption/desorption, sedimentation, sediment–water exchange, and chemical, photochemical, and biological transformation reactions) can be defined interactively. The results of a simulation are graphically presented, and they may be stored on a file. Finally, interactively defined models can be saved for future use.

Library files (that can be built up by the user) allow fast access to both compound-specific and system (i.e., lake)-specific data. Compounds are characterized by all pertinent physicochemical parameters and reactivities. The lakes are described in terms of morphometric (volume, lake area as a function of depth, etc.), hydraulic (rate of throughflow, depth of epilimnion, vertical mixing rates as a function of depth, etc.), and physical and chemical parameters (T, pH, concentration of dissolved and particulate species, particule settling velocities, etc.). Both the spatial and the temporal variation of all these parameters can be taken into account.

MASAS supports an iterative form of model development. Based on initially simple models, models of increasing complexity can be constructed. Various types of mod-

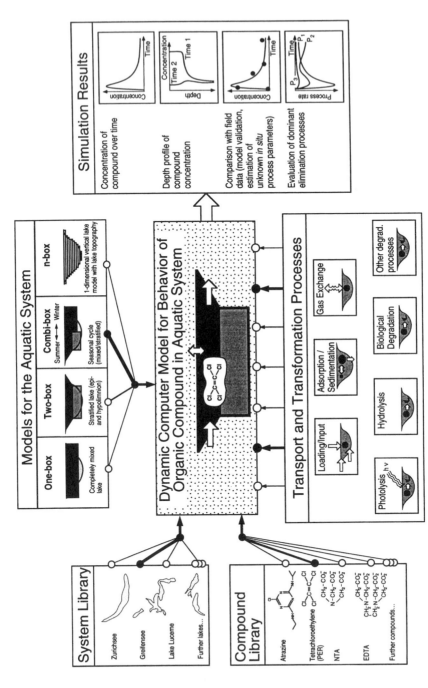

Figure 6.13. General concept of the "MASAS" system showing its major elements. **Left:** Libraries for storage of system specific and compound specific data. **Top:** Models of different spatial complexity. **Bottom:** Modules for the description of individual processes. **Center:** Program with which the model is built and which calculates the simulation results; **Right:** Simulation results. With the indicated setting (bold arrows), the user has chosen a combi-box model for PER in Greifensee with loading and gas exchange as the sole processes (see Figure 6.17). Flushing as a transport process is automatically included in the model, and therefore it is not shown.

els (one-box, two-box, "combi-box," n-box) with different degrees of spatial resolution are available to describe the vertical variation in the concentration of the substance being studied. Processes can be described at various levels of sophistication. Information fields show the user which parameters are missing and which routines are available to estimate these missing parameters.

The program with a user interface consisting of menus, standard dialogue boxes, and interactive text and graphic windows has been developed on an Apple Macintosh personal computer (Ulrich et al., 1995). Because of its user-friendliness, MASAS has not only been proven to be a useful tool in research and consulting, it is also well suited for teaching purposes.

"Back-of-the-Envelope" Calculations

Before applying (more) sophisticated models to a given problem, it is useful to start out with some simple back-of-the-envelope calculations. Using the already available data, one can often get a feeling for the relative importance of the various processes. Furthermore, the results of such simple calculations can be used as a starting point for building more complex models.

In the case of a stratified lake, for example, as a first approximation the epilimnion may be looked at as a well mixed box that is isolated from the hypolimnion. In addition, each of the processes can be expressed as a (pseudo-) first-order reaction exhibiting a characteristic rate constant (see Figure 6.14). By doing so, the relative im-

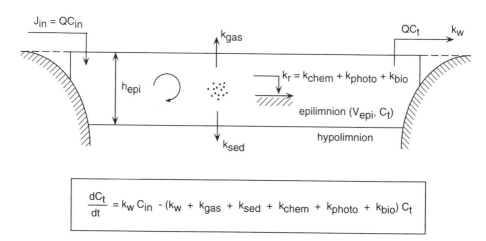

where C_t = total concentration of compound in the epilimnion
   C_{in} = average input concentration of compound (defined as J_{in} / Q)
   J_{in} = average load of compound (mass per time)
   k = characteristic first-order rate constant for a given process
       (i.e., export by water (flushing), gas exchange, sedimentation,
       and chemical, photochemical and biological transformation)

Figure 6.14. One-box model for the (well-mixed) epilimnion of a lake. The model is used for some back of-the envelope calculations by making the following assumptions: (1) total input of water and compound into epilimnion, (2) no water exchange between epilimnion and hypolimnion, and (3) all processes can be expressed as (pseudo-) first-order reactions.

portance of each process can immediately be seen from a comparison of the various characteristic (first-order) rate constants. Since flushing from the epilimnion is equally important for all compounds (dissolved and particulate species), the characteristic rate constant, k_w, for this process can be taken as a reference point:

$$k_w = \frac{Q}{V_{\text{epi}}} \tag{6.3}$$

where Q is the average flow rate (e.g., in $m^3 \ d^{-1}$), and V_{epi} is the volume of the epilimnion. Note that in our case, it is assumed that the total water and compound input occurs into the epilimnion. The k_w value for the epilimnion of Greifensee is about 0.01 d^{-1}. The estimation and discussion of the rate constants for the pertinent processes of the various model compounds will be given in the following sections.

Dynamic Behavior of the Model Compounds in the Lake

NTA

Nitrilotriacetic acid (NTA) is a widely used complexing agent that enters the aquatic environment primarily via wastewater discharges. In Switzerland, NTA is used predominantly as a phosphate substitute in detergents (phosphate has been banned from detergents since 1986). It has been found to biodegrade both under aerobic and anaerobic conditions.

At ambient pH values in natural waters, NTA is present predominantly as anionic metal complex (see Figure 6.11). Therefore, gas exchange can be neglected. Also, NTA can be considered persistent with respect to (abiotic) chemical degradation. Furthermore, based on the results of various batch and column experiments of several authors, one can conclude that NTA adsorbs only very weakly to particles, and therefore sedimentation can be excluded as a major removal process from the water column.

In summary, based on these considerations, it was assumed that microbial degradation was the most important elimination process for NTA in lakes. The major goal of this study was to check this hypothesis and to determine the *in situ* rate for the degradation process.

Input Measurements. During the time period of the field investigation, the average daily NTA input to Greifensee estimated from numerous samples was 3.0 kg d^{-1} (range from 0.5–6.0 kg d^{-1}). During storm-water events, spot samples incidated much higher loads (by up to a factor of 5). In general, during stratification of the lake (May–November), it can be assumed that the major part of the input occurred into the epilimnion of the lake (Ulrich, 1991).

Concentration Profiles and Removal Rate of NTA from the Water Column. Figure 6.15 shows some representative vertical concentration profiles of NTA, as well as the temperature profiles indicating whether the lake was stratified or well-mixed at the various sampling dates. Significantly higher concentrations were always observed in the epilimnion as compared to the hypolimnion.

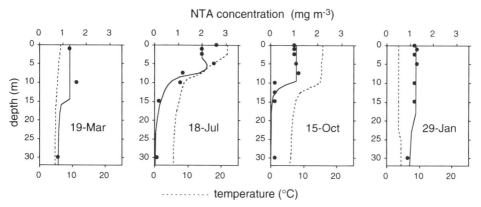

Figure 6.15. Some selected vertical concentration profiles of NTA determined at the deepest point in Greifensee. Also shown are the temperature profiles (dotted lines) indicating the stratification of the lake. The solid lines represent the results of model calculations using an *n*-box (*n* = 18) model.

Some important conclusions can be drawn by applying simple box models (e.g., two-box for epilimnion/hypolimnion) to evaluate the field data. First, in addition to flushing, an elimination rate with a characteristic first-order rate constant of between 0.02–0.04 d^{-1} for the epilimnion is calculated. Furthermore, the model calculation indicates that NTA was also eliminated from the hypolimnion, but at a smaller rate. Quantification of this rate is, however, more difficult, since the concentrations of NTA in the hypolimnion were close to the detection limit of the analytical method. The results of the two-box model calculations were confirmed by application of a more refined *n*-box (*n* = 18, vertical mixing rates derived from temperature data) model, which, as shown by the solid lines in Figure 6.15, reproduces the vertical distribution of NTA quite satisfactorily (for more details see Ulrich et al., 1994). Based on this model, an average rate constant for elimination of NTA other than by flushing of 0.035 d^{-1} is obtained for the epilimnion of Greifensee.

Interpretation of the NTA Results and Conclusions. The seasonal variation in the distribution of NTA in Greifensee could be modeled successfully with an *n*-box model that includes time-variable input, flushing, and a first-order elimination process. The rate at which NTA was removed from the epilimnion did not vary significantly with season, which excludes photolytic transformation and leaves microbial degradation as the most likely elimination process. The average characteristic reaction rate constant of 0.035 d^{-1} determined for the epilimnion in Greifensee is of the same order of magnitude as found for NTA degradation in estuarine and marine waters, but small (up to more than 10 times lower) when compared to elimination rates reported for rivers where NTA concentrations are usually much higher [Giger et al. (1991) and references cited by Ulrich (1991)].

The NTA study is a nice illustration of a case in which an anticipated behavior of an anthropogenic compound in a natural system has been confirmed by a field investigation, and in which an *in situ* reaction rate could be determined for a process that is very difficult to quantify from laboratory data (i.e., microbial transformation). Note that the model developed for NTA in Greifensee has also been applied successfully to describe the dynamic behavior of this compound in other Swiss lakes.

PER

With a global annual consumption on the order of 10^6 tons, PER has to be considered worldwide as one of the most important solvents. In Switzerland, PER is primarily used for degreasing metals (60–70%) and for dry-cleaning (20–30%). Over the past 10 years, the consumption of PER in Switzerland has decreased by about a factor of five, and various measures have been taken to diminish losses to the environment, particularly to natural waters.

In the water column of lakes, PER is present primarily in dissolved form as is evident from the following calculation. Assuming hydrophobic partitioning as the sole sorption process, and using a K_{oc} value for PER of about 500 L kg_{oc}^{-1} and a particulate organic matter concentration, [POC], of 2×10^{-6} kg_{oc}^{-1} L^{-1} in the epilimnion of Greifensee, the maximum fraction of PER in particulate form, f_p, is estimated to be

$$f_p = \frac{K_{oc}[POC]}{1 + K_{oc}[POC]} = 0.001 \qquad (6.4)$$

As discussed elsewhere (Ulrich et al., 1994), the characteristic first-order rate constant, k_{sed} (Figure 6.14), for the loss of a compound from the epilimnion can be expressed by

$$k_{sed} = \frac{v_s}{h_{epi}} f_p \qquad (6.5)$$

where v_s is the average particle settling velocity (maximum value 2.5 m · d^{-1}) and h_{epi} (= 5 m) is the average epilimnion depth. Insertion of these values into Eq. (6.5) yields a k_{sed} of 0.0005 d^{-1}. Hence, as was assumed above for NTA, sedimentation should not be an important process for PER which is consistent with earlier findings (Schwarzenbach et al., 1979). In contrast to NTA, however, PER exhibits a rather large Henry constant (K_H^* = dimensionless Henry constant ≈ 1 at 25°C), and therefore volatilization from the lake to the atmosphere must be considered.

In lakes, gas exchange to the atmosphere can be described by a two-film model (for details see Schwarzenbach et al., 1993, Chapter 10). In the case of volatile organic compounds such as PER, gas exchange is, however, only controlled by the liquid film, and the flux of the compound between the lake and the air can be simply expressed as

$$\text{Flux (e.g., in mass m}^{-2}\text{ d}^{-1}) = v_w (C_w - C_L/K_H^*) \qquad (6.6)$$

where v_w is the liquid-film mass transfer coefficient (e.g., in m d^{-1}), C_w (= $(1 - f_p)$ $C_t \approx C_t$) is the concentration in the water in dissolved form, and C_L/K_H^* is the concentration that would be established in the water at equilibrium with the actual concentration, C_L, of the compound in the air. When assuming that $C_L/K_H^* \ll C_w$—that is, when neglecting the PER concentration in the air—a characteristic first-order rate constant for loss from the epilimnion by gas exchange can be defined (Schwarzenbach et al., 1993, Chapter 15):

$$k_{gas} = \frac{v_w}{h_{epi}} (1 - f_p) \qquad (6.7)$$

Using empirical correlations between liquid-film mass transfer coefficients for oxygen and average wind speeds above the lake, an average v_W value for PER in Greifensee of 0.15 m d^{-1} is estimated (Ulrich et al., 1994). This estimated v_W value corresponds well with experimental data obtained from an earlier mass balance study for another volatile organic compound, 1,4-dichlorobenzene, in Lake Zürich (Schwarzenbach et al., 1979). With an average epilimnion depth of 5 m, our simple back-of-the-envelope calculation (Figure 6.14) then yields an average characteristic rate constant, k_{gas}, of 0.03 d^{-1}, which means that volatilization should be about three times more important for elimination of PER from the epilimnion of Greifensee as compared to export by the outflow. Since PER is considered to be quite persistent to chemical, photochemical, and biological transformation in lakes, volatilization and flushing are the only elimination processes incorporated into the model.

Input Measurements for PER. As mentioned earlier, measurements for quantification of the input of PER to Greifensee were carried out only during the first (1982–1983) and during the last (1990) year of the study. Based on the rather small data set available, it is not possible to give reliable numbers for the PER input during these time periods. Rough estimates can, however, be made. During the first year, the estimated load varied between about 0.1 and 0.6 kg d^{-1}, while in 1990 the average "measured" input of PER was only on the order of 0.01 kg d^{-1}. One of the goals of the modeling efforts was, therefore, also to provide a better picture of the "input history" of PER in Greifensee.

Concentration Profiles and Model Calculations for PER. Figure 6.16 shows some of vertical concentration profiles of PER determined during the first year of the investigation. The solid lines represent the results of model calculations with an n-box ($n = 32$) model using the first concentration profile as the starting point. Note that the only fitting parameter used in the model calculations was the average PER load for a given time period. These calculated values are also given in Fig. 16. All other parameters including gas exchange rate, flushing rate, and vertical mixing rates were estimated (gas exchange, see above) and/or derived independently. As can be seen, over the one-year period shown in Figure 16, the calculated and measured concentration profiles match very well. Similarly good results are also obtained when extending the time period to several years. In addition, after an extraordinary input of a large quantity of PER into the epilimnion of the lake in 1985, the model enabled the estimation of the quantity of PER discharged to the lake, as well as a description of the changes in the vertical distribution of PER observed after the incident (for details see Ulrich et al., 1994).

For evaluating the history of the total PER load to Greifensee between 1982 and 1990, a simple combi-box model (i.e., two-box model during stratification, one-box model during the mixing periods) can be used. Figure 6.17 shows the time course of the PER load obtained from fitting in the total PER content of the lake. Also shown is the change in the total PER content in the lake over the 9 years. Note that the accidental input of PER in 1985 has not been incorporated into the combi-box model calculations.

There are two striking features that are apparent from Figure 6.17. First, the model simulation suggests that between 1982 and 1990, the PER load to Greifensee decreased steadily, overall by a factor of about 10. This result is consistent with the continuous decrease of the total PER consumption in Switzerland during this time period (Ulrich et al., 1994), and it also complies with the input measurements at the beginning and at the end of the study period (see above). Second, the relatively sharp peak in the total PER content of the lake after the incident in early summer 1985 illustrates that even

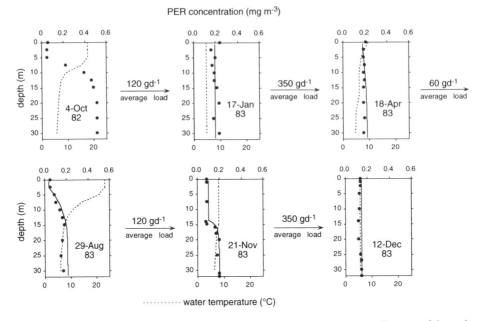

Figure 6.16.   Selected vertical concentration profiles of PER determined during the first year of the study at the deepest point in Greifensee. Also shown are the temperature profiles and the calculated average PER loads between two consecutive sampling dates. The solid lines represent the results of model calculations using an *n*-box (*n* = 32) model.

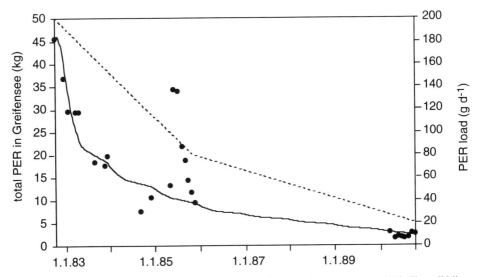

Figure 6.17.   Time course of the total PER content of Greifensee between 1982 and 1990. The solid line represents the result of a model calculation using a combi-box model. The dots indicate the values calculated from measured profiles. Note that the accidental input in 1985 has not been incorporated into the model. Also shown (dotted line) is the time course of the PER load obtained from fitting the general trend of the total PER content of the lake.

after a substantial additional input (the PER content of the lake increased by more than a factor of three) of a volatile organic compound such as PER to the epilimnion of a lake, normal conditions resume within a relatively short period of time (i.e., within a few months). Note that if the additional input of PER had occurred into the hypolimnion of the lake, the recovery time would have taken much longer.

In summary, the PER example demonstrates that, by using a film model for quantification of gas exchange and by using typical vertical mixing rates, the dynamic behavior of a persistent volatile organic compound in a lake can be modeled successfully over a long time period. In addition, the example shows that in cases in which the rates of all elimination processes are known or can be predicted, a reasonable estimate of the (average) load of a given compound to the lake can be obtained.

Atrazine

Atrazine is worldwide one of the most widely used herbicides. Since 1989, in Switzerland, the use of atrazine has been restricted to application on corn fields between January and June, with a maximum amount of 1.5 kg per hectare and year.

There is a vast literature concerning the fate of atrazine in soils and aquifers (e.g., Adams and Thurman, 1991; Agertved et al., 1992; Gaynor et al., 1992). Much less is, however, known about the dynamic behavior of this compound in surface waters, particularly in lakes. In a preliminary study, Buser (1990) measured the concentrations of atrazine and other triazine herbicides in various Swiss lakes. From his data he concluded that atrazine should be quite persistent in the water column of the lakes investigated. Kolpin and Kalkhoff (1993) found, however, a significant elimination of atrazine in a small stream in Iowa. They attributed the observed disappearance to (indirect) photolysis, most probably to the reaction of atrazine with OH radicals that are formed in natural waters primarily from the photolysis of nitrate (Zepp et al., 1987). The major sinks for OH radicals in surface waters include natural organic matter, bicarbonate, and carbonate (Larson and Zepp, 1988).

For the epilimnion of Greifensee, using typical concentrations of nitrate, natural organic matter, and (bi)carbonate, Schwarzenbach et al. (1993, Chapter 13) calculated a 24-h-averaged steady-state OH-radical concentration of 3×10^{-18} M for a sunny midsummer day. Assuming a second-order rate constant for the reaction of atrazine with OH radicals of 6×10^9 M^{-1} s^{-1} (Zepp et al., 1987), a characteristic rate constant, k_{photo}, of about 0.002 d^{-1} is obtained, which is small when compared to the flushing rate k_w.

Furthermore, when considering that atrazine exhibits a rather small K_{oc} value (i.e., $K_{oc} \approx 100$ L kg_{oc}^{-1}) and a very small Henry constant $K^*_H < 10^{-6}$), elimination by sedimentation or gas exchange can be neglected (see above). Finally, the rates reported for hydrolysis and microbial transformation are in general smaller than k_w for the epilimnion of Greifensee (Ulrich, 1991). Thus, as a starting point for the model calculations, a conservative behavior was assumed for atrazine.

Input Measurements for Atrazine. Great efforts were made to quantify the input of atrazine to Greifensee. The detailed load function is presented elsewhere (Ulrich et al., 1994). The total atrazine load determined experimentally for the one-year period was on the order of 21 kg, of which 12.4 kg were introduced to the lake during and right after the application period (May, June).

Experimental Data, Model Calculations, and Some Conclusions for Atrazine. Figure 6.18 shows some selected atrazine profiles taken over a period of one year. Again, the solid lines represent the results of model calculations using an n-box ($n = 32$) model. For the model calculations, the experimental input data were used except for the period during the high water event at the end of June, where, as indicated in Figure 6.19 by the increase in the total atrazine content of the lake, the total load was obviously underestimated by about a factor of 1.5. Thus, for this event, a corrected input value was used. The result of this model calculation is shown by the dotted line in Figure 6.19.

From a comparison between model calculations and experimental data, some interesting conclusions can be drawn. First, except for the epilimnion during the months of July and August, the concentration profiles could be described very well with a model that assumes flushing to be the only removal process. Only for the period between the end of June and the end of August, an additional elimination from the epilimnion with a characteristic rate constant of 0.003 d^{-1} must be incorporated into the model to fit the data. Considering that this elimination rate is small and that it lies within the range of the rates estimated for photolysis (see above) or reported for hydrolysis and microbial transformation of atrazine in natural waters, identification of the process(es) responsible for the additional removal of atrazine from the epilimnion of Greifensee is impossible. Nevertheless, the results of this field investigation are a clear

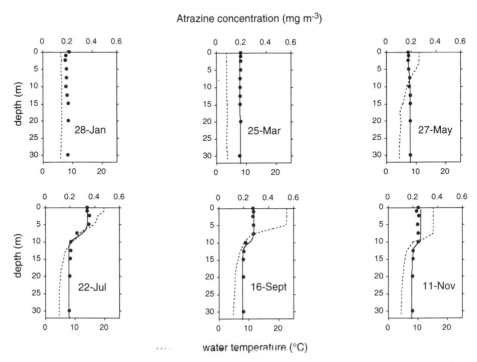

Figure 6.18. Selected vertical concentration profiles of atrazine and temperature (dotted line) determined over a one-year period at the deepest point in Greifensee. The solid lines represent the results of model calculations using an n-box ($n = 32$) model and experimental input data.

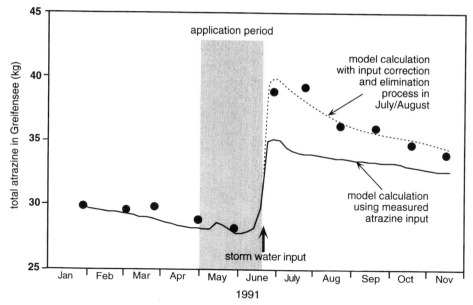

Figure 6.19.   Time course of the total atrazine content of Greifensee over a one-year period. The solid line indicates the model calculation using the storm-water input estimated from measurements in the effluents of the treatment plants and in the major tributaries. The dotted line represents the model calculation with input correction and *in situ* elimination during the months of July and August.

confirmation of the assumption that atrazine is quite conservative in the water column of deep lakes such as Greifensee.

A second important finding of this study is that over 30% of the total annual load of atrazine to the lake occurred during one single high-water event right after the application period. This result is consistent with the results of other studies, in which it was demonstrated that export of atrazine from soils by runoff after storm events is inversely related to the period of time between atrazine application and the storm event (Klaine et al., 1988). Thus, particularly in small lakes located in agricultural areas, concentrations of atrazine (and possibly also of other herbicides) may increase significantly after a storm event during or immediately after the application period. Because of the conservative behavior of atrazine in the water column, the "recovery" of a lake from such a pulse input will be determined primarily by the rate of flushing of the lake.

This last example demonstrates that, with a reasonable number of field measurements, it is possible to get a good picture of the time course of the input of an organic chemical to a lake. The example also shows that, in such a case, convincing evidence for the persistence of a compound in a given aquatic system can be obtained.

CONCLUSION

The goal of this chapter was to illustrate some aspects of environmental organic chemistry and to provide some insights into the general approaches taken in this field. As has hopefully become evident, modern environmental organic chemistry does not just consist of measuring environmental concentrations of organic pollutants, although it should be stressed that analytical chemistry plays an important role. Furthermore, when considering the great complexity of the problems encountered when trying to assess the dynamic behavior of organic contaminants in the environment and, consequently, considering the need for multidisciplinary approaches to solve these problems, it seems obvious that environmental organic chemistry is more than just the application of basic chemistry to environmental issues. Nevertheless, the future development of environmental (organic) chemistry hinges on a much stronger involvement of basic chemists. It would be most fruitful if this challenging and important field would draw more interest in the chemistry community, all the more considering that knowledge of the whole life cycle of a given chemical forms an important base for the future development and use of synthetic chemicals in our society.

References

Achman, D. R., Hornbuckle, C., and Eisenreich, S. J.,1993, Volatilization of polychlorinated biphenyls from Green Bay, Lake Michigan. *Environ. Sci. Technol.*, 27, 75.

Adams, C. D., and Thurman, E. M., 1991, Formation and transport of deethylatrazine in the soil and vadose zone. *J. Environ. Qual.*, 20, 540.

Agertved, J., Rügge, K., and Barker, J. F., 1992, Transformation of the herbicides MCPP and atrazine under natural aquifer conditions. *Ground Water*, 30, 4, 500.

Ballschmiter, K., 1992, Transport and fate of organic compounds in the global environment. *Angew. Chem. Int. Ed. Engl.*, 31, 487.

Berg, M., Müller, S. R., and Schwarzenbach, R. P.,1995, Simultaneous determination of triazines including atrazine and their major metabolites hydroxyatrazine, desethylatrazine, and deisopropylatrazine in natural waters. *Anal. Chem.*, 67, 11, 1860–1865.

Brownawell, B. J., Chen, M., Collier, J. M., and Westall, J. J., 1990, Adsorption of organic cations to natural materials. *Environ. Sci. Technol.*, 24, 1234.

Brusseau, M. L., 1995, The effect of nonlinear sorption on transformation of contaminants during transport in porous media. *J. Contam. Hydrol.* 17, 277–291.

Buser, H. R., 1990, Atrazine and other s-triazine herbicides in lakes and in rain in Switzerland. *Environ. Sci. Technol.*, 24, 1049.

Cirpka, O., Reichert, P., Wanner, O., Müller. S. R., and Schwarzenbach, R. P., 1993, Gas exchange at river cascades: field experiments and model calculations. *Environ. Sci. Technol.*, 27, 2086–2097.

Davis, J. A., and Kent, D. B., 1990, Surface complexation modelling in aqueous geochemistry. In: Hochella, M. F., Jr., White, A. F., eds., *Mineral–Water Interface Geochemistry*, Mineralogical Society of America, Chelsea, MI, Chapter 5.

Dunnivant, F., Schwarzenbach, R. P., and Macalady, D. L., 1992, Reduction of substituted nitrobenzenes in aqueous solutions containing natural organic matter. *Environ. Sci. Technol.*, 26, 2153.

Eisenreich, S. J., Willford, W. A., and Strachan, W. M. J., 1989, The role of atmospheric deposition in organic contaminant cycling in the Great Lakes. In: Allen, D., ed., *Intermedia Pollutant Transport: Modelling and Field Measurements*, Plenum, New York.

Fesch, C. M., Simon, W., Reichert, P., Haderlein, S. B., and Schwarzenbach, R. P., 1997, Nonlinear sorption and non-equilibrium solute transport in aggregated porous media. *J. Contam. Hydrol.*, submitted.

Foster, R. E., 1969, *Organic Charge Transfer Complexes*, Academic Press, New York.

Gaynor, J. D., Mac Tavish, D. C., and Findlay, W. I., 1992, Surface and subsurface transport of atrazine and alachlor from a Brookston Clay Loam under continuous corn production. *Arch. Environ. Contam. Toxicol.*, 23, 240.

Giger, W., Schaffner, Ch., Kari, F. G., Ponusz, M., Reichert, P., and Wanner, O., 1991, Occurrence and behavior of NTA and EDTA in Swiss rivers. *Nachrichten der EAWAG*, No. 32D, p. 27 (in German).

Haderlein, S. B., and Schwarzenbach, R. P., 1993, Adsorption of substituted nitrobenzenes and nitrophenols to mineral surfaces. *Environ. Sci. Technol.*, 27, 316–326.

Haderlein, S. B., and Schwarzenbach, R. P., 1995, Abiotic reduction of nitroaromatic compounds in the subsurface. In: J. Spain, ed., *Biodegradation of Nitroaromatic Compounds*, Plenum, New York, pp. 199–225.

Haderlein S. B., Weissmahr K. W., and Schwarzenbach, R. P., 1996, Specific adsorption of nitroaromatic explosives and pesticides to clay minerals. *Environ. Sci. Technol.*, 30, 612–622.

Heijman, C. G., Grieder, E., Holliger, C., and Schwarzenbach, R. P., 1995, Reduction of nitroaromatic compounds coupled to microbial iron reduction in laboratory aquifer columns, *Environ. Sci. Technol.*, 29, 775–783.

Imboden, D. M., and Schwarzenbach, R. P., 1985, Spatial and temporal distribution of chemical substances in lakes: modeling concepts. In: W. Stumm, ed., *Chemical Processes in Lakes*, John Wiley & Sons, New York, pp. 1–30.

Klaine, S. J., Hinman, M. L., Winkelmann, D. A., Sauser, K. R., Martin, J. R., and Moore, L. W., 1988, Characterization of agricultural nonpoint pollution: pesticide migration in a West Tennessee watershed. *Environ. Toxicol. Chem.*, 7, 609.

Klausen, J., Tröber, S. P., Haderlein, S. B., and Schwarzenbach, R. P., 1995, Reduction of substituted nitrobenzenes by Fe(II) in aqueous mineral suspensions. *Environ. Sci. Technol.*, 29, 2396–2404.

Kolpin, D. W., Kolkhoff, S. J., 1993, Atrazine degradation in a small stream in Iowa. *Environ. Sci. Technol.*, 27, 134.

Larson, R. A., and Zepp, R. G., 1988, Reactivity of the carbonate radical with aniline derivatives, *Environ. Toxicol. Chem.*, 7, 265.

Rügge, K., Hofstetter, T. B., Haderlein, S. B., Bjerg, P. L., Knudsen, S., Zraunig, C., Mosbek, H., and Christensen, T. H., 1997. Characterization of predominant reductants in an anaerobic leachate-affected aquifer by nitroaromatic probe compounds, *Environ. Sci. Technol.*, submitted.

Schwarzenbach, R. P., and Imboden, D. M., 1984, Modelling concepts for hydrophobic organic pollutants in lakes, *Ecological Modelling*, 22, 171.

Schwarzenbach, R. P., Molnar-Kubrica, E., Giger, W., and Wakeham, S. G., 1979, Distribution, residence time, and fluxes of tetrachloroethylene and 1,4-dichlorobenzene in lake Zurich, Switzerland. *Environ. Sci. Technol.*, 13, 1367.

Schwarzenbach, R. P., Gschwend, P. M., and Imboden, D. M., 1993, *Environmental Organic Chemistry*, John Wiley & Sons, New York, 681 pp.

Schwarzenbach, R. P., Gschwend, P. M., and Imboden, D. M., 1995, *Environmental Organic Chemistry. Illustrative Examples, Problems, and Case Studies*, John Wiley & Sons, New York.

Spalding, R. F., and Fulton, J. W., 1988, Groundwater munition residues and nitrate near Grand Island, Nebraska, USA. *J. Contam. Hydrol.*, 2, 139.

Stumm, W., and Morgan, J. J., 1981, *Aquatic Chemistry*, John Wiley & Sons, New York.

Stumm, W., Schwarzenbach, R. P., and Sigg, L., 1983, From environmental analytical chemistry to ecotoxicology—a plea for more concepts and less monitoring and testing. *Angew. Chem. Int. Ed. Engl.*, 22, 380.

Ulrich, M., 1991, Modeling of chemicals in Lakes—Development and Application of User-Friendly Simulation Software (MASAS and CHEMSEE) on Personal Computers. Ph.D. thesis ETH. No. 9632. The program including a copy of the Ph.D. thesis are available for a nominal fee. Contact: M. Ulrich, Swiss Federal Institute for Water Resources and Water Pollution Control (EAWAG), 8600 Dübendorf, Switzerland.

Ulrich, M. M., Müller, S. R., Singer, H. P., Imboden, D. M., and Schwarzenbach, R. P., 1994, Input and dynamic behavior of the organic pollutants tetrachloroethene, atrazine and NTA in a lake: a study combining mathematical modeling and field measurements. *Environ. Sci. Technol.* 28, 1674–1685.

Ulrich, M. M., Imboden, D. M., and Schwarzenbach, R. P., 1995, MASAS—a user friendly simulation tool for modeling the fate of anthropogenic substances in lakes. *Environ. Software*, 10, 177–198.

Wanner, O., Egli, T., Fleischmann, T., Lanz, K., Reichert, P., and Schwarzenbach, R. P., 1989, Behavior of the insecticides disulfoton and thiometon in the Rhine River; a chemodynamic study, *Environ. Sci. Technol.*, 23, 1232.

Weissmahr, K. W., 1996, *Mechanism and Environmental Significance of Electron Donor–Acceptor Interactions of Nitroaromatic Compounds with Clay Minerals*. Ph.D. thesis No. 11631, Swiss Federal Institute of Technology (ETH), Zurich.

Weissmahr, K. W., Haderlein, S. B., Schwarzenbach, R. P., Hauy, R., and Nüesch, R., 1997, *In situ* spectroscopic investigations of adsorption mechanisms of nitroaromatic compounds at clay minerals. *Environ. Sci. Technol.*, submitted.

Zachara, J. M., Ainsworth, C. C., and Smith, S. C., 1990, The sorption of *N*-heterocyclic compounds on reference and subsurface smectite clay isolates, *J. Contam. Hydrol.*, 6, 281.

Zepp, R. G., Hoigné, J., and Bader, H.,1987, Nitrate-induced photooxidation of trace organic chemicals in water, *Environ. Sci. Technol.*, 21, 443.

CHAPTER 7

Correlation Analysis of Environmental Reactivity of Organic Substances

PAUL G. TRATNYEK

ABSTRACT. Quantitative structure–activity relationships (QSARs) are ubiquitous in environmental chemistry because they provide a powerful tool for assimilating properties of the many environmentally important organic substances. The abundance of QSARs also reflects the ease with which statistically significant correlations can be obtained. This is evidenced by the correlation matrix of substituted phenol properties, which shows that many variable combinations exhibit patterns that suggest significant relationships. However, rather few of these apparent relationships yield useful QSARs. QSARs are most valuable when they are based on appropriate predictor variables, an explicit mechanistic model, and appropriate statistics. These criteria are met to varying degrees by the many published QSARs that describe the rates of oxidation of substituted phenols. A problematic application in which QSARs are often employed is predicting the reactivity of compounds outside the scope of their respective training sets. The abundance of data on correlation analysis of phenol oxidation makes it possible to explore the limits and possibilities of this activity.

INTRODUCTION

Assessing the environmental impact of an organic substance requires consideration of its physical transport and chemical transformation. Both transport and transformation can be the net result of many contributing processes, each of which is determined by properties of the substance as well as environmental factors. Therefore, preliminary impact assessment for just one compound may require values for dozens of physical and chemical properties. [For example, consider the input parameters for computer models like MASAS (Ulrich, 1991) and EXAMS (Burns et al., 1982).] Measured values of these properties are the preferred input parameters, but they are often not available, despite recent efforts to develop comprehensive databases (Akland and Waters, 1983; Howard et al., 1991; Kollig and Kitchens, 1993). Furthermore, this lack of data is not likely to be overcome because of the expense of measuring large numbers of substance properties, the continued growth in numbers of manufactured chemicals, and the expanding range of substances subject to environmental regulation (Donaldson, 1992).

The demand for data on the properties of environmental chemicals has led to the development of a wide range of estimation methods. In some cases, these methods take the form of qualitative or semiquantitative rules-of-thumb, but most interest is in relationships that provide quantitative estimates of unknown parameters. Quantitative predictions can be obtained by computation from fundamental molecular structure theory, or by correlation analysis. Despite recent advances in the former approach (Karickhoff et al., 1991), traditional correlation analysis is still the most widely used tool for predicting properties of environmental chemicals. Correlation analysis typically involves regression of available property data for a series of related compounds with one, or several, convenient descriptor variables. The resulting relationship is then used in reverse to estimate values of the property for compounds that were not included in the original data set.

The intent of this review is to survey the principles of correlation analysis as it applies to the reactivity of environmental organic contaminants. The emphasis will be on the practical aspects of formulation and interpretation. To provide continuity while illustrating the widest possible range of correlation techniques, the review will focus on examples involving substituted phenols. This class of compounds is ubiquitous in the environment, both as natural products and as contaminants. Substituted phenols also represent one of the prototypical systems of substrates for correlation analysis and have been extensively studied for this purpose. Although this review is not meant to be comprehensive, it does deal systematically with most aspects of correlation analysis as it is applied in environmental chemistry.

There are relatively few published reviews focusing on the application of correlation analysis to the fate of organic contaminants. Several surveys emphasize chemical transformations (Mill, 1979; Brezonik, 1990, 1994), and a few emphasize correlation of biodegradation rates (Larson, 1990; Parsons and Govers, 1990); however, only one edited volume seems to focus on the practical aspects of formulating and interpreting correlations (Lyman et al., 1982). In contrast, there are numerous monographs, as well as review articles, that treat various aspects of correlation analysis from the perspective of physical organic chemistry (Leffler and Grunwald, 1963; Wells, 1968; Shorter, 1973; Fuchs and Lewis, 1974; Chapman and Shorter, 1978; Wold and Sjöström, 1978; Shorter, 1982; Exner, 1988). It is these works that have most influenced the present discussion. The largest body of literature on correlation analysis concerns toxicology (Hermens, 1986, 1989; Nirmalakhandan et al., 1988) and pharmacology (Testa and Kier, 1991; Hansch and Leo, 1995; Leo et al., 1995). The practice of correlation analysis in these two areas is highly evolved, but the methods used to describe chemical reactivity are largely the same as those used by physical chemists. This review emphasizes the physical chemistry perspective, since it focuses on the chemical *reactivity* of contaminants in the environment.

BACKGROUND CONCEPTS

Terminology

Many different combinations of substance properties have provided correlations that are useful in environmental chemistry. Based on the nature of these properties, a variety of terminology has been developed to identify various classes of correlations. For example, structure–activity relationships (SARs) have been distinguished from property–activity relationships (PARs), both of which have been distinguished from structure–property relationships (SPRs). Careful classification along these lines certainly has heuristic value (Brezonik, 1990), but few researchers adhere to these distinctions rigorously. Instead, only a few terms are commonly used and they are often applied to a wider range of correlation types than strict use of each expression would allow. One example is the term *Linear Free-Energy Relationship* (LFER), which originally referred to a specific type of correlation practiced by physical organic chemists but has come to represent the whole field of correlation analysis in organic chemistry (Shorter, 1973). Similarly, *Quantitative Structure–Activity Relationship* (QSAR) was originally coined for use in drug design, but it is now commonly used to refer to many types of correlations used in the pharmaceutical, toxicological, and environmental sciences. In the present discussion, only the latter two expressions are used, and both are used in accord with their more general definitions. The expression *correlation analysis* is used to represent the broadest scope, including LFERs, QSARs, and a host of related techniques.

Purpose of Correlation Analysis

Despite the similarity in methods and results found in the many disciplines employing correlation analysis, there is a wide range of views on the fundamental basis for their interpretation. This range of views is represented by three distinct interpretations as formulated by Sjöström and Wold (1981). They classify correlation analyses as (i) fundamental expressions of causal relationships, (ii) locally valid linearizations of complex relationships, or (iii) empirical models of similarity. Failure to recognize these differences in purpose has led to disagreement and misunderstanding among the many practitioners of correlation analysis.

The physical organic chemist's primary goal in developing LFERs is usually to better understand the fundamental aspects of how chemical structure determines reactivity. To this end, the most satisfactory LFERs are those that can be interpreted in the most explicitly mechanistic terms (i.e., as causal relationships). In contrast, the growing interest of environmental regulators in QSARs is motivated primarily by the need to predict critical properties of environmental contaminants (Donaldson, 1992) or to prioritize testing for potentially toxic substances (Eriksson et al., 1990). QSARs for this purpose are often developed to have maximum statistical power, with little attention paid to their physical interpretation (i.e., they are empirical models of similarity). The intermediate possibility is that QSARs and LFERs do describe causal relationships, but are approximations that are valid only over a limited range of variables. This interpretation has benefits from the perspective of chemometrics (Sjöström and Wold, 1981; Wold and Sjöström, 1986), but does not entirely reflect the purposes of correlation analysis as practiced by physical chemists (Kamlet and Taft, 1985). One goal of this review is to establish a balance between statistical and mechanistic criteria that reflects the mixed purposes of environmental scientists.

MODEL FORMULATION FOR CORRELATION ANALYSIS

Correlation analysis involves discovering relationships between properties of a series of closely related compounds. The compounds must be related by a common structure or reaction center that is responsible for the property of primary interest. The target property is the dependent or "response" variable. The series of individual compounds on which the correlation is based (the "training set") are distinguished by secondary modifications, usually substituents. These modifications are characterized by one or more independent variables—referred to as "descriptor variables"—that can be conveniently measured, or otherwise determined. When correlation of the response and descriptor variables yields a consistent relationship, the resulting fitted equation can be used as a quantitative model for comparing and predicting properties of related compounds.

Response Variables

Appropriate definition of the response variable is the first step in developing meaningful correlations of chemical properties. Ideally, the response variable reflects a well-defined characteristic, or set of characteristics, of the substances to be correlated. This means that all external effects on the response variable must be rigorously identified and either factored out or held constant. Clear formulation of response variables becomes more challenging as system complexity increases, and can be very difficult for environmental systems.

The difficulties in formulating a response variable are illustrated by the various ways in which reaction rates can be quantified for environmental applications. The kinetics of most environmental transformations can be described by a rate-limiting bimolecular interaction between the substrate of interest, P, and a specified reactant found in the environment, E. The corresponding rate law

$$-\frac{d[\text{P}]}{dt} = k[\text{P}][\text{E}] \qquad (7.1)$$

defines the second-order rate constant, k. The advantage of this approach is that values of k will be independent of environmental conditions (apart from the effects of temperature and ionic strength, which are often negligible) (Hoigné, 1990). Second-order rate constants are the response variable in most of the examples of correlations discussed below.

Very often, however, environmental reactions are not sufficiently well characterized to permit identification and quantification of specific environmental reactants. Under these circumstances, the term for E cannot be explicitly resolved, so the kinetics can only be expressed in terms of a pseudo-first-order rate law

$$-\frac{d[\text{P}]}{dt} = k_{\text{obs}}[\text{P}] \qquad (7.2)$$

where the rate constant k_{obs} is the product of k and [E]. Good correlations are often obtained using k_{obs} as the response variable when the data come from a series of replicate experiments, thereby ensuring that reaction conditions are generally consistent. However, such correlations do not provide a basis for quantitative reaction rate predictions under other environmental conditions and therefore are generally less useful than those based on second-order rate constants. Occasionally, an appropriate kinetic model cannot be identified, and rates are expressed as a change in concentration of P over an arbitrary time interval, or as time for an arbitrary change in concentration. Statistically satisfactory correlations can be obtained even with these parameters as response variables, but the generality of such relationships can be highly uncertain.

Rate constants for use as response variables in correlation analysis can be further refined by taking into account equilibrium speciation of the reactants P and E. Oxidation of substituted phenols in aqueous solution is a reaction that illustrates the importance of including speciation. Dissociation of the phenolic hydroxyl group results in an equilibrium mixture of the parent compound and its dissociated form, the phenoxide (or phenolate) anion. The undissociated phenol and the phenoxide anion react as independent species with very different rate constants, resulting in an apparent dependence of k on pH. At least two modifications of the rate law will take this effect into account (e.g., Scurlock et al., 1990; Tratnyek and Hoigné, 1991); but both require determination of separate rate constants, k_{ArOH} and k_{ArO-}, for the two phenolic species. Therefore, correlation analysis should be performed on these two sets of rate constants separately in order to maintain maximum generality.

Descriptor Variables

The range of descriptor variables that have been applied in correlation analysis is extensive (Chignell, 1983; Franke, 1984; Grüber and Buß, 1989; Eriksson et al., 1993), and new alternatives are continually being proposed (e.g., Kier and Hall, 1990; Stanton and Jurs, 1990). This proliferation makes it difficult to determine the best set of descriptor variables for a new correlation. Some guidance can be obtained by consideration of the various criteria by which descriptor variables are classified. Table 7.1 shows many of the commonly used descriptor variables, organized around two types of criteria, one theoretical and the other practical.

The rows in Table 7.1 reflect physical aspects of the reaction mechanism: (A) uptake to the reaction site, (B) electronic interactions at the reaction site, and (C) steric interactions at the reaction site. Descriptor variables from one or more of these categories often can be excluded based on the nature of the response variable. For example, most of the data discussed below are rate constants measured in well-mixed homogeneous solutions, so descriptor variables that reflect substrate transport across cell membranes (Type A) are clearly not relevant. Similarly, correlations are often developed from data on a series of compounds that are selected, in part, to ensure good correlations by minimizing the likelihood of steric effects on the response variable. In these cases, descriptors of steric interactions (Type C) should not be used.

The columns in Table 7.1 show a practical distinction between three different approaches to defining descriptor variables: (I) substituent constants, (II) molecular descriptors, and (III) reaction properties. The most commonly used descriptors are constants defined by substituent effects on a reference reaction (Type I). Substituent constants are usually designated σ and are applied to correlations in the form of the Hammett equation or its various extensions (Eq. (7.5)). The various σ constants reflect primarily electronic effects on reactivity; however, substituent-based scales also exist for describing steric interactions, E_s, and hydrophobicity, π. Alternatively, a descriptor variable can be a property of the substrate molecule that is available, readily measurable, or can be calculated by independent means (Type II). A familiar example of this is the use of pK_a values as a descriptor variable to correlate hydrolysis rates with the Brönsted equation Eq. (7.9). Finally, a descriptor variable can consist of reaction

TABLE 7.1
A Classification of Representative Descriptor Variables

	(I) Substituent constants	(II) Molecular descriptors	(III) Reaction properties
(A) Uptake to the reaction site	π	K_{ow}, log P, t_R, T_m, SA	k_{other}
(B) Electronic effects at the reaction site	σ, $\sigma-$, σ^+, σ^*, BS, ν	pK_a, $E_{1/2}$, EA, IP, E_{LUMO}, E_{HOMO}, $\Delta l \delta l_{xy}$	k_{other}, ΔG, V_{max}, K_m
(C) Steric effects at the reaction site	E_s	γ_{vdw}, $N\chi$, $N\chi N$	k_{other}

π = hydrophobic fragment constant, K_{ow} and P = octanol-water partition coefficient, t_R = HPLC retention time, T_m = melting temperature, SA = molecular surface area (accessible or total), k_{other} = rate constant for another reactant or another medium, σ = Hammett substituent constants (superscripts indicate various scales), BS = bond strength, ν = bond stretching frequencies, $E_{1/2}$ = polarographic half-wave potential, EA = electron affinity, IP = ionization potential, E_{LUMO} = energy of the lowest unoccupied molecular orbital, E_{HOMO} = energy of the highest occupied molecular orbital, $\Delta l \delta l_{xy}$ = superdelocalizability, ΔG = free energy of reaction, V_{max} and K_m half-maximum velocity and constant from Michaelis–Menton kinetics, E_s = steric substituent constants, γ_{vdw} = van der Waals radii, $N\chi$ and $N\chi N$ connectivity indices.

properties (Type III), with different values for each combination of E and the various substrates P. An example of this is the correlation of rate constants for one reaction with those of another, or for one reaction measured in two different media.

The advantage of the substituent-oriented approach is that constants for a limited number of substituents can be combined to provide values of the descriptor variable for new, more complex substrate molecules. However, not all substances can be adequately represented as the sum of independent substituents. Correlations based on molecular properties are not limited by uncertainties over the additivity of substituent effects because values of their descriptor variables are determined on whole molecules, thereby incorporating the effects of interacting substituents.

Functional Relationships

In general terms, all correlations involve a functional relationship between a response variable, represented by the vector **Y**, and one or several descriptor variables, denoted by the vector **X**. The relationship can be summarized as

$$\mathbf{Y} = \mathbf{X}\beta + \epsilon \qquad (7.3)$$

where the vector β represents fitted coefficients, and the residual term, ϵ, denotes all of the variability in **Y** that is not described by the model, including missing terms as well as random experimental error. Although correlations of chemical properties come in many forms, the vast majority are simple univariate, linear relationships. These can be described by

$$y_i = \beta_0 + \beta_1 x_i + \epsilon_i \qquad (7.4)$$

where β_0 and β_1 are fitted coefficients, x_i is a single descriptor variable, and ϵ_i is unexplained variability in y_i. The subscript i identifies individual members of a closely related series of substances from which the correlation is derived. The correlation model described by Eq. (7.4) is the same as the simple linear regression model found in standard texts on statistics (Draper and Smith, 1981), and a great deal of insight into the proper formulation of correlation analyses can be found in these sources. Multivariate linear models are often proposed, but the burden of defending the physical and statistical validity of the model goes up greatly with each additional descriptor variable. As a result, established bivariate models are encountered only occasionally, and trivariate models very rarely.

Many linear correlation models involve log-transformed response and/or descriptor variables. In most cases, these formulations were developed empirically by visual inspection of the various graphing options, and theoretical justifications were developed only later. Currently, linear regression on log-transformed data is widely used in correlation analysis despite the many statistical advantages of applying nonlinear regression to the untransformed data (de Levie, 1986). This is, in part, because the various σ scales are *defined* by the semilogarithmic Hammett equation (eq. 7.5), and therefore the use of nonlinear regression techniques would be inconsistent with prior practice.

In correlation analysis, data that do not provide a linear correlation to a semilogarithmic model often indicate an important effect that has been neglected, and these deviations can provide useful mechanistic information when interpreted in this way (Schreck, 1971). There are, however, a few fundamentally nonlinear models on which some correlations are based. The Marcus equation [Eq. (7.13), below] is an example of such a model. It was derived from theory and has a roughly parabolic form. Fitting data to the model requires the use of nonlinear regression methods, of which the Levenberg–Marquardt algorithm is the most frequently employed (Press et al., 1986). In practice, the biggest uncertainty in using iterative nonlinear regression procedures in correlation analysis is that of reaching anomalous local minima, which are determined by the choice of initial parameter values. Fortunately, the fitting parameters in theoretically derived correlations usually have physical interpretations that can be used to judge the reasonableness of the regression results.

FITNESS OF A CORRELATION

Statistical Criteria

The fitness of a correlation depends on statistical and mechanistic criteria. Therefore, it is important that results of each regression calculation are reported with appropriate statistical parameters for judging the goodness of fit between the data and the model. One parameter that should always be included is a measure of the uncertainty in the fitted coefficients. These values are usually reported in parentheses as $\pm$ one standard deviation [e.g., see Eqs. (7.6) and (7.7)]. If the uncertainty is large relative to the value of a fitted coefficient, a t-test should be performed to determine if the coefficient is significantly different from zero. This is particularly important for judging *ad hoc* multivariate models that often result from QSAR analyses of environmental data.

Judging the overall fitness of the regression equation can be done by a variety of criteria, and there is no consensus that any one parameter is preferred (Shorter, 1982).

Most authors report only a few common parameters, but include just enough informa-tion that other criteria can be derived. The reported values usually consist of the stan-dard deviation about the regression, s; the correlation coefficient, r; and the number of points included in the regression, n. The value of s reflects the scatter of the data about the fitted line or, in other words, the precision of the regression equation. It is directly useful in judging possible outliers, and it forms the basis for many other goodness of fit parameters. One of these is r, which should approach unity for data that fit well to a linear model. There are, however, many pitfalls in the overreliance on r in judging the results of correlation analysis (Davis and Pryor, 1976; Shorter, 1982; Tiley, 1985). Although various modifications and alternatives to r have been proposed, the most re-liable goodness of fit criterion is still inspection of the data in graphical form.

The basic statistical parameters for evaluating multivariate correlations are analo-gous to those described above for univariate models. These include the standard devi-ations about each fitted parameter, the standard deviation for the regression equation as a whole, and the multiple correlation coefficient, R. Again, there is debate over which statistical parameters are most appropriate for validating the form of a model (Shorter, 1982). Evaluation of multivariate correlations is further complicated by the variety of regression methods that may be used (Draper and Smith, 1981) and the lack of an accepted graphical method of evaluating the results (Thioulouse et al., 1991).

Mechanistic Criteria

The fitness of a correlation that has a form derived from theory is primarily determined by the adequacy of the existing theory. On the other hand, most QSAR analyses are based on an *ad hoc* mixture of statistical and mechanistic considerations and therefore should be judged on both types of criteria. Various multivariate statistical techniques can be used to determine the optimum combination of descriptor variables, and many of these techniques are being used in the development of QSARs (Shorter, 1982). How-ever, the reported result should represent a parameterization that covers all of the im-portant factors influencing reactivity without significant overlap between the descrip-tor variables. The use of overlapping, or partially redundant, descriptor variables is one of the most common pitfalls in multivariate QSAR analysis. The statistical manifesta-tion of mechanistic redundancy in the descriptor variables is that the variables are co-variant with one another, a condition known as *multicollinearity* (Draper and Smith, 1981). Figure 7.1 shows that there are good correlations *between* many of the descriptor variables commonly used to describe chemical reactivity, and thus they exhibit a high degree of multicollinearity. Additional evidence for multicollinearity among common descriptors of chemical reactivity can be found in the literature (Simpson et al., 1965; Grüber and Buß, 1989). Clearly, a multivariate correlation including descriptor vari-ables that are covariant with one another is of dubious validity.

EXAMPLES WITH ENVIRONMENTAL APPLICATION

Hammett Equation and its Extensions

By the mid-1930s, many studies had reported linear correlations between various properties of substituted benzenes, which led to the general method of correlation analy-

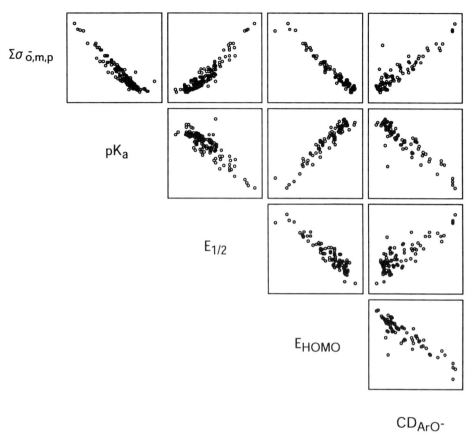

Figure 7.1. Scatter plot matrix of common molecular descriptors for substituted phenols. The data include the sum of Hammett substituent constants for all ring positions, $\Sigma\sigma_{o,m,p}^-$; pK_a; half-wave oxidation potential, $E_{1/2}$; energy of the highest occupied molecular orbitals, E_{HOMO}; and the calculated charge density at the oxygen of the phenoxide moiety, CD_{ArO^-}.

sis proposed by Hammett (1937). The approach used data for a reference reaction to define values of a parameter, σ, for each substituent on the aromatic center. Values of σ proved to be sufficiently characteristic of individual substituents and independent of the reference reaction to be useful in correlating other properties and other reaction types. The continued application of the approach taken by Hammett has led to many refinements, which have been discussed thoroughly by others (Jaffé, 1953; Shorter, 1982; Exner, 1988). This review emphasizes only those points with the most general relevance to correlation analysis of environmental applications.

The Hammett equation, as it has been known, is a univariate model based on the observed linear correlation between the log of a response variable and the descriptor variable σ. The response variable is usually the rate constant k_R, or the equilibrium constant K_R, for a series of substituted substrates. For k_R, it can be written in a form consistent with Eq. (7.4)

$$\log k_R = \rho\sigma + \log k_0 \qquad (7.5)$$

where the fitting parameters are ρ, which reflects the sensitivity of the reaction rate to substituent effects, and log k_0, which reflects the rate constant for a hypothetical substrate that exhibits no effect of substituents. The interpretation of k_0 is a common source of confusion and is discussed further in the section on interpreting intercepts.

Sigma constant scales

Values of the various substituent constants, σ, are defined using experimental data obtained from a reference reaction (Jaffé, 1953; Exner, 1988). A variety of σ scales are available, reflecting differences in the way substituents exert their influence on the reference reaction. Hammett's original scale was defined using the dissociation of benzoic acids in water at 25°C as the reference reaction. By assigning $\rho = 1$ for this reaction, and $\sigma = 0$ for benzoic acid, values of σ_m and σ_p could be obtained for substituents in the meta and para positions, respectively. This σ scale reflects the net effect of substituents on the electron density of the aromatic ring: A positive value indicates a net electron-withdrawing effect, and a negative value indicates an electron-donating effect.

Formally, the net electronic effect of a substituent can be divided into inductive and resonance interactions. Inductive effects involve delocalization of electrons that is conducted through σ bonds, and resonance effects are communicated through π bonds. Various σ scales have been developed that distinguish these two effects (e.g., σ^o, σ_I, and σ_R), but their proper use implies a level of analysis that is not often sought in environmental contexts, so they are considered outside the scope of this review. There are environmentally relevant reaction types, however, which exhibit substituent effects that are different from those of the reference reaction on which the original σ scale was developed. The major examples of this are due to through-resonance, whereby certain substituents stabilize charge in the parent molecule by a resonance interaction that results in charge delocalization. Through-resonance has been included in the Hammett approach by defining special σ scales using alternative reference reactions. Substituents that donate electron density by through-resonance (e.g., alkyl, methoxy, and halogen groups) are described by the σ^+ scale, whereas through-resonance accepting substituents (e.g., nitro, acetyl, and cyano groups) are described by the σ^- scale. The latter effect is particularly pronounced for phenoxide anions, and the reference reaction on which the σ^- scale is defined is acid dissociation of substituted phenols in water. Therefore, it is not surprising that σ^- values give the best Hammett correlation for oxidation rate constants of phenoxide anions (Figures 7.2 and 7.3).

The choice of σ, σ^-, or σ^+ scales should always be verified by comparing the resulting correlations, where possible, in graphical form. Unfortunately, there is variation among reported σ values even within a particular scale. These inconsistencies are usually due to variety in the choice of reference reaction or in experimental conditions such as solvent, ionic strength, or temperature. Several investigators have compiled reported σ values and attempted to identify "best" values for each σ scale (Wold and Sjöström, 1972; Hansch et al., 1973, 1991; Exner, 1978; Hansch and Leo, 1979; Shorter, 1994). Although these exercises are not likely to affect the general result of a correlation analysis, σ values for certain substituents can be quite variable. For these substituents, the selection of σ values may account for the appearance of individual outliers and other anomalies. Among substrates of environmental interest, charged substituents—such as carboxylate or sulfonate—are most likely to exhibit this effect.

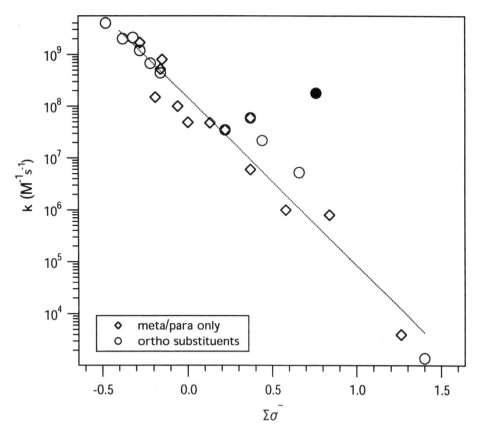

Figure 7.2. Correlation between the second-order rate constants for reaction of substituted phenoxide anions (k_{ArO^-}) with chlorine dioxide and the sum of σ^- substituent constants. The line corresponds to Eq. (7.6) and is for regression on all data except vanillin (●). (Original data from Tratnyek and Hoigné, 1994a.)

Additivity and proximity effects

An advantage of the substituent constant approach to correlation analysis is that σ values are often found to be additive for multiple substituents in meta or para positions (Shorter, 1982; Exner, 1988). Therefore, it is possible to correlate a variety of di- and trisubstituted compounds by calculating net substituent effects from a limited set of σ values. Additivity of substituent constants is an important principle in environmental correlation analysis, because the substances of interest are often heavily substituted compared to the simple model compounds traditionally used in LFER development.

In practice, additivity of substituent constants occasionally extends to ortho substituents, which then can be correlated by using the corresponding values of σ_p. This is nicely exemplified by the extensive data set for oxidation of substituted phenoxide anions by ClO$_2$ (Hoigné and Bader, 1993; Tratnyek and Hoigné, 1994a). The sum of σ^- values is adequate to correlate second-order reaction rate constants, k_{ArO^-}, for a wide range of substituent patterns (Figure 7.2). The only outlier in this case is the phenoxide of vanillin (4-hydroxy-3-methoxybenzaldehyde), which behaves anomolously because of its high degree of intramolecular hydrogen bonding (Sadekov et al., 1970). Linear regression on all other data gives a satisfactory Hammett correlation:

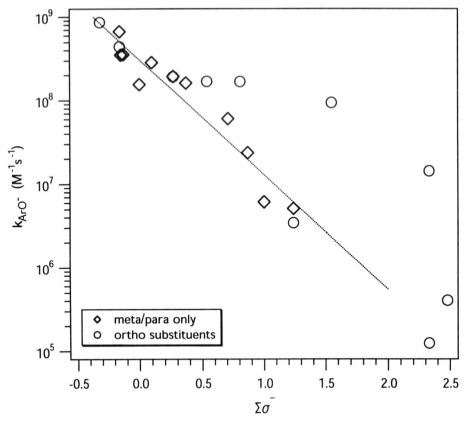

Figure 7.3. Correlation between the second-order rate constants for reaction of substituted phenoxide anions (k_{ArO^-}) with singlet oxygen and the sum of σ^- substituent constants. The line corresponds to Eq. (7.7) and reflects only phenoxides without ortho substituents. (Original data from Tratnyek and Hoigné, 1991.)

$$\log k_{ArO^-} = 8.2(\pm 0.2) - 3.2(\pm 0.4)\Sigma\sigma^-_{o,m,p},$$

$$n = 23 \qquad s = 0.39 \qquad r = 0.97 \tag{7.6}$$

Results of this calculation have been presented in the conventional form for QSAR analyses, with the uncertainties in the fitted parameters given as standard deviations in parentheses, and other criteria for judging the goodness of fit in the line below. Shown here are the number of points included in the regression, n; the standard deviation about the regression line, s; and the correlation coefficient, r. The line determined by Eq. (7.6) has been superimposed on the data in Figure 7.2.

Simple correlations like the one in Figure 7.2 are very appealing, but it is important to recognize that substituent constants will not be additive if there are interaction effects. Ortho substituents often exhibit interaction effects due to their proximity to the reaction center, and therefore their effect tends to be case-specific and not easily described by correlation analysis. The data in Figure 7.3 illustrate this situation for oxidation of phenoxide anions by the first electronically excited state of molecular oxygen, singlet oxygen, or $^1O_2$ (Tratnyek and Hoigné, 1991). In this case, it is apparent that additivity of σ constants does not extend to substituents in the ortho position, even though meta- and para-substituted phenoxides correlate well to the Hammett equation.

$$\log k_{\text{ArO}-} = 8.5(\pm 0.1) - 1.4(\pm 0.1) \Sigma \sigma^-_{m,p}$$

$$n = 11 \qquad s = 0.22 \qquad r = 0.96 \tag{7.7}$$

Comparison of the $^1O_2$ case with the correlation of ClO_2 reaction rates discussed above suggests that the two reactions occur by different mechanisms. The observation that ortho-substituted phenoxides fit a simple Hammett correlation for reaction with ClO_2 is consistent with evidence that this reaction involves an electron transfer mechanism (Tratnyek and Hoigné, 1994a). In contrast, $^1O_2$ reacts with phenoxides, in part, by cycloaddition (Tratnyek and Hoigné, 1994b). The encounter between reactants that is necessary for cycloaddition is more likely to be affected by ortho substituents, which is consistent with the observed deviations from the correlation line.

Multiparameter extensions of the Hammett equation

A great deal of research has been devoted to overcoming the limitations that the "ortho effect" imposes on the application of Eq. (7.5) (e.g., Charton, 1971; Fujita and Nishioka, 1976). Unfortunately, there does not seem to be any prospect for a generally applicable σ scale for substituents in the ortho position. One alternative is to allow multivariate models that include additional terms that take into account specific interaction effects. The approach of Fujita and Nishioka (1976) correlates ortho-substituted substrates with an extended form of the Hammett equation that contains two additional terms to correct for polar and steric proximity effects. In the environmental literature, this approach has been used to accommodate the ortho effect on the oxidation of substituted phenols by $^1O_2$ (Tratnyek and Hoigné, 1991) and chromate (Elovitz and Fish, 1994).

The Fujita–Nishioka treatment is one of many multiparameter extensions of the original Hammett correlation (Shorter, 1978). Despite their potential effectiveness at correlating experimental data, few of these models will ever be widely used in environmental applications, mainly due to the burden of defending the statistical and physical significance of multivariate correlations. One exception to this is the Taft equation (Taft, 1956), which was developed to quantitatively separate inductive and resonance contributions to the substituent effects described by Hammett's σ scale. The separation can be achieved using Taft's σ^* scale. This measure of substituent effects was defined using aliphatic substrates, which lack multiple bonds separating substituents from the reaction center so they cannot be subject to resonance effects. The resulting model has two terms: the first for inductive effects includes σ^* and a fitting parameter ρ^*, and the second for steric effects, with the steric effect constants E_s and a fitting parameter δ.

$$\log k/k_0 = \rho^* \sigma^* + \delta E_s \tag{7.8}$$

The Taft equation and its associated σ scales form the preferred basis for correlating the environmental reactivity of aliphatic compounds such as halogenated methanes and ethanes (Roberts et al., 1991).

Interpreting rho values

For Hammett correlations of rate constants, the slope of the fitted line gives the reaction constant ρ, which can provide insight into the reaction mechanism and its rate-limiting factors. Positive values of ρ result from reactions that are facilitated by decreased electron density at the reaction center, whereas negative ρ values indicate that the reaction is favored by substituents that increase electron density. Nucleophilic substitution is a common example of the former, and electrophilic substitution mechanisms result in the latter. For reactions involving free radicals, the sign of ρ is usually negative, but exceptions to this are relatively common. Absolute magnitudes of ρ range from 0 to about 4, with larger values indicating a greater sensitivity to substituent effects. The most significant factor in determining the magnitude of ρ is the distance of separation between the substituent and the reaction center of substrate molecules, but other factors such as solvent and temperature also contribute (Jaffé, 1953; Shorter, 1973; Fuchs and Lewis, 1974; Exner, 1988).

In the two examples of phenoxide oxidation discussed above, the fitted lines [Eqs. (7.6) and (7.7)] give reaction constants $\rho = -3.2 \pm 0.4$ for ClO_2 and -1.4 (± 0.1) for $^1O_2$. The signs of these fitted parameters are negative as expected for oxidations, where the reaction is favored by increased electron density at the reaction center. The magnitude of ρ for oxidation by ClO_2 indicates a strong substituent effect, as expected for a reaction where electron transfer is rate limiting (Tratnyek and Hoigné, 1994a). For reaction of phenoxides with $^1O_2$, the modest magnitude of ρ suggests a rate-limiting step that is relatively insensitive to substituent effects, perhaps formation of the precursor complex that precedes oxidation (Tratnyek and Hoigné, 1991).

Interpreting intercepts

In most respects, it is preferable to obtain Hammett correlations by fitting log k to σ and allowing both slope and intercept to be fitting parameters, as implied by Eq. (7.5). This is in contrast to the alternative method of performing regression on response variable data normalized to the measured value for the parent (unsubstituted) substrate, i.e. log k/k_0 (Shorter, 1982). In the former case, log k_0 is obtained as a fitting parameter and can then be compared to the experimental value. Usually, the two values are not significantly different. For example, in Eq. (7.6) and (7.7), the calculated intercepts suggest values of k_0 that are not statistically different from the measured values of k_{ArO^-} for phenol [(4.9 $\pm$ 0.5) $\times$ 10^8 $M^{-1}s^{-1}$ for ClO_2 (Tratnyek and Hoigné, 1994a) and (1.5 $\pm$ 0.1) $\times$ 10^8 $M^{-1}s^{-1}$ for $^1O_2$ (Tratnyek and Hoigné, 1991)]. When there is a significant difference between k_0 from regression and k observed for the unsubstituted parent molecule, it indicates that replacement of hydrogen on the parent molecule with any other substituent, has an effect on the reaction rate.

Brønsted Equation and Correlations to pK_A

An alternative to correlations based on substituent constants is to use a measurable property of whole molecules as the independent (descriptor) variable (Table 7.1, Type II). The Brønsted relationship is of this type (Brønsted 1928; Bell 1978). It was first developed for the base-catalyzed decomposition of nitramide, but the treatment has proven to be effective at describing a wide range of reactions (Duboc 1978). For a reaction subject to general acid catalysis, the Brønsted relationship can be written

$$\log k = \alpha \log K_A + C \tag{7.9}$$

where K_A represents the acid-dissociation constants for a series of catalysts, and α and C are fitting parameters. As with the Hammett relationship, the Brønsted equation was developed empirically, but has been interpreted theoretically by many subsequent investigators (e.g., Kresge, 1973; Pross, 1984).

In environmental applications, there has been little use of the Brønsted correlation in its original form because general acid or base catalysis is normally not significant in environmental systems (Perdue and Wolfe, 1982). Instead, Eq. (7.9) provides a basis for the various empirical linear correlations between $\log k$ and pK_a that are commonly reported in the environmental literature. Examples include the hydrolysis of organophosphate esters, organophosphorothionate esters, and carbamates, all of which have phenolic or naphtholic leaving groups (Wolfe et al., 1978; Wolfe, 1980). The hydroxyl moiety of each leaving group imparts a pK_a, which has proven to be an excellent descriptor of experimentally determined second-order alkaline hydrolysis rate constants. Correlations of this type have been interpreted as evidence that the reaction occurs by an elimination mechanism (Williams, 1973a,b).

Another example of a Brønsted-type correlation is provided by a series of studies on the microbial oxidation of substituted phenols and catechols (Paris et al., 1982, 1983). Formally, "second-order" rate constants, k_b, were obtained for these reactions by dividing pseudo-first-order disappearance rate constants by measured cell densities. One set of data was obtained using a pure culture of *Pseudomonas putida* (Paris et al., 1982). The reported values of $\log k_b$ for this reaction were shown to correlate with pK_a only after excluding *p*-methoxyphenol and *p*-cyanophenol as outliers. After excluding these outliers, regression on the remaining data gives a statistically satisfactory fit to a model in the form of the Brønsted-type equation.

$$\log k_b = -17.2(\pm 0.4) + 0.58(\pm 0.05)pK_a,$$

$$n = 8 \qquad s = 0.13 \qquad r = 0.97 \tag{7.10}$$

Presumably, this correlation is successful because both k_b and pK_a are similarly influenced by electron density of the aromatic ring, which, in turn, is determined by the effects of substituents. [A similar correlation for these k_b values was obtained to the shift in IR stretching frequencies for the phenolic O−H bonds (Collette, 1992).] That pK_a alone appears to be a satisfactory descriptor variable can be attributed to the relative simplicity of the pure culture systems. In contrast, a follow-up study using five natural water samples as media (Paris et al., 1983) did not find good correlations between k_b and pK_a.

Despite their appeal, correlations with pK_a as a descriptor variable often are not among the most satisfactory. This is exemplified by the data for phenoxide oxidation rates by ClO_2 and $^1O_2$ discussed above. The correlations to σ constants (Figures 7.2 and 7.3) are much better than to values of pK_a (not shown), which make the pK_a correlations appear unsatisfactory by comparison. Figure 7.1 shows that pK_a is substantially covariant with many of the other common descriptor variables for substituted phenols. Therefore, it appears that relatively modest differences in descriptor variables can account for large changes in the fitness of a QSAR.

Correlations to Redox Potential

Molecular properties commonly used as descriptor variables in correlation analysis include various measures of substrate redox potential. There are many possibilities in this case, including one-electron reduction potentials for single-electron transfers, even-electron potentials for equilibria between starting materials and stable products, polarographic electrode potentials, and electron affinities or ionization potentials. Ideally, the parameter values should apply to conditions similar to those under which the data to be correlated were obtained. Unfortunately, most organic redox reactions do not produce reversible half-reactions from which equilibrium Nernstian potentials can be determined. Instead, potentials are either (i) estimated from other thermodynamic data using the Born–Haber approach (Curtis, 1991) or (ii) approximated from measurements using polarographic or spectroscopic techniques (Eberson, 1987).

The one-electron reduction potential, E', is often the most appropriate descriptor variable for correlating rates of reactions that involve electron transfer, since it is usually a single electron-transfer step that is rate limiting (Eberson, 1987). For a narrow range of potentials, the expected linear relationship can be expressed as

$$\log k = (\alpha / 0.059)E' + \beta \tag{7.11}$$

for 25°C and E' in volts (Tratnyek et al., 1991). In this form, the slope of the observed correlation, α, is a unitless measure of the sensitivity of $\log k$ to changes in reduction potential of the substrate. This approach has been used in the correlation of reduction rates of nitroaromatic compounds by natural organic matter (Dunnivant et al., 1992) and the phenoxides of hydroquinones selected to model natural organic matter (Schwarzenbach et al., 1990). These correlations have been interpreted mechanistically by analogy to Marcus theory. However, it is important to note that linear correlations to E' are not, alone, compelling evidence for an electron transfer mechanism. The major limitation to practical use of Eq. (7.11) is the unavailability of E' values, especially for compounds of environmental interest.

Half-wave potentials, $E_{1/2}$, are an example of a descriptor variable that is directly measurable, in this case by polarography (Zuman, 1967). Values of $E_{1/2}$ are generally available because they have been widely tabulated (Meites and Zuman, 1979), and some $E_{1/2}$ data have been correlated to give QSARs that can be used to estimate unreported values (Suatoni et al., 1961; Jovanovic et al., 1991). Unfortunately, most polarography is done in nonaqueous solutions, and under conditions that are not otherwise environmentally representative. To the extent that the various methods of determining values of $E_{1/2}$ can be used to obtain a self-consistent set of molecular properties, however, they may be quite suitable for correlation analysis. Once again, the correlations are often linear, and numerous examples have been reported in the chemical literature.

One set of $E_{1/2}$ values that has proven to be particularly useful in correlating rates of environmentally significant reactions is that of Suatoni et al. (1961). They reported values measured by anodic voltametry for 41 phenols and 32 anilines. This data set has been used frequently in the recent environmental literature to correlate various types of biological response (Kaiser et al., 1984) and reaction rate constants (Stone, 1987; Laha and Luthy, 1990; Tratnyek and Hoigné, 1991; Tratnyek et al., 1991; Elovitz and Fish, 1994; Tratnyek and Hoigné, 1994a,b). As an example, the rate constants for reductive dissolution of manganese oxides by oxidation of phenols correlate linearly with Suatoni's $E_{1/2}$ values for the phenols (Figure 7.4). Regression on all of the data shown gives the following QSAR:

$$\log k = 6.1(\pm 0.9) - 9.7(\pm 1.4)E_{1/2},$$

$$n = 10 \quad s = 0.41 \quad r = 0.92 \qquad (7.12)$$

where the values of k, in $M^{-1}s^{-1}$, are derived from data on Mn(III/IV) oxides (Stone, 1987). In a subsequent study, rate constants for aniline oxidation were measured and found to fall on roughly the same line as the phenols (Laha and Luthy, 1990).

Suatoni et al. (1961) also showed that the effects of monosubstitutions could be used to estimate $E_{1/2}$ values for disubstituted phenols, and that the measured values correlate to the Hammett equation, allowing estimation of additional substituent effects from their γ constants. Several recent studies have made use of both measured and estimated values of $E_{1/2}$ to correlate oxidation rate constants for substituted phenols. These include the reactions with ClO_2 (Tratnyek and Hoigné, 1994a) and $^1O_2$ (Tratnyek and Hoigné, 1991; Tratnyek et al. 1994a) discussed above, and the reduction of

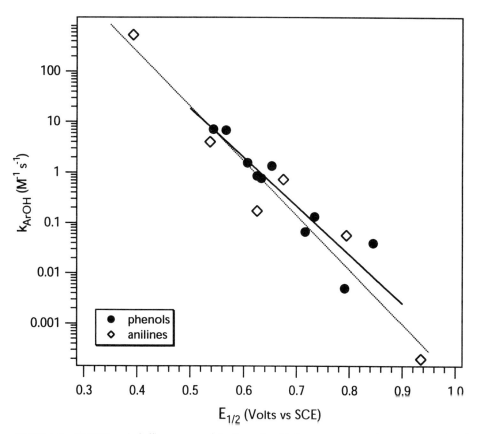

Figure 7.4. Correlation to half-wave potentials, $E_{1/2}$, of oxidation rate constants for substituted phenol and anilines by manganese oxides. The solid regression line ([Eq. (7.12)] and the solid points are for phenols at pH 4.4 (Stone, 1987). Open diamonds show that data for anilines follow a very similar correlation (dotted line) (Laha and Luthy, 1990).

chromate (Elovitz and Fish, 1994). The success of such nested correlation analyses indicates remarkable robustness, but special care must be taken to evaluate them critically as the assumptions and conditions under which they were developed become increasingly remote from the conditions to which they are being applied.

Correlations to ΔG of Reaction and the Marcus Theory

The two general types of correlations described above—correlations to substituent constants and to molecular properties—rely on descriptor variables that reflect only the homologous series of reactants, P, and not the common reactant, E. In contrast, reaction properties (Table 7.1, Type III) reflect both P and E. One such reaction property is the free energy of reaction, ΔG. The basis for correlation between values of k and ΔG has been analyzed extensively and used to rationalize the success of LFERs and QSARs in general (Godfrey, 1978; Agmon, 1981; Hoffmann, 1981).

For reactions that proceed by an outer-sphere one-electron transfer mechanism, Marcus theory predicts a quadratic relationship between log k and ΔG (Marcus, 1975; Eberson, 1987; Bolton and Archer, 1991). Several environmentally oriented studies of electron transfer reactions involving metals have been interpreted using this approach (Hoffmann, 1981; Wehrli, 1990). For electron transfer reactions involving organic substrates, the Marcus relationship is written

$$k = \frac{k_d}{1 + \dfrac{k_d}{K_d Z} \exp\left\{\left[W + \dfrac{\lambda}{4}\left(1 + \dfrac{\Delta G^0}{\lambda}\right)^2\right] /RT\right\}} \tag{7.13}$$

where k and ΔG^0, are the response and descriptor variables, respectively (Eberson, 1982, 1985). The remaining terms in Eq. (7.13) are the diffusion-limited reaction rate constant, k_d; the equilibrium constant for precursor complex formation, K_d; the universal collision frequency factor, Z; a correction for electrostatic work, W; the ideal gas constant, R; absolute temperature, T; and the reorganization energy, λ. Although the fundamental implications of this relationship have been studied in detail, its practical applications are still emerging.

In an environmental context, Eq. (7.13) may be applied in several ways. Eberson has suggested that it may aid in screening families of compounds for toxicity (Eberson, 1985). The premise is that any xenobiotic substance with redox potential that is sufficiently high or low will undergo nonspecific electron transfer reactions with cell constituents, resulting in free radicals that cause cell damage. Application of electron transfer theory can help identify the combinations of reactants that are likely to produce toxicity by this mechanism. An illustration of this notion can be found in the formation of radical oxidation products from phenols and amines by reaction with horseradish peroxidase. In this case, data from Job and Dunford (1976) were found to correlate with Eq. (7.13), suggesting that the compounds studied were sufficiently reducing to undergo electron transfer with the enzyme. Other examples have appeared subsequently (Macdonald et al., 1989; Netto et al., 1991).

To apply Eq. (7.13), values of ΔG must be available for conditions similar to those under which values of k are measured. These free energies may be estimated from the one-electron redox potentials for the two half-reactions that make up the rate-limiting step, if the necessary potentials are available. Again, the situation for phenoxide oxidation by ClO_2 and $^1O_2$ is typical. Although E_{ClO_2/ClO_2^-} and $E_{^1O_2/O_2^-}$ are readily available, $E_{ArO^-/ArO\cdot}$ are difficult to obtain because the latter couples are not usually reversible. A solution is to approximate values of $E_{ArO^-/ArO\cdot}$ from half-wave potentials, $E_{1/2}$. The relationship between polarographic potentials and Nernstian redox potentials is complex in theory and complicated by various experimental uncertainties, so their use in place of rigorously defined thermodynamic values must be made with these limitations in mind. However, the approach can be highly successful, as illustrated for the phenoxide oxidation by ClO_2 (Figure 7.5). Nonlinear regression of the data to equation 7.13 gives a fit that is statistically satisfactory, and the fitted value $\lambda = 30.1 \pm 0.6$ kcal/mole is consistent with independent determinations of the reorganization energies λ_{ClO_2/ClO_2^-} and $\lambda_{ArO^-/ArO\cdot}$ (Tratnyek and Hoigné, 1994a). This result is strong evidence that the mechanism of reaction involves rate-limiting single-electron transfer, and it is consistent with the mechanistic interpretation of Figure 7.2 given above. In addition, the fitted equation provides a nonlinear theoretically based QSAR that exhibits no outliers and is valid over almost seven orders of magnitude in k_{ArO^-}.

In contrast to the situation for chlorine dioxide, the data for phenoxide oxidation with $^1O_2$ (recall Figure 7.3) do not conform to the Marcus theory. These data have been included on Figure 7.5 but do not fit Eq. (7.13). This result suggests that processes

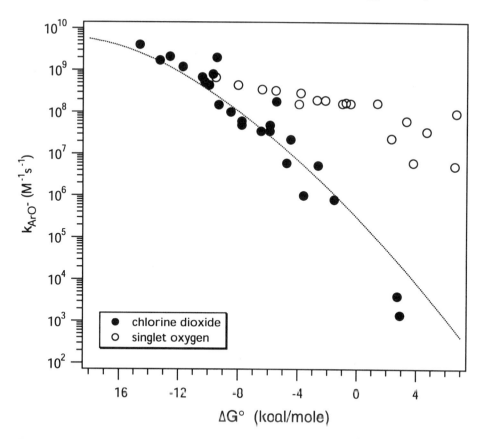

Figure 7.5. Marcus correlation for phenoxide oxidation by ClO_2 (Tratnyek and Hoigné, 1994a) and $^1O_2$ (Tratnyek and Hoigné, 1991; Tratnyek and Hoigné, 1994b). The line is from Eq. (7.13) fitted to data for ClO_2.

such as precursor complex formation (Tratnyek and Hoigné, 1991) or cycloaddition (Tratnyek and Hoigné, 1994b) may be rate limiting instead of electron transfer.

Correlations between Rate Constants

The focus of this review has been primarily on correlating reaction rate constants as the response variable, but values of k can also serve as descriptor variables in correlation analysis. Unlike correlations to ΔG, correlations between values of k are usually developed on an *ad hoc* basis and are evaluated on statistical criteria rather than mechanistic criteria. For example, values of k_{ArO^-} and k_{ArOH} for phenol oxidation by ClO_2, $^1O_2$, or O_3 do exhibit similar trends due to substituents (Figure 7.6). However, these trends are not easily interpreted. Clearly, there are outliers, mostly due to intramolecular H-bonding effects, and the correlations are noisy and nonlinear. A case where better correlations are obtained is illustrated by the relationships between aqueous- and gas-phase rate constants for the reaction of organic substrates with hydroxyl radical (Haag and Yao, 1992). In this case, linear correlations are obtained for some

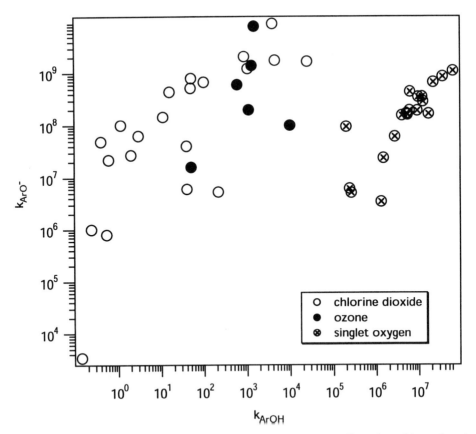

Figure 7.6. Scatter plot correlating measured values of k_{ArO^-} to corresponding values of k_{ArOH} for oxidation of substituted phenols by various oxidants. ClO_2 data from Tratnyek and Hoigné (1994a), $^1O_2$ data from Tratnyek and Hoigné (1991); Tratnyek and Hoigné (1994b), and O_3 data from Hoigné and Bader (1983).

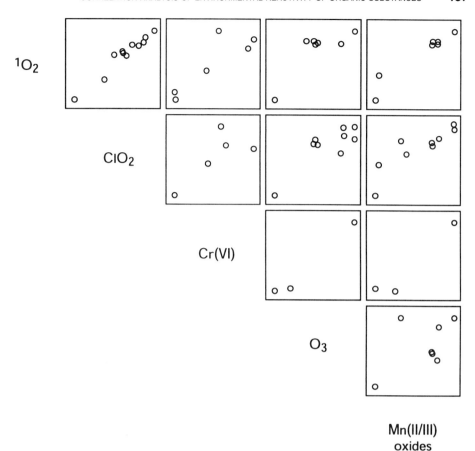

Figure 7.7. Scatter plot correlation matrix of log second-order rate constants for phenol oxidation. Data for k_{ArO^-} are included for $^1O_2$ (Tratnyek and Hoigné, 1991; Tratnyek and Hoigné, 1994b) and ClO_2 (Tratnyek and Hoigné, 1994a), and values of k_{ArOH} are used for chromate (Elovitz and Fish, 1994), O_3 (Hoigné and Bader, 1983), and manganese oxides (Stone, 1987).

very diverse sets of substrates as long as the primary mechanism of reaction is constant; for example, reaction with hydroxyl radical occurs by either H-abstraction or addition to unsaturated sites.

For well-studied families of substrates, such as the substituted phenols, it is possible to correlate k's for reaction with one environmental factor, to k's for reaction with another factor. Although such correlations may have predictive value, their main benefit is in evaluating the quality and mechanistic implications of previously reported data. Figure 7.7 provides a correlation matrix of scatter plots based on five sets of phenol oxidation rate constants. Data that have been included are a mixture of values for log k_{ArO^-} and log k_{ArOH}, corresponding to the species for which the largest number of reliable data are available. Several combinations give satisfactory correlations, particularly the values of k_{ArO^-} for $^1O_2$ and ClO_2. Linear regression on these data gives

$$\log k(^1O_2) = 5.5 \ (\pm 0.2) - 0.36 \ (\pm 0.03) \log k(ClO_2),$$

$$n = 10 \qquad s = 0.16 \qquad r = 0.975 \tag{7.14}$$

where $k(^1O_2)$ represents k_{ArO^-} for oxidation by singlet oxygen, and $k(ClO_2)$ represents k_{ArO^-} for oxidation by chlorine dioxide. All of the phenoxides for which both rate constants are available are included in the correlation. The agreement implies that substituent effects are parallel for phenoxide oxidation by ClO_2 and $^1O_2$, even though Figure 7.5 indicates that only the rate of reaction with ClO_2 is controlled by outer-sphere electron transfer. This result is consistent with the possibility that the rate of oxidation by $^1O_2$ is limited by the precursor formation step.

APPLICATION OF QSARS

Correlation analysis for regularities in a series of rate constants serves at least three purposes: validation, prediction, and mechanistic interpretation. In practice, most QSARs serve multiple purposes, with each use benefiting the others. This is particularly true in applications such as environmental chemistry, where the overall quality of a correlation should be judged with all three purposes in mind.

Validation

Correlation analysis helps in judging the reliability of individual values because regular correlations between the properties of molecules are the norm, making small numbers of outliers easy to identify. For example, data presented in this review show that phenoxide anions often give smooth correlations to appropriate descriptors. The only common outliers are substrates that exhibit a high degree of intramolecular hydrogen bonding, although there are some descriptor variables that incorporate this effect. Rate data with relatively large uncertainties, such as the values of k_{ArOH} for $^1O_2$ and ClO_2, do not correlate as well.

Prediction

Fitting an equation to a good correlation yields a quantitative structure–activity relationship (QSAR) that can be used to estimate rate constants for additional substrates. Under the best circumstances, this will involve only interpolation using an equation developed for the system of interest. However, predictions are often made by modest extrapolation with respect to the descriptor variable or with respect to other characteristics of the system. Justification of these extrapolations requires substantial insight into the mechanistic basis for correlation.

Mechanism

Finally, correlations of rate data can often be interpreted in terms of reaction mechanisms. In general, simple correlations imply a single mechanism, similarities between correlations imply similarities in mechanism, and deviations from either of these rules imply differences in mechanism. The appropriateness of the mechanism implied by correlation analysis lends further support to the reliability of the experimental data and helps define the limits of safe extrapolation based on the resulting QSARs.

References

Agmon, N., 1981, From energy profiles to structure–reactivity correlations. *Int. J. Chem. Kinetics* 13:333–365.

Akland, A. H., and Waters, M. D., 1983, Chemical and toxicological data bases for assessment of structure–activity relationships. In: L. Goldberg, ed., *Structure–Activity Correlation as a Predictive Tool in Toxicology: Fundamentals, Methods, and Applications*, Vol. 2, Hemisphere Publishing Corporation, Washington, D.C., pp. 23–48.

Bell, R. P., 1978, The Brønsted equation—its first half-century. In: N. B. Chapman and J. Shorter, eds., *Correlation Analysis in Chemistry: Recent Advances*, Plenum, New York, pp. 55–84.

Bolton, J. R., and Archer, M. D., 1991, Basic electron-transfer theory. In: J. R. Bolton, N. Mataga, and G. McLendon, eds., *Electron Transfer in Inorganic, Organic, and Biological Systems*, Advances in Chemistry Series, No. 228, American Chemical Society, Washington, D.C., pp. 7–23.

Brezonik, P. L. ,1990, Principles of linear free-energy and structure–activity relationships and their applications to the fate of chemicals in aquatic systems. In: W. Stumm, ed., *Aquatic Chemical Kinetics: Reaction Rates of Process in Natural Waters*, Wiley-Interscience, New York, pp. 113–143.

Brezonik, P. L., 1994, Prediction methods for reaction rates and compound reactivity. In: *Chemical Kinetics and Process Dynamics in Aquatic Systems*, Lewis, Boca Raton, FL, pp. 553–641.

Brønsted, J. N., 1928, Acid and basic catalysis. *Chem. Rev.*, 5, 231–338.

Burns, L. A., Cline, D. M., and Lassiter, R. R. 1982, *Exposure Analysis Modeling System (EXAMS): User Manual and System Documentation*. Final Report. EPA 600/3-82-023, U.S. Environmental Protection Agency, Washington, D.C., 456 pp.

Chapman, N. B., and Shorter, J., eds., 1978, *Correlation Analysis in Chemistry: Recent Advances*, Plenum, New York, 546 pp.

Charton, M. ,1971, The quantitative treatment of the ortho effect. *Progr. Phys. Org. Chem.*, 8, 235–317.

Chignell, C. F., 1983, Overview of molecular parameters that relate to biological activity in toxicology. In: L. Goldberg, ed., *Structure–Activity Correlation as a Predictive Tool in Toxicology*, Hemisphere, Washington, D.C., pp. 61–74.

Collette, T. W., 1992, Infrared spectroscopy-based property–reactivity correlations for predicting environmental fate of organic chemicals. *Environ. Toxicol. Chem.*, 11(7), 981–991.

Curtis, G. P., 1991, *Reductive Dehalogenation of Hexachloroethane and Carbon Tetrachloride by Aquifer Sand and Humic Acid*, Ph.D. thesis, Stanford University, 228 pp.

Davis, W. H. J., and Pryor, W. A., 1976, Measures of goodness of fit in linear free energy relationships. *J. Chem. Educ.*, 53(5), 285–287.

de Levie, R , 1986, When, why, and how to use weighted least squares. *J. Chem. Educ.*, 63(1), 10–15.

Donaldson, W. T., 1992, The role of property–reactivity relationships in meeting the EPA's needs for environmental rate constants. *Environ. Toxicol. Chem.*, 11(7), 887–891.

Draper, N. R., and Smith, H., 1981, *Applied Regression Analysis*, John Wiley & Sons, New York, 709 pp.

Duboc, C., 1978, The correlation analysis of nucleophilicity. In: N. B. Chapman and J. Shorter, eds., *Correlation Analysis in Chemistry*, Plenum, New York, pp. 313–355.

Dunnivant, F. M., Schwarzenbach, R. P., and Macalady, D. L., 1992, Reduction of substituted nitrobenzenes in aqueous solutions containing natural organic matter. *Environ. Sci. Technol.*, 26(11), 2133–2141.

Eberson, L., 1982, Electron-transfer reactions in organic chemistry. *Adv. Phys. Org. Chem.*, 18, 79–185.

Eberson, L., 1985, The Marcus theory of electron transfer, a sorting device for toxic compounds. *Adv. Free Rad. Biol. Med.*, 1, 19–90.

Eberson, L., 1987, *Electron Transfer Reactions in Organic Chemistry*, Springer-Verlag, Berlin, 234 pp.

Elovitz, M. S., and Fish, W., 1994, Redox interactions of Cr(VI) and substituted phenols: kinetic investigation. *Environ. Sci. Technol.*, 28(12), 2161–2169.

Eriksson, L., Jonsson, J., Hellberg, S., Lindgren, F., Skagerberg, B., Sjöström, M., Wold, S., and Berglind, R., 1990, A strategy for ranking environmentally occurring chemicals. Part III. Multivariate quantitative structure–activity relationships for halogenated aliphatics. *Environ. Toxicol. Chem.*, 9(11), 1339–1351.

Eriksson, L., Verhaar, H. J. M., Sjöström, M., and Hermens, J. L. M., 1993, Multivariate characterization and modelling of the chemical reactivity of epoxides. Part II: Extension to di- and trisubstitution. *Quant. Struct.–Act. Relat.*, 12: 357–366.

Exner, O., 1978, A critical compilation of substituent constants. In N. B. Chapman and J. Shorter, eds., *Correlation Analysis in Chemistry: Recent Advances*. Plenum, New York, pp. 439–540.

Exner, O., 1988, *Correlation Analysis of Chemical Data*. Plenum, New York, 275 pp.

Franke, R., 1984, QSAR parameters. In: J. K. Seydel, ed., *QSAR and Strategies in the Design of Bioactive Compounds*, Verlag Chemie, Weinheim, pp. 59–78.

Fuchs, R., and Lewis, E. S., 1974, Linear free-energy relations. In: E. S. Lewis, ed., *Investigation of Rates and Mechanisms of Reactions*, Vol. VI, Part I, Wiley-Interscience, New York, pp. 777–824.

Fujita, T., and Nishioka, T., 1976, The analysis of the ortho effect. *Progr. Phys. Org. Chem.*, 12, 49–89.

Godfrey, M., 1978, Theoretical models for interpreting linear correlations in organic chemistry. In: N. B. Chapman and J. Shorter, eds., *Correlation Analysis in Chemistry: Recent Advances*, Plenum, New York, pp. 85–117.

Grüber, C. and V. Buß, 1989, Quantum-mechanically calculated properties for the development of quantitative structure-activity relationships (QSAR's). pK_a-values of phenols and aromatic and aliphatic carboxylic acids. *Chemosphere*, 19(10/11), 1595–1609.

Haag, W. R., and Yao, C. C. D., 1992, Rate constants for reaction of hydroxyl radicals with several drinking water contaminants. *Environ. Sci. Technol.*, 26(5), 1005–1013.

Hammett, L. P., 1937, The effect of structure upon the reactions of some organic compounds. Benzene derivatives. *J. Am. Chem. Soc.*, 59(1), 96–103.

Hansch, C., and Leo, A., 1979, *Substituent Constants for Correlation Analysis in Chemistry and Biology*, John Wiley & Sons, New York, 339 pp.

Hansch, C., and Leo, A., 1995, *Exploring QSAR: Fundamentals and Applications in Chemistry and Biology*, Vol. 1, American Chemical Society, Washington, D.C., 880 pp.

Hansch, C., Leo, A., and Taft. R. W., 1991, A survey of Hammett substituent constants and resonance field parameters. *Chem. Rev.*, 91(2), 165–195.

Hansch, C., Leo, A., Unger, S. H., Kim, K. H., Nikaitani, D., and Lien, E. J., 1973, "Aromatic" substituent constants for structure–activity correlations. *J. Med. Chem.*, 16(11), 1207–1216.

Hermens, J. L. M., 1986, Quantitative structure–activity relationships in aquatic toxicology. *Pesticide Sci.*, 17, 287–296.

Hermens, J. L. M., 1989, Quantitative structure–activity relationships of environmental pollutants. In: O. Hutzinger, ed., *Handbook of Environmental Chemistry*. Springer, Berlin, pp. 111–162.

Hoffmann, M. R., 1981, Thermodynamic, kinetic, and extrathermodynamic considerations in the development of equilibrium models for aquatic systems. *Environ. Sci. Technol.*, 15(3), 345–353.

Hoigné, J., 1990, Formulation and calibration of environmental reaction kinetics: oxidations by aqueous photooxidants as an example. In: W. Stumm, ed., *Aquatic Chemical Kinetics: Reaction Rates of Processes in Natural Waters*, Wiley-Interscience, New York, pp. 43–70.

Hoigné, J., and Bader, H., 1983, Rate constants of reactions of ozone with organic and inorganic compounds in water—II. *Water Res.*, 17(2), 185–194.

Hoigné, J., and Bader, H., 1993, Kinetics of reactions involving chlorine dioxide (OClO) in water. I. Inorganic and organic compounds. *Water Res.*, 28(1), 45–55.

Howard, P. H., Boethling, R. W., Jarvis, W. F., Meylan, W. M., and Michalenko, E. M., 1991, *Handbook of Environmental Degradation Rates*, Lewis, Chelsea, MI, 725 pp.

Jaffé, H. H., 1953, A reexamination of the Hammett equation. *Chem. Rev.*, 53, 191–261.

Job, D., and Dunford, H. B., 1976, Substituent effect on the oxidation of phenols and aromatic amines by horseradish peroxidase compound I. *Eur. J. Biochem.*, 66, 607–614.

Jovanovic, S. V., M., Tosic, and Simic, M.G., 1991, Use of the Hammett correlation and σ^+ for calculation of one-electron redox potentials of antioxidants. *J. Phys. Chem.*, 95(26), 10824–10827.

Kaiser, K. L. E., Dixon, D. G., and Hodson, P. V., 1984, QSAR studies on chlorophenols, chlorobenzenes and para-substituted phenols. In: K. L. E. Kaiser, ed., *QSAR in Environmental Toxicology*, Reidel, Dordrecht, pp. 189–205.

Kamlet, M. J., and Taft, R. W., 1985, Linear solvation energy relationships. Local empirical rules—or fundamental laws of chemistry? A reply to the chemometicians. *Acta Chem. Scand.*, B39(8), 611–628.

Karickhoff, S. W., McDaniel, V. K., Melton, C., Vellino, A. N., Nute, D. E., and Carreira, L. A., 1991, Predicting chemical reactivity by computer. *Environ. Toxicol. Chem.*, 10, 1405–1416.

Kier, L. B., and Hall, L. H., 1990, An electrotopological-state index for atoms in molecules. *Pharm. Res.*, 7(8), 801–807.

Kollig, H. P., and Kitchens, B. E., 1993, FATE, the environmental fate constants information database. *J. Chem. Inf. Comput. Sci.*, 32(1), 131–134.

Kresge, A. J., 1973, The Brønsted relation—recent developments. *Chem. Soc. Rev.*, 2(4), 475–503.

Laha, S. and Luthy, R. G., 1990, Oxidation of aniline and other primary aromatic amines by manganese dioxide. *Environ. Sci. Technol.*, 24(3), 363–373.

Larson, R. J., 1990, Structure–activity relationships for biodegradation of linear alkylbenzenesulfonates. *Environ. Sci. Technol.*, 24(8), 1241–1246.

Leffler, J. E., and Grunwald, E., 1963, *Rates and Equilibria of Organic Reactions*, Dover, New York, 458 pp.

Leo, A., Hansch, C., and Hoekman, D., 1995, *Exploring QSAR: Hydrophobic, Electronic, and Steric Constants*, Vol. 2, American Chemical Society, Washington, D.C., 880 pp.

Lyman, W. J., Reehl, W. F., and Rosenblatt, D. H., eds., 1982, *Handbook of Chemical Property Estimation Methods*, McGraw-Hill, New York.

Macdonald, T. L., Gutheim, W. G., Martin, R. B., and Guengerich, F. P., 1989, Oxidation of substituted *N,N*-dimethylanilines by cytochrome P-450—estimation of the effective oxidation potential of cytochrome P-450. *Biochemistry*, 28(5), 2071–2077.

Marcus, R. A., 1975, Electron transfer in homogeneous and heterogeneous systems. In: *The Nature of Seawater*, Dahlem Konferenzen, Berlin, pp. 477–503.

Meites, L., and Zuman, P., 1979, *CRC Handbook Series in Organic Electrochemistry*, five volumes, Chemical Rubber Company, Cleveland, OH.

Mill, T., 1979, *Structure Reactivity Correlations for Environmental Reactions*, EPA 560/11-79-103, U.S. Environmental Protection Agency, Washington, D.C.

Netto, L., Ferreira, A. M. D., and Augusto, O., 1991, Iron(III) binding in DNA solutions—complex-formation and catalytic activity in the oxidation of hydrazine derivatives. *Chemico-Biological Interactions*, 79(1), 1–14.

Nirmalakhandan, N., and Speece, R.E., 1988, Structure–activity relationships. *Environ. Sci. Technol.*, 22(6), 606–615.

Paris, D. F., Wolfe, N. L., and Steen, W. C., 1982, Structure–activity relationships in microbial transformation of phenols. *Appl. Environ. Microbiol.*, 44(1), 153–158.

Paris, D. F., Wolfe, N. L., Steen, W. C., and Baughman, G. L., 1983, Effect of phenol molecular structure on bacterial transformation rate constants in pond and river samples. *Appl. Environ. Microbiol.*, 45(3), 1153–1155.

Parsons, J. R., and Govers, H. A. J., 1990, Quantitative structure–activity relationships for biodegradation. *Ecotox. Environ. Saf.*, 19, 212–227.

Perdue, E. M., and Wolfe, N. L., 1982, Modification of pollutant hydrolysis kinetics in the presence of humic substances. *Environ. Sci. Technol.*, 16(12), 847–852.

Press, W. H., Flannery, B. P., Teukolsky, S. A., and Vetterling, W. T., 1986, *Numerical Recipes. The Art of Scientific Computing.* Cambridge University Press, New York, 818 pp.

Pross, A., 1984, Relationship between rates and equilibria and the mechanistic significance of the Brønsted parameter. A qualitative valence-bond approach. *J. Org. Chem.*, 49(10), 1811–1818.

Roberts, A. L., Jeffers, N. L., Wolfe, N.L., and Gschwend, P. M., 1991, Structure–activity relationships in dehydrohalogenation reactions of polychlorinated and polybrominated alkanes. *Crit. Rev. Environ. Sci. Technol.*, 23(1), 1–39.

Sadekov, I. D., Minkin, V. I., and Lutskii, A. E., 1970, The intramolecular hydrogen bond and the reactivity of organic compounds. *Russ. Chem. Rev.*, 39(3), 179–195.

Schreck, J. O., 1971, Nonlinear Hammett relationships. *J. Chem. Educ.*, 28(2), 103–107.

Schwarzenbach, R. P., Stierli, R., Lanz, K., and Zeyer, J., 1990, Quinone and iron porphyrin mediated reduction of nitroaromatic compounds in homogeneous aqueous solution. *Environ. Sci. Technol.*, 24(10), 1566–1574.

Scurlock, R., Rougee, M., and Bensasson, R. V., 1990, Redox properties of phenols, their relationships to singlet oxygen quenching and to their inhibitory effects on benzo(*a*)pyrene-induced neoplasia. *Free Rad. Res. Commun.*, 8(4–6), 251–258.

Shorter, J., 1973, *Correlation Analysis in Organic Chemistry: An Introduction to Linear Free-Energy Relationships.* Clarendon Press, Oxford, 119 pp.

Shorter, J., 1978, Multiparameter extensions of the Hammett equation. In: N. B. Chapman and J. Shorter, eds., *Correlation Analysis in Chemistry: Recent Advances*, Plenum, New York, pp. 119–173.

Shorter, J., 1982, *Correlation Analysis of Organic Reactivity with Particular Reference to Multiple Regression*, John Wiley & Sons, Chichester, England, 235 pp.

Shorter, J., 1994, Values of σ_m and σ_p base on the ionization of substituted benzoic acids in water at 25°C. *Pure Appl. Chem.*, 66(12), 2451–2468.

Simpson, H. N., Hancock, C. K., and Meyers, E. A., 1965, The correlation of the electronic spectra, acidities, and polarographic oxidation half-wave potentials of 4-substituted 2-chlorophenols with substituent constants. *J. Org. Chem.*, 30(8), 2678–2683.

Sjöström, M., and Wold, S., 1981, Linear free energy relationships. Local empirical rules—or fundamental laws of chemistry. *Acta Chem. Scand.*, B35(8), 537–554.

Stanton, D. T., and Jurs, P. C., 1990, Development and use of charged partial surface area structural descriptors in computer-assisted quantitative structure–property relationship studies. *Anal. Chem.*, 62(21), 2323–2329.

Stone, A. T., 1987, Reductive dissolution of manganese(III/IV) oxides by substituted phenols. *Environ. Sci. Technol.*, 21(10), 979–988.

Suatoni, J. C., Snyder, R. E., and Clark, R. O., 1961, Voltammetric studies of phenol and aniline ring substitution. *Anal. Chem.*, 33(13), 1894–1897.

Taft, R. W., Jr., 1956, Separation of polar, steric, and resonance effects in reactivity. In: M. S. Newman, ed., *Steric Effects in Organic Chemistry*, Chapman and Hall, New York, pp. 556–675.

Testa, B., and Kier, L. B., 1991, The concept of molecular structure in structure–activity relationship studies and drug design. *Med. Res. Rev.*, 11(1), 35–48.

Thioulouse, J., Devillers, J., Chessel, D., and Auda, Y., 1991, Graphical techniques for multidimensional data analysis. In: J. Devillers and W. Karcher, eds., *Applied Multivariate Analysis in SAR and Environmental Studies*, Kluwer, Dordrecht, pp. 153–205.

Tiley, P. F., 1985, The misuse of correlation coefficients. *Chem. Brit.*, Feb, 162–163.

Tratnyek, P. G., and Hoigné, J., 1991, Oxidation of substituted phenols in the environment: a QSAR analysis of rate constants for reaction with singlet oxygen. *Environ. Sci. Technol.*, 25(9), 1596–1604.

Tratnyek, P. G., and Hoigné, J., 1994a, Kinetics of reactions of chlorine dioxide (OClO) in water. II. Quantitative structure–activity relationships for phenolic compounds. *Water Res.*, 28(1), 57–66.

Tratnyek, P. G., and Hoigné, J., 1994b, Photooxidation of 2,4,6-trimethylphenol in natural waters and laboratory systems: kinetics of reaction with singlet oxygen. *J. Photochem. Photobiol. A. Chem.*, 84(2), 153–160.

Tratnyek, P. G., Hoigné, J., Zeyer, J., and Schwarzenbach, R., 1991, QSAR analyses of oxidation and reduction rates of environmental organic pollutants in model systems. *Sci. Total Environ.*, 109/110, 327–341.

Ulrich, M., 1991, *Modeling of chemicals in lakes—Development and application of user-friendly simulation software (MASAS & CHEMSEE)*, Ph.D. thesis, Swiss Federal Institute of Technology, Zürich, 213 pp.

Wehrli, B., 1990, Redox reactions of metal ions at mineral surfaces. In: W. Stumm, ed., *Aquatic Chemical Kinetics: Reaction Rates of Processes in Natural Waters*, Wiley-Interscience, New York, pp. 311–336.

Wells, P. R., 1968, *Linear Free Energy Relationships*. Academic Press, London, 116 pp.

Williams, A., 1973a, Alkaline hydrolysis of substituted phenyl *N*-phenylcarbamates. Structure–reactivity relationships consistent with an ElcB mechanism. *J. Chem. Soc. Perkin II*, (6), 808–812.

Williams, A., 1973b, Participation of an elimination mechanism in alkaline hydrolyses of alkyl *N*-phenylcarbamates. *J. Chem. Soc. Perkin*, II(9), 1244–1247.

Wold, S., and Sjöström, M., 1972, Statistical analysis of the Hammett equation. I. Methods and model calculations. *Chemica Scripta* 2(2), 49–55.

Wold, S., and Sjöström, M., 1978, Linear free energy relationships as tools for investigating chemical similarity—theory and practice. In: N. B. Chapman and J. Shorter, eds., *Correlation Analysis in Chemistry: Recent Advances*, Plenum, New York, pp. 1–54.

Wold, S., and Sjöström, M., 1986, Linear free energy relationships. Local empirical rules—or fundamental laws of chemistry? A reply to Kamlet and Taft. *Acta Chem. Scand.*, B40(4), 270–277.

Wolfe, N. L., 1980, Organophosphate and organophosphorothionate esters: application of linear free energy relationships to estimate hydrolysis rate constants for use in environmental fate assessment. *Chemosphere*, 9(9), 571–579.

Wolfe, N. L., Zepp, R. G., and Paris, D. F., 1978, Use of structure-reactivity relationships to estimate hydrolytic persistence of carbamate pesticides. *Water Res.*, 12, 561–563.

Zuman, P., 1967, Physical organic polarography. *Prog. Phys. Org. Chem.*, 5, 81–206.

Photolysis of Organics in the Environment

JASON GEDDES AND GLENN C. MILLER

ABSTRACT. Photolysis of organic compounds in the environment is an important environmental transformation pathway. Sunlight and environmentally available reactants transform organic compounds at rates which are often competitive with other transformation pathways. The rates of transformation are dependent on both the chemical and physical properties of the compound being examined, as well as the properties of each environmental compartment in which the compound may reside. Direct photolysis occurs when the compound absorbs sunlight ($\lambda > 290$ nm) directly and undergoes a transformation. Indirect reactions require some other substance to absorb light which initiates a series of reactions which transforms the organic compound. Light screening, oxidant concentrations, pH, sensitizers, and sunlight intensity all affect the rates of photolysis and complicate interpretation of experimental results.

INTRODUCTION

Sunlight-induced transformation of compounds is common in a variety of environmental compartments. The rates and products of these photochemical reactions can vary substantially, depending on a variety of physical and chemical factors. Reactions that occur in water may not be observed in air; photolysis of contaminants on soils is dramatically affected by photochemical quenching reactions and light screening; on leaf surfaces, observed photolysis rates are affected by transport into protected parts of leaves. The complexity of various environmental compartments has precluded a generalized model for photochemical transformations. However, understanding of basic photochemical principles governing light absorption and excited state reactivity is necessary prior to beginning to disentangle the diverse and more complicated reactions that occur in heterogeneous compartments such as particles, environmental waters, and leaves.

One relatively simple scenario for which photochemical processes have been modeled successfully is the direct photolysis of contaminants in water. Direct photolysis occurs when a chemical absorbs radiation and a series of processes occurs to the excited state molecule which eventually convert that chemical to products. When the concentration of the contaminant is low and the environmental reagent concentrations are constant (i.e., O_2), direct photolysis reactions can be reliably predicted for most substances. Although these simple conditions are relatively rare, the chemical and physical aspects of this less complicated case exist in nearly all of the other, more complicated environmental compartments.

DIRECT PHOTOLYSIS

Light Absorption

Any sunlight-induced transformation begins with absorption of sunlight. The first law of photochemistry, the Grotthus–Draper Law, states that *only the light that is absorbed by a molecule can be effective in producing photochemical change in the molecule.* The result of this absorption is an electronically excited molecule that can subsequently undergo various chemical and physical processes. For a single absorbing component in transparent media, the absorption of light follows the Beer–Lambert Law:

$$I = I_0 10^{-\epsilon P l}$$

where I_0 is the incident radiation intensity, I is the transmitted radiation intensity, ϵ is the molar extinction coefficient, P is the molar concentration, and l is the pathlength of light through the medium (Finlayson-Pitts and Pitts, 1986). The absorbance (A) of the solution is expressed as follows:

$$A = -\log I/I_0 = \epsilon P l$$

Molar extinction coefficients, which represent the intrinsic absorptivity of each chemical at each wavelength, can be determined on commonly available ultraviolet-visible spectrometers and are necessary for modeling photolysis kinetics. Rearrangement of the Beer–Lambert Law gives

$$\epsilon = A/P l$$

Such measurements are often complicated when absorption spectra of very nonpolar compounds are being determined in water. Solubility limitations may require the use of a cosolvent such as methanol or acetonitrile. Care must be taken to ensure that the cosolvents are not altering the spectra by spectral shifts to either higher or lower wavelengths. Gas-phase spectra of organic compounds often are shifted to lower wavelengths, compared to solution phase spectra (Hackett et al., 1995).

Rates of direct photolysis are dependent not only on the absorption spectra of contaminants, but also on the overlap with the spectrum of sunlight striking the earth's surface. Sunlight wavelengths shorter than 290 nm are almost completely absorbed by atmospheric ozone and are not available for photochemical reactions in the troposphere (Figure 8.1). The ultraviolet wavelengths (290–400 nm) are attenuated more strongly than the visible wavelengths as they pass through the atmosphere, primarily due to molecular scattering, but also by absorption by atmospheric gases, such as ozone (Peterson, 1976). Sunlight intensity varies seasonally, diurnally, and as a function of lat-

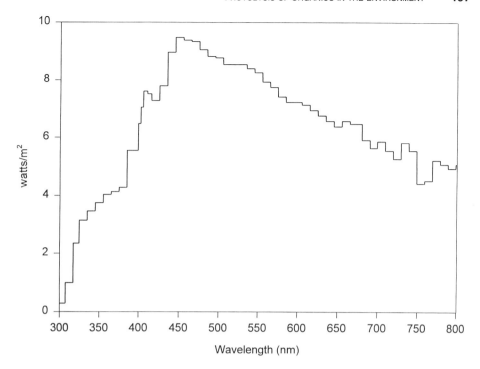

Figure 8.1.   Tropospheric solar radiation intensity as a function of wavelength.

itude, and the variation is substantial for the short UV-B wavelengths (280–320 nm), which are particularly important because these wavelengths are absorbed the most strongly (in general) by anthropogenic chemicals. Changes in intensity of sunlight wavelengths are, however, predictable and can be calculated using available solar models (i.e., GCSOLAR, US EPA, 1988).

Once a compound absorbs a photon of light energy, several pathways are available for loss of that energy (Figure 8.2), and photochemical transformation is only one of these processes. Deactivation of the molecule can occur by luminescence, physical quenching, energy transfer, electron transfer, and photoionization as well as photo-transformation (Calvert and Pitts, 1966). The relative importance of each of these pathways depends on the characteristics of the chemical and on the chemical environment in which it resides.

Luminescence occurs when a quantum of light is emitted during the transition to the ground electronic state and some residual vibrational excitation is lost via collision. Luminescence can come from either singlet states (fluorescence), where all of the electrons are paired, or triplet states (phosphorescence), where two electrons are unpaired. Physical quenching occurs when excited-state energy of the compound is dissipated by collisional deactivation, most often with the solvent.

Energy transfer can occur if a suitable acceptor exists that allows the electronic energy to be transferred, resulting in the deactivation of the molecule that absorbed the light and the excitation of the acceptor. Molecular oxygen is an example of an important quencher. Transformation of the chemical can also occur by electron transfer,

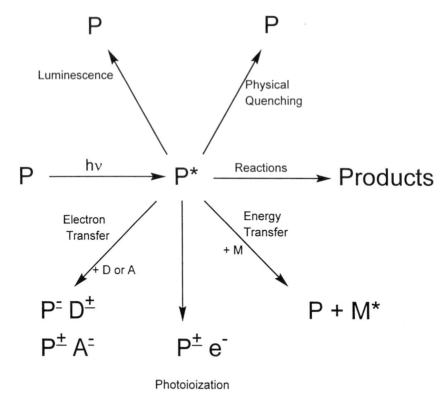

Figure 8.2. Pathways for deactivation of excited-state molecules.

photoionization, homolysis, or excited state reactions with available environmental reagents.

The contribution of each of the nonreactive processes indicated in Figure 8.2 to the overall rate of deactivation is dependent on the individual reaction rate constants, which vary from being very rapid, on the order of 10^9 s^{-1}, to relatively slow, on the order of 10^3 s^{-1}. Photolysis reactions must have first-order rate constants on the order of 10^5 to 10^9 s^{-1} in order to compete with photophysical deactivation processes.

These competitive deactivation schemes are expressed in the second law of photochemistry. The Stark–Einstein–Bodenstein Law states that *the absorption of light by a molecule is a one-quantum process, so that the sum of the primary-process quantum yields (ϕ) must be unity*. Thus, the fraction of light absorption events that results in transformation is termed the quantum yield for reaction (ϕ_r) and is equal to the number of molecules that are transformed divided by the number of photons absorbed. (Finlayson-Pitts and Pitts, 1986).

Photolysis in Aqueous Systems

Photolysis of contaminants in water has received extensive attention, and many comprehensive reviews of the various types of reactions encountered in pure and natural aquatic environments are in the literature (Crosby, 1976; Crosby et al., 1976; Hautala, 1978; Miller et al., 1979; Perry et al., 1978; Smith et al., 1978; Ware et al., 1980; Roof, 1980; Zafiriou et al., 1984; Zepp, 1978a,b; Zepp and Baughman, 1978; Zepp and Cline, 1977; Zepp and Schlotzhauer, 1979). The interest in aquatic photolysis stems from public concern over the potential for human exposure to the chemicals which can enter the hydrosphere from airborne drift, wet and dry deposition, surface runoff, subsurface runoff, and injection into ground-water systems.

The large amount of information on aqueous systems is also due in part to the well-defined physical and photochemical properties of water. Generally, if the quantum yield and absorption spectrum of a compound are known, as well as the light transmission characteristics of the water column, direct photolysis rates may be calculated at specific latitudes and dates for compounds in pure water and in waters containing dissolved light-absorbing materials (Zepp and Cline, 1977; Zepp and Baughman, 1978).

The rate of transformation in pure water by direct photolysis is given by

$$-d[P]/dt = k_P[P] = \phi_r I_a$$

where k_P is the first-order direct photolysis rate constant, $[P]$ is the concentration, I_a is the sunlight absorption rate, and ϕ_r is the quantum yield for reaction. The quantum yield for reaction is simply the number of molecules which absorb light and are transformed divided by the total number of molecules which absorb light. In dilute solution the clear-sky rate constant for sunlight absorption (I_a) is calculated by

$$I_a = \Sigma \epsilon_\lambda L_\lambda$$

where ϵ_λ is the molar extinction coefficient of the chemical at wavelength λ, and L_λ is a value proportional to the day-average sunlight intensity in a defined wavelength interval λ. This relationship is applicable only for pure nonabsorbing solutions when the concentration of the compound in water is sufficiently low that essentially none of the light is absorbed by the compound. For higher concentrations and when other absorbing material is present, the equation is more complicated (Miller and Zepp, 1979).

Values for L_λ have been calculated for solar radiation from 280 to 800 nm at various latitudes and have been compiled by Mill et al. (1981). These calculations are straightforward and can be performed using a simple calculator. Alternatively, the microcomputer program GCSOLAR (US EPA, 1988) allows rapid calculation of photolysis rates as a function of time of day, time of year, and latitude and allows changes in other parameters such as ozone and water clarity. A simplified version of GCSOLAR has been built into the Exposure Analysis Modeling System (EXAMS) (Burns, 1990), which accepts input of the physical properties of organic compounds and allows comparison of photolysis rates of transformation with other transformation pathways. Once a photolysis first-order rate has been determined, a photolysis half-life ($t_{1/2}$) may be calculated as follows:

$$t_{1/2} = 0.693/k_p$$

One of the very useful aspects of the GCSOLAR program is that it allows an estimation of the seasonal variability of direct photolysis. Zepp and Cline (1977) demonstrated that the summer and winter intensities differ by up to a factor of four for wavelengths longer than 320 nm and up to a factor of 36 for the UV-B (295–320 nm) region. A tenfold or greater seasonal variation in photolysis half-lives can then be expected due to the sunlight intensity differences.

Quantum yields often vary with solvents, and thus extrapolation of direct photolysis quantum yields to other solvents or systems is generally invalid (Miller et al., 1979). Although quantum yields for reactions are usually independent of wavelength, this assumption is useful only for small changes in wavelength and any attempt to extrapolate the rate of photolysis in the environment from data collected using 254 nm irradiation can introduce significant errors into estimates of environmental fate.

Translation of direct photolysis rates to environmental conditions requires a thorough understanding of all competing degradation factors in aquatic ecosystems. Within natural waters, humic substances, suspended sediments, algae, and natural inorganic materials can have a substantial influence on the rate of direct photolysis. These materials will generally attenuate sunlight through the water column and reduce direct photolysis rates. Miller and Zepp (1979) have demonstrated that suspended sediments can affect photolysis rates through light attenuation and scattering. Their results indicate that more than 99% light attenuation from incident radiation can occur at depths of 25 cm for solutions containing natural sediment concentrations of 17 mg/L to 106 mg/L.

INDIRECT PHOTOLYSIS

Indirect photolysis reactions occur when sunlight is absorbed by some substance that initiates a series of reactions that eventually consume the compound. In the environment, indirect photolysis can result from absorption of sunlight by a ground-state sensitizer (^1S). The sunlight changes the ground-state sensitizer to an excited singlet state ($^1S*$). The excited singlet state is a short-lived species and can undergo molecular intersystem crossing to result in a longer-lived triplet sensitized state ($^3S*$). The various deactivation processes can occur, and, with certain restrictions, this energy can be transferred to the compound of interest resulting in product formation.

The energy of $^3S*$ can be transferred to ground-state molecular oxygen to form reactive singlet oxygen ($^1O_2*$). This singlet oxygen contains 22 kcal/mol more energy than its triplet ground state and will react with electron-rich substances to form a variety of products (Foote, 1976). If the sensitized reactant undergoes photoionization, electron transfer, or fragmentation, reactive free radicals are generated which generate various reactive oxidation intermediates which can consume other compounds through nonspecific oxidative processes.

Although humic substances attenuate sunlight and can substantially retard direct photolysis, several studies have demonstrated that indirect photolysis rates can be enhanced by the presence of these UV-absorbing materials in natural waters. Zepp et al. (1976) demonstrated that the photolysis rate of methoxychlor in several natural waters was dramatically enhanced compared to rates found in distilled waters.

Photolysis of natural organics in water can result in free radical production. Mill et al. (1980) have shown that irradiation of natural water containing cumene causes the formation of products that are characteristic of oxidation by alkylperoxy and hydroxyl radicals. These researchers suggest that alkylperoxy radical formation can be sufficiently high that, for certain highly reactive chemicals, oxidation by photochemically generated peroxy and hydroxyl radicals can be an important degradative process in natural waters. Zepp et al. (1977, 1981) demonstrated that in natural waters, humic substances also photosensitize the formation of singlet oxygen. The rate of singlet oxygen production in these studies suggests that photooxidation of a variety of electron-rich organic compounds by singlet oxygen can be environmentally important. Polycyclic aromatic compounds are especially prone to photosensitized oxygenation.

PHOTOCHEMICAL REACTIONS

The presence of reactive nucleophiles, oxidizing or reducing agents, natural sensitizers, quenchers, and other environmental reactants will direct the photochemical fate of a compound in its environmental compartment. The rate and variety of photoproduct formation depend not only on the kinds of environmental reactants present and their concentrations, but also on the photochemical processes acting on the organic compound. Several reviews are available which provide a detailed discussion of the nature and extent of photochemical transformation of many organics (Calvert and Pitts, 1966; Kan, 1966; Turro, 1978). Several reviews are also available that discuss generation of photoproducts in pesticides (Miller and Crosby, 1983; Moilanen et al., 1975; Sundstrom and Ruzo, 1978; Zepp, 1979; Zepp et al., 1984).

Photoproducts observed from direct photolysis are often the same compounds observed for other nonbiological and biological pathways. Oxidation, reduction, hydrolysis, and rearrangements are common. In addition, specialized reactions that occur only for excited-state molecules generate products unique to photolysis reactions.

Photooxidation is the most common photolysis pathway for both direct and indirect reactions (see Figure 8.3). Molecular ground-state oxygen ($^3O_2$) is found in abundance in most environmental compartments and will react with free radicals to form other oxidants. These oxidants can then react to form products through such processes as hydrogen atom extraction, nitrogen oxidation, side-chain oxidation, hydroxyl radical addition to aromatics, and addition to double bonds.

Other photoreactions produce products characteristic of hydrolysis or nucleophilic substitutions. Examples include elimination reaction of carbaryl (Crosby et al., 1976), photohydrolysis of parathion (Grunwell and Erickson, 1973), and nitrogen photosubstitution in the presence of nucleophiles (Nakagawa and Crosby, 1974) (see Figure 8.4). Photoproducts resulting from most photosubstitutions are attributed exclusively to excited-state reactions and are not observed in other transformation pathways.

Photoreduction is also commonly observed in halogens and nitro substituents, particularly where hydrogen sources are available (Crosby and Wong, 1977) (see Figure 8.5). Dehalogenation is another important photochemical pathway for other chlorinated and brominated aromatic compounds, including PCBs and PBBs (Sundstrom and Ruzo, 1978). Photoreduction of nitro groups to amines has been demonstrated for parathion (Ruzo, 1982) and various nitrophenyl ethers (Ruzo et al., 1980).

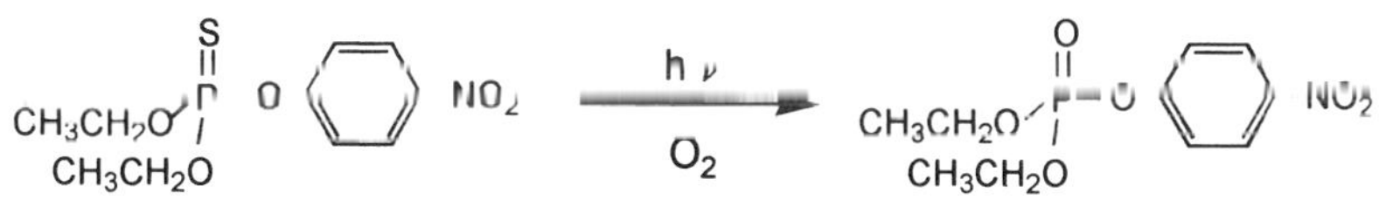

Figure 8.3. Photooxidation of ethyl parathion.

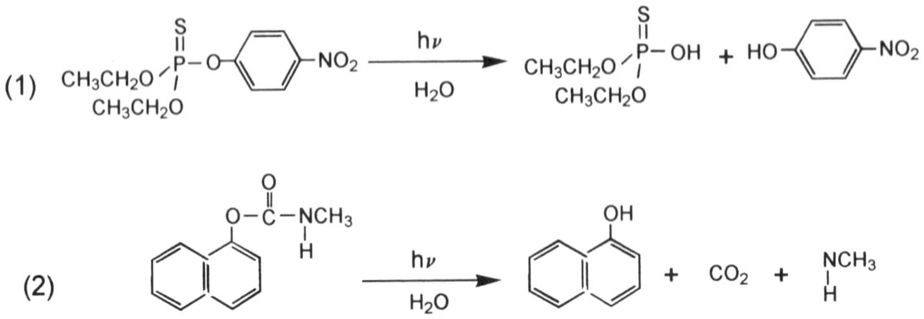

Figure 8.4. Photohydrolysis of ethyl parathion (1) and carbaryl (2).

Due to the complexity of possible pathways, it is easiest to state photochemical fate in terms of disappearance of the organic compound. Determination of the rate constant (k), half-life $(t_{1/2})$, and quantum yield (ϕ) are based on pseudo-first-order degradation and disappearance. These characteristics reflect on the combined effects of all photochemical processes other than deactivation of the excited state.

ATMOSPHERIC PHOTOLYSIS

The lower troposphere serves as an important transport environment and sink for medium-weight organic compounds (de Voogt and Jannson, 1993; Woodrow et al., 1990). These compounds can enter the atmosphere from direct emission, formation in the atmosphere, and volatilization from plant, soil, and water surfaces. Once in the atmosphere, these compounds can be transported globally and redeposited to the earth through adsorption and wet and dry deposition (McClure, 1976). There is considerable evidence that DDT and PCBs can undergo long-range transport and global cycling; both have been found in many pristine settings, including the polar regions, that have never received direct pesticide application.

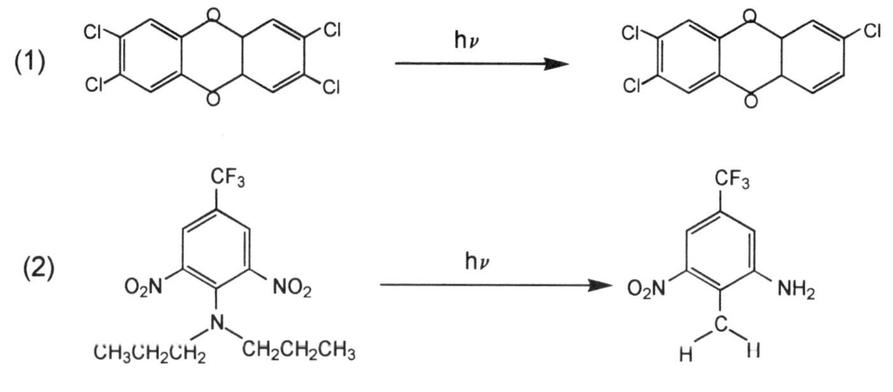

Figure 8.5. Photoreduction of 2,3,7,8-TCDD (1) and trifluralin (2).

However, physical processes are considered of minor importance for removal of reactive compounds from the lower troposphere (Seiber et al., 1983). Atmospheric photodegradation, whether acting by direct photolysis or by indirect processes such as hydroxyl radical and ozone photooxidation, have been reported as important processes in degradation (Majewski and Capel, 1995; Soderquist et al., 1975; Woodrow et al., 1978).

Although the atmosphere is a generally well-defined homogenous system, attempts at modeling kinetics of atmospheric gas-phase photolysis have been only partially successful. This is primarily a result of difficulties in establishing stable vapor concentrations for low-volatility compounds. Compounds with low vapor pressures tend to sorb onto the walls of the reaction vessels. In small vessels, wall reactions tend to predominate due to the large area-to-volume ratio (Lemaire et al., 1981).

Thus, except for compounds with relatively high vapor pressure ($>10^{-2}$ mm Hg), gas-phase experiments conducted in sealed containers generally provide only qualitative information on transformation products. Their utility is limited when attempting to establish relevant environmental photolysis rates and photoproduct distributions. Woodrow et al. (1978) described a field technique to assess atmospheric photolysis rates using a photochemically stable compound with similar physical characteristics as a tracer in order to create a ratio of compound to tracer. In this case the mixture was released via aerial application, and the ratio of the photolabile pesticide to the stable tracer was determined over time. Mongar and Miller (1988) used a similar technique and a large outdoor chamber to determine the atmospheric photolysis rate of trifluralin.

Methylisothiocyanate (MITC) is a volatile transformation product of the fumigant, metam sodium. Because of a large spill of metam sodium into the Sacramento River in 1990, the atmospheric chemistry of MITC was examined in two studies. Alvarez and Moore (1994) showed that the primary photolysis product of MITC is methylisocyanide and is generated with a quantum yield of near 1.0. Geddes et al. (1995) showed that MITC has a half-life in the atmosphere in midsummer of slightly less than one day and revealed several other photolysis products, including methylisocyanate. Because of its relatively high volatility, complicating wall reactions were less of a concern. These studies also revealed that photolysis is a major transformation process in the atmosphere.

Oxidants responsible for indirect degradation are also available in the troposphere. Atmospheric reactivity and residence times for 46 compounds found to react with ozone and hydroxyl radicals have been reported (Cupitt, 1980). Hydroxyl radical reactions dominate the transformation of these substances and are most likely the primary mode for oxidative degradation of compounds in air. Very little data are available on this subject due to the difficulty in performing these experiments. However, chemical oxidation by hydroxyl radicals or ozone is probably the major degradative pathway for compounds that do not undergo direct photolysis. Atkinson (1986) has refined methods for measuring hydroxyl radical reaction rates with a variety of organic compounds. These methods also utilize a reference compound for which the rate constant for reaction with OH is well known. Although the majority of this work has been restricted to compounds with relatively high vapor pressures, examples are available (i.e., thiocarbamates) which use this technique for an important class of agricultural chemicals (Kwok et al., 1992).

PHOTOLYSIS ON SOIL SURFACES

Organic compounds on soil and plant surfaces are exposed to the strong environmental forces of sunlight, oxygen, and heat for a period of time before being absorbed, transformed to other products, or evaporated from those surfaces. Photolysis of organic compounds on soil and leaf surfaces is likely to be a major degradative process. In contrast to aqueous and air photolysis, soils are heterogeneous and have variable organic, mineral, moisture, and texture properties. Each of these physical characteristics has the potential to affect photolysis and to cause differences in rates and product distributions.

Light attenuation by soils is the primary factor limiting photolysis of surface deposits. Both the organic (humic) fraction and the inorganic fraction can attenuate light (Miller and Zepp, 1979) and can limit the photic zone of soils to the top 0.5 mm (Hebert and Miller, 1990). The depth dependence of photolysis varies with soil properties; the soils lower in organic content are generally more transparent. In all soils examined, however, photolysis occurred only at the surface, and therefore any transport processes that move the compound to and from the surface will have a substantial effect on the importance of soil photolysis. The effect of transport has been demonstrated in two recent studies. When hydrocarbon films were present on soil, 2,3,7,8-tetrachlorodibenzo-p-dioxin (TCDD) was shown to degrade much more extensively than on unamended soils (Kieatiwong et al., 1990). Similarly, the relatively water-soluble herbicide, napropamide, was shown to move with water upward in the soil column from a depth of 3 cm to soil surface where it underwent photolysis (Donaldson and Miller, 1996). In both cases, transport resulted in a substantially greater contribution of photolysis to the overall loss of the compound, compared to when the compounds were stationary.

Even though direct photolysis processes in soils are generally retarded compared to those in transparent media, photolysis can be a major transformation process since surface applied, adsorbed, and wet and dry deposited chemicals will exist in greatest concentration on these surfaces. Photochemical conversion of compounds susceptible to indirect photolysis can also be substantially enhanced as a result of the production of reactive oxidants on irradiated soils. Gohre and Miller (1983) found that singlet oxygen is produced on soils at environmentally significant rates and can be particularly important for conversion of sulfides to sulfoxides on soil surfaces.

At present there are no successful predictive models for assessing the kinetics of either direct or indirect photolysis on soil surfaces. This is due in part to the inherent complexity of soils and is also a result of problems encountered when trying to develop an environmentally relevant methodology for modeling soil photolysis (Guth et al., 1976).

Perhaps the most significant health-related problem associated with pesticide soil photolysis results from thiophosphate insecticide photooxidation into its more toxic oxon moiety. Paraoxon formation in agricultural fields after parathion application has been implicated as the principal toxic constituent in poisoning incidents among farm workers in California citrus groves (Nigg and Allen, 1979).

EXPERIMENTAL CONSIDERATIONS

Techniques to examine the environmental photochemistry of medium-weight organics have evolved over the past 30 years, and a large body of information has developed on the most successful methods for predicting how organic compounds will interact with sunlight in the environment. In many cases, however, the data are difficult to apply to real environmental situations. A particular problem is extrapolation of rate data that have been obtained using uncalibrated laboratory lamps. Clear-sky sunlight intensity at reported latitudes and times of year can be compared between laboratories and applied toward field conditions. Whenever obtainable, these data are the most reliable, especially if an actinometer system is used to measure the sunlight intensity. Alternatively, xenon arc lamps equipped with sunlight filters that remove wavelengths below 290 nm provide a useful approximation to sunlight. When properly calibrated, results from these systems can be compared between laboratories and applied to field situations by comparison of solar intensity measured with actinometers or modern radiometers.

Particularly in early studies, many reports of photolysis utilized lamps which emit wavelengths below 290 nm. Irradiation of compounds at wavelengths below 290 nm often results in photolysis rates and product distributions atypical of environmental conditions. Additionally, many organics simply do not absorb sunlight and thus do not undergo direct photolysis in sunlight. Unfiltered mercury arc lamp studies are particularly suspect because mercury lamps primarily emit 254-nm radiation. Even for compounds that do not absorb sunlight, mercury arc lamps can produce products not found when the chemical is exposed to sunlight (Miller and Crosby, 1978).

Another concern often overlooked is the medium in which the compound is exposed to light. If water is the system, use of methanol, hexane, or any other organic solvent will very likely provide misleading results. An example is the photolysis of 3,4-dichloroaniline in water, yielding 2-chloro-5-aminophenol as a photoproduct. In acetonitrile, the product is not observed and the photolysis rate is over 100 times slower (Miller et al., 1979). Any photolysis experiment performed in another medium should be validated in the medium of the environmental compartment of interest before environmental significance is assigned to the results.

The same considerations must be made when dealing with soil and plant surfaces. Soils contain a heterogeneous variety of dissolved and particulate organic substances, living organisms, colloidal suspensions, and mineral particulates that can screen sunlight or participate in energy transfer in light-induced transformations. The surfaces of the plants are composed of a hydrocarbon outer coating and internally contain pigments which can screen the light. Thus, use of silica gel, glass plates, or other simple surfaces to predict photolysis on these environmental surfaces will not provide dependable predictive information.

The physical and chemical nature of most environmental compartments as they affect photolysis remains sufficiently complicated so as to preclude reliable predictions of photolysis in the environment. Although we can predict direct photolysis in solution, extrapolation of photolysis rates and products to other environmental compartments is uncertain. However, the 25+ years of research on environmental photochemistry of contaminants has revealed that photolysis is important under a wide variety of conditions, and continued studies will refine our understanding of how individual compounds will react in each environmental compartment.

References

Alvarez, R. A., and Moore, C. B., 1994, Quantum yield for production of CH_3NC in the photolysis of CH_3NCS. *Science*, 263, 205–207.

Atkinson, R., 1986, Kinetics and mechanism of the gas-phase reactions of the hydroxyl radical with organic compounds under atmospheric conditions. *Chem. Rev.*, 85, 69–201.

Burns, L. A., April 1990, *EXAMS: An Exposure Analysis Modeling System*, US Environmental Protection Agency, EPA/600/3-89/084, Athens, GA.

Calvert, J. G., and Pitts, J. N., 1966, *Photochemistry*, John Wiley & Sons, New York.

Crosby, D. G., 1976, Photochemistry of benchmark pesticides. In: *A Literature Survey of Benchmark Pesticides*, report by the Department of Medical and Public Affairs of the George Washington University Medical Center, Washington, D.C., under Contract 68-01-2889 for the Office of Pesticide Programs, US Environmental Protection Agency.

Crosby, D. G., Leitis, E., and Winterlin, W. H., 1976, Photodecomposition of carbamate insecticides. *J. Agric. Food Chem.*, 13, 204–207.

Crosby, D. G., and Wong, A. S., 1977, Environmental degradation of 2,3,7,8-tetrachlorodibenzo-*p*-dioxin. *Science*, 195, 1337–1338.

Cupitt, L. T., 1980, *Fate Of Toxic And Hazardous Materials In The Air Environment*. EPA-600/53-80-084. US Environmental Protection Agency, Athens, GA.

de Voogt, P., and Jannson, B., 1993, Vertical and long-range transport of persistent organics in the atmosphere. *Rev. Environ. Contamin. Toxicol.*, 132, 1–27.

Donaldson, S. G., and Miller, G. C., 1996, Coupled transport and photodegradation of napropamide in soils undergoing evaporation from a shallow water table. *Environ. Sci. Technol.*, 30, 924–930.

Finlayson-Pitts, B. J., and Pitts, J. N., 1986, *Atmospheric Chemistry: Fundamentals and Experimental Techniques*, John Wiley & Sons, New York.

Foote, C. S., 1976, Photosensitized oxidation and singlet oxygen. In: W. A. Pryor, ed., *Free Radicals in Biology*, Vol. 2, Academic Press, New York, pp. 84–133.

Geddes, J. D., Miller, G. C., and Taylor, G. E., Jr., 1995, Gas phase photolysis of methyl isothiocyanate. *Environ. Sci. Technol.*, 29, 2590–2594.

Gohre, K., and Miller, G. C., 1983, Singlet oxygen generation on soil surfaces. *J. Agric. Food Chem.*, 31, 1104–1108.

Grunwell, J. R., and Erickson, R. H., 1973, Photolysis of parathion (*O,O*-diethyl-*O*-[4-nitrophenyl] thiophosphate). *J. Agric. Food Chem.*, 21, 929–931.

Guth, J. A., Burkhard, N., and Eberle, D. O., 1976, Experimental approaches for studying the persistence of pesticides in soils. In: *Persistence of Pesticides and Herbicides*, Proceedings British Crop Protection.

Hackett, M., Wang, H., Miller, G. C., and Bornhop, D. J., 1995, Ultraviolet-visible detection for capillary gas chromatography and combined ultraviolet–visible–mass spectrometry using a remote flow cell, *J. Chrom. A*, 695, 243–257.

Hautala, R. R., 1978, *Surfactant Effects on Pesticide Photochemistry in Soil and Water*, EPA-600/3-78-060, US Environmental Protection Agency, Washington, D.C.

Hebert, V. R., and Miller, G. C., 1990, Depth dependence of direct and indirect photolysis on soil surfaces, *J. Agric. Food Chem.*, 38, 913–918.

Kan, R. O., 1966, *Organic Photochemistry*, McGraw-Hill, New York.

Kieatiwong, S., Nguyen, L. V., Hebert, V. R., Hackett, M., Miller, G. C., Miille, M. J., and Mitzel, R., 1990, Photolysis of 2,3,7,8-TCDD on soils, *Environ. Sci. Technol.*, 24, 1575 1580.

Kwok, E., Atkinson, R., and Arey, J., 1992, Gas phase atmospheric chemistry of selected thiocarbamates. *Environ. Sci. Technol.*, 26, 1798–1807.

Lemaire, J., Campell, I., Hulpe, H., Guth, J. A., Merz, W., Philp, J., and Von Waldow, C.,

1981, An assessment of test methods for photodegradation of chemicals in the environment. In: L. Turner, ed. *European Chemical Industry Ecology and Toxicology Center*, Technical Report No. 3, Brussels, Belgium, pp. 1–61.

Majewski, M. S., and Capel, P. D., 1995, *Pesticides in the Atmosphere: Distribution, Trends and Governing Factors*. U.S. Geological Survey, Open-File Report 94-506, Sacramento, CA.

McClure, V. E., 1976, Transport of heavy chlorinated hydrocarbons in the atmosphere. *Environ. Sci. Technol.*, 10, 1223–1229.

Mill, T., Hendry, D. G., and Richardson, H., 1980, Free radical oxidants in natural waters. *Science*, 207, 815–820.

Mill, T., Mabey, W. R., Bomberger, D. C., Chou, T. W., Henry, D. G., and Smith, J. H., 1981, Laboratory protocols for evaluating the fate of organic chemicals in air and water, Contract No. 68-03-2227. U.S. Environmental Protection Agency, Athens, GA, pp. 1–329.

Miller, G. C., and Zepp, R. G., 1979, Photoreactivity of aquatic pollutants sorbed on suspended soil and sediments. *Environ. Sci. Technol.*, 13, 860–863.

Miller, G. C., and Crosby, D. G., 1983, Pesticide products: generation and significance. *J. Toxicol. Clin. Toxicol*, 19, 707–735.

Miller, G. C., Miille, M. J., Crosby, D. G., Sontum, S., and Zepp, R. G., 1979, Photosolvolysis of 3,4-dichloroaniline in water. *Tetrahedron*, 35, 1797–1799.

Moilanen, K. W., Crosby, D. G., Soderquist, C. J., and Wong, A. S., 1975, Dynamic aspects of pesticide photodecomposition. In: R. Haque and V. Freed, eds., *Environmental Dynamics of Pesticides*, Plenum Press, New York, pp. 45–60.

Mongar, K., and Miller, G. C., 1988, Vapor phase photolysis of trifluralin in an outdoor chamber, *Chemosphere*, 17, 2183–2188.

Nakagawa, M., and Crosby, D. G., 1974, Photonucleophilic reactions of nitrofen. *J. Agric Food Chem.*, 22, 930–933.

Nigg, H. N., and Allen, J. C., 1979, A comparison of time-weather models for predicting parathion disappearance under California conditions. *Environ. Sci. Technol.*, 13, 231–233.

Perry, F. M., Day, E. W., Jr., Magadanz, H. E., and Sanders, D. G., 1978, Fate of nitrosamines in the environment: photolysis in natural waters. Presented at the 176th National Meeting, American Chemical Society, Miami Beach, FL, September.

Peterson, J. T., 1976, Calculated actinic fluxes (290–700 nm) for air pollution photochemistry applications, EPA-600/4-76-025. Research Triangle Park, NC: US Environmental Protection Agency, pp. 1–55.

Roof, A. A. M., 1980, Aquatic photochemisty. In: O. Hutzinger, ed., *Handbook of Environmental Chemsitry*, Vol. 2b, Springer-Verlag, New York.

Ruzo, L. O., 1982, Photochemical reactions of the synthetic pyrethroids. In: *Progress in Pesticide Chemistry*, D. H. Hutson and T. R. Roberts, eds., John Wiley & Sons, New York, pp. 1–33.

Ruzo, L. O., Lee, J. K., and Zabik, M. J., 1980, Solution-phase photodecomposition of several substituted diphenyl ether herbicides. *J. Agric. Food Chem.*, 28, 1289–1292.

Seiber, J. N., Kim, Y. H., Wehner, T., and Woodrow, J. E., 1983, Analysis of xenobiotics in air. In: J. Miyamoto and P. C. Kearney, eds., *Pesticide Chemistry: Human Welfare and the Environment*, Pergamon Press, Oxford, pp. 1–13.

Smith, J. H., Mabey, W. R., Bohonos, N., Holt, B. R., Lee, S. S., Chou, T. W., Bomberger, D. C., and Mill, T., 1978, *Environmental Pathways of Selected Chemicals in Fresh-*

water Systems: Part Ii. Laboratory Studies, EPA-600/7-78-074, US Environmental Protection Agency, Washington, D.C.

Soderquist, C. J., Crosby, D. G., Moilanen, K. W., Seiber, L. N., and Woodrow, J. E., 1975, Occurrence of trifluralin and its photoproducts in air. *J. Agric. Food Chem.*, 23, 304–309.

Sundstrom, G., and Ruzo, L. O., 1978, Photochemical transformation of pollutants in water. *Permagon Ser. Environ. Sci.*, 1, 205–222.

Turro, N. J., 1978, *Modern Molecular Photochemistry*. Benjamin-Cummings, Menlo Park, CA.

US EPA, 1988, GSCOLAR, VERSION 1.1. Available Through the Center For Exposure Assessment Modeling, USEPA Environmental Research Laboratory, College Station Road, Athens, GA.

Ware, G. W., Crosby, D. G., and Giles, J. W., 1980, Photodecomposition of DDA. *Arch. Environ. Contam. Toxicol.*, 9, 135–46.

Woodward, R. B., 1941, Structure and the absorption spectra of unsaturated ketones. *J. Am. Chem. Soc.*, 63, 1123.

Woodrow, J. E., Crosby, D. G., Mast, T., Moilanen, K. W., and Seiber, J. N., 1978, Rates of transformation of trifluralin and parathion vapors in the air. *J. Agric. Food Chem.*, 26, 1312–1316.

Woodrow, J. E., McChesney, M. M., and Seiber, J. N., 1990, Modeling the volatilization of pesticides and their distribution into the atmosphere. In: D. A. Kurtz, ed., *Long Range Transport of Pesticides*, Lewis, Chelsea, MI.

Zafiriou, O. C., Joussot-Dubien, J., Zepp, R. G., and Zika, R., 1984, Photochemistry of natural waters. *Environ. Sci. Technol.*, 18, 358a–371a.

Zepp, R. G., 1978a, Assessing the photochemistry of organic pollutants in aquatic environments. Paper presented before the Division of Environmental Chemistry, American Chemical Society, Miami Beach, FL, September.

Zepp, R. G., 1978b, Quantum yields for reaction of pollutants in dilute aqueous solution. *Environ. Sci. Technol.*, 12, 327–329.

Zepp, R. G., 1979, Photochemistry in aquatic environments. In: R. Haque, ed., *Dynamics, Exposure and Hazard Assessment of Toxic Chemicals*, Ann Arbor Science, Ann Arbor, MI, Chapter 9.

Zepp, R. G., Wolfe, N. L., Gordon, J. A., and Baughman, G. L., 1976, Dynamics of 2,4-D esters in surface waters; hydrolysis, photolysis and vaporization. *Environ. Sci. Technol.*, 9, 1144–1150.

Zepp, R. G., and Cline, D. M., 1977, Rates of direct photolysis in the aqueous environment. *Environ. Sci. Technol.*, 11, 359–366.

Zepp, R. G., Wolfe, N. L., Baughman, G. L., and Hollis, R. C., 1977, Singlet oxygen in natural waters. *Nature*, 267, 421–423.

Zepp, R. G., and Baughman, G. L., 1978, Prediction of photochemical transformation of pollutants in the aquatic environment. In: O. Hutzinger, I. H. Van Lelyveld, and B. C. J. Zoeteman, eds., *Aquatic Pollutants: Transformation And Biological Effects*, Pergamon Press, Oxford, pp. 237–243.

Zepp, R. G., and Schlotzhauer, P. F., 1979, Photoreactivity of selected aromatic hydrocarbons in water. In: P. W. Jones and P. Leber, eds., *Polynuclear Aromatic Hydrocarbons*, Ann Arbor Science Publishers, Ann Arbor, MI pp. 141–158.

Zepp, R. G., Baughman, G. L., and Scholtzhauer, P. F., 1981, Comparison of photochemical behavior of various humic substances in water. II. Photosensitized oxygenations. *Chemosphere*, 10, 119–126.

Zepp, R. G., Schlotzhauer, P. F., Simmons, M. S., Miller, G. C., Baughman, G. L., and Wolfe, N. L., 1984, Dynamics of pollutant photoreactions in the hydrosphere. *J. Fresenius Z. Anal. Chem.*, 319, 119–125.

CHAPTER 9

Microbiologically Mediated Reactions in Aquatic Systems

PETRA M. RADEHAUS

ABSTRACT. This chapter describes a select group of chemical processes that are mediated by microorganisms in aquatic systems. Typical mechanisms used by microbes to degrade organic pollutants and the dependence of microbially mediated degradation rates on physicochemical conditions are described. A discussion of the impacts of nonbiological processes on biological reactions, and vice versa, is also included. Finally, an overview is presented describing the major interactions of microorganisms with inorganic chemicals in their aquatic habitats.

AQUATIC SYSTEMS AS BIOSPHERES FOR MICROORGANISMS

Life as we know it could not exist without water, and wherever there is water, there is almost certainly life. Where there is life, microorganisms are likely to be found. Microorganisms, a diverse group of single-celled organisms consisting of bacteria, fungi, algae, and protozoa, require water in order to carry out metabolic processes (Chapelle, 1993). Nourishment supplied to cells must be dissolved in order to be taken into the cells through the cell membranes. The polar chemical properties of water make it an ideal biological solvent because many biologically important molecules (polysaccharides, proteins, etc.) are themselves polar and thus readily dissolved (Brock et al., 1994).

The very small size of the microbial cells is particularly significant because it gives them a high surface-to-volume ratio and thus enables them to have a more efficient exchange of nutrients and metabolic products with their surroundings. This is of major importance because many natural waters are extremely dilute in inorganic and organic nutrients, and thus microbes must grow under conditions of severe nutrient limitation (Sterritt and Lester, 1988). Aquatic habitats for microorganisms can be divided into freshwater and marine habitats. Freshwater habitats include lakes, ponds, swamps, springs, streams, rivers, and aquifers. Marine habitats include oceans and the estuarine habitats that occur at the interface between freshwater and marine ecosystems (Atlas and Bartha, 1992). Other aquatic systems can be streams of pollutants in water such as sewage, wastewater, or leachates from landfills.

Although these aquatic systems have substantial differences, they also have some common characteristics. All can be roughly divided, at least conceptually, into a sediment fraction, which is suspended or settled in a bottom layer, and a liquid fraction. This zonation has an influence on the microbial populations distributed at various densities throughout these layers. Most sediment layers are stratified into an upper oxidized (aerobic) zone and a lower reduced (anaerobic) zone, each having a distinctive bacterial flora (Scow, 1982). The variety and quantity of microorganisms in a water body also depends on ecological factors, such as the physical properties of the aqueous solutions (density, viscosity, refractive index, temperature, light, pH) and the chemical composition (presence and concentration of organic and inorganic compounds, oxygen, and other gases) of the aquatic medium. The ecological conditions have a strong impact on the number of microorganisms found per unit volume of water; for example, the number of bacteria can be as low as one per milliliter in rainwater or subsoil systems (e.g. groundwater, springs, artesian wells) to 10^8 per milliliter in heavily polluted water (Grady and Lim, 1980).

Due to the enormous variety of metabolic reactions performed, microorganisms can substantially alter—either directly or indirectly—the chemical and physical properties of their environment. Organic pollutants may be directly oxidized or reduced by intra- or extracellular enzymes or may be indirectly altered by abiotic reactions with molecules in the environment. Abiotic reactions are defined as those processes that do not feature the direct involvement of the metabolic processes of organisms. However, microorganisms may play an indirect role in these processes (Macalady et al., 1986) The major impacts microorganisms have on aquatic systems include the following:

1. Algae, photosynthetic bacteria, and cyanobacteria are capable of fixing inorganic carbon and are the predominant producers of organic matter that supports the rest of the food chain (consumers and decomposers) in bodies of water (Manahan, 1993). The consumers of the organic matter include herbivorous aquatic animals and the carnivores which feed on them. When the consumer organisms die and decay, they constitute a source of food for the decomposer organisms.

2. Aquatic microorganisms (decomposers) play a key role in nutrient cycling. They mineralize organic chemicals and biomass; that is, they convert these compounds to inorganic substances (H_2O, CO_2, mineral salts) and thus make nutrients available to the primary producers once again (Sterritt and Lester, 1988).

3. Microorganisms mediate most of the oxidation–reduction processes in the hydrosphere. They oxidize reduced substances of inorganic origin such as divalent iron to trivalent iron, ammonium to nitrites and nitrates, hydrogen sulfide to sulfates, and so on. Likewise, under reducing conditions, they reduce sulfates to sulfides, carbon dioxide to methane, and so on. The reduced forms of iron and sulfur may then react as reductants of electron transfer mediators such as quinone-type compounds and transition metal complexes In abiotic reduction reaction these mediators pass electrons to compounds, particularly those of anthropogenic origin, which may be unreactive toward reduced iron or sulfur. In this way, microbes can provide the bulk reactants for otherwise abiotic processes (Dunnivant et al., 1992).

4. Aquatic microorganisms are essential for the major biogeochemical cycles in nature such as the carbon, nitrogen, sulfur, phosphorus, and iron cycles (Brock et al., 1994; Ehrlich, 1990).

5. Microorganisms are directly responsible for the degradation and detoxification of many pollutants in aquatic systems. They often play a vital role in minimizing the impacts of pollutants in natural environments and are essential for water purification and secondary wastewater treatment (Grady and Lim, 1980; Kemmer, 1988; Sayles and Suidan, 1993).

BIODEGRADATION IN AQUATIC SYSTEMS

The section details a few examples of the many metabolic activities of microbes in aquatic systems. The emphasis is on the microbial degradation of toxic chemicals in aquatic systems and on the direct and indirect effects of microbes on the degradation of hazardous pollutants. Biodegradation reactions are of major interest because of the negative impact a growing world population has on nature and the remediation solutions urgently needed to clean-up the environment. As used here, "biodegradation" refers to any biologically mediated structural alteration of a compound. "Mineralization" refers to the conversion of an organic substrate to inorganic products.

During the production, transport, and use of toxic chemicals, these compounds are often accidentally introduced into aquatic environments. Xenobiotics, synthetic chemicals that pose a threat to living organisms, are of special concern because they are not readily biodegradable by microorganisms. Such recalcitrant compounds, which include chlorinated solvents or pesticides (polychlorinated biphenyls, DDT, lindane, pentachlorophenol, etc.), may persist in aquatic systems over extremely long time periods. Biodegradation rates are strongly influenced by the metabolic capabilities of the degrading microorganisms. Some chemicals such as aliphatic petroleum compounds may be readily biodegradable, but due to their low water solubility they may be not available to microorganisms. Thus not only the metabolic capabilities of microorganisms are of importance for bioremediation, but also physical and chemical properties of the chemicals of concern.

From an ecological point of view, it is necessary to include the number and physiological state of microorganisms, the seasonal state of microbial development, and other possible phenomena as controlling factors for biodegradation processes (Toth and Tomascovicova, 1989). An often-neglected aspect is the interaction of degrading and nondegrading microorganisms sharing the same ecological niche. Predator–prey interactions between bacteria and protozoa, for example, have an impact on the biodegradation of pollutants (Schmidt et al., 1992). It should also be stressed that nonbiological (abiotic) factors play an important role in the transformation of pollutants. Each of these influences on biodegradation rates and mechanisms is discussed in more detail below.

Typical Microbially Mediated Degradation Mechanisms

The most important reaction types involved in the first steps of biodegradation sequences include hydrolysis, hydroxylation, dehalogenation, demethylation or other dealkylations, methylation or acetylation, nitro reduction, ether cleavage, and conversion of nitrile to an amide (Alexander, 1994; Toth and Tomasovicova, 1989; Scow, 1982). The microbial enzymes catalyzing these reactions include oxidoreductases, transferases, hydrolases, lyases, isomerases, and ligases (Glaser, 1989). Once a single degradation step has occurred, pollutants may or may not be detoxified. Detoxification refers to the conversion of a toxicant to innocuous metabolites, which is the most common effect of biodegradation. However, several types of biodegradation reactions described below may actually yield products far more toxic to some organisms than the original compound (Alexander, 1994).

Hydrolysis. Cleavage of a bond by the addition of water is a common means by which microorganisms inactivate toxicants (Alexander, 1994; Chapter 5, this volume). Functional groups that are dissociated include esters, anhydrides, amides and other acyl derivatives (Glaser, 1989). An example is the bacterial hydrolysis of the organophosphate ester bond in the insecticide parathion (*O,O*-diethyl-*O-p*-nitrophenyl phosphorothioate). Tchelet et al. (1993) found out that, depending on the bacterial species, metal cations can have inhibitory or stimulatory effects on the activity of the involved parathion hydrolases.

Hydroxylation. The addition of OH to an aromatic or aliphatic molecule is called hydroxylation. The first step in the microbial oxidative degradation pathway of *n*-alkanes is the hydroxylation of a terminal or subterminal methyl group (Chapelle, 1993). In the case of aromatic hydrocarbons, ring fission requires a dihydroxylation reaction and the subsequent formation of a *cis*-dihydrodiol. The reaction is carried out by a membrane-bound enzyme system. Further oxidation leads to the formation of catechols that are substrates for another deoxygenase that catalyzes ring fission (Chapelle, 1993). In the case of aerobic oxidation, the oxygen inserted into ring structures comes exclusively from molecular oxygen. But even under anaerobic conditions, hydroxylation reactions can take place. Studies by Grbic-Galic and Vogel (1987) tracing the ^{18}O composition of intermediate degradation products suggest the oxygen inserted into the ring structures may come from water. Under Fe(III)-reducing conditions, found in many ground-water systems, the oxygen may be derived from Fe(III) oxyhydroxides (Lovely and Lonergan, 1990).

Dehalogenation. Some microorganisms are capable of dehalogenation reactions. They can replace a halogen (X) by hydrogen (= reductive dehalogenation)

$$RX \rightarrow RH$$

and/or by an hydroxyl group (hydrolytic dehalogenation)

$$RX \rightarrow ROH$$

Under anaerobic conditions, reductive dehalogenation reactions are the most important processes for the degradation of alkyl halides, such as trichloroethene and tetrachloroethene (DiStefano et al., 1992). In electron donor–poor ground-water systems, reductive dehalogenations tend to be incomplete, resulting in the accumulation of the mutagenic and carcinogenic compound vinyl chloride (Chapelle, 1993). Reductive dehalogenation reactions may also be important microbial processes involved in degrading aryl halides, such as hexachlorobenzene (Fathepure et al., 1988) or chlorinated phenolic compounds (Zhang and Wiegel, 1990; Larsen et al., 1991; Hendriksen and Ahring, 1993: Armenante et al., 1993). The dechlorinated degradation products are less toxic, less likely to bioaccumulate, and more susceptible to further microbial attack (Fathepure et al., 1988). Studies on pentachlorophenol (PCP) show that the chlorine atoms are sequentially removed from the aromatic ring. One sequence of products observed in biologically active anaerobic soil-sewage sludge mixtures was PCP → 2,3,4,5-tetra-chlorophenol → 3,4,5-trichlorophenol → 3,5 or 3,4-dichlorophenol → 3-chlorophenol (Mikesell and Boyd, 1988).

Aerobic microbial PCP metabolism involves dechlorination (Radehaus and Schmidt, 1992) and hydroxylation either ortho or para to the phenolic hydroxyl group, forming a catechol or quinone, respectively (Rochkind-Dubinsky et al., 1987). ^{18}O-labeling experiments done with *Rhodococcus chlorophenolicus* indicate that the first hydroxyl group replacing the chlorine on the PCP ring is derived from water molecules and not from molecular oxygen (Apajalahti and Salkinoja-Salonen, 1987). Apajalahti and Salkinoja-Salonen (1987) also showed that the sequence of complete PCP dechlorination reactions in *Rhodococcus chlorophenolicus* consists of two hydrolytic and three reductive reaction steps.

Demethylation or Other Dealkylation Reactions. Many heavily used herbicides and pesticides are polluting aquatic systems, and considerable interest exists in the development of technologies to eliminate these biocides. Most of them contain methyl or other alkyl substituents, often linked to nitrogen or oxygen atoms (*N*- or *O*-alkyl substitution). A limited number of microorganisms have been reported to mediate specific steps in the degradation of these biocides, particularly *N*- or *O*-dealkylation, which frequently results in loss of toxicity. An example is the dealkylation of the *s*-triazine herbicide atrazine (2-chloro-4-ethylamino-6-isopropylamino-1,3,5-triazine) by fungi (Kaufman and Blake, 1970), by *Nocardia* strains (Giardina et al., 1980; Nagy et al., 1995), a *Rhodococcus* strain (Behki et al., 1993), and a *Pseudomonas* species (Behki and Khan, 1988). The dealkylation of the atrazine molecule results in the loss of its *N*-ethyl or *N*-isopropyl groups. These alkyl side chains are used by some microorganisms as source of carbon (Radosevich et al., 1995). A cytochrome P-450 enzyme system is involved in the *N*-dealkylation of atrazine (Behki et al., 1993; Nagy et al., 1995). In nutrient-poor aquifer systems, the degradation of atrazine seems to be limited to deethylation reactions (Chapelle, 1993).

Methylation and Acetylation. Some microbes have been shown to use methylation and acetylation reactions as an alternative to degradation for detoxification of their environment. The hydroxy group of phenolic compounds is methylated (Suzuki, 1978, 1983; Häggblom et al., 1988), but the carbon skeleton of the substrates remains unaltered:

$$R–OH \rightarrow R–OCH_3$$

Methoxybenzenes resulting from these biotransformation reactions are more resistant toward microbial attack and have a high potential for bioaccumulation, but seem to be less toxic to microbes than the untransformed compounds. Cserjesi and Johnson (1972) assumed that the methylation of the phenolic hydroxy group inactivates the reactive binding group. Microorganisms were found to *O*-methylate chlorinated phenols, guaiacols (= 2-methoxyphenols), syringols (2,6-dimethoxyphenols), and hydroquinones

(Häggblom et al., 1988). The lower solubility of pentachloroanisole (PCA), the methylated form of PCP, in water (at least 10 times lower than that of PCP) may also contribute to its lower toxicity (see also the paragraph on "solubility" below).

Nitro Reduction. Nitroaromatic compounds are organic chemicals in which the nitro group ($-NO_2$) is bonded to one or more carbons in a benzene ring. Nitro aromatic moieties are found in compounds used as dyes, pesticides, herbicides (2,4-dinitro-o-cresol, dinoseb = 2-*sec*-butyl-4,6-dinitrophenol), plasticizers, explosives [2,4,6-trinitrotoluene (TNT)], and solvents. The presence of the nitro group makes these compounds more resistant to biodegradation than the unsubstituted analogs. However, some microorganisms are capable of reductive removal of the nitro substituents from nitroaromatic compounds (Stevens et al., 1991; Kaake et al., 1994), a pathway which involves initial reduction of the nitro substituent to an amino group:

$$RNO_2 \rightarrow RNH_2$$

Some of the metabolites detected during microbial degradation of 2,4,6-trinitrotoluene were 2-amino-4,6-dinitrotoluene and 4-amino-2,6-dinitrotoluene (Mondecar et al., 1994). Abiotic reduction of nitroaromatic compounds is an important competing pathway and may dominate in many anaerobic aquatic systems (Chapter 5, this volume).

Ether Cleavage. Phenoxy herbicides contain ether linkages ($C—O—C$), and the cleavage of these linkages destroys the phytotoxicity of the molecule (Alexander, 1994). A typical example for this reaction type is the microbial conversion of herbicide 2,4-dichlorophenoxyacetic acid (2,4-D) to 2,4-dichlorophenol (Top et al., 1995).

Conversion of Nitrile to an Amide. Numerous microorganisms are capable of degrading nitriles (organic cyanides) by converting them to the corresponding amide (Thompson et al., 1988; Knowles and Wyatt, 1992). The necessary nitrile hydratases can hydrolyze a wide variety of structurally diverse nitriles:

$$R—C \equiv N \rightarrow R-CO-NH_2$$

The microbial conversion of the plant growth inhibitor 2,6-dichlorobenzonitrile to 2,6-dichlorobenzamide, inactivates the molecule (Ashton and Crafts, 1981; Alexander, 1994).

DEPENDENCE OF THE DEGRADATION RATE OF ORGANIC COMPOUNDS ON PHYSICO-CHEMICAL CONDITIONS

Biodegradation rates are influenced by (a) the physicochemical properties of the chemical of concern and (b) the conditions in the microbial environment. The structure, water solubility and vapor pressure of the chemical, the pH, the temperature, and the availability of electron acceptors have a strong impact on the biodegradation of recalcitrant compounds in aquatic systems. Some molecules are recalcitrant to degradation because the number, length, or location of functional groups impede enzyme attack (Godsy, 1994). Thus the addition of chemical substituents to organic molecules can transform them from biodegradable compounds to persistent compounds. Such substituents include amino, methoxy, sulphonyl, nitro, and chloro groups, a range of meta substituents on the benzene ring, ether linkages and carbon-chain branching (Sterritt and Lester, 1988).

Physicochemical Factors that Influence Biodegradation Rates

Vapor Pressure. The vapor pressure of a substance depends on its molecular structure as well as the temperature of the environment. An increase in temperature increases the equilibrium vapor pressure, resulting in an increased volatilization rate. Volatilization may change the number of constituents in a mixture of chemicals and thus decrease the availability of certain components of the mixture to microorganisms. A typical example is an oil spill in the sea, where hydrocarbons rise to the surface, come into contact with air, and some of the low molecular weight hydrocarbons volatilize. Wolfe et al. (1994) studied the fate of the 35,5000 metric tons of petroleum hydrocarbons spilled from the Exxon Valdez tank vessel. They estimated that approximately 20% of the spilled oil evaporated and underwent photolysis in the atmosphere (see "photolysis" below).

Volatilization can also be a major fate of many alkyl halides. Evaporative losses of alkyl halides, such as dichloromethane or trichloroethene, from soils and surface water generally exceed 80% in less than 10 days (Chapelle, 1993; Levin and Gealt, 1993). This is due to the high vapor pressure of these compounds. In addition, the relatively low octanol–water partition coefficients of these chemicals result in their relatively low affinity for sorption onto organic matter and minerals (Chapelle, 1993; Chapter 5, this volume).

pH Values. The pH is an important factor influencing microbial growth rates, the biodegradation of pollutants, and other microbially mediated reactions (Armenante et al., 1993; Lallai et al., 1988). The hydrogen ion activity is governed by metabolic compounds produced by microorganisms and is controlled especially by the CO_3^{2-}:HCO_3^-:CO_2 equilibria. Other conditions being equal, a higher pH favors primary production through the availability of CO_2 in the forms of HCO_3^- and CO_3^{2-}. Most bacteria show optimal growth in the neutral to slightly basic pH range, while fungi prefer a slightly acidic pH of 5–6. The presence of carbonate minerals (limestone, dolomite, shell material) in an aqueous system, such as aquifers or near beaches, buffers pH changes that would otherwise result from biological production of carbon dioxide or other acids and bases. Extremes of pH inhibit microbial degradation processes; microbial oxidation is most rapid between pH 6 and 8 (Scow, 1982). For ionizable compounds such as pentachlorophenol, the apparent octanol–water partition coefficient varies as a nonlinear function with the pH of the aqueous solution (Kaiser and Valdmanis, 1982).

Redox Potential. The redox potential, termed E_h, is extremely important in considerations of the biodegradation of organic and inorganic contaminants (Water Science and Technology Board et al., 1993). Biodegradation is more extensive when high concentrations of oxygen or other electron acceptors are available. Usually, heavily contaminated sites are anoxic because ongoing microbial respiration has depleted all available O_2. The resulting anoxic conditions tend to favor different electron acceptors, with the most oxidized compounds (higher E_h) being used first (Godsy, 1994).

This means that NO_3^- (nitrate) is utilized (denitrification) after the available oxygen is depleted. Once nitrate is depleted, Fe^{3+} (iron reduction) is utilized. If Fe^{3+} is absent or depleted, SO_4^{2-} is utilized (sulfate reduction). Finally, when all the sulfate is reduced, CO_2 (carbon dioxide) is used as an electron acceptor and is reduced to CH_4 (methane). Thus the composition of bacterial populations changes, depending on the availability of the various electron acceptors. Contaminants in an aquifer may serve as electron donors or electron acceptors, depending on their chemistry. For example, halogenated compounds are highly oxidized compared to their nonhalogenated counterparts, and thus tend to accept electrons and to be reductively dehalogenated. Such reductive dehalogenation reactions are influenced by the presence of other electron acceptors, such as nitrate or sulfate (Häggblom and Young, 1995; Häggblom et al., 1993). Reductive dechlorination is more frequently observed under methanogenic conditions where these anions are less available (Godsy, 1994).

Temperature. Temperature is an important factor controlling microbial metabolic activity and thus the rates of organic matter decomposition in water. Most microorganisms are capable of degradation at most ambient temperatures. However, the temperature can influence biodegradation rates by changing a contaminant's physical properties, bioavailability, or toxicity to microorganisms (Larsen et al., 1991). As pointed out above, an increase in temperature may increase the equilibrium vapor concentration, resulting in an increased volatilization rate. Temperature changes can have similar indirect effects on biodegradation rates through the temperature-dependence of other physicochemical processes (Godsy, 1994).

Solubility. Water-insoluble compounds are thought to persist longer than those that are water-soluble. Alexander (1973) suggests the following reasons for this behavior:

1. Inability of the compound to reach the reaction site in the microbial cell
2. A reduced rate of reaction when biodegradation is regulated by the rate of solubilization
3. The inaccessibility of insoluble compounds because of increased adsorption or trapping in inert material

Typical hydrocarbons are immiscible with water and build a separate liquid phase in an aqueous system. They are therefore called non-aqueous-phase liquids (NAPL). The water solubility of each component of a complex hydrocarbon mixture can vary from ideal conditions due to cosolvency effects. Interaction between components can either enhance the solubility of a given component (cosolvency) or reduce the solubility of that component by a kind of salting-out process (Groves, 1988). Microbial degradation of hydrocarbons is a multiphase reaction, involving gas, water-soluble hydrocarbons, water-insoluble hydrocarbons, water, dissolved salts, solids, and microorganisms.

The fact that the first step in hydrocarbon catabolism involves a membrane-bound oxygenase makes it essential for microorganisms to come into direct contact with the hydrocarbon substrate (Rosenberg et al., 1992). Microbial growth always proceeds at the hydrocarbon–water interface. To enhance cell–hydrocarbon contact, bacteria either directly adhere to the petroleum or produce extracellular oil emulsifying agents. Microorganisms of the first class have a high cell-surface hydrophobicity and can adhere to the oil phase. The chemical at or near the point of contact with the organism passes through the cell surface into the cytoplasm.

Organisms in the second class do not have a high cell-surface hydrophobicity, do not adhere to oil, produce emulsifying agents (biosurfactants such as lipoproteins, emulsans, etc.), and take up emulsified hydrocarbons (Alexander, 1994). Biosurfactants or bioemulsifiers are surface-active and are amphiphilic in nature. Hence, these molecules concentrate at the oil–water or air–water interface. The formation of an ordered molecular film at the interface lowers the interfacial energy and causes the unique properties of the surfactant (Phale et al., 1995). Emulsifier production is induced by growth on hydrocarbons. Biosurfactants produced by microorganisms show substrate specificity. For example, the biosurfactant emulsan is specific for a mixture of aliphatic and aromatic hydrocarbons only (Rosenberg et al., 1979). The Biosur-Pm produced by *Pseudomonas maltophila* CSV89 showed maximum emulsification activity toward 1-naphthaldehyde and good activity toward substituted aromatic hydrocarbons (Phale et al., 1995). Mutants which do not produce the emulsifier grow poorly on hydrocarbons. Besides extracellular biosurfactants, cell-bound biosurfactants are proposed to be involved in adhesion to hydrophobic surfaces and transport of hydrocarbons (Bar-Ness et al., 1988).

Influence of Nonbiological (Abiotic) Mechanisms on Biodegradation Reactions

As mentioned above, pollutants can be degraded through both biotic and abiotic reactions. Although abiotic degradation reactions are not the topic of this chapter, they should be mentioned because it is apparent that chemical and biological oxidation and reduction processes are strongly coupled. Microorganisms may provide the bulk reactants for abiotic reactions in which microorganisms are not directly involved. Nevertheless, such abiotic degradation reactions are important from a biological point of view because they have an impact on microbially mediated reactions. Abiotic transformation of pollutants can lead to products that are more readily or sometimes less biodegradable (Sunda and Kleber, 1994).

Abiotic Reduction of Organic Pollutants. Abiotic reduction of organic pollutants include reactions such as the reductive dehalogenation of polyhalogenated hydrocarbons (Hashsham et al., 1995), the reduction of nitroaromatic compounds (Tratnyek and Macalady, 1989; Wolfe et al., 1986; Dunnivant et al., 1992), and the reduction of aromatic azo compounds (Weber and Wolfe, 1987). The most abundant natural reductants, often designated as "bulk" reductants, present in reducing environments include reduced inorganic forms of iron and sulfur, such as aqueous iron(II), iron(II) carbonates, iron(II) sulfides, and hydrogen sulfide. These bulk reductants transfer electrons to mediators, such as quinone-type compounds found in natural organic matter, which pass these electrons to the pollutants. The mediators may be reduced again by the bulk reductants (Dunnivant et al., 1992; Chapter 5, this volume).

Hydrolysis. Abiotic hydrolyses of anthropogenic organic chemicals, especially pesticides, have been extensively investigated. Most of these studies focus on factors that enhance or retard the hydrolytic reactions, such as the effects of natural organic matter, such as humic substances, in aquatic systems (Macalady et al., 1989), the effects of surface acidity of clays or of transition metal cations (Tchelet et al., 1993). In the hydrolytic process, an organic molecule, RX, reacts with water to form a new carbon oxygen bond with the water molecule and cleave the $C-X$ bond in the original molecule. The net reaction is commonly a direct replacement of X by OH. Important classes of organic molecules that undergo hydrolytic reactions include aliphatic halides; esters of carboxylic, phosphoric, phosphonic, sulfonic, and sulfuric acids; carbamates; epoxides; amides; and nitriles.

Photolysis. When aqueous solutions of chemical compounds are exposed to sunlight, photodegradation or photodecomposition reactions may occur. A typical example is the photolysis of the herbicide alachlor [2-chloro-2′,6′-diethyl-N(methoxymethyl)-acetanilide]. Irradiation with UV light dechlorinates the molecule and forms a number of intermediates (Fang, 1977; Somich et al., 1988). Solutions of alachlor that were subjected to photolysis and ozonation were rapidly mineralized, whereas the microbial degradation of untreated parent compounds was very slow (Somich et al., 1988).

Another example is the photolysis of pentachlorophenol (PCP), a pesticide that is widely used for wood preservation. PCP absorbs sunlight readily in the ultraviolet region of sunlight ($\lambda_{max} = 320$ nm). The photochemical reaction leads to degradation products such as chloranilic acid (2,5-dichloro-3,6-dihydroxybenzoquinone), tetrachlororesorcinol, and several complex chlorinated benzoquinones. As Crosby et al. (1972) reported, the initial and rate-limiting reaction is the photonucleophilic replacement of PCP chlorine atoms by hydroxyl groups. The chlorine atoms can also be displaced by chlorophenoxide ions and may lead to the generation of extremely toxic chlorinated dibenzo-p-dioxins (Wong and Crosby, 1978). The above examples clearly show that in light-penetrable waters, photolysis can inhibit or support microbial degradation.

As previously discussed, photolysis is important not only for dissolved molecules, but also for those that evaporate from the aqueous system. Evaporated monoaromatics, along with naphthalenes and substituted naphthalenes, can be assigned a mean half-life in air of approximately one day, compared to approximately two days for biphenyl, acenaphthene, fluorene, phenanthrene, and anthracene and seven days for the 4- and 5-ring polycyclic aromatic hydrocarbons (Wolfe et al., 1994).

Sorption of Chemicals and Microbes to Particulate Matter. Many organic contaminants are hydrophobic and tend to sorb onto particulate matter (mineral colloids, humic materials, bacterial surfaces). Only a small fraction of the compounds present in an aqueous environment may actually be in the bulk water phase. Most evidence indicates the uptake of compounds by microorganisms proceeds via the liquid phase. Consequently, a process such as sorption that reduces the solution concentration tends to reduce the biotransformation rate (Stevens et al., 1990). However, this is only true if the rate of desorption of contaminants from the particulates to water is slower than the rate of intrinsic biodegradation (Bouwer et al., 1994).

Microorganisms have a tendency to adsorb onto particulate matter suspended in aqueous systems and to build a biofilm on surfaces (Toth and Tomasovicova, 1989). Long ago, ZoBell (1943) proposed that bacteria attached to surfaces gain a growth advantage in low nutrient waters by utilizing organic molecules and nutrients adsorbed at the surfaces as energy source. Solid surfaces may also retard the diffusion of exoenzymes away from the microorganisms, thereby promoting the assimilation of nutrients (ZoBell, 1943). It is also likely that microbial metabolites accumulate within biofilms (Marshall and Goodman, 1994). This may be of importance if a sufficiently high level of a metabolite is necessary to induce the expression of genes encoding further metabolic processes.

Most particulate surfaces in natural aqueous systems possess or acquire a net negative surface charge. Counterions (cations, including protons) are attracted to the surface. This can lead to a lower pH in the microenvironment near the surface which may have an effect on gene expression and can cause alterations of the transmembrane potentials of bacteria adhering to the surfaces (Marshall and Goodman, 1994).

HOW MICROBES CHANGE THE INORGANIC CHEMISTRY OF AQUEOUS SYSTEMS

Changes in Solubility

As mentioned above, microbes not only change structures of organic molecules through their metabolic activities, they also react directly or indirectly with inorganic compounds. Due to microbial activities, inorganic chemicals are oxidized or reduced. Consequently, these inorganic molecules behave differently in aqueous systems. Sometimes, insoluble inorganic compounds may become soluble; in other instances, compounds precipitate or volatilize due to microbial activities. Biotransformations of iron and manganese are notable in this regard.

A less obvious example is that soluble uranium(VI) is reduced by some sulfate-reducing and Fe(III)-reducing bacteria to the insoluble U(IV) mineral uraninite. This microbial reduction of soluble U(VI) to insoluble U(IV) may play an important role in the geochemical cycle of uranium and may also serve as a mechanism for the bioremediation of uranium-contaminated water (Lovely and Philips, 1992). An other example for the action of microorganisms on metal compounds and for the microbially mediated change of solubility is the production of acid-mine waters resulting from the oxidation of pyrite (FeS_2) or other metal sulfides to sulfuric acid (Manahan, 1993).

Microbial Transformations of Nitrogen

Nitrogen levels in the environment are affected by an interacting web of processes, including the nitrogen fixation, the nitrification, the ammonification, the nitrate reduction to ammonium, and denitrification reactions (Kuenen and Robertson, 1988; Carter et al., 1995). What follows is a short description of the reactions involved.

Nitrogen Fixation. Most nitrogen is stored in the atmosphere in molecular form, N_2, which is chemically inert and not suitable for most living organisms. A number of specialized organisms, including *Azotobacter*, *Rhizobium*, and *Cyanobacteria*, can fix nitrogen—that is, convert it to organic forms (Grady and Lim, 1980). Because this reaction is highly energy-consuming, the amount of nitrogen fixation that takes place in oligotrophic aqueous systems is very limited (Chapelle, 1993).

Ammonification. Organic nitrogen compounds are mineralized by microorganisms, and the nitrogen is released as ammonia (NH_3) without any change of oxidation state. This decomposition reaction is called ammonification (Grady and Lim, 1980).

Nitrification. A portion of the ammonia in the biosphere is oxidized first to nitrite (NO_2^-) and then to nitrate (NO_3^-) by two highly specialized kinds of bacteria. This conversion, nitrification, does not alter the availability of the nitrogen to plants and algae since they can assimilate nitrate as well as ammonia. Once the nitrate is in the cell, it is reduced to ammonia for incorporation into organic compounds. Discharges of ammonia in wastewater treatment plant effluents can lead to nitrification with resultant oxygen depletion in the receiving stream. Therefore, nitrification is often carried out in biochemical operations within the plant (Grady and Lim, 1980).

Dissimilatory Nitrate Reduction. Under conditions where oxygen is deficient, bacteria can use nitrate in place of oxygen as a final electron acceptor in metabolism. Nitrate can be reduced to ammonium (nitrate ammonification) or to nitrogen gas (denitrification). Denitrification occurs in a number of steps, each of which is catalyzed by different microorganisms, in the order $NO_3 \rightarrow NO_2 \rightarrow N_2O \rightarrow N_2$ (Chapelle, 1993). Nitrogen in wastewater is removed via denitrification processes (Grady and Lim, 1980). Anaerobic biodegradation of petroleum hydrocarbons in polluted soil or ground water is often indicated by an increased release of nitrogen gas which indicates that nitrate is used as electron acceptor (Levin and Gealt, 1993). Recently, several bacterial strains isolated from soils and freshwater sediments were shown to reduce nitrate under aerobic conditions (Carter et al., 1995). These bacteria use a periplasmatic nitrate reductase for nitrate respiration.

Microbial Transformations of Sulfur

Sulfur undergoes a number of chemical transformations in nature mediated exclusively by microorganisms. The sulfur cycle involves a variety of sulfur species, such as inorganic soluble sulfates, insoluble sulfates, soluble sulfide, gaseous hydrogen sulfide, insoluble sulfides, biologically bound sulfur, elemental sulfur, polysulfides, and sulfur in synthetic organic compounds (Manahan, 1993).

Sulfate Reduction. Many microorganism use sulfate as sulfur source for biosynthesis, a process which is called assimilative sulfate reduction. If sulfate is used as an electron acceptor for energy-generating reactions, the process is called dissimilative sulfate reduction (Brock et al., 1994). Dissimilative sulfate reduction is performed by the sulfate-reducing bacteria, a group of anaerobes which are widely distributed in nature. The end product of sulfate reduction is hydrogen sulfide (H_2S, or HS^-). At the high sulfate concentrations found in seawater (20–30 mM), sulfidogenesis is the major terminal electron-accepting process during degradation of organic matter in marine sediments and is therefore central to carbon turnover (Capone and Kiene, 1988). Sulfates as electron acceptors have been shown to support anaerobic degradation of halogenated aromatic compounds in freshwater and estuarine sediments (Häggblom et al., 1993) and of toluene and xylene in aquifer sediments (Edwards et al., 1992). Häggblom and Young (1995) found that biodegradation of chlorophenol in estuarine sediments is dependent on sulfidogenesis and did not proceed in the absence of sulfate. This example clearly shows how microbial changes of inorganic chemicals have an impact on organic molecules in nature, and vice versa.

Sulfide Oxidation. Reduced sulfur compounds, hydrogen sulfide and elemental sulfur, can be used by a variety of chemolithotrophic bacteria as a source of energy. The reaction

$$2 \, H_2S + O_2 \rightarrow 2H_2O + 2 \, S$$

yields 83 kcal of energy, and the reaction

$$2S + O_2 + 2 \, H_2O \rightarrow 2H_2SO_4$$

yields 236 kcal of energy (Chapelle, 1993).

Important aquatic habitats for sulfide-oxidizing bacteria are "black smokers," hot hydrothermal vents on the sea floor of tectonically active regions. The chemolithotrophic bacteria grow on the expense of inorganic energy sources emitted from the vents and provide the base of a food chain that includes grazing protozoa, filter feeders, crabs, and fish (Chapelle, 1993; Brock et al., 1994).

Degradation of Organic Sulfur Compounds. When organic compounds containing reduced sulfur undergo decomposition, the sulfur is released as H_2S (putrefaction and desulfurylation). This decomposition reaction is a significant source of sulfides in fresh water. In the sulfate-rich marine environment, dissimilatory sulfate reduction is the main source of hydrogen sulfide (Brock et al., 1994). The sulfides are rapidly oxidized to sulfate under aerobic conditions. Under anaerobic conditions, however, the hydrogen sulfide can cause problems because of odor and toxicity.

Microbial Transformations of Phosphorus

Phosphorus occurs in nature in the form of organic and inorganic phosphates and is essential to microorganisms for the formation of nucleic acids and phospholipids. Phosphorus is the most common limiting nutrient in fresh water, particularly for the growth of algae. Bacteria are more effective than algae in taking up phosphate from water, accumulating it as excess cellular phosphorus that can be released to support bacterial growth if the supply of phosphorus becomes limiting (Manahan, 1993). Studies of aquatic ecosystems show that bacteria take up not only water-soluble phosphorus (Holtan et al., 1988) but also water-insoluble, hydrophobic organic phosphates, such as the cell membrane phospholipids from dying cells (Lemke et al., 1995). Microorganisms can utilize organic phosphates through the action of cell enzymes called phosphatases. These mineralization reactions which release inorganic phosphorus from organic forms are important because they supply algae with the necessary orthophosphates and deactivate highly toxic organophosphate esters, such as the insecticide parathion (Tchelet et al., 1993).

Microbial Transformations of Iron Minerals

Ferrous iron (Fe^{2+}) is relatively soluble in water and is therefore quite mobile. Ferric iron (Fe^{3+}), on the other hand, tends to form insoluble Fe(III) oxyhydroxides and is therefore relatively immobile (Chapelle, 1993). In aerobic aquatic systems with a near-neutral pH, ferrous iron rapidly oxidizes nonbiologically to the ferric state.

Bacterial Iron Oxidation. At an acid pH, ferrous iron oxidation occurs much less rapidly and is available for microbial oxidation (Chapelle, 1993). Thus, most iron-oxidizing bacteria, such as *Thiobacillus ferrooxidans*, are obligately acidophilic. They are able to use carbon dioxide as a carbon source and either ferrous iron or reduced sulfur compounds as electron donors. Another iron-oxidizing bacterium is the archae-bacterium *Sulfolobus* which lives in hot, acid springs at temperatures up to the boiling point of water. At near-neutral pH, iron is only soluble at interfaces between anoxic and oxic conditions. Here we find iron-oxidizing bacteria such as *Gallionella ferruginea*, *Sphaerotilus natans*, and *Leptothrix ochracea* which form characteristic iron deposits (Brock, 1994).

Bacterial Iron Reduction. Ferric iron can be reduced to the ferrous state by a variety of microorganisms and leads to the solubilization of iron, an important geochemical process. In coastal sediments, the reduced iron has been shown to precipitate sulfides generated by dissimilatory sulfate reduction; that is, microbial iron reduction has an impact on the geochemistry of sulfur (Hines et al., 1991). The reduction of iron is also coupled to the oxidation of a wide variety of organic compounds. In these metabolic processes, ferric iron serves as an electron acceptor. As with nitrate reduction, this is a major form of anaerobic respiration (Brock et al., 1994). Lovely et al. (1990) reported the existence of microbial populations in deep subsurface aquifers that are capable of coupling the oxidation of organic matter to carbon dioxide with the reduction of Fe^{3+}. As Crawford (1991) points out, the existence of ground-water microorganisms capable of coupling the anaerobic reduction of Fe^{3+} to the oxidation of organic compounds may allow the bioremediation of ground water contaminated with hydrocarbon pollutants. The existence of such microorganisms has been confirmed by Lovely and Lonergan (1990), who describe the first microorganism known to couple the oxidation of toluene, phenol, and *p*-cresol to the reduction of Fe^{3+}.

Microbial Transformations of Manganese

Manganese (Mn) can exist in the oxidation states 0, +2, +3, +4, +6, and +7. However, in nature only the +2, +3, and +4 oxidation states are commonly found. Of these, only Mn(II) can occur as a free ion in solution. Manganese(III) can occur in solution only when it is complexed (Ehrlich, 1990). As in the case of iron, the most important geomicrobial interactions with manganese are those which lead to precipitation of dissolved manganese in an insoluble phase or solubilization of insoluble forms of manganese. These reactions frequently, but not always, involve oxidations or reductions of manganese (Ehrlich, 1990). Microbial reduction or oxidation of Mn may be enzymatic or nonenzymatic. Nonenzymatic microbial redox reactions of manganese are caused indirectly by microbially produced metabolic products that react with the manganese (Bromfield, 1979).

Microbial Manganese Oxidation. Many microorganisms in natural waters and sediments are capable of oxidizing manganese(II) to manganese oxides. Manganese oxides can be found in large quantities in concretions (nodules) or crusts on the ocean floors. The average nodule contains more than 20% manganese, 6% iron, and about 1% each of copper and nickel (Press and Siever, 1986). The oxidation state of manganese in nodules is mainly +4 (Ehrlich, 1990). The Mn(IV) oxides in the nodules have a strong capacity to scavenge cations including divalent manganese, cobalt, copper, and nickel. Therefore the surface of the nodules is an optimal habitat for those bacteria that oxidize divalent manganese ions. Microbes isolated from nodules oxidize divalent manganese only if it is first bound to Mn(IV) oxides or ferromanganese (Ehrlich, 1982). The newly produced Mn(IV) oxides again act as cation scavenging sites, and in this way nodules keep growing.

Some manganese-oxidizing bacteria deposit manganese oxides on their surfaces or on an extracellular polymer sheath. The manganese oxide coating may protect the cells against toxic metals, oxygen species, or ultraviolet light (Ghiorse, 1984). Sunda and Kleber (1994) reported that microbially produced Mn oxides lyse complex humic substances. Humic substances represent one of the largest organic reservoirs in natural water, sediments, and soils. In general, microorganisms cannot use these recalcitrant complex molecules as a carbon source. However, the oxidation of the refractory pools of natural organic matter by Mn oxides may produce a variety of low-molecular-weight carbon compounds that can be used as substrates for microbial growth.

Microbial Manganese Reduction. Some bacteria have the capacity to reduce manganese(IV) to manganese(II) aerobically, while others reduce manganese only under strict anaerobic conditions. Under anaerobic conditions the microbes, mostly chemoorganotrophs, are not restricted to Mn(IV) but can also use Fe(III), nitrate, fumarate, glycine, iodate, sulfite, and thiosulfate as electron acceptors. The organisms are also quite versatile in regard to the electron donors used for manganese oxide reduction. Depending on the species, butyrate, propionate, lactate, succinate, acetate, pyruvate, formate, and H_2 may be used as electron donors (Brock et al., 1994).

Tetravalent manganese from ferromanganese concretions (nodules) or crusts found on ocean floors can be enzymatically reduced to soluble divalent manganese in a respiratory process by some bacteria native to the concretions (Ehrlich and Brierley, 1990). Co, Cu, and Ni are present in variable concentrations in these crusts and are solubilized as the manganese is reduced. However, only little of the iron in the concretions is solubilized (Ehrlich et al., 1973). The number of Mn(IV) reducers growing on the nodules was found to be much greater than the number of Mn(II) oxidizers.

Microbial Interactions With Other Minerals

The above-described interactions of microorganisms with nitrogen, sulfur, phosphorus, iron, and manganese are only a few examples that show how microbes influence geochemical cycles. This chapter can only cover a limited number of microbial processes. However, it should be mentioned that microbes also have important impacts on the chemistry of carbonates, silicon, mercury, chromium, selenium, tellurium, and so on (Ehrlich, 1990).

CONCLUSIONS

The conditions in aquatic systems are constantly shaped by microbially mediated reactions. Microbes play an important role in the biogeochemical cycling of organic and inorganic compounds. Scientists studying these complex reactions in aquatic systems must understand that chemical, physical, and biological reactions are interacting and that only an interdisciplinary research approach will make it possible to learn more about these reactions.

Most of the knowledge about aqueous systems was gained from bench-scale experimentation. However, field conditions are more complex, involving many reactants, inhibiting and activating chemicals, competing substrates, and many other factors. Microbial processes are difficult to evaluate, and therefore only a few *in situ* investigations have been performed. Further field investigations by scientists and engineers will be necessary to get a deeper understanding of microbially mediated reactions and the many factors that have an impact on the reactions. These future studies will certainly give us more fascinating insights into microbial activities in aquatic systems. This knowledge will allow us to better understand and optimize already existing technologies and to develop new technologies in the field of engineering, mining, and environmental remediation.

References

Alexander, M., 1973, Nonbiodegradable and other recalcitrant molecules. *Biotechnol. Bioeng.* 15, 611–647.

Alexander, M., 1994, *Biodegradation and Bioremediation*, Academic Press, San Diego, CA.

Apajalahti, J. H. A., and Salkinoja-Salonen, M. S., 1987, Complete dechlorination of tetrachlorohydroquinone by cell extracts of pentachlorophenol-induced *Rhodococcus chlorophenolicus. J. Bacteriol.*, 169, 5125–5130.

Armenante, P. M., Kafkewitz, D., Jou, C.-J., and Lewandowski, G., 1993, Effect of pH on the anaerobic dechlorination of chlorophenols in a defined medium. *Appl. Microbiol. Biotechnol.*, 39, 772–777.

Ashton, F. M., and Crafts, A. S., 1981, *Mode of Action of Herbicides*, John Wiley & Sons, New York.

Atlas, R. M., and Bartha, R., 1992, *Microbial Ecology: Fundamentals and Applications*, Benjamin-Cummings, Redwood City, CA.

Bar-Ness, R., Avrahamy, N., Matsuyama, T., and Rosenberg, M. 1988, Increased cell surface hydrophobicity of a *Serratia marcescens* NS38. *J. Bacteriol.*, 170, 4361–4364.

Behki, R., Topp, E., Dick, W., and Germon, P., 1993, Metabolism of the herbicide atrazine by *Rhodococcus* strains. *Appl. Environ. Microbiol.*, 59, 1955–1959.

Behki, R. M., and Khan, S. U., 1988, Degradation of atrazine by *Pseudomonas*: N-dealkylation and dehalogenation of atrazine and its metabolites. *J. Agric. Food Chem.*, 34, 746–749.

Bouwer, E., Durant, N., Wilson, L., Zhang, W., and Cunningham, A., 1994, Degradation of xenobiotic compounds *in situ*: capabilities and limits. *FEMS Microbiol. Rev.*, 307 317.

Brock, T. D., Madigan, M. T., Martinko, J. M., and Parker, J., 1994, *Biology of Microorganisms*, Prentice-Hall, Englewood Cliffs, NJ.

Bromfield, S. M., 1979, Manganous ion oxidation at pH values below 5.0 by cell-free substances from *Streptomyces* sp. cultures. *Soil Biol. Biochem.*, 11, 115–118.

Capone, D. G., and Kiene, R. P., 1988, Comparison of microbial dynamics in marine and freshwater sediments: contrasts in anaerobic carbon metabolism. *Limnol. Oceanogr.*, 33, 725–749.

Carter, J. P., Hsiao, Y. H., Spiro, S., and Richardson, D. J., 1995, Soil and sediment bacteria capable of aerobic nitrate respiration. *Appl. Environ. Microbiol.*, 61, 2852–2858.

Chapelle, F. H., 1993, *Ground-Water Microbiology and Geochemistry*, John Wiley & Sons, New York.

Crawford, R. L., 1991, Bioremediation of groundwater pollution. *Curr. Opin. Biotechnol.*, 2:436–439.

Crosby, D. G., Moilanen, K. W., Nakagawa, M., and Wong, A. S., 1972, Photonucleophilic reactions of pesticides. In: F. Matsumura, G. M. Boush, and T. Misato, eds., *Environmental Toxicology of Pesticides*, Academic Press, New York, pp. 423–431.

Cserjesi, A. J., and Johnson, E. L., 1972, Methylation of pentachlorophenol by *Trichoderma virgatum. Can. J. Microbiol.* 18:45–49.

DiStefano, T. D., Gossett, J. M., and Zinder, S. H., 1992, Hydrogen as an electron donor for dechlorination of tetrachloroethene by an anaerobic mixed culture. *Appl. Environ. Microbiol.*, 58, 3622–3629.

Dunnivant, F. M., Schwarzenbach, R. P., and Macalady, D. L., 1992, Reduction of substituted nitrobenzenes in aqueous solutions containing natural organic matter. *Environ. Sci. Technol.*, 26, 2133–2141.

Edwards, E. A., Wills, L. E., Reinhard, M., and Grbic-Galic, D., 1992, Anaerobic degradation of toluene and xylene by aquifer miroorganisms under sulfate-reducing conditions. *Appl. Environ. Microbiol.*, 58, 794–800.

Ehrlich, H. L., 1982, Enhanced removal of Mn^{2+} from seawater by marine sediments and clay minerals in the presence of bacteria. *Can. J. Microbiol.*, 28, 1389–1395.

Ehrlich, H. L., 1990, *Geomicrobiology*, Marcel Dekker, New York.

Ehrlich, H. L., and Brierley, C. L., 1990, *Microbial Mineral Recovery*, McGraw-Hill, New York.

Ehrlich, H. L., Yang, S. H., and Mainwaring, J. D., 1973, Bacteriology of manganese nodules. VI. Fate of copper, nickel, cobalt, and iron during bacterial and chemical reduction of the manganese (IV). *Z. Allg. Mikrobiol.*, 13, 39–48.

Fang, C.-H., 1977, Effects of soils on the degradation of herbicide alachlor under light. *J. Chin. Agric. Chem. Soc.*, 15:53–59.

Fathepure, B. Z., Tiedje, J. M., and Boyd, S. A., 1988, Reductive dechlorination of hexachlorobenzene to tri- and dichlorobenzenes in anaerobic sewage sludge. *Appl. Environ. Microbiol.*, 54:372–330.

Ghiorse, W. C., 1984, Biology of iron- and manganese-depositing bacteria. *Annu. Rev. Microbiol.*, 38:515–550.

Giardina, M. C., Giardi, M. T., and Filacchioni, G., 1980, 4-Amino-2-chloro-1,3,5-triazine: a new metabolite of atrazine by a soil bacterium. *Agric. Biol. Chem.*, 44, 2067–2072.

Glaser, J. A., 1989, Enzyme systems and related reactive species, p. 9.61–9.73. In: H. M. Freeman, ed., *Standard Handbook of Hazardous Waste Treatment and Disposal*, McGraw-Hill, New York.

Godsy, E. M., 1994, Microbiological and geochemical degradation processes. In: *Symposium on Intrinsic Bioremediation of Ground Water*, Denver, Colorado, August 30–

September 1, 1994, Office of Research and Development, U.S. Environmental Protection Agency, Washington, D.C. (EPA/540/R-94/515).

Grady, C. P., and Lim, H. C., 1980, Biological wastewater treatment: theory and applications. In: *Pollution Engineering and Technology*, Vol. 12. P. M. Cheremisinoff, ed., Marcel Dekker, New York, pp. 197–228.

Grbic-Galic, D., and Vogel, T. M., 1987, Transformation of toluene and benzene by mixed methanogenic cultures. *Appl. Environ. Microbiol.*, 53, 254–260.

Groves, F. R., 1988, Effect of cosolvents on the solubility of hydrocarbons in water. *Environ. Sci. Technol.*, 22, 282–286.

Häggblom, M. M., and Young, L. Y., 1995, Anaerobic degradation of halogenated phenols by sulfate-reducing consortia. *Appl. Environ. Microbiol.*, 61, 1546–1550.

Häggblom, M. M., Nohynek, L. J., and Salkinoja-Salonen, M. S., 1988, Degradation and *o*-methylation of chlorinated phenolic compounds by *Rhodococcus* and *Mycobacterium* strains. *Appl. Environ. Microbiol.*, 54, 3043–3052.

Häggblom, M. M., Rivera, M. D., and Young, L. Y., 1993, Influence of alternative electron acceptors on the anaerobic biodegradability of chlorinated phenols and benzoic acids. *Appl. Environ. Microbiol.*, 59, 162–1167.

Hashsham, S. A., Scholze, R., and Freedman, D. L., 1995, Cobalamin-enhanced anaerobic biotransformations of carbon tetrachloride. *Environ. Sci. Technol.*, 29, 2856–2863.

Hendriksen, H. V., and Ahring, B. K., 1993, Anaerobic dechlorination of pentachlorophenol in fixed-film and upflow anaerobic sludge blanket reactors using different inocula. *Biodegradation*, 3, 399–408.

Hines, M. E., Bazylinski, D. A., Tugel, J. B., and Lyons, W. B., 1991, Anaerobic microbial biogeochemistry from two basins in the Gulf of Maine: evidence for iron and manganese reduction. *Estuarine Coastal Shelf Sci.*, 32, 313–324.

Holtan, H., Kamp-Nielsen, L., and Stuanes, A. O., 1988, Phosphorus in soil, water and sediment: an overview. *Hydrobiologia*, 170, 19–34.

Kaake, R. H., Crawford, D. L., and Crawford, R. L., 1994, Optimization of an anaerobic bioremediation process for soil contaminated with the nitroaromatic herbicide dinoseb (2-*sec*-butyl-4,6-dinitrophenol). In: R. E. Hinchee, D. B. Anderson, F. B. Metting, and G. D. Sayles, eds., *Applied Biotechnology for Site Remediation*, Lewis, Boca Raton, FL, pp. 337–341.

Kaiser, K. L. E., and Valdmanis, I., 1982, Apparent octanol/water partition coefficients of pentachlorophenol as a function of pH. *Can. J. Chem.*, 60, 2104–2106.

Kaufman, D. D., and Blake, J., 1970, Degradation of atrazine by soil fungi. *Soil Biol. Biochem.*, 2, 73–80.

Kemmer, F. N., 1988, *The Nalco Water Handbook*, McGraw-Hill, New York.

Knowles, C. J., and Wyatt, J. W., 1992, The degradation of cyanide and nitriles. In: J. C. Fry, G. M. Gadd, R. A. Herbert, C. W. Jones, and I. A. Watson-Craik, eds., *Microbial Control of Pollution*, 48th Symposium of the Society for General Microbiology, University of Cardiff, Cambridge University Press, Cambridge, pp. 113–128.

Kuenen, J. G., and Robertson, L. A., 1988, Ecology of nitrification and denitrification. In: J. A. Cole and S. J. Ferguson, eds., *The Nitrogen and Sulphur Cycles*, Cambridge University Press, Cambridge, pp. 161–218.

Lallai, A., Mura, G., Miliddi, R., and Mastinu, C., 1988, Effect of pH on growth of mixed cultures in batch reactors. *Biotechnol. Bioeng.*, 31, 130–134.

Larsen, S., Hendriksen, H. V., and Ahring, B. K., 1991, Potential for thermophilic (50°C) anaerobic dechlorination of pentachlorophenol in different ecosystems. *Appl. Environ. Microbiol.*, 57, 2085–2090.

Lemke, M. J., Churchill, P. F., and Wetzel, R. G., 1995, Effect of substrate and cell surface hydrophobicity on phosphate utilization in bacteria. *Appl. Environ. Microbiol.*, 61, 913–919.

Levin, M. A., and Gealt, M. A., 1993, *Biotreatment of Industrial and Hazardous Waste.* McGraw-Hill, New York.

Lovely, D., and Lonergan, D., 1990, Anaerobic oxidation of toluene, phenol and p-cresol by the dissimilatory iron-reducing microorganism, GS-15. *Appl. Environ. Microbiol.*, 56, 1858–1864.

Lovely, D. R., and E. J. P., Phillips, 1992, Reduction of uranium by *Desulfovibrio desulfuricans. Appl. Environ. Microbiol.*, 58, 850–856.

Lovely, D., Chapelle, F., and Phillips, E., 1990, Fe(III)-reducing bacteria in deeply buried sediments of the Atlantic coastal plain. *Geology*, 18, 954–957.

Macalady, D. L., Tratnyek, P. G., and Grundl, T. J., 1986, Abiotic reduction reactions of anthropogenic chemicals in anaerobic systems: a critical review. *J. Contam. Hydrol.*, 1, 1–28.

Macalady, D. L., Tratnyek, P. G., and Wolfe, N. L., 1989, Influence of natural organic matter on the abiotic hydrolysis of organic contaminants in aqueous systems. In: I. H. Suffet and P. MacCarthy, eds., *Aquatic Humic Substances: Influence on Fate and Treatment of Pollutants*, Advances in Chemistry Series, No. 219, American Chemical Society, Washington, D.C., pp. 323–332.

Manahan, S. E., 1993. *Fundamentals of Environmental Chemistry*, Lewis, Boca Raton, FL.

Marshall, K. C., and Goodman, A. E., 1994, Effects of adhesion on microbial cell physiology. *Colloids and Surfaces B: Biointerfaces*, 2, 1–7.

Mikesell, M. D., and Boyd, S. A., 1988, Enhancement of pentachlorophenol degradation in soil through induced anaerobiosis and bioaugmentation with anaerobic sewage sludge. *Environ. Sci. Technol.*, 22, 1411–1414.

Mondecar, M., Bender, J., Ross, J., George, W., and Preslan, J., 1994, Removal of 2,4,5-trinitrotoluene from contaminated water with microbial mats. In: R. E. Hinchee, D. B. Anderson, F. B. Metting, and G. D. Sayles, eds., *Applied Biotechnology for Site Remediation*, Lewis, Boca Raton, FL, pp. 342–345.

Nagy, I., Compernolle, F., Ghys, K., Vanderleyden, J., and de Mot, R., 1995, A single cytochrome P-450 system is involved in degradation of the herbicide EPTC (*S*-ethyl dipropylthiocarbamate) and atrazine by *Rhodococcus* sp. strain N186/21. *Appl. Environ. Microbiol.*, 61, 2056–2060.

Phale, P. S., Savithri, H. S., Rao, N. A., and Vaidyanathan, C. S., 1995, Production of biosurfactant "Biosur-Pm" by *Pseudomonas maltophila* CSV89: characterization and role in hydrocarbon uptake. *Arch. Microbiol.*, 163, 424–431.

Press, F., and Siever, R., 1986, *Earth*, W. H. Freeman, New York.

Radehaus, P. M., and Schmidt, S. K., 1992, Characterization of a novel *Pseudomonas* sp. that mineralizes high concentrations of pentachlorophenol. *Appl. Environ. Microbiol.*, 58, 2879–2885.

Radosevich, M., Traina, S. J., Hao, Y.-L., and Tuovinen, O. H., 1995, Degradation and mineralization of atrazine by a soil bacterial isolate. *Appl. Environ. Microbiol.*, 61, 297–302.

Rochkind-Dubinsky, M. L., Sayler, G. S., and Blackburn, J. W., 1987, *Microbiological Decomposition of Chlorinated Aromatic Compounds, Microbiology Series*, Vol. 18, A. I. Laskin, and R. I. Mateles, eds., Marcel Dekker, New York.

Rosenberg, E., Perry, A., Gibson, D. T., and Gutnick, D. L., 1979, Emulsifier of *Arthrobacter* RAG-1: specificity of hydrocarbon substrate. *Appl. Environ. Microbiol.*, 37, 409–413.

Rosenberg, E., Legmann, R., Kushmaro, A., Taube, R., Adler, E., and Ron, E. Z., 1992, Petroleum bioremediation—a multiphase problem. *Biodegradation*, 3, 337–350.

Sayles, G. D., and Suidan, M. T., 1993, Biological treatment of industrial and hazardous wastewater. In: M. A. Levin and M. A. Gealt, eds., *Biotreatment of Industrial and Hazardous Waste*, McGraw-Hill, New York, pp. 245–267.

Schmidt, S. K., Smith, R., Sheker, D., Hess, T. F., Silverstein, J., and Radehaus, P. M., 1992, Interactions of bacteria and microflagellates in sequencing batch reactors exhibiting enhanced mineralization of toxic organic chemicals. *Microb. Ecol.*, 23, 127–142.

Scow, K. M., 1982, Rate of biodegradation. In: W. J. Lyman, W. F. Reehl and D.H. Rosenblatt, eds., *Handbook of Chemical Property Estimation Methods. Environmental Behavior of Organic Compounds*, McGraw-Hill, New York, pp. 9.1–9.85.

Somich, C. J., Kearney, P. C., Muldoon, M. T., and Elsasser, S., 1988, Enhanced soil degradation of alachlor by treatment with ultraviolet light and ozone. *J. Agric. Food Chem.*, 36, 1322–1326.

Sterritt, R. M., and Lester, J. N., 1988, *Microbiology for Environmental and Public Health Engineers*, E. & F. N. Spon, London.

Stevens, T. O., Crawford, R. L., and Crawford, D. L., 1990, Biodegradation of dinoseb (2-*sec*-butyl-4,6-dinitrophenol) in several Idaho soils with various dinoseb exposure histories. *Appl. Environ. Microbiol.*, 56, 133–139.

Stevens, T. O., Crawford, R. L., and Crawford, R. L., 1991, Selection and isolation of bacteria capable of degrading dinoseb (2-*sec*-butyl-4,6-dinitrophenol). *Biodegradation*, 2, 1–13.

Sunda, W. G., and Kleber, D. J., 1994, Oxidation of humic substances by manganese oxides yields low-molecular-weight organic substrates. *Nature*, 367, 62–64.

Suzuki, T., 1978, Enzymatic methylation of pentachlorophenol and its related compounds by cell-free extracts of *Mycobacterium* sp. isolated from soil. *J. Pesticide Sci.*, 3, 441–443.

Suzuki, T., 1983, Methylation and hydroxylation of pentachlorophenol by *Mycobacterium* sp. isolated from soil. *J. Pesticide Sci.*, 8, 419–428.

Tchelet, R., Levanon, D., Mingelgrin, U., and Henis, Y., 1993, Parathion degradation by a *Pseudomonas* sp. and a *Xanthomonas* sp. and by their crude enzyme extracts as affected by some cations. *Soil Biol. Biochem.*, 25, 1665–1671.

Thompson, L. A., Knowles, C. J., Linton, E. A., and Wyatt, J. M., 1988, Microbial biotransformations of nitriles. *Chem. Br.*, 900–902.

Top, E. M., Holben, W. E., and Forney, L. J., 1995, Characterization of diverse 2,4-dichlorophenoxyacetic acid-degradative plasmids isolated from soil by complementation. *Appl. Environ. Microbiol.*, 61, 1691–1698.

Toth, D., and Tomasovicova, D., 1989, *Microbial Interactions with Chemical Water Pollution*, Ellis Horwood, Chichester.

Tratnyek, P. G., and Macalady, D. L., 1989, Abiotic reduction of nitro aromatic pesticides in anaerobic laboratory systems. *J. Agric. Food Chem.*, 37, 248–254.

Water Science and Technology Board, Commission on Engineering and Technical Systems, National Research Council, 1993, *In Situ Bioremediation. When Does It Work?* National Academy of Sciences, Washington, D.C.

Weber, E. J., and Wolfe, N. L., 1987, Kinetic studies of the reduction of aromatic azo compounds in anaerobic sediment/water systems. *Environ. Toxicol. Chem.*, 6, 911–916.

Wolfe, N. L., Kitchens, B. P., Macalady, D. L., and Grundl, T. J., 1986, Physical and chemical factors that influence the anaerobic degradation of methyl parathion in sediment systems. *Environ. Toxicol. Chem.*, 5, 1019–1026.

Wolfe, D. A., Hameedi, M. J., Galt, J. A., Watabayashi, G., et al., 1994, The fate of the oil spilled from the *Exxon Valdez. Environ. Sci. Technol.*, 28, 561–568.

Wong, A. S., and Crosby, D. G., 1978, Photolysis of pentachlorophenol in water. In: K. R. Rao, eds., *Pentachlorophenol: Chemistry, Pharmacology, and Environmental Toxicology*, Plenum Press, New York, pp. 19–25.

Zhang, X., and Wiegel, J., 1990, Sequential anaerobic degradation of 2,4-dichlorophenol in freshwater sediments. *Appl. Environ. Microbiol.*, 56, 1119–1127.

ZoBell, C. E., 1943, The effect of solid surfaces upon bacterial activity. *J. Bacteriol.*, 46, 39–56.

PART II

ATMOSPHERIC ENVIRONMENTAL CHEMISTRY

CHAPTER 10

Oxidant Formation in the Troposphere

JOHN W. BIRKS

ABSTRACT. In the troposphere, ozone concentrations have approximately doubled over the past century, almost certainly as a result of increasing emissions of carbon monoxide (CO), hydrocarbons, and oxides of nitrogen. Ozone and other oxidants derived from ozone are damaging to biological tissues and thus have detrimental effects on natural ecosystems, crops, and human health. Ozone is a secondary pollutant produced as a byproduct of the oxidation of CO and hydrocarbons in the presence of oxides of nitrogen. The level of nitric oxide (NO) may be thought of as a "chemical switch" that changes the pathways by which CO and hydrocarbons are oxidized from ones that naturally destroy ozone to ones that form ozone. Ozone formation may be greatly enhanced in rural areas as a result of the reactions of biogenic hydrocarbons such as isoprene and the monoterpenes with oxides of nitrogen which are transported from cities. This chapter describes the chemistry of ozone formation precursors and discusses the role of oxidant formation in acid precipitation.

INTRODUCTION

Ozone, O_3, is one of the most important trace gases in the atmosphere. In the stratosphere it serves as a protective shield against biologically harmful ultraviolet radiation at wavelengths between approximately 200 and 320 nm. Ultraviolet light in the UV-B region, 280–320 nm, is particularly damaging to life because of its strong absorption by proteins, nucleic acids, and other biological molecules. Hence the great concern for ozone depletion in the stratosphere resulting from anthropogenic emissions of halocarbons and other ozone depleting chemicals.

By contrast, in the troposphere the principal concern is the production of enhanced levels of ozone and other oxidants derived from ozone. The identification of ozone formation in photochemical smog was first made by Haagen-Smit and co-workers and described in a classic series of papers in the 1950s (Haagen-Smit et al., 1951, 1952, 1953, 1954, 1955, 1959). Oxidant production in the troposphere is enhanced by anthropogenic emissions of the pollutants carbon monoxide, hydrocarbons, and the oxides of nitrogen. Ironically, human pollution of the atmosphere tends to decrease ozone in the stratosphere where high concentrations of ozone are desirable and to increase ozone in the troposphere where ozone is undesirable because of its toxicity to plants and animals.

The toxicity of ozone results from the weakness of its $O-O$ chemical bond, which has a bond energy of only 143 kJ/mol, compared to the $O-O$ bond strength of 498 kJ/mol in O_2. Thus, ozone is predicted to be much more reactive, since it can more easily enter into chemical reactions in which the $O-O$ bond is broken and an oxygen atom is donated to another molecule—that is, an "oxidation" reaction. Such reactions are damaging to biological tissues, and exposure to high levels of ozone is harmful to human health, agricultural crops, and natural ecosystems. Atmospheric reactions of ozone result in the production of many other oxidants, including the hydroxyl radical (OH), the hydroperoxyl radical (HO_2), hydrogen peroxide (H_2O_2), peroxynitric acid (HO_2NO_2), the nitrate radical (NO_3), organic peroxides (ROOR), hydroperoxides (ROOH), peroxy acids (R(CO)OOH), and peroxyacyl nitrates (PANs). Furthermore, many of these oxidants are more damaging to plant and animal life than ozone itself.

The fact that tropospheric ozone has increased over the past century is evident from a comparison of ozone measurements made at Montsouris (near Paris) in the 1870s and 1880s (Volz and Kley, 1988) with those made at other Northern Hemisphere locations in the 1980s (Logan, 1985, 1988). Figure 10.1 shows the seasonal dependence of those measurements (Houghton et al., 1990, p. 29). In the 1880s the mixing

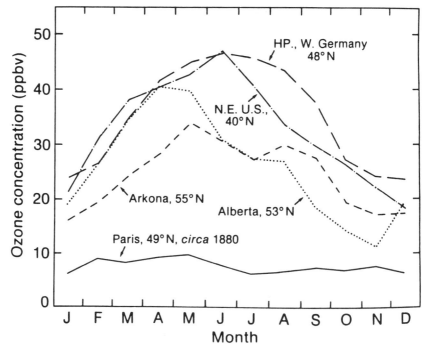

Figure 10.1. Comparison of monthly mean surface ozone mixing ratios from the 1970s and 1980s with observations at Montsouris (near Paris), France in the 1880s. Reprinted with permission from IPCC, 1990. Climate Change, The IPCC Scientific Assessment. Eds. J. T. Houghton, G. J. Jenkins & J. J. Ephraums. Cambridge University Press, UK, 365 pp.

ratio of ozone (fraction of air molecules that are ozone) averaged less than 10 ppbv (ppbv = part per billion by volume = 10^{-9} fraction of air molecules) throughout the year and had no significant seasonal trend, suggesting that most of this ozone resulted from transport from the stratosphere. By contrast, in the 1970s and 1980s, surface ozone ranged from winter mixing ratios of 15–25 ppbv to summer values of 30–45 ppbv. The strong seasonal variation is highly suggestive of a photochemical source of ozone that has increased severalfold over the past century.

MECHANISM OF OZONE PRODUCTION IN THE TROPOSPHERE

Ozone in the troposphere results from two sources: (1) transport of ozone downward from the stratosphere and (2) direct photochemical production within the troposphere. Transport from the stratosphere accounts for a background level of ozone in the troposphere of less than 10 ppbv. Levels of ozone much higher than this background, measured in the tens of ppbv in rural continental areas and up to hundreds of ppbv in heavily polluted urban areas, are due to photochemical production within the troposphere. The necessary ingredients for photochemical production of ozone in the troposphere are a carbon source (CO and/or hydrocarbons), oxides of nitrogen ($NO_x = NO + NO_2$), and sunlight.

The pivotal role of NO_x in ozone production is illustrated by considering the atmospheric oxidation of CO under conditions of low and high NO_x concentration:

Low NO_x	High NO_x
$CO + OH \rightarrow CO_2 + H$	$CO + OH \rightarrow CO_2 + H$
$H + O_2 + M \rightarrow HO_2 + M$	$H + O_2 + M \rightarrow HO_2 + M$
$HO_2 + O_3 \rightarrow OH + 2\,O_2$	$HO_2 + NO \rightarrow OH + NO_2$
	$NO_2 + h\nu \rightarrow NO + O$
Net: $CO + O_3 \rightarrow CO_2 + O_2$	$O + O_2 + M \rightarrow O_3 + M$
	Net: $CO + 2\,O_2 \rightarrow CO_2 + O_3$

In both reaction sequences, carbon monoxide oxidation is initiated by reaction with hydroxyl radicals, OH; in fact, OH is the only species known to react with CO in the troposphere. In the second reaction of each sequence, the hydrogen atom product rapidly combines with molecular oxygen in the atmosphere via a termolecular reaction to form the hydroperoxyl radical, HO_2. Here, M is any molecule, principally N_2 and O_2, and is necessary to remove kinetic energy and thereby stabilize the addition product, HO_2. The two reaction sequences diverge at the third reaction where HO_2 reacts either with O_3 (at low NO_x) or with NO (at high NO_x). The net results of the two reaction sequences, either destruction of ozone at low NO_x or formation of ozone at high NO_x, depends on the fate of HO_2. In relatively unpolluted air—that is, in areas remote from urban pollution—oxides of nitrogen are sufficiently low in concentration that the HO_2 radicals react preferentially with ozone and the oxidation of CO to CO_2 results in net ozone destruction, whereas in polluted atmospheres the HO_2 reacts preferentially with NO, and ozone is produced in the oxidation of CO.

For atmospheric oxidation of hydrocarbons an analogous branching occurs; for example, for methane, two reaction sequences are as follows:

Low NO_x	High NO_x
$CH_4 + OH \rightarrow CH_3 + H_2O$	$CH_4 + OH \rightarrow CH_3 + H_2O$
$CH_3 + O_2 + M \rightarrow CH_3O_2 + M$	$CH_3 + O_2 + M \rightarrow CH_3O_2 + M$
$CH_3O_2 + HO_2 \rightarrow CH_3OOH + O_2$	$CH_3O_2 + NO \rightarrow CH_3O + NO_2$
$CH_3OOH + h\nu \rightarrow CH_3O + OH$	$CH_3O + O_2 \rightarrow CH_2O + HO_2$
$CH_3O + O_2 \rightarrow CH_2O + HO_2$	$HO_2 + NO \rightarrow OH + NO_2$
	$(NO_2 + h\nu \rightarrow NO + O)$ Twice
Net: $CH_4 + O_2 \rightarrow CH_2O + H_2O$	$(O + O_2 + M \rightarrow O_3 + M)$ Twice

$$\text{Net: } CH_4 + 4\,O_2 \rightarrow CH_2O + H_2O + 2\,O_3$$

The oxidation of methane, like CO, is initiated by OH radicals. As for CO oxidation, the first two reaction steps are identical and the branch point for ozone formation versus ozone destruction occurs at the third reaction, where, in this case, the methylperoxy radical, CH_3O_2, may react with either HO_2 or NO. If the nitric oxide concentration is sufficiently high, the oxidation of methane to formaldehyde, CH_2O, produces two ozone molecules. In the presence of nitric oxide, the further oxidation of CH_2O to CO produces up to two more ozone molecules, and the subsequent oxidation of CO to CO_2, as described above, produces yet another ozone. Thus, a total of five ozone molecules can be produced in the photochemical oxidation of methane. Although no ozone is destroyed in the oxidation of methane to formaldehyde at low NO_x concentrations, subsequent oxidation of CH_2O in the absence of NO_x does result in ozone destruction. The photochemical oxidation of higher hydrocarbons, as described in Figure 10.2, is more complex, but in all cases ozone may be either produced or destroyed, depending on the relative mixing ratios of nitric oxide and ozone.

For this reason, nitric oxide may be thought of as a "chemical switch" for oxidant formation. We can easily calculate the nitric oxide concentration required for CO oxidation to "switch" from destroying ozone to producing ozone. This occurs when the rates of the $HO_2 + O_3$ and $HO_2 + NO$ reactions are equal:

$$k_{NO+HO_2}[NO][HO_2] = k_{O_3+HO_2}[O_3][HO_2] \tag{10.1}$$

$$[NO] = [O_3]\frac{k_{O_3+HO_2}}{k_{NO+HO_2}} \tag{10.2}$$

The rate constant for the $NO + HO_2$ reaction (Howard and Evenson, 1977) is ~4000 times the rate constant for the $O_3 + HO_2$ reaction. Thus, the critical nitric oxide concentration for switching from ozone destruction to production is 1/4000 of the O_3 concentration. For a "clean air" mixing ratio of ~20 ppbv, this corresponds to a nitric oxide mixing ratio of ~5 pptv (pptv = parts per trillion by volume = 10^{-12} fraction of air molecules).

It is sometimes more useful to express this critical concentration in terms of NO_x—that is, the sum of NO and NO_2 concentrations. In "clean air," NO, NO_2, and O_3 rapidly establish a photochemical steady state described by the following three reactions:

$$NO + O_3 \rightarrow NO_2 + O_2$$

$$NO_2 + h\nu \rightarrow NO + O$$

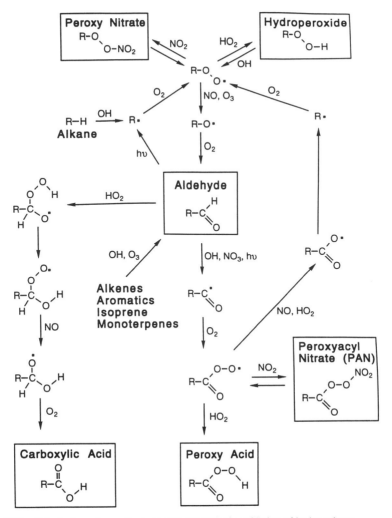

Figure 10.2.   Principal reaction pathways in the atmospheric oxidation of hydrocarbons.

$$O + O_2 + M \rightarrow O_3 + M$$

The third reaction, the association of O atoms with molecular oxygen, is so fast (natural lifetime, τ, of 12×10^{-3} s) compared to the photolysis of NO_2 ($\tau \sim 120$ s) that we may consider the photolysis of NO_2 to result in nearly instantaneous production of ozone. At steady state, the first two reactions occur at equal rates and the ratio of NO and NO_2 concentrations is given by

$$\frac{[NO_2]}{[NO]} = \frac{k_{NO+O_3}}{j_{NO_2}} [O_3] \tag{10.3}$$

Here, k_{NO+O_3} is the second-order rate coefficient for the reaction of NO with O_3, and j_{NO_2} is the first-order rate constant for photolysis of NO_2 to produce NO and O products. For a "clean air" ozone mixing ratio of $\sim$20 ppbv, the ratio of NO_2 to NO is calculated to be $\sim$1.0. The critical NO_x concentration for switching from ozone destruction to ozone production is therefore approximately twice the critical NO concentration (estimated above to be ~5 pptv) or ~10 pptv. However, even in only slightly polluted air masses the [NO$_2$]/[NO] ratio is often larger than calculated from this expression because other oxidants, especially HO$_2$ and RO$_2$ radicals (discussed below), also oxidize NO to NO$_2$, resulting in a critical NO$_x$ level for ozone production that is higher than the "clean air" value of ~10 pptv. For example, if the ozone mixing ratio is 40 ppbv and the [NO$_2$]/[NO] ratio is 1.5, the NO$_x$ "chemical switch" occurs at ~25 pptv; for the heavily polluted case of 200 ppbv ozone and a [NO$_2$]/[NO] ratio of 2.0, the critical NO$_x$ level for ozone production is calculated to be ~150 pptv.

Mixing ratios of NO$_x$ generally exceed the critical value for ozone production in urban and many rural areas where there are large contributions from fossil fuel combustion. In fact, NO$_x$ may range up to tens of ppbv in polluted regions. However, over vast regions remote from large NO$_x$ sources, such as over much of the oceans and at high latitudes, NO$_x$ mixing ratios are approximately equal to or less than the critical level for ozone production. A major global concern is that future increases in NO$_x$ emissions will result in elevated concentrations of NO$_x$ in these remote regions and thereby flip the "chemical switch" from a state of oxidant destruction to one of oxidant formation.

COMPARISON OF THE EFFECT OF NO$_X$ ON STRATOSPHERIC AND TROPOSPHERIC OZONE

In the natural stratosphere, NO$_x$, derived mostly from N$_2$O emitted by the biosphere (Crutzen, 1971), is responsible for more than half of the continuous *in situ* ozone destruction. This is a result of the following catalytic cycle (Crutzen, 1970, 1971; Johnston, 1971):

$$\begin{array}{c} NO + O_3 \rightarrow NO_2 + O_2 \\ \underline{NO_2 + O \rightarrow NO + O_2} \\ Net: O_3 + O \rightarrow 2O_2 \end{array}$$

In this cycle of reactions, NO$_x$ catalyzes the combination of oxygen atoms with ozone to form molecular oxygen. The reaction is said to be catalyzed by NO$_x$ because NO$_x$ is not consumed in the reaction; the sequence of reactions can cycle many times with the result that part-per-billion levels of NO$_x$ can affect part-per-million levels of ozone. There is a direct reaction between O and O$_3$, but the indirect NO$_x$-catalyzed reaction of O with O$_3$ is much faster throughout most of the stratosphere.

The O atoms in the NO$_x$ catalytic cycle are derived from ozone photolysis,

$$O_3 + h\nu \rightarrow O_2 + O$$

which occurs readily in both the stratosphere and troposphere—mainly as a result of visible light absorption in the Chapuuis band of ozone, which has a maximum near 600 nm. The lifetime of ozone with respect to photolysis is about 30 minutes throughout the troposphere and stratosphere (Brasseur and Solomon, 1984). In the troposphere, the principal fate of oxygen atoms is association with molecular oxygen to reform ozone:

$$O + O_2 + M \rightarrow O_3 + M$$

The steady state concentration of O atoms is found by equating its rate of production in the photolysis of O_3 with its rate of loss in the above reaction and is given by

$$[O] = \frac{j_{O_3}\,[O_3]}{k_{O+O_2+M}\,[O_2[M]}\tag{10.4}$$

The much higher partial pressure of oxygen and much lower concentration of ozone in the troposphere results in several orders of magnitude lower concentration of oxygen atoms in the troposphere as compared to the stratosphere. Because of the low oxygen atom concentration, the NO_x catalytic cycle is unimportant in the troposphere. Instead, NO_x catalyzes ozone production in the oxidation of CO and hydrocarbons, as described earlier. The oxidation of CH_4 (and CO derived from CH_4) is important in the stratosphere as well, and there is an altitude in the lower stratosphere (of the order of 18 km but dependent on latitude) where ozone production due to photochemical oxidation just balances ozone destruction via the NO_x catalytic cycle. At lower altitudes, smog chemistry with ozone production dominates, while at higher altitudes the net effect of NO_x is ozone reduction. Thus, projected effects of NO_x emissions by proposed future fleets of supersonic aircraft flying in the stratosphere are highly dependent on flight altitude (Johnston, 1971; Crutzen, 1972).

OXIDANTS DERIVED FROM OZONE

Several powerful oxidants are produced in the atmosphere from ozone, and some of these are much more reactive than ozone itself. The most reactive and most important oxidant derived from ozone is the hydroxyl radical. Hydroxyl radicals are produced as the result of the photolysis of ozone by ultraviolet light:

$$O_3 + h\nu \rightarrow O_2(^1\Delta_g) + O(^1D_2), \qquad \lambda \leq 310 \text{ nm}$$

$$O(^1D_2) + H_2O \rightarrow 2\,OH$$

Photolysis of ozone below about 310 nm produces electronically excited $O(^1D_2)$ ("O singlet D") oxygen atoms and electronically excited $O_2(^1\Delta_g)$ ("O_2 singlet delta"). These atomic and molecular oxygen species differ from their ground states in that all of their electrons are spin-paired; in their ground states, O and O_2 both have two unpaired electrons occupying different orbitals. The term symbols $^1D_2$ and $^1\Delta_g$ are spectroscopic notations specifying the atomic and molecular electronic configurations and need not be further discussed here [see Herzberg (1944, 1950) for a detailed description]. The energy required to pair electrons in these species provides electronic potential energies of 190 and 94 kJ/mol for $O(^1D_2)$ and $O_2(^1\Delta_g)$, respectively. This extra potential energy allows these excited-state species to undergo many reactions that are unavailable to their ground states. For example, in the reaction shown above $O(^1D_2)$ reacts at a near-collisional rate with water molecules to produce two hydroxyl radicals. The corresponding reaction of ground-state $O(^3P)$ ("O triplet P") is endothermic by 70 kJ/mol and therefore does not occur to any significant extent in the atmosphere. Once formed, the principal fates of $O(^1D_2)$ and $O_2(^1\Delta_g^-)$ are return to their ground states [$O(^3P)$ and $O_2(^3\Sigma_g^-)$, respectively] by collisions with N_2 and O_2:

$$O(^1D_2) + M \rightarrow O(^3P) + M$$

$$O_2(^1\Delta_g) + M \rightarrow O_2(^3\Sigma_g^-)$$

Due to rapid collisional deactivation and reaction with water vapor, the steady-state concentration of $O(^1D_2)$ in the troposphere is extremely low, less than 1 molecule/cm^3. Approximately 5–10% of the $O(^1D_2)$ produced reacts with water to form OH, and despite its very low concentration, $O(^1D_2)$ is the primary source of OH in "clean" tropospheric air. No significant role for the low concentrations of $O_2(^1\Delta_g)$ present in the troposphere has been identified.

Hydroxyl radicals play an extremely important role in the troposphere of cleansing the atmosphere of a wide range of chemical species emitted to the troposphere by the natural biosphere and as a result of human activities. The lifetime of a chemical species with respect to the OH reaction is given by

$$\tau = \frac{1}{k[\text{OH}]_{\text{ave}}} \tag{10.5}$$

where k is the second-order reaction rate constant. The concentration of hydroxyl radicals varies over many orders of magnitude, and its average concentration is still not well established but is in the range 10^5 to 10^6 molecule/cm^3. Some examples of chemical species whose tropospheric concentrations are limited largely by reaction with OH are listed in Table 10.1, along with estimates of their lifetimes with respect to reaction with OH.

More than 90% of CO and CH_4 destruction occurs via reaction with OH, with the ozone formation/destruction consequences described earlier. Hydrocarbons larger than methane (nonmethane hydrocarbons) have much faster reaction rates with OH and thus much shorter atmospheric lifetimes. The faster reaction rates result from weaker $C-H$ bonds and/or the availability of $C=C$ double bonds for addition of OH. As a result of the higher reactivity and shorter lifetimes of nonmethane hydrocarbons, these compounds play a more important role in oxidant formation on a local to regional scale, while methane and CO contribute to oxidant formation on a global scale.

Note in Table 10.1 that hydroxyl radicals provide for rapid oxidation of NO_2 and SO_2, forming nitric and sulfuric acids in the atmosphere, as discussed below. The lifetimes of hydrochlorofluorocarbons (HCFCs), which are being phased in as commercial replacements for chlorofluorocarbons (CFCs), also are determined by the average tropospheric OH concentration. The "ozone depletion potentials" of HCFCs are reduced relative to CFCs because they contain $C-H$ bonds that are susceptible to attack by OH in the troposphere. Reaction of CFCs with OH is negligible since the abstraction of either Cl or F is endothermic (positive enthalpy change); for example, for CFC-12 the reactions and enthalpies are

$$\text{CF}_2\text{Cl}_2 + \text{CH} \rightarrow \text{CF}_2\text{Cl} + \text{HOCl}, \qquad \Delta H^0_{298} = +113 \text{ kJ/mol}$$

$$\text{CF}_2\text{Cl}_2 + \text{CH} \rightarrow \text{CFCl}_2 + \text{HOF}, \qquad \Delta H^0_{298} = +264 \text{ kJ/mol}$$

while for HCFC-22 the abstraction of a hydrogen atom by OH to form water is exothermic (negative enthalpy change):

$$\text{CHF}_2\text{Cl} + \text{CH} \rightarrow \text{CF}_2\text{Cl} + \text{H}_2\text{O}, \qquad \Delta H^0_{298} = -64.9 \text{ kJ/mol}$$

The fully halogenated CFCs are chemically inert in the troposphere and remain in the atmosphere long enough to be transported to the stratosphere where they are degraded by short-wavelength UV photolysis and reaction with $O(^1D_2)$. The estimated lifetimes due to reaction with OH of some HCFCs are provided in Table 10.1. Although HCFCs do have much shorter lifetimes in the troposphere than CFCs, future increases in emissions, particularly as use of refrigeration and air conditioning is expanded in develop-

TABLE 10.1
Estimated Lifetimes of Atmospheric Species with Respect to Reaction with the Hydroxyl Radical

Species	Lifetime[a]
Carbon Monoxide/Hydrocarbons	
Carbon monoxide, CO	2.5 mo
Methane, CH_4	9.3 yr
Ethane, C_2H_6	2.8 mo
Propane, C_3H_8	18 d
Ethylene, $H_2C=CH_2$	2.2 d
Biogenic Sulfur Compounds	
Hydrogen sulfide, H_2S	3.9 d
Methane thiol, CH_3SH	6.4 hr
Dimethyl sulfide, CH_3SCH_3	3.7 d
Dimethyl disulfide, CH_3SSCH_3	1.0 hr
Acid Rain Precursors	
Sulfur dioxide, SO_2	17 d
Nitrogen dioxide, NO_2	1.2 d
CFCs, HCFCs, and Halons	
CCl_3F (CFC-11)	>18,000 yr
CCl_2F_2 (CFC-12)	>13,000 yr
$CHClF_2$ (HCFC-22)	12.5 yr
$CHCl_2F$ (HCFC-21)	1.9 yr
CH_3Cl (Methyl chloride)	1.3 yr
CH_3Br (Methyl bromide)	1.9 yr
CH_2ClF (HCFC-31)	1.2 yr
CH_3CCl_3 (Methyl chloroform)	5.9 yr
CHF_2CF_3	1.4 yr
$CHFClCF_3$ (HCFC-124)	5.7 yr
CH_2ClCF_2Cl	3.5 yr
CH_2ClCF_3	4.3 yr
CH_3CFCl_2 (HCFC-141b)	9.7 yr
CH_3CF_2Cl (HCFC-142b)	18.1 yr

[a]*Assumes an average hydroxyl radical concentration of 6.5×10^5 cm^{-3}, average temperature of 288 K, and pressure of 1 atmosphere. Rate constants from: Chemical Kinetics and Photochemical Data for Use in Stratospheric Modeling, Evaluation No. 10, NASA, Jet Propulsion Laboratory Publication 92-20, 1992.*

ing countries, could also cause significant degradation of the ozone layer. For this reason, HCFCs can only be considered as an interim solution to the ozone depletion problem (Birks et al., 1992).

The cutoff near 310 nm for $O(^1D_2)$ production and therefore OH production is within the biologically damaging UV-B region (280–320 nm). Thus, a large fraction of the UV radiation that "leaks" through the stratospheric ozone layer, causing damage to plants and animals, also is responsible for hydroxyl radical production in the troposphere. This results in important couplings between the chemistries of the stratosphere and troposphere. For example, ozone depletion in the stratosphere may result in increased oxidant formation in the troposphere. Also, as discussed above and summarized in Table 10.1, the lifetimes of many species in the troposphere are determined by the OH radical concentration, and the fluxes of those species to the stratosphere are therefore affected by depletion of the ozone layer. Again, methane is important because its oxidation in the stratosphere is responsible for a large fraction of the water vapor (from which stratospheric OH is derived) and H_2 content.

Some other oxidants derived from ozone include peroxy radicals (HO_2 and RO_2), hydrogen peroxide (H_2O_2), organic peroxides (ROOH, ROOR'), peroxy nitrates ($ROONO_2$), peroxyacyl nitrates ($R(C=O)OONO_2$), and the nitrate radical (NO_3). An important source of HO_2 radicals is the reaction of OH with O_3:

$$OH + O_3 \rightarrow HO_2 + O_2$$

The photooxidation of CO and hydrocarbons also produces HO_2 radicals via (a) the formation of hydrogen atoms which attach to oxygen and (b) the formation of alkoxy radicals ($R\text{-}CH_2O$) which donate hydrogen atoms to oxygen:

$$H + O_2 + M \rightarrow HO_2 + M$$

$$R\text{-}CH_2O + O_2 \rightarrow R\text{-}CHO + HO_2$$

As discussed above, HO_2 radicals are key species in oxidant formation and may either react with and thereby destroy O_3 or oxidize NO to NO_2 and make ozone, depending on the $[NO]/[O_3]$ ratio.

Another important role of HO_2 is the production of the oxidant H_2O_2 via the self-reaction:

$$HO_2 + HO_2 \rightarrow H_2O_2 + O_2$$

This is the only gas-phase source of H_2O_2 in the atmosphere. Hydrogen peroxide is highly soluble in water and is rapidly taken up by cloud and fog droplets. As described below, reaction of SO_2 with H_2O_2 in cloud droplets is one of the principal methods of oxidation of SO_2 to sulfuric acid, H_2SO_4.

Yet another important role for HO_2 radicals is the reaction with aldehydes to form organic acids, for which the reaction mechanism is thought to be

Because of their high polarity and resulting low vapor pressures, carboxylic acids readily condense to form aerosol particles, which subsequently are removed from the atmosphere by rainout and dry deposition. Thus, the reaction of HO_2 with aldehydes represents an important sink for carbon in the atmosphere. The fraction of carbon removed from the atmosphere by deposition of organic aerosols is not well known, and the rate of removal of organic acids and other oxygenated hydrocarbons from the atmosphere is a critically important research problem.

Organoperoxy radicals, RO_2, play an analogous role to HO_2 in oxidant formation. They are produced as a result of hydroxyl radical abstraction of hydrogen atoms from hydrocarbons and other organic species, RH:

$$RH + OH \rightarrow R + H_2O$$

$$R + O_2 + M \rightarrow RO_2 + M$$

As for HO_2, organoperoxy radicals may react with O_3 or NO with the net effect of either destroying or producing ozone. Disproportionation and cross reaction with HO_2 serves as a source of organic peroxides and hydroperoxides in the atmosphere:

$$RO_2 + R'O_2 \rightarrow ROOR' + O_2$$

$$RO_2 + HO_2 \rightarrow ROOH + O_2$$

Peroxy acids are formed in the atmosphere by reaction of HO_2 with peroxyacyl radicals. Peroxyacyl radicals are produced from aldehydes as a result of hydroxyl radical abstraction followed by oxygen addition:

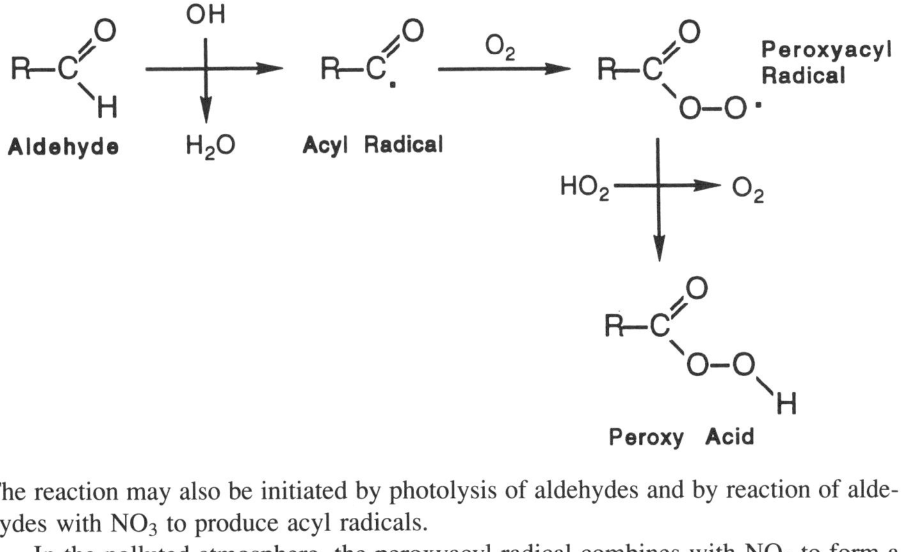

The reaction may also be initiated by photolysis of aldehydes and by reaction of aldehydes with NO_3 to produce acyl radicals.

In the polluted atmosphere, the peroxyacyl radical combines with NO_2 to form a class of especially strong oxidants, peroxyacyl nitrates (Stephens, 1969):

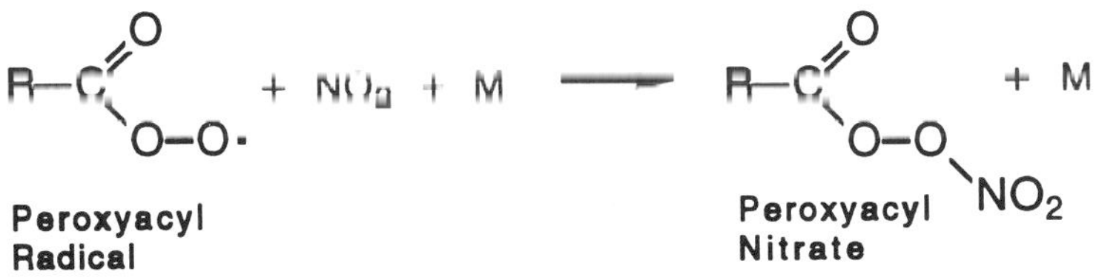

Peroxy*acetyl* nitrate, or "PAN," is derived from acetaldehyde and is the most abundant of these compounds. Peroxyacyl nitrates derived from higher aldehydes are referred to generically as "PANs." PANs are strong eye irritants and are more damaging to vegetation (stronger phytotoxins) than is ozone (Wayne, 1991). As discussed below, they play an important role in transport of NO_x from urban to rural areas and thus enhance oxidant formation on a regional basis.

During the night, hydroxyl radical concentrations are extremely low due to the lack of UV light to photolyze O_3 and produce them. However, another highly reactive oxidant, the nitrate radical, reaches high concentrations at night. The nitrate radical is produced in the reaction of NO_2 with O_3:

$$NO_2 + O_3 \rightarrow NO_3 + O_2$$

Because of strong absorption at visible wavelengths, NO_3 is rapidly photolyzed during the daytime,

$$NO_3 + h\nu \rightarrow NO_2 + O$$

$$\rightarrow NO + O_2$$

so that its daytime concentration is extremely low. During the night, the principal fate of NO_3 is to combine with NO_2 to form N_2O_5 via a reversible reaction:

$$NO_2 + NO_3 + M \leftrightarrow N_2O_5 + M$$

This thermal equilibrium is maintained during the night and provides a continuous source of NO_3 that decreases as oxides of nitrogen are converted to nitric acid via the reactions

$$N_2O_5 + H_2O \rightarrow 2\ HNO_3$$

$$NO_3 + RH \rightarrow HNO_3 + R$$

Figure 10.3 summarizes the atmospheric reactions of oxides of nitrogen. The concentrations of reactive nitrogen species vary over several orders of magnitude in the atmosphere because of the very short and highly variable lifetime of NO_x (approximately one day in relatively clean air and less in polluted air). The final oxidation product, nitric acid, is removed from the atmosphere by wet and dry deposition. Thus, even the removal of NO_x from the atmosphere can be harmful by contributing to the environmental problems associated with acid rain.

Hydroxyl and nitrate radicals play complementary roles, OH being an active oxidant during the daytime and NO_3 at night. Although the NO_3 reaction rate constants are generally smaller than those for OH, its nighttime mixing ratio (~100 pptv) can be as much as 10^3 times the daytime mixing ratio of OH (~0.1 pptv). Oxidation reactions of NO_3 include the abstraction of hydrogen atoms from hydrocarbons, as shown above, and addition to double bonds, for which the detailed reaction mechanism is not well understood. The abstraction by NO_3 of allylic hydrogens from compounds containing double bonds is particularly facile due to the relatively weak $C-H$ bond (~335 kJ/mol compared to ~380–420 kJ/mol for most other $C-H$ bonds). The oxidation by NO_3 of natural hydrocarbons such as isoprene and the various terpenes, all of which contain double bonds, is very fast. As summarized in Table 10.2, the lifetimes of these biogenic hydrocarbons may be only a few minutes at night due to fast reactions with NO_3, as compared to several hours during the daytime when both OH and O_3 make significant contributions to the total destruction rate.

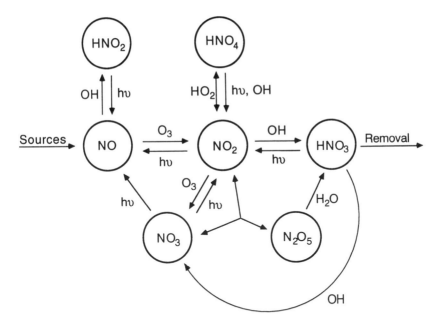

Figure 10.3. Principal tropospheric reactions of the oxides of nitrogen.

INHIBITION OF OXIDANT FORMATION BY NO$_X$ AT HIGH CONCENTRATIONS AND THE EXPORT OF NO$_X$ FROM URBAN TO RURAL AREAS

As discussed above, for mixing ratios exceeding a critical level, NO$_x$ has a catalytic effect on oxidant formation in the troposphere. At the high ppbv levels encountered in urban air pollution, however, NO$_x$ can actually reduce oxidant formation by acting as a free radical scavenger. This effect has important implications to the development of policy for air pollution abatement. Being a radical species itself, NO$_2$ can combine with other radicals to form stable compounds. One of the most important of these is the combination with hydroxyl radicals to form nitric acid:

TABLE 10.2
Comparison of Lifetimes of Isoprene and Monoterpenes with Respect to Reactions with OH, O$_3$, and NO$_3$[a]

	Lifetime Due to Reaction with		
Species	OH $(6.5 \times 10^5$ cm$^{-3})$	O$_3$ $(7.5 \times 10^{11}$ cm$^{-3})$	NO$_3$ $(2.5 \times 10^8$ cm$^{-3})$
Isoprene	4.3 hr	26 hr	1.9 hr
α-Pinene	8.1 hr	4.4 hr	13 min
β-Pinene	5.5 hr	18 hr	27 min
Δ^3 Carene	4.7 hr	3.1 hr	6.3 min
d-Limonene	2.8 hr	35 min	4.8 min

[a]Adapted from Table 14.5 in Finlayson-Pitts and Pitts (1986). p. 988.

$$NO_2 + OH + M \rightarrow HNO_3 + M$$

Besides serving as the principal means for removal of NO$_x$ from the atmosphere and contributing to acid precipitation, this reaction also reduces the hydroxyl radical concentration. Other radical termination reactions include (a) the combination with peroxy radicals, HO$_2$ and RO$_2$, to form peroxynitric acid and organoperoxy nitrates,

$$NO_2 + HO_2 + M \rightarrow HOONO_2 + M$$

$$NO_2 + RO_2 + M \rightarrow ROONO_2$$

and (b) the reaction (discussed above) of peroxyacyl radicals with NO$_2$ to form PANs.

Smog chambers are large reaction vessels irradiated with simulated sunlight and used to investigate the effects of different reactants on the formation of ozone and other reaction products (Finlayson-Pitts and Pitts, 1986, pp. 380–400). The inhibition of ozone formation at high NO$_x$ concentrations has been simulated in smog chambers (Finlayson-Pitts and Pitts, 1977), as illustrated in Figure 10.4. This figure shows the effect of initial NO$_x$ concentration on ozone levels after six hours of photochemistry for a range of initial concentrations of hydrocarbons. As expected, higher carbon loadings result in higher ozone concentrations, and the maximum in the curve shifts to higher NO$_x$ levels as the carbon loading increases. Note that at NO$_x$ levels beyond the maxima, a decrease in NO$_x$ concentration results in an increase in ozone concentration. Thus, reducing NO$_x$ emissions can actually cause ozone and other oxidants to increase.

One method of reducing the concentrations of oxidants in cities would be to reduce hydrocarbon emissions while allowing NO$_x$ to increase—for example, by operating automobile engines at higher temperatures where CO and hydrocarbons are more completely oxidized but more nitric oxide is produced. A disadvantage of this approach

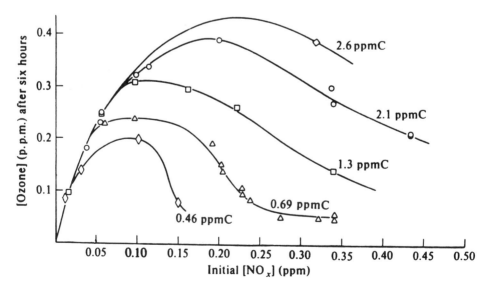

Figure 10.4. Ozone mixing ratios reached after 6-hour irradiations of NO$_x$ with a surrogate mixture of hydrocarbons simulating ambient air for various initial hydrocarbon and NO$_x$ concentrations. Reprinted with permission from Finlayson-Pitts, B. J. and Pitts, J. N., Jr., *Adv. Environ. Sci. Technol.* 7, 75 (1977). Copyright © 1977 John Wiley & Sons, Inc.

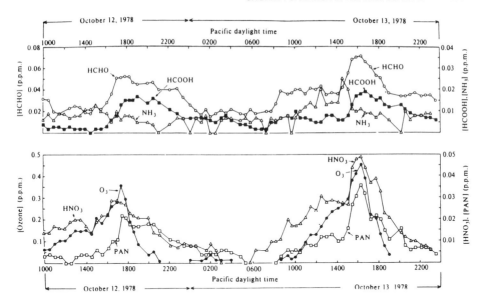

Figure 10.5.   Variation in concentrations of O_3, HNO_3, PAN, formaldehyde, formic acid, and ammonia on two successive polluted days in Southern California. Reprinted with permission from Tuazon, E. C., Winer, A. C. and Pitts, J. N., Jr., *Environ. Sci. Technol.*, 15, 1232 (1981). Copyright 1981 American Chemical Society.

is that the peroxy nitrates and PANs that build up and act to terminate photochemical reactions locally are thermally unstable and provide a mechanism for export of NO_x to rural areas. There, these compounds may reversibly decompose into the radicals that form them and act as accelerators for oxidant formation. As a result, this approach to emissions regulation could result in reduced ozone levels in urban areas at the expense of much higher ozone levels in rural areas where damage to crops and forests would increase.

It should be noted that the distance over which NO_x is transported by this mechanism is highly temperature-dependent because of the activation energies for the thermal decomposition of the sequestering compounds. For example, the lifetime of PAN is estimated to be 14 days at $-10°C$, 1.7 hours at $20°C$, and 5 minutes at $40°C$.

DIURNAL TRENDS OF OXIDANTS AND SMOG CHAMBER STUDIES

The general mechanisms of oxidant formation described above have been elucidated through a combination of field studies, smog chamber experiments, and direct measurements of reaction rate coefficients and product distributions for most of the key chemical reactions. Figure 10.5 displays observations of the diurnal (day and night) variations of ozone, formaldehyde, formic acid, PAN, and nitric acid for two successive days of high pollution in southern California (Tuazon et al., 1981). The concentrations of all of these species drop to low levels at night and build up again the following day as the result of automobile and industrial sources of NO_x and hydrocarbons, reaching a maximum in the late afternoon. Note that on these two days, ozone reached the very high levels of 350–450 ppbv and greatly exceed the standard set by the U.S. Environmental Protection Agency of 120 ppbv. (The U.S. EPA defines a region to be in violation of the air quality standard whenever two days in a calendar year register a maximum 1-hour-averaged ozone mixing ratio in excess of 120 ppbv.)

A smog chamber run starting with ~500 ppbv of the hydrocarbon propene, ~420 ppbv nitric oxide, and ~70 ppbv NO$_2$ is shown in Figure 10.6. This figure demonstrates the classic oxidation of an olefin to its corresponding aldehydes, formaldehyde and acetaldehyde, with the production of high levels of ozone (~450 ppbv). NO$_2$ builds up as an intermediate due to the NO + O$_3$ reaction and decays again as it is converted to PAN (resulting from the further reaction of acetaldehyde) and nitric acid (not shown). This and other laboratory simulations verify the general mechanism of oxidant formation resulting from the interaction of hydrocarbons, NO$_x$, and sunlight.

SOURCES OF HYDROCARBONS, CARBON MONOXIDE AND NO$_X$

A quantitative explanation of present oxidant levels in the troposphere and the prediction of future oxidant trends relies on a knowledge of the natural and anthropogenic sources of hydrocarbons, CO and NO$_x$. Methane and carbon monoxide are important

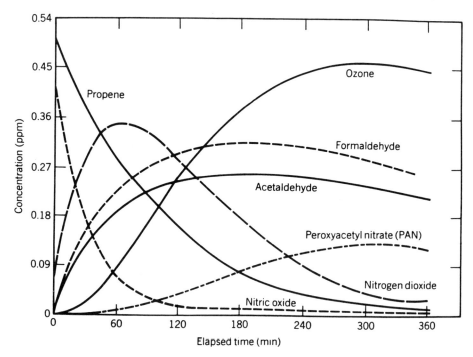

Figure 10.6. Typical primary and secondary pollutant profiles for a smog chamber irradiation of a propene–NO$_x$ mixture. Adapted from Pitts, J. N., Jr., Lloyd, A. C. and Sprung, J. L., *Chem. Br.* 11, 247 (1975).

contributors to a "background" oxidant level on a regional and global basis, while non-methane hydrocarbons (NMHC), because of their much higher reactivities, are responsible for high oxidant levels above this background in urban areas where the sources are dominated by anthropogenic emissions and in rural areas where vegetation may contribute to high levels of NMHC. The biogenic sources of hydrocarbons and NO_x are especially difficult to quantify due to the wide variety of species responsible, the large number of specific compounds emitted, the high degree of diurnal and seasonal variability, the highly complex distribution of soil and vegetation types across the Earth's surface, and a lack of understanding of the complex processes that control emissions. Nevertheless, flux measurements have been made for a wide range of compounds and attempts made to develop global emissions inventories. The current best estimates of these inventories are discussed here for methane, NMHC, isoprene, monoterpenes, and NO_x.

Methane

The present methane concentration in the atmosphere is approximately 1.7 ppmv in the Southern Hemisphere and 1.8 ppmv in the Northern Hemisphere (Steele et al., 1987). The methane concentration has increased at a rate of 0.8–1.0% per year over the past decade, with the rate of increase declining in recent years. A sharp gradient between the Northern and Southern Hemispheres is seen in the "flying carpet" diagram of Figure 10.7 and is explained by a larger source of CH_4 in the Northern Hemisphere (Steele et al., 1987). Also evident in Figure 10.7 is the seasonal variability of CH_4 aris-

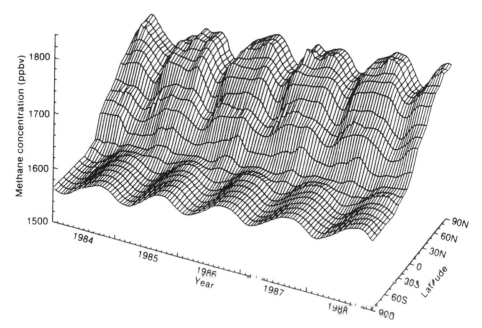

Figure 10.7. Measured methane mixing ratios as a function of latitude and time. Reprinted with permission from IPCC, 1990. Climate Change, The IPCC Scientific Assessment. Eds. J. T. Houghton, G. J. Jenkins & J. J. Ephraums. Cambridge University Press, UK, 365 pp.

ing from seasonal variability in the biogenic source strength and in the hydroxyl radical concentration. Estimates of various anthropogenic and biogenic sources and sinks of methane (Houghton et al., 1990, pp. 18–22) are summarized in Table 10.3. The largest sources of methane arise from anaerobic bacteria present in natural wetlands and rice paddies and in the digestive tracks of ruminants (animals that chew their cud) such as cattle. Most, if not all, of the observed increase in the atmospheric burden of methane over the past decade is believed to be anthropogenic and due to increased rice cultivation and cattle production. Other significant anthropogenic sources include (a) loss of methane to the atmosphere in natural gas production and transmission through pipelines, (b) biomass burning in tropical regions, (c) landfills, and (d) coal mining. Termites are a significant source of methane (Zimmerman et al., 1982), and termite emissions could possibly increase as a result of deforestation. Oceans and freshwaters are a relatively minor source of methane. Vast quantities of methane hydrate is present in permafrost and may be released as global temperatures increase, thereby providing a positive feedback to global warming (Kvenvolden, 1988). The methane sink due to reaction with OH is estimated to be 300–450 Tg C/yr, and there is a much smaller uptake of 10–35 Tg C/yr by soils. Although the uncertainties in individual sources of methane are quite large, the total source of methane to the atmosphere may be estimated to be 340–520 Tg C/yr based on a mass balance that takes into account the known sinks and the measured increase of 30–35 Tg C/yr.

TABLE 10.3
Estimated Annual Sources and Sinks of Methane to the Atmosphere[a]

Source/Sinks	Emission Estimates, Tg C/yr	
	IPCC[a]	IGAC-GEIA[b]
Source		
Fuel combustion		28
Natural wetlands (bogs, swamps, tundra, etc.)	75–150	100–115
Rice paddies	20–130	40–70
Enteric fermentation (animals)	50–75	80
Gas drilling, venting, transmission	20–40	25–42
Biomass Burning	15–60	45
Termites	10–75	20
Landfills	15–50	20–25
Coal mining	15–40	35–50
Oceans	4–15	10
Freshwaters	1–20	
CH$_4$ hydrate destabilization	0–75	
Wastewater treatment and animal waste		30
Other (industrial, residential waste burning, peat mining, geothermal, CH$_4$ hydrate destabilization, freshwaters)		9
Total sources	340–520	442–524
Sinks		
Removal by soils	10–35	
Reaction with OH in the atmosphere	300–450	
Total sinks	310–485	
Annual atmospheric increase	30–35	

[a]Adapted from Table 1.2 in Houghton et al. (1990).
[b]IGAC-GEIA Methane Emission Working Group Report (1993).

Nonmethane Hydrocarbons

The principal anthropogenic source of nonmethane hydrocarbons to the atmosphere arises from the production, refining, transport, and incomplete combustion of fossil fuels. The U.S. Environmental Protection Agency estimates that emissions of nonmethane hydrocarbons increased from 15 Tg C/yr^{-1} in 1940 to 25 Tg C/yr^{-1} in 1970 and decreased to 18.7 Tg C/yr^{-1} in 1990 (U.S. EPA, 1991). Transportation, industrial and chemical processing, and solvent evaporation make the largest contributions.

The emission of hydrocarbons by vegetation is much larger in magnitude but more dispersed, so that its contribution to urban smog formation is relatively small. These hydrocarbons may, however, react with oxides of nitrogen imported from urban centers to increase ozone and other oxidant concentrations on a regional basis. Natural hydrocarbon emissions are made up of isoprene and a variety of monoterpenes, of which α-pinene is the most abundant. In addition, there are emissions of sesquiterpenes, alkanes, alkenes, alcohols, aldehydes, ketones, and acetates (Winer et al., 1992). The various monoterpenes, such as those summarized in Figure 10.8, have the formula $C_{10}H_{16}$ and consist of two C_5H_8 isoprene units. For example, myrcene, which is emitted by loblolly pine, California black sage, and redwood trees, is a dimer of isoprene:

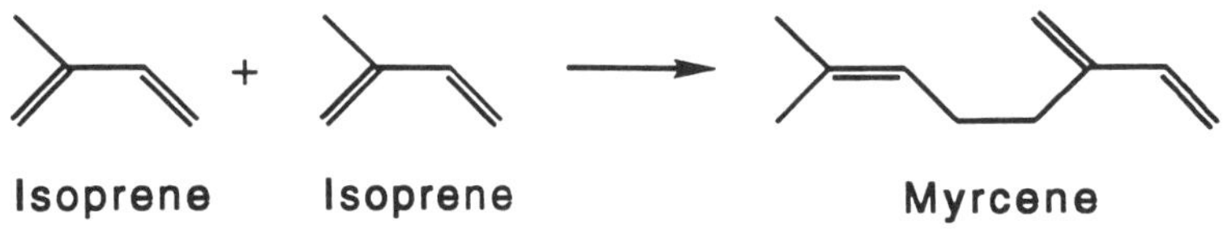

| Isoprene | Isoprene | Myrcene |

Because of their higher molecular weight, monoterpenes have much lower vapor pressures than isoprene, and their release to the atmosphere is highly temperature dependent. Note that isoprene and the monoterpenes contain C=C double bonds, which make them highly reactive toward OH, O_3, and NO_3. Thus, the lifetimes of these nonmethane hydrocarbons (see Table 10.2) range from a few minutes to a few hours.

Recent estimates (Guenther et al., 1995) place the annual rate of emission of isoprene to the atmosphere at 420 Tg C/yr and emission rate of terpenes at 130 Tg C/yr, of which α-pinene comprises about 50%. In addition, plants emit oxygenated hydrocarbons such as alcohols, ethers, aldehydes, ketones, and carboxylic acids, for which current estimates are about 380 Tg C/yr (Guenther et al., 1995). On a global basis these natural sources of ~830 Tg C/yr of nonmethane organic compounds exceeds anthropogenic emissions by at least an order of magnitude.

Figure 10.8. Structures of some common monoterpenes emitted by vegetation.

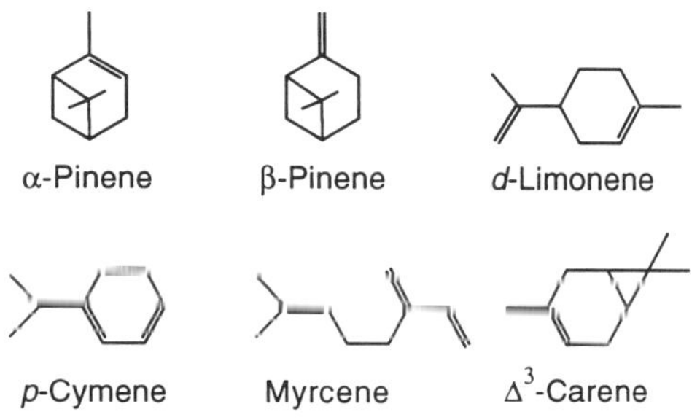

α-Pinene   β-Pinene   d-Limonene

p-Cymene   Myrcene   Δ³-Carene

Carbon Monoxide

Natural and anthropogenic sources of CO to the atmosphere are summarized in Table 10.4 (World Meteorological Organization, 1985). By far the largest natural sources are oxidation of methane (some of which is actually anthropogenic) and non-methane hydrocarbons, with much smaller contributions from natural wildfires and oceans. The major anthropogenic sources are fossil fuel combustion, biomass burning, and forest clearing in the tropics. Oxidation of anthropogenic hydrocarbons and the burning of wood as a fuel are minor anthropogenic contributions. The natural and anthropogenic contributions to atmospheric CO are estimated to be approximately equal and sum to $\sim$1100 Tg C/yr.

Oxides of Nitrogen

Current estimates of natural and anthropogenic sources of NO_x to the atmosphere are approximately equal, as summarized in Table 10.5. The natural source of $\sim$30 Tg N/yr of NO_x is thought be due mostly to lightning, with smaller contributions from soils, ammonia oxidation, and transport downward from the stratosphere (Penner, 1992). It should be noted, however, that the lightning source of NO_x is highly uncertain, as is its vertical distribution. The estimated anthropogenic contribution of $\sim$34 Tg N/yr arises from fossil fuel combustion (Penner et al., 1991) and biomass burning (Dignon and Penner, 1991).

Because of its short atmospheric lifetime, of the order of one day, NO_x concentrations vary over approximately four orders of magnitude, being much higher in concentration near its sources. The high degree of variability of NO_x makes its atmospheric distribution extremely difficult to characterize at any point in time; this results in a dilemma for atmospheric scientists in that prediction of future states of the atmosphere are extremely sensitive to the three-dimensional distribution of NO_x.

TABLE 10.4
Estimated Annual Emissions of CO to the Atmosphere[a]

Source	Tg C/yr
Natural	
Oxidation of CH4	260
Oxidation of biogenic non-methane hydrocarbons	250
Natural wildfires	10
Oceans	20
Subtotal, natural	540
Anthropogenic	
Fossil fuel combustion	190
Biomass burning	110
Forest clearing in tropical regions	160
Oxidation of anthropogenic hydrocarbons	40
Wood used as a fuel	20
Subtotal, anthropogenic	520
Total, all sources	1060

[a]*Adapted from World Meteorological Organization, 1985, Atmospheric Ozone 1985, Global Ozone Research and Monitoring Project, Report No. 16, Vol. 1, p. 106.*

TABLE 10.5
Estimated Annual Emissions of NO_X to the Atmosphere

Source	Tg N/yr
Natural	
Lightning	20
Stratospheric injection	2
Ammonia oxidation	3
Soil emissions	5
Subtotal, natural	30
Anthropogenic	
Biomass burning	13
Fossil fuel combustion	21
Subtotal, anthropogenic	34
Total, all sources	64

ROLE OF OXIDANT FORMATION
IN ACID PRECIPITATION

Sulfur dioxide is a primary pollutant whose principal source is the combustion of sulfur-containing coal, while nitric oxide is produced from N_2 and O_2 in all high-temperature combustion processes. Nitric and sulfuric acids are the final oxidized products resulting from NO_x and SO_2 emissions. The acidities of fogs, clouds, precipitation, and dry deposition are highly dependent on the rates of oxidation of these species. These acids have low vapor pressures and condense in the atmosphere to form aerosol particles, which are responsible for visibility reduction—that is, the "brown clouds" of our cities. The aerosols also scatter solar radiation back to space and have a net cooling effect on climate. Recently, it has been demonstrated that nitrate and sulfate aerosols may have partially offset global warming that otherwise would have resulted from the increase in atmospheric concentrations of greenhouse gases during the past century.

In the gas phase, both NO_x and SO_2 are oxidized to their corresponding acids upon reaction with the OH radical. For NO_x the reactions are

$$NO + O_3 \rightarrow NO_2 + O_2$$

$$NO_2 + OH + M \rightarrow HNO_3 + M$$

For SO_2 the gas-phase mechanism is

$$SO_2 + OH + M \rightarrow HSO_3 + M$$

$$HSO_3 + O_2 \rightarrow SO_3 + HO_2$$

$$SO_3 + H_2O \rightarrow H_2SO_4$$

It is now recognized that the extent of heterogeneous oxidation of SO_2 within cloud droplets is approximately equal to that of gas-phase oxidation by hydroxyl radicals (Finlayson-Pitts and Pitts, 1986, pp. 691–693). At a pH less than ~5, H_2O_2 is the principal oxidant in cloud water. Because of its high solubility in water, H_2O_2 is rapidly scavenged from the gas phase by cloud and fog droplets. It has recently been shown that H_2O_2 may be produced by photochemical reactions within cloud droplets as well (Zuo and Hoigné, 1993; Faust et al., 1993). The detailed reaction mechanism for SO_2 oxidation by H_2O_2 is much more complex, but the net stoichiometric reaction may be written

$$SO_2 + H_2O_2 \rightarrow H_2SO_4$$

One should bear in mind that sulfur in the $+4$ oxidation state, S(IV), is present in solution as the hydrate $SO_2 \cdot (H_2O)$ and as the anions HSO_3^- (bisulfite) and SO_3^{2-} (sulfite), with HSO_3^- being the principal form for the atmospherically relevant range of pH 2–6. As a strong acid, sulfuric acid, for which sulfur exists in a $+6$ oxidation state, will be ionized in an aqueous solution to form sulfate (SO_4^{2-}) and a small amount of bisulfate, (HSO_4^-).

The limiting reagent in cloud droplets is frequently H_2O_2, rather than SO_2 (Chaimedes, 1984; Gunz and Hoffmann, 1990). As a result, the oxidizing capacity of the atmosphere strongly affects the severity of acid deposition. This has important policy implications in that reducing SO_2 emissions from sources such as power plants could have a very limited effect on sulfuric acid deposition, a phenomenon often referred to as the "nonlinear effect." Another consequence is that increased oxidant formation in the atmosphere is predicted to increase the severity of acid precipitation.

RESEARCH NEEDS FOR QUANTITATIVE PREDICTION OF OXIDANT FORMATION

Our understanding of oxidant formation in the troposphere has evolved through knowledge and insight gained in laboratory studies of reaction kinetics and photochemistry, field measurements of chemical fluxes and concentrations, and computer modeling. Both laboratory and field studies have benefited greatly from the continued development of new analytical techniques that are more sensitive and free from interferences. Similarly, the ability to model the atmosphere has advanced with improvements in computer technology.

Clearly, there is much yet to be learned about chemical emissions to, transformations within, and deposition from the atmosphere. This will not be enough, however; a fully predictive capability will require that oxidant formation be considered in the context of other problems such as ozone depletion, climate change, and the integrity of ecosystems. We will need to go beneath the surface—of a leaf, of soil, of the ocean—and encompass the whole of biological sciences, ranging from the biochemistry of living cells to ecosystem dynamics, before this small but important part of the whole, the Earth System, is fully understood.

References

Birks, J. W., Calvert, J. G., and Sievers, R. E., eds., 1992, *The Chemistry of the Atmosphere: Its Impact on Global Change, CHEMRAWN VII, Perspectives and Recommendations*, International Union of Pure and Applied Chemistry and the American Chemical Society, Published by the Agency for International Development, pp. 21–22.

Brasseur, G., and Solomon, S., 1984, *Aeronomy of the Middle Atmosphere*, Reidel, Dordrecht, pp. 154–159.

Chaimedes, W. L., 1984, The photochemistry of a remote marine stratiform cloud. *J. Geophys. Res.*, 89, 4739.

Crutzen, P. J., 1970, The influence of nitrogen oxides on the atmospheric ozone content. *J. R. Meteor. Soc.*, 96, 320.

Crutzen, P. J., 1971, Ozone production rates in an oxygen-hydrogen-nitrogen oxide atmosphere. *J. Geophys. Res.*, 76, 7311.

Crutzen, P. J., 1972, SST's—a threat to the earth's ozone shield. *Ambio*, 1, 41.

Dignon, J., and Penner, J. E., 1991, Biomass burning: a source of nitrogen oxides in the atmosphere. In: J. Levine, ed., *Global Biomass Burning*, MIT Press, Cambridge, pp. 370–375.

Faust, B. C., Anastasio, C., J., Allen, M. and Arakaki, T., 1993, Aqueous-phase photochemical formation of peroxides in authentic cloud and fog waters. *Science*, 260, 73.

Finlayson-Pitts, B. J., and Pitts, J. N., Jr., 1977, The chemical basis of air quality: kinetics and mechanisms of photochemical air pollution and application to control strategies. *Adv. Environ. Sci. Technol.*, 7, 75.

Finlayson-Pitts, B. J., and Pitts, J. N., Jr., 1986, *Atmospheric Chemistry: Fundamentals and Experimental Techniques*, Wiley, New York.

Gunz, D. W., and Hoffmann, M. R., 1990, Atmospheric chemistry of peroxides: a review. *Atmos. Environ.*, A24, 1601.

Guenther, A., Hewett, C.N., Erickson, D., Fall, R., Geron, C., Graedel, T., Harley, P., Klinger, L., Lerdau, M., McKay, W. A., Pierce, T., Scholes, B., Steinbrecher, R., Tallamraju, R., Taylor, J., and Zimmerman, P., 1995, A global model of natural volatile organic compound emissions, *J. Geophys. Res.*, 100, 8873–8892.

Haagen-Smit, A. J., 1952, Chemistry and physiology of Los Angeles smog. *Indust. Eng. Chem.*, 44, 1342.

Haagen-Smit, A. J., and Fox, M. M., 1954, Photochemical ozone formation with hydrocarbons and automobile exhaust. *J. Air Pollut. Control Assoc.*, 4, 105.

Haagen-Smit, A. J., and Fox, M. M., 1955, Automobile exhaust and ozone formation. *AEA Trans.*, 63, 575.

Haagen-Smit, A. J., Darley, E. F., Zaitlin, M., Hull H., and Noble, W., 1951, Investigation on injury to plants from air pollution in the Los Angeles area, *Plant Physiol.*, 27, 18.

Haagen-Smit, A. J., Bradley, C. E., and Fox, M. M., 1953, Ozone formation in photochemical oxidation of organic substances. *Indust. Eng. Chem.*, 45, 2086.

Haagen-Smit, A. J., Brunelle, M. F., and Haagen-Smit, J. W., 1959, Ozone cracking in the Los Angeles area. *Rubber Chem. Technol.*, 32, 1134.

Herzberg, G., 1944, *Atomic Spectra and Atomic Structure*, 2nd ed., Dover, New York.

Herzberg, G., 1950, *Spectra of Diatomic Molecules*, 2nd ed., Van Nostrand, Princeton, NJ.

Houghton, J. T., Jenkins, G. J., and Ephraums, J. J., eds., 1990, *Climate Change: The IPCC Scientific Assessment, Intergovernmental Panel on Climate Change*, Cambridge University Press, Cambridge.

Howard, C. J., and Evenson, K. M., 1977, Kinetics of the reaction of HO_2 with NO." *Geophys. Res. Lett.*, 4, 437.

Johnston, H. S., 1971, Reduction of stratospheric ozone by nitrogen oxide catalysts from supersonic transport exhaust. *Science*, 173, 517.

Kvenvolden, K. A., 1988, Methane hydrates and global climate. *Global Biogeochem. Cycles*, 2, 221.

Logan, J. A., 1985, Tropospheric ozone: seasonal behavior, trends and anthropogenic influences. *J. Geophys. Res.*, 90, 10463.

Logan. J. A., 1988, The ozone problem in rural areas of the United States. In: I. S. A. Isaksen, ed., *Tropospheric Ozone: Regional and Global Scale Interactions*, North American Treaty Organization Advanced Study Institutes, Reidel, Dordrecht, Holland, pp. 327–344.

Penner, J. E., 1992, Tropospheric chemistry and climate change. In: *Atmospheric and Geophysical Sciences Program Report, 1990–1991*, Lawrence Livermore National Laboratory, Report UCRL-51444-90/91, pp. 75–83.

Penner, J. E., Atherton, C. S., Dignon, J., Ghan, S. J., Walton, J. J., and Hameed, S., 1991, Tropospheric nitrogen: a three-dimensional study of sources, distribution, and deposition. *J. Geophys. Res.*, 96, 959.

Pitts, J. N., Jr., Lloyd, A. C., and Sprung, J. L., 1975, Ecology, energy and economics, *Chem. Br.*, 11, 247.

Steele, L. P., Fraser, P. J., Rasmussen, R. A., Khalil, M. A. K., Conway, T. J., Crawford, A. J., Gammon, R. H., Masarie, K. A., and Thoning, K. W., 1987, The global distribution of methane in the troposphere. *J. Atmos. Chem.*, 5, 125.

Stephens, E. R., 1969, The formation, reactions, and properties of peroxyacyl nitrates (PANs) in photochemical air pollution. *Adv. Environ. Sci.*, 1, 119.

Tuazon, E. C., Winer, A. C. and Pitts, J. N., Jr., 1981, Trace pollutant concentrations in a multiday smog episode in the California south coast air basin by long path length Fourier transform infrared spectroscopy. *Environ. Sci. Technol.*, 15, 1232.

U.S. Environmental Protection Agency, 1991, *National Air Pollutant Emission Estimates 1940–1990*, Publication No. EPA-450/4-91-026.

Volz, A., and Kley, E., 1988, Evaluation of the Montsouris series of ozone measurements made in the nineteenth century. *Nature*, 332, 240.

Wayne, R. P., 1991, *Chemistry of Atmospheres*, 2nd ed., Clarendon Press, Oxford, p. 261.

Winer, A. C., Arey, A. J., Atkinson, R., Aschman, S., Long, W., and Morrison, C., 1992, Emission rates of organics from vegetation in California's Central Valley. *Atmos. Environ.*, 26A, 2647–2660.

World Meteorological Organization, 1985, *Atmospheric Ozone 1985*, Global Ozone Research and Monitoring Project, Report No. 16, Vol. I, p. 106.

Zimmerman, P. R., Greenberg, J. P., Wandiga, S. O., and Crutzen, P. J., 1982, Termites: a potentially large source of atmospheric methane, carbon dioxide and molecular hydrogen. *Science*, 218, 563.

Zuo, Y., and Hoigné, J., 1993, Evidence for photochemical formation of H_2O_2 and oxidation of SO_2 in authentic fog water. *Science*, 260, 71.

The Tropospheric Aerosol and Its Role in Atmospheric Chemistry

SONIA M. KREIDENWEIS

ABSTRACT. The characteristics of aerosol in different regions of the troposphere are reviewed. Particles serve as an important sink of oxidized species in many trace gas chemical cycles. The tropospheric aerosol is also linked to the hydrologic cycle, since wet deposition is a primary removal mechanism for particulate matter, and since the chemical and physical nature of aerosol particles serving as cloud condensation and ice nuclei can influence the properties of clouds, including precipitation rates. Aerosols also have a more direct effect on tropospheric chemistry via heterogeneous chemistry occurring in or on aerosol particles and in cloud droplets, especially aqueous-phase chemical reactions that are important in oxidizing some tropospheric trace gases. Heterogeneous reactions have also been implicated in affecting the oxidative capacity of the atmosphere.

INTRODUCTION

The sources, sinks, chemical and physical nature, and chemical roles of aerosol in the troposphere and stratosphere are markedly different. In general, the tropospheric aerosol is more complex chemically and exhibits greater spatial and temporal inhomogeneity; this complexity arises in part from the link between certain aerosols and the hydrologic cycle. Aerosol in the stratosphere has a relatively long lifetime, a consequence of the lack of wet removal processes and the stable thermal structure of the stratosphere which inhibits dynamic mixing processes.

Since particles in the lower atmosphere have a relatively short lifetime, on the order of days to a week, aerosol characteristics are strongly influenced by local sources and sinks. It is possible, however, to define broad categorizations of aerosol that generally exhibit similar properties. The aerosol near the Earth's surface within the planetary boundary layer, which is the turbulent region generally comprising the lowest 200–2000 m of the atmosphere, can be broadly classified as marine, continental, or polar. Aerosol in the free troposphere above the boundary layer has different chemical and physical properties. The free troposphere is removed from the exchanges of heat, mass, and momentum that occur within the boundary layer, and thus the aerosol in mid- and upper levels of the troposphere is less directly influenced by surface sources and sinks of particles and gases.

In the following sections, the chemical constituents of the atmospheric aerosol are briefly reviewed, and the role of particulate matter in tropospheric chemical cycles, including the hydrological cycle, are described. Heterogeneous chemistry involving aerosol particles is treated in the final section.

DESCRIPTION OF THE MAIN CONSTITUENTS OF THE TROPOSPHERIC AEROSOL

Continental Aerosol

The continental aerosol has components derived from soils, minerals, biogenic sources, combustion, and gas-to-particle conversion, which describes the process by which gaseous species (which may be of natural or anthropogenic origin) are converted to condensable vapors and incorporated into particles. Gas-to-particle conversion is the mechanism by which aerosol particles serve as sinks in the chemical cycles of several important trace gases in the troposphere. A distinction is made between primary and secondary aerosol: The former term describes those species that have been directly emitted to the atmosphere (e.g., windblown dust), and the latter term refers to aerosol which has formed in the atmosphere by physical or chemical transformations. An example of a secondary aerosol is sulfate which forms downwind of an industrial source, through the chemical conversion of sulfur gases to condensable species. The continental aerosol may be further classified into rural (remote), urban, and desert dust storm aerosol. Characteristics of each category are described in the reviews by Jaenicke (1988) and d'Almeida et al. (1991).

In general, higher number concentrations (on the order of several thousand cm^{-3}) than those observed in clean marine regions are observed in regions influenced by anthropogenic continental sources. Concentrations may be even higher in urban locations due to the proximity of industrial and transportation sources. The chemical composition of aerosol of anthropogenic origin differs from other types of aerosol, having, among others, components arising from combustion of fossil fuels (e.g., soot, trace metals, and partially oxidized organic matter).

Marine Aerosol

Fitzgerald (1991) has compiled a review of the characteristics of particles within the marine boundary layer (MBL) which are of marine origin. This aerosol is often considered to define the "background", that is, it is representative of air masses that have resided over the ocean for a time period (on the order of 10 days) sufficient to minimize influences from continental anthropogenic gaseous and particulate emissions. Near the east coast of the United States, measurements (Hoppel et al., 1990) indicate that total particle concentrations are typically near 6000 cm^{-3}, reflecting continental influences; the background concentrations range from less than 100 cm^{-3} to a mean of about 200 cm^{-3} throughout the tropical trade wind regions. Similar marine background concentrations have been measured over the Pacific Ocean (Hoppel and Frick, 1990; Parungo et al., 1987). Sea salt is the most important component of particles larger than 0.5 μm, with a variable fraction of mineral dust of continental origin (Raemdonck et al., 1986). Particles with radius less than 0.3 μm comprise 90–95% of the number, but only about 5% of the aerosol mass, and consist primarily of non-sea-salt sulfate. The sulfate is presumably generated from oxidation of reduced sulfur compounds, principally dimethylsulfide (DMS), emitted from the ocean.

Polar Aerosol

The phenomenon known as Arctic haze has been studied intensively since the early 1970s. Field programs have resulted in numerous publications reporting observations and analyses of data (see, for example, Table 4.4 in d'Almeida et al. (1991), which presents a survey of studies carried out in the Arctic; and Harriss et al. (1992, 1994), which overview the Arctic Boundary Layer Expeditions, ABLE 3A and 3B). The present understanding is that the seasonal haze, which peaks between February and April, results from transport of polluted midlatitude air masses, primarily from Eurasia (Barrie, 1986). Meteorological conditions that favor this springtime transport (i.e., the location of the polar front) are shifted at other times of the year to prevent intrusion of anthropogenic pollutants.

The Antarctic troposphere has been less well-studied. Aerosol in this remote, continental region has primarily mineral, sea-salt, and natural sulfate components, and the variability of number concentration and composition appears to be less than that of the Arctic aerosol. Gras (1993) summarizes particle measurements at Mawson, Antarctica, from 1985 to 1990, which show a marked seasonal cycle in aerosol number concentration and size distribution. A recent study has found evidence for new sulfate particle production in the Antarctic (Ito, 1993), presumably by homogeneous nucleation from the gas phase, which is favored in clean regions where low particle surface area inhibits aerosol scavenging of condensable vapors.

Aerosol in the Free Troposphere

There is recent evidence that particles in the free troposphere have significantly different characteristics than particles found at lower levels. The calculations of Delmas (1992) and the measurements of Clarke (1992) suggest that these particles have a high sulfur content and an acidic character. Airborne observations of ultrafine particles (diameter less than 0.01 μm) suggest that sporadic but significant new particle formation from the gas phase may occur in the free troposphere (Clarke, 1992). The observed ultrafine aerosol concentrations are occasionally very high (greater than 40,000 cm^{-3}), and the small size of these particles results in long atmospheric lifetimes and potentially large transport distances.

Clarke (1993) also found that the presence of ultrafine aerosol in the Pacific midtroposphere was strongly correlated with convection, particularly evident above the cloudy region of the equatorial intertropical convergence zone (ITCZ). Ikegami et al. (1994) observed high concentrations of aerosol particles, including a large contribution from sea salt, in the upper troposphere over the tropical Pacific Ocean. Both of these observations suggest that deep convection plays a large role in transporting aerosol and aerosol precursors from the boundary layer to the mid- and upper troposphere.

Cirrus clouds in the upper troposphere are composed largely of ice crystals. These clouds and other atmospheric ice particles result from both homogeneous freezing of droplets and haze particles (Sassen and Dodd, 1988; Heymsfield and Sabin, 1989) and heterogeneous nucleation on ice nuclei (Vali, 1985; Rogers and DeMott, 1991). It appears that important species which may nucleate ice include sulfate particles, soot (De-Mott, 1990), and, in some instances, ice formation is inhibited by particles which enter the troposphere from the stratosphere (Sassen, 1992).

Cloud Condensation Nuclei and Clouds

A subset of the tropospheric aerosol is classified as CCN, or cloud condensation nuclei. Designation as a CCN refers to the particle's capability for taking up water and growing to cloud droplet size (greater than approximately 5 μm), a process called activation. Whether or not a particle becomes activated depends upon its size and chemical composition and the water vapor supersaturation the particle experiences. More hygroscopic and larger particles act as CCN at lower supersaturations. Those particles which take up water, but do not activate, form haze and are important to both climate (Penner et al., 1994) and visibility (Malm et al., 1994). Several decades ago, sea salt was thought to be an important source of CCN, but subsequent measurements suggest that both maritime and continental CCN are primarily composed of sulfate species; it has also been suggested that organic species are a significant component (Novakov and Penner, 1993). Studies of biomass burning have shown that this process may be an important source of CCN (Crutzen and Andreae, 1990). A comprehensive review of the state of knowledge of cloud condensation nuclei is provided by Hudson (1993).

THE ROLE OF AEROSOL IN BIOGEOCHEMICAL CYCLES

The tropospheric chemical cycle of a species has four components: sources, chemical transformations, transport, and removal processes. In general, reduced compounds are emitted and are oxidized in the atmosphere. The oxidation products are often low-vapor-pressure or soluble species that partition to the aerosol phase and are removed in precipitation or by dry deposition. Thus, aerosol particles play a major role in defining the removal pathways and rates for many species. Here and in the following sections, the tropospheric sulfur cycle is used to illustrate the role of aerosol in tropospheric chemistry.

Figure 11.1 depicts a simplified version of the tropospheric sulfur cycle. Reduced sulfur gases, such as dimethylsulfide and hydrogen sulfide, are emitted from natural and anthropogenic sources and are oxidized in the troposphere to sulfur dioxide (and other sulfur species). Sulfur dioxide, which can be a primary as well as a secondary pollutant, is further oxidized, both in the gas and in the aqueous phase, to sulfuric acid or, more generally, to sulfate, SO_4^{2-}. These final oxidation products have very low vapor pressures and will either form new particles or condense onto existing particles

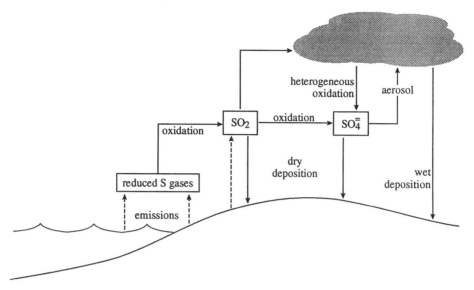

Figure 11.1. A simplified schematic of the tropospheric sulfur cycle.

or hydrometeors, which are eventually removed from the troposphere. The rate of removal, and hence the lifetime of sulfate species, depends upon (a) the rates at which aerosol particles are incorporated into precipitating clouds or are scavenged by falling precipitation and (b) the rates at which particles dry-deposit on the Earth's surface. Dry deposition is generally effective only in the lowest levels of the troposphere, whereas precipitation can remove species from higher levels and is a stronger, although sporadic, removal mechanism.

Deposited species represent a flux to the surface that may in turn affect emissions of other species. For example, the hydrogen ion in rainwater, H^+, is a key weathering reagent (Butcher, 1992). Another example is the wet and dry deposition of mineral-containing aerosol to the ocean surface. This process can represent an important flux of mineral nutrients to the ocean surface (Duce et al., 1991) and thereby impacts ocean productivity and the release of biogenic trace gases from surface waters.

THE ROLE OF AEROSOL IN THE HYDROLOGIC CYCLE

The role of tropospheric CCN in cloud formation establishes a fundamental link between particles, trace gas chemical cycles, and the hydrologic cycle. Particles influence the formation and microphysical properties of clouds, and introduce a variety of chemical species into cloud water. Since precipitation is a primary aerosol mass removal mechanism, the hydrologic cycle determines the strength of aerosol sinks and thus the lifetime of aerosol species. If a cloud reevaporates instead of precipitating, particulate matter is resuspended rather than removed from the troposphere. Interestingly, the aerosol itself—particularly the number concentration of CCN—may affect cloud formation, cloud microphysical properties, and precipitation efficiency (Albrecht, 1989), and thus particles may alter their own sink mechanism (Baker and Charlson, 1990).

Studies examining the interactions between aerosol and clouds have been particularly concerned with inadvertent anthropogenic modification of clouds (Leaitch et al., 1992). The relationship between aerosol mass and number is nonlinear (e.g., Kreidenweis et al., 1991), so it is not clear to what extent number concentrations will increase as airborne anthropogenic aerosol mass increases. It appears likely that increases in number concentration would affect cloud properties by increasing cloud droplet number concentration and decreasing mean cloud droplet size, if the liquid water content remains approximately constant (Twomey, 1991; Twomey et al., 1984; Hudson and Frisbie, 1991). Evidence for this effect has been deduced from observational studies of pollutant effects on cloud droplet number concentration (Hudson, 1991; Han et al., 1994), including the phenomena known as ship tracks (Coakley et al., 1987; Radke et al., 1989).

Ship tracks occur in satellite images as long-lived, bright bands, embedded within marine stratocumulus layers. They result from the injection of combustion products, including particles, from the ship's stack into the boundary layer. These contaminants are incorporated into the stratus cloud and alter its microphysical and radiative properties. The cloud within a ship track is characterized by smaller mean droplet size, higher droplet number concentration, and higher reflectivity than the surrounding layer, with the altered properties persisting for remarkably long periods of time. A related consequence of the altered cloud microphysical state is that precipitation processes are inhibited by the shift to smaller drop diameters, leading to longer cloud lifetimes and perhaps increased fractional cloudiness if anthropogenic perturbations on the global scale behave similarly (Albrecht, 1989).

CHEMICAL IMPACT OF THE ATMOSPHERIC AEROSOL

Haze, fog, and cloud droplets serve as sites for heterogeneous chemistry, both on the surfaces of particles and in the liquid phase. Many chemical processes which would occur slowly or not at all in the gas phase proceed rapidly in solution or on the surfaces of atmospheric particles. Indeed, modeling studies (Graedel and Weschler, 1981; Chameides, 1984; Schwartz, 1986; Jacob, 1986) and observations (Hegg and Hobbs, 1982; Gunz and Hoffmann, 1990; Zuo and Hoigné, 1993) suggest that heterogeneous reactions are significant components of global atmospheric chemistry and are important oxidation pathways for some tropospheric trace gases, such as SO_2. The composition of particulate matter determines the amount of liquid water in a haze particle in equilibrium with the ambient relative humidity. Aerosol chemical composition and size also influence (a) the pH and thus the solubility of various gases in cloud water, (b) the chemical composition of the aqueous phase and acidity of subsequent precipitation, and (c) the rates of reaction in the liquid phase.

Effects of Aerosol Species on Solubility of Trace Gases

The solubility of gases is a strong function of temperature, increasing as temperature decreases. The apparent solubility for a species that dissociates in solution, however, also depends upon the pH of the solution, as seen from the expression for the effective Henry's Law constant for sulfur dioxide (SO_2), $H^*_{S(IV)}$:

$$H^*_{S(IV)} = H_{SO_2}[1 + K_{s1}/[H^+] + K_{s1}K_{s2}/[H^+]^2] \tag{11.1}$$

A complete derivation of Eq. (11.1) may be found in the text by Seinfeld (1986). In Eq. (11.1), H_{SO_2} is the Henry's Law constant for SO_2 and is an equilibrium constant describing the partitioning of SO_2 between the gas and aqueous phases. K_{s1} and K_{s2} are the first and second dissociation constants for $SO_2 \cdot H_2O$ and give the expected concentrations of HSO_3^- and SO_3^{2-} ions as a function of pH (pH = $-\log 10[H^+]$). The effective Henry's Law constant describes the equilibrium between gas-phase SO_2 and the total free S(IV) in solution, which is the sum of dissolved SO_2, HSO_3^- and SO_3^{2-}. Since the term in parentheses is always greater than one, the apparent solubility is always greater than the physical solubility, but decreases as $[H^+]$ becomes large (indicating a low pH, acidic solution). This example demonstrates how a fundamental chemical property, the solubility of a gas in atmospheric liquid water, is affected by the chemical nature of the aerosol which has been incorporated into the aqueous solution and which determines the acidity of that solution prior to aqueous-phase reactions.

Heterogeneous Chemistry in Cloud, Fog and Haze Droplets

The important role of sulfur species in atmospheric chemistry and global change has resulted in numerous modeling, laboratory, and observational studies aimed at elucidating the components of the sulfur cycle and its linkages to atmospheric processes. Such studies point to the importance of aqueous chemistry in sulfur oxidation, because gas-phase mechanisms appear to be too slow to account for observed lifetimes. Results from a global model of the sulfur cycle (Langner and Rodhe, 1991) suggest that the aqueous conversion pathway may account for as much as 84% of the atmospheric SO_2-to-sulfate transformation. Smaller-scale, coupled chemistry/cloud models (Hegg and Hobbs, 1987; Chaumerliac et al., 1987; Hegg et al., 1989; Barth et al., 1992) have been used to demonstrate that a large percentage of precipitated sulfate can be attributed to conversion occurring in-cloud, although in some instances [e.g., in continental background conditions as in the simulations of Roelofs (1992)] sulfate from aerosol incorporated either as CCN or by scavenging below-cloud may be the dominant source. Sulfate and nitrate (NO_3^-) ions account for most of the acidity in precipitation (Calvert et al., 1985). Nitrate is an oxidation product of nitrogen oxides, and its removal from the atmosphere in the form of particulate matter is a major sink in the atmospheric NO_x chemical cycle, analogous to the role of sulfate in the atmospheric S cycle.

Only a small percentage of clouds precipitate (Pruppacher and Jaenicke, 1995). The aqueous chemistry occurring in nonprecipitating cloud cycles increases the mass and modifies the chemical composition of the CCN that is returned to the atmosphere upon evaporation of the cloud. Current understanding is that in-cloud sulfate production is the most likely mechanism for transfer of aerosol mass from smaller to larger particles (Hoppel, 1989), and the soluble mass added to the particle by this pathway allows it to serve as a CCN at a lower supersaturation in the next cloud cycle in which it participates (Bower and Choularton, 1993). The aqueous-phase oxidation of S(IV) to S(VI) is primarily by dissolved oxygen, ozone (O_3), and hydrogen peroxide (H_2O_2). The S(VI) species which are produced include sulfuric acid (H_2SO_4), bisulfate ion (HSO_4^-), and sulfate ion (SO_4^{2-}). Concentrations of the individual S(VI) species depend upon the appropriate aqueous equilibria and, for sulfuric acid, upon gas–liquid partitioning, although sulfuric acid is so highly soluble that its concentration in the gas phase is assumed negligible if liquid water is present. The trace metal ions Fe(III), Mn(II), and Cu(II), which can enter droplets through scavenged particles, appear to be effective catalysts in promoting the rate of aqueous-phase oxidation of SO_2 by dissolved oxygen (Grgic et al., 1992; Berglund and Elding, 1995). For the conditions typically assumed for "remote" regions, rates of oxidation by ozone or hydrogen peroxide are faster and are assumed to be the dominant pathways. However, trace metal ion chemistry may be expected to be important in regions where the aerosol composition has been strongly influenced by industrial sources, such as smelters, which release trace metals into the atmosphere (Williams et al., 1988), or where crustal metal sources are significant.

Many studies of the catalytic effect of particulate carbon upon sulfur oxidation have been published; a listing of some experimental studies of soot-catalyzed aqueous sulfur oxidation is found in Table 5.10 of the text by Seinfeld (1986). Chughtai et al. (1993) examined the individual and combined effects of soot and transition metal oxides on SO_2 oxidation and concluded that the combination of soot and metal oxides is an effective catalyst. There is no general agreement in proposed rate expressions, and further work is needed in this area, particularly since observational evidence points to the likelihood that this is an important mechanism in the atmosphere. For example, samples of aerosols from the plume of the Kuwait oil fires indicated that about 70% of the particles were CCN (Hudson and Clarke, 1992). Electron micrographs and chemical analyses frequently showed a core of soot embedded in sulfate (Sheridan et al., 1992; Okada et al., 1992). The sulfate component apparently converted nonhygroscopic soot aerosol to efficient CCN and thereby significantly reduced the lifetime of smoke particles in the atmosphere.

Although most studies reported in the literature have focused upon the role of cloud and, to some extent, fog droplets in converting SO_2 to sulfate, some studies have indicated that significant sulfate production may occur in haze formed on sea-salt aerosols. Sea salt is found primarily in the coarse particle mode (diameter >2.5 μm) and, as a consequence of its ionic composition, forms alkaline, buffered solution droplets (Pszenny et al., 1982). The resulting high pH favors rapid S(IV)-to-S(VI) oxidation by ozone. Measurements and modeling studies suggest that 5–25% of the total non-sea-salt sulfate in the marine boundary layer may be formed via this sea-salt aerosol mechanism (Luria and Sievering, 1991; Sievering et al., 1992). In addition to being an important component of the marine sulfur cycle, this pathway may influence the size distribution of marine aerosols: The non-sea-salt sulfate thus formed is associated with coarse-mode particles that have a large dry removal rate and are rapidly cycled near the surface. Thus, sulfur gas emissions that are cycled via this pathway cannot contribute to the growth of fine-mode particles (diameter <2.5 μm) or to the generation of new particles by homogeneous nucleation (Chameides and Stelson, 1992).

There is reason to expect that aqueous chemistry can vary from drop to drop within the same cloud. After a cloud has formed, the droplet size distribution is usually not very broad, so that, for example, a soluble 0.1-μm aerosol particle is grown into a cloud drop that is approximately the same size as that grown from a 1.0-μm particle. The masses represented by these two aerosol particles, however, differ by three orders of magnitude, and thus the aqueous concentrations of released ionic species differ markedly (Twohy et al., 1989). Chemically distinct particles will also produce drops with chemically distinct compositions (e.g., NaCl vs. $(NH_4)_2SO_4$). It is therefore expected that the chemistry occurring in individual cloud droplets should vary significantly, and, indeed, the overall conversion of S(IV) to S(VI) computed for a "bulk" cloudwater chemistry differs substantially from that computed by explicitly considering the distribution of aerosol mass in the droplets (Ayers and Larson, 1990; Lin and Chameides, 1991; Ogren and Charlson, 1992). Field experiments designed to detect drop-size-dependent variations in chemistry have provided some supporting observations (Ogren et al., 1989; Collett et al., 1994; Rao and Collett, 1995), although inhomogeneities are in some cases less pronounced than expected. Postulated reasons for decreased variability include mixing of droplets by turbulent motion in clouds and evaporation of some droplets.

Effects of Heterogeneous Chemistry on the Oxidative Capacity of the Troposphere

In addition to the effect upon oxidation of sulfur, trace-metal-catalyzed reactions on aerosol particles may be important for the destruction of the peroxy radical ($HO_2\cdot$) and the production of H_2O_2. Recent studies have also shown that the presence of metals and organic species in cloud water promotes aqueous-phase photochemical production of H_2O_2 (Zuo and Hoigné, 1993; Anastasio et al., 1994). The existence of aqueous-phase oxidant production pathways implies that the availability of oxidants produced in the gas phase and subsequently dissolved into droplets may not limit aqueous-phase oxidation of sulfur, as has been often assumed (Anastasio et al., 1994). Model studies suggest that such heterogeneous reactions may significantly affect tropospheric HO_x chemistry (Ross and Noone, 1991).

Reactions of N_2O_5 and NO_3 on wet aerosol surfaces have been invoked to account for observed rapid nighttime losses of NO_x reservoir species. Inclusion of these reactions in a tropospheric chemical model (Dentener and Crutzen, 1993) led to improved simulations of observed nitrate wet deposition patterns. Additionally, the heterogeneous reactions lowered computed ozone concentrations by 25% for some regions and seasons. The effect upon ozone arises from the role that tropospheric NO_x species play in determining the concentrations of the important atmospheric oxidants O_3 and $OH\cdot$.

Heterogeneous reactions occurring on atmospheric aerosol particles have been implicated in the destruction of lower tropospheric ozone at polar sunrise in the Arctic (Barrie et al., 1988). The ozone-destruction reactions involve halogens, particularly Br (Fan and Jacob, 1992). The Polar Sunrise Experiment, conducted in 1992, was designed to collect data to elucidate the chemistry involved in this phenomenon (Barrie et al., 1994). Observations indicated that heterogeneous reactions occurring on acidic sulfate aerosols, analogous to those implicated in stratospheric ozone depletion, may be important in tropospheric ozone destruction and in the partitioning of bromine, nitrogen, and other chemical species in the Arctic troposphere.

CONCLUSION

The aerosol in the troposphere exhibits substantial spatial and temporal variability in physical and chemical characteristics. Although this chapter has focused particularly on their role in the global sulfur cycle, aerosol species are important intermediates in many atmospheric chemical cycles and are linked to the hydrologic cycle by their role as cloud condensation nuclei. Aerosol particles affect the formation and evolution of hazes and clouds, and their chemical composition strongly impacts the rates of uptake and aqueous chemical conversion of gases. Aqueous chemical processes in turn shape the aerosol size spectra and chemical composition, and precipitation serves as a primary removal process for atmospheric particulate matter.

Aerosol, haze, and cloud particles serve as sites for heterogeneous reactions that may be important in determining the oxidative capacity of the troposphere, including the destruction of lower tropospheric ozone at polar sunrise in the Arctic. Additionally, species transported by aerosol particles, such as organic pollutants and trace metal ions, serve as catalysts for heterogeneous reactions. The global-scale effects of heterogeneous chemistry involving aerosol appear to be significant, although present estimates of global impacts must rely on limited knowledge of the nature of the interactions.

References

Albrecht, B. A., 1989, Aerosols, cloud microphysics, and fractional cloudiness. *Science*, 245, 1227.

Anastasio, C., Faust, B. C., and Allen, J. M., 1994, Aqueous phase photochemical formation of hydrogen peroxide in authentic cloud waters. *J. Geophys. Res.*, 99, 823i.

Ayers, G. P., and Larson, T. V., 1990, Numerical study of droplet size dependent chemistry in oceanic, wintertime stratus cloud at southern mid-latitudes. *J. Atmos. Chem.*, 11, 143.

Baker, M. B., and Charlson, R. J., 1990, Bistability of CCN concentrations and thermodynamics in the cloud-topped boundary layer. *Nature*, 345, 142.

Barrie, L. A., 1986, Arctic air pollution: an overview of current knowledge. *Atmos. Environ.*, 20, 643.

Barrie, L. A., Bottenheim, J. W., Schnell, R. C., Crutzen, P. J., and Rasmussen, R. A., 1988, Ozone destruction and photochemical reactions at polar sunrise in the lower Arctic atmosphere. *Nature*, 334, 138.

Barrie, L. A., Bottenheim, J. W., and Hart, W. R., 1994, Polar Sunrise Experiment 1992 (PSE 1992): Preface. *J. Geophys. Res.*, 99, 25,313.

Barth, M. C., Hegg, D. A., and Hobbs, P. V., 1992, Numerical modeling of cloud and precipitation chemistry associated with two rainbands and some comparisons with observations. *J. Geophys. Res.*, 97, 5825.

Berglund, J., and Elding, L. I., 1995, Manganese catalysed autooxidation of dissolved sulfur dioxide in the atmospheric aqueous phase. *Atmos. Environ.*, 29, 1379.

Bower, K. N., and Choularton, T. W., 1993, Cloud processing of the cloud condensation nucleus spectrum and its climatological consequences. *Q. J. R. Meteor. Soc.*, 119, 655.

Butcher, S. S., 1992, In: S. S. Butcher, R. J. Charlson, G. H. Orians, and G. V. Wolfe, eds., *Global Biogeochemical Cycles*, Academic Press, New York, p. 235.

Calvert, J. G., Lazrus, A., Kok, G. L., Heikes, B. G., Walega, J. G., Lind, J., and Cantrell, C. A., 1985, Chemical mechanisms of acid generation in the troposphere. *Nature*, 317, 27.

Chameides, W. L., 1984, The photochemistry of a remote marine stratiform cloud, *J. Geophys. Res.*, 89, 4739–4755.

Chameides, W. L., and Stelson, A. W., 1992, Aqueous-phase chemical processes in deliquescent sea-salt aerosols: a mechanism that couples the atmospheric cycles of S and sea salt. *J. Geophys. Res.*, 97, 20,565.

Chaumerliac, N., Richard, E., Pinty, J.-P., and Nickerson, E. C., 1987, Sulfur scavenging in a mesoscale model with quasi-spectral microphysics: two-dimensional results for continental and maritime clouds. *J. Geophys. Res.*, 92, 3114.

Chughtai, A. R., Brooks, M. E., and Smith, D. M., 1993, Effect of metal oxides and black carbon (soot) on SO_2 / O_2 / H_2O reaction systems. *Aer. Sci. Tech.*, 19, 121.

Clarke, A. D. 1992, Atmospheric nuclei in the remote free-troposphere. *J. Atmos. Chem.*, 14, 479.

Clarke, A. D., 1993, Atmospheric nuclei in the Pacific midtroposphere: Their nature, concentration, and evolution. *J. Geophys. Res.*, 98, 20,633.

Coakley, J. A., Bernstein, R. L., and Durkee, P. A., 1987, Effect of ship-stack effluents on cloud reflectivity. *Science*, 237, 1020.

Collett, J. L., Jr., Bator, A., Rao, X., and Demoz, B. B., 1994, Acidity variations across the cloud drop size spectrum and their influence on rates of atmospheric sulfate production. *Geophys. Res. Lett.*, 21, 2393.

Crutzen, P. J., and Andreae, M. O., 1990, Biomass burning in the tropics: Impact on atmospheric chemistry and biogeochemical cycles, *Science*, 250, 1669.

d'Almeida, G. A., Koepke, P., and Shettle, E. P., 1991, *Atmospheric Aerosols: Global Climatology and Radiative Characteristics*. A. Deepak Publishing, Hampton, Virginia, 561 pp.

Delmas, R. J., Free tropospheric reservoir of natural sulfate 1992. *J. Atmos. Chem.*, 14, 261.

DeMott, P. J., 1990, Studies of condensation and ice nucleation by soot aerosols. *J. Appl. Meteor.*, 29, 1072.

Dentener, F. J., and Crutzen, P. J., 1993, Reaction of $N2_{25}$ on tropospheric aerosols: impact on the global distributions of NO_x, O_3, and OH, *J. Geophys. Res.*, 98, 7149.

Duce, R. A., Liss, P. S., Merrill, J. T., Atlsa, E. L., Buat-Menard, P., Hucks, B. B., Miller, J. M., Prospero, J. M., Arimoto, R., Church, T. M., Ellis, W., Galloway, J. N., Hansen, L., Jickells, T. D., Knap, A. H., and Reinhardt, K. H., 1991, The atmospheric input of trace species to the world ocean. *Global Biogeochemical Cycles*, 5, 193.

Fan, S.-M., and Jacob, D. J., 1992, Surface ozone depletion in Arctic spring sustained by bromine reactions on aerosols. *Nature*, 359, 522.

Fitzgerald, J. W., 1991, Marine aerosols: a review. *Atmos. Environ.*, 25A, 533.

Graedel, T. E., and Weschler, C. J., 1981, Chemistry within aqueous atmospheric aerosols and raindrops. *Rev. Geophys. Space Phys.*, 19, 505.

Gras, J. L., 1993, Condensation nucleus size distribution at Mawson, Antarctica: seasonal cycle. *Atmos. Environ.*, 21A, 1417.

Grgic, I., Hudnik, V., Bizjak, M., and Levec, J., 1992, Aqueous S(IV) oxidation, II. Synergistic effects of some metal ions. *Atmos. Environ.*, 26A, 571.

Gunz, D. W., and Hoffmann, M. R., 1990, Atmospheric chemistry of peroxides: a review. *Atmos. Environ.*, 24(A), 1601.

Han, Q., Rossow, W. B., and Lacis, A. A., 1994, Near-global survey of effective droplet radii in liquid water clouds using ISCCP data. *J. Climate*, 7, 465.

Harriss, R. C., Wofsy, S. C., Bartlett, D. S., Shipham, M. C., Jacob, D. J., Hoell, J. M., Jr., Bendura, R. J., Drewry, J. W., McNeal, R. J., Navarro, R. L., Gidge, R. N., and Rabine, V. E., 1992, The Arctic Boundary Layer Expedition (ABLE 3A): July–August 1988. *J. Geophys. Res.*, 97, 16,383.

Harriss, R. C., Wofsy, S. C., Hoell, J. M., Jr., Bendura, R. J., Drewry, J. W., McNeal, R. J., Pierce, D., Rabine, V., and Snell, R. L., 1994, The Arctic Boundary Layer Expedition (ABLE 3B): July–August 1990. *J. Geophys. Res.*, 99, 1635.

Hegg, D. A., and Hobbs, P. V., 1987, Comparisons of measured sulfate production due to ozone oxidation in clouds with a kinetic rate equation. *Geophys. Res. Lett.*, 14, 719.

Hegg, D. A., Rutledge, S. A., Hobbs, P. V., Barth, M. C., and Hertzmann, O., 1989, The chemistry of a mesoscale rainband. *Q. J. R. Meteorol. Soc.*, 115, 867.

Hegg., D. A., and Hobbs, P. V., 1982, Measurements of sulfate production in natural clouds. *Atmos. Environ.*,16, 2663.

Heymsfield, A. J., and Sabin, R. M., 1989, Cirrus crystal nucleation by homogeneous freezing of solution droplets. *J. Atmos. Sci.*, 46, 2252.

Hoppel, W. A., 1989, The role of nonprecipitating cloud cycles and gas-to- particle conversion in the maintenance of the submicron aerosol size distribution over the tropical oceans. In: P. V. Hobbs and M. P. McCormick, eds., *Aerosols and Climate*, A. Deepak Publishing, Hampton, VA, p. 9.

Hoppel, W. A., and Frick, G. M., 1990, Submicron aerosol size distributions measured over the tropical and south Pacific. *Atmos. Environ.*, 24A, 645.

Hoppel, W. A., Fitzgerald, J. W., Frick, G. M., Larson, R. E., and Mack, E. J., 1990, Aerosol size distributions and optical properties found in the marine boundary layer over the Atlantic Ocean. *J. Geophys. Res.*, 95, 3659.

Hudson, J. G., 1991, Observations of anthropogenic CCN. *Atmos. Environ.*, 25A, 2449.

Hudson, J. G., 1993, Cloud condensation nuclei, *J. Appl. Meteor.*, 32, 596.

Hudson, J. G., and Frisbie, P. R., 1991, Cloud condensation nuclei near marine stratus. *J. Geophys. Res.*, 96, 20795.

Hudson, J. G., and Clarke, A. D., 1992, Aerosol and cloud condensation nuclei measurements in the Kuwait plume. *J. Geophys. Res.*, 97, 14533.

Ikegami, M., Okada, K., Zaizen Y., and Makino, Y., 1994, Sea-salt particles in the upper tropical troposphere. *Tellus*, 46B, 142.

Ito, T., 1993, Size distribution of Antarctic submicron aerosols. *Tellus*, 45B, 145.

Jacob, D. J., 1986, Chemistry of OH in remote clouds and its role in the production of formic acid and peroxymonosulfate. *J. Geophys. Res.*, 91, 9807.

Jaenicke, R., 1988, Aerosol physics and chemistry. In: D. Fischer, ed., *Zahlenwerte und Funktionen aus Naturwissenschaften und Technik*, Landolt-Börnstein New Series V, Springer-Verlag, pp. 391–456.

Kreidenweis, S. M., Penner, J. E., Yin, F., and Seinfeld, J. H., 1991, The effects of dimethylsulfide upon marine aerosol concentrations. *Atmos. Environ.*, 25A, 2501.

Langner, J., and Rodhe, H., 1991, A global three-dimensional model of the tropospheric sulfur cycle. *J. Atmos. Chem.*, 13, 225.

Leaitch, W. R., Isaac, G. A., Strapp, J. W., Banic, C. M., and Wiebe, H. A., 1992, The relationship between cloud droplet number concentrations and anthropogenic pollution: observations and climatic implications. *J. Geophys. Res.*, 98, 2463.

Lin, X., and Chameides, W. L., 1991, Model studies of the impact of chemical inhomogeneity on SO_2 oxidation in warm stratiform clouds. *J. Atmos. Chem.*, 13, 109, 1991.

Luria, M., and Sievering, H., 1991, Heterogeneous and homogeneous oxidation of SO_2 in the remote marine atmosphere, *Atmos. Environ.*, 25A, 1489.

Malm, W. C., Sisler, J. F., Huffman, D., Eldred R. A., and Cahill, T. A., 1994, Spatial and seasonal trends in particle concentration and optical extinction in the United States. *J. Geophys. Res.*, 99, 1347.

Novakov, T., and Penner, J. E., 1993, Large contribution of organic aerosols to cloud condensation nuclei concentrations. *Nature*, 365, 823.

Ogren, J. A., and Charlson, R. J., 1992, Implications for models and measurements of chemical inhomogeneities among cloud droplets, *Tellus*, 44B, 208.

Ogren, J. A., Heintzenberg, J., Zuber, A., Noone, K. J., and Charlson, R. J., 1989, Measurements of the size-dependence of solute concentrations in cloud droplets. *Tellus*, 41B, 24.

Okada, K., Ikegami, M., Uchino, O., Nikaidou, Y., Zaizen, Y., Tsutsumi, Y., and Makino, Y., 1992, Extremely high proportions of soot particles in the upper troposphere over Japan. *Geophys. Res. Lett.*, 19, 921.

Parungo, F. P., Nagamoto C. T., Rosinski J., and Haagenson P. L., 1987, Marine aerosols in Pacific upwelling regions. *J. Aerosol Sci.*,18, 277.

Penner, J. E., Charlson, R. J., Hales, J. M., Laulainen, N. S., Lefier, R., Novakov, T., Ogren, J., Radke, L. F., Schwartz, S. E., and Travis, L., 1994, Quantifying and minimizing uncertainty of climate forcing by anthropogenic aerosols. *Bull. Am. Meteorol. Soc.*, 75, 137.

Pruppacher, H. R., and Jaenicke, R., 1995, The processing of water vapor and aerosols by atmospheric clouds, a global estimate. *Atmos. Res.*, 38, 283.

Pszenny, A. A. P., MacIntyre, F., and Duce, R. A.,1982, Sea-salt and the acidity of marine rain on the windward coast of Samoa. *Geophys. Res. Lett.*, 9,751.

Radke, L. F., Hobbs, P. V., Coakley, J. A., and King, M. D., 1989, Direct and remote sensing observations of the effects of ships on clouds. *Science*, 24, 1146.

Raemdonck, H., Maenhaut W., and Andreae M. O., 1986, Chemistry of marine aerosols over the tropical and equatorial Pacific. *J. Geophys. Res.*, 91, 8623.

Rao, X., and Collett, J. L., Jr., 1995, Behavior of S(IV) and formaldehyde in a chemically heterogeneous cloud. *Environ. Sci. Technol.*, 29, 1023.

Roelofs, G. J. H., 1992, Drop size dependent sulfate distribution in a growing cloud. *J. Atmos. Chem.*, 14, 109.

Rogers, D. C., and DeMott, P. J., 1991, Advances in laboratory cloud physics, 1987–1990. *Rev. Geophys.*, 29 (Suppl. Vol. 1), 80.

Ross, H. B., and Noone, K. J., 1991, A numerical investigation of the destruction of peroxy radical by Cu ion catalysed reactions on atmospheric particles. *J. Atmos. Chem.*, 21, 121.

Sassen, K., 1992, Evidence for liquid-phase cirrus cloud formation from volcanic aerosols: climatic implications. *Science*, 257, 516.

Sassen, K., and Dodd, G. C., 1988, Homogeneous nucleation rate for highly supercooled cirrus cloud drops. *J. Atmos. Sci.*, 45, 1357.

Schwartz, S. E., 1986, Mass transport considerations pertinent to aqueous phase reactions of gases in liquid-water clouds. In: Jaeschke, W., ed., *Chemistry of Atmospheric Systems*, NATO ASI Series, Vol. G6, Springer-Verlag, Berlin, p. 467.

Seinfeld, J. H., 1986, *Atmospheric Chemistry and Physics of Air Pollution*, John Wiley & Sons, New York.

Sheridan, P. J., Schnell, R. C., Hofmann, D. J., Harris, J. M., and Deshler, T., 1992, Electron microscope studies of aerosol layers with likely Kuwaiti origins over Laramie, Wyoming during Spring 1991. *Geophys. Res. Lett.*, 19, 389.

Sievering, H., Boatman, J., Gorman, E., Kim, Y., Anderson, L., Ennis, G., Luria, M., and Pandis, S., 1992, Removal of sulphur from the marine boundary layer by ozone oxidation in sea-salt aerosols. *Nature*, 360, 571.

Twohy, C. H., Austin, P. H., and Charlson, R. J., 1989, Chemical consequences of the initial diffusional growth of cloud droplets: a clean marine case. *Tellus*, 41B, 51.

Twomey, S., 1991, Aerosols, clouds and radiation. *Atmos. Environ.*, 25A, 2435.

Twomey, S., Piepgrass, M., and Wolfe, T. L., 1984, An assessment of the impact of pollution on global cloud albedo. *Tellus*, 36B, 356.

Vali, G., 1985, Atmospheric ice nucleation: a review. *J. Rech. Atmos.*, 19, 105.

Williams, P. T., Radojevic, M., and Clarke, A. G., 1988, Dissolution of trace metals from particles of industrial origin and its influence on the composition of rainwater. *Atmos. Environ.*, 22, 1433.

Zuo, Y., and Hoigné, J., 1993, Evidence for photochemical formation of H_2O_2 and oxidation of SO_2 in authentic fog water. *Science*, 260, 71.

CHAPTER 12

Why Carbon Dioxide from Fossil Fuel Burning Won't Go Away

PIETER P. TANS

ABSTRACT. The carbon dioxide added to the combined atmosphere–ocean–biosphere as a result of the combustion of fossil fuels will not leave that system, which we call the mobile reservoirs, for an extremely long time. The extra carbon will keep moving back and forth between the mobile reservoirs in response to climatic and biological processes. Therefore, future anthropogenic greenhouse forcing of the earth's climate will be controlled primarily by the total amount and the rate of fossil fuel burning. Humankind may have a measure of control on the storage of carbon by the terrestrial biosphere.

INTRODUCTION

Carbon dioxide in the atmosphere influences the earth's radiative balance and, hence, its surface temperature and climate. In this chapter we will concern ourselves with the question of how the atmospheric concentration of CO_2 is determined and how it may change as a result of human activities. Carbon at the surface of the earth is held in reservoirs that differ in capacity by many orders of magnitude. Figure 12.1 shows the major carbon reservoirs and the rates at which carbon is transferred from one form of storage to another. Two features immediately attract attention. The oceans hold much more carbon than the atmosphere, while their carbon inventory in turn pales in comparison with the amount present in sedimentary rocks. The second feature concerns the fluxes of carbon between the various reservoirs. The exchange between the atmosphere and the biosphere and between the atmosphere and the oceans is vigorous, whereas other fluxes are much smaller. Looking at the relative sizes of the reservoirs of carbon, it might at first sight seem surprising that the burning of fossil fuels by our industrialized society could be a problem because there appears to be plenty of capacity for the uptake of CO_2 from the atmosphere into other reservoirs, especially the oceans. We will see that the earth's capability to absorb from the atmosphere the excess CO_2 produced from fossil fuels is limited on a timescale of interest to human society.

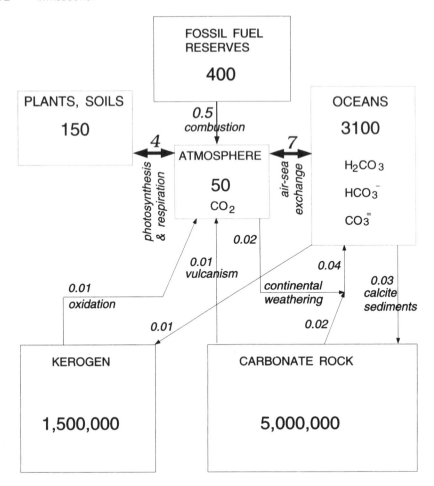

Figure 12.1. Diagram of the global carbon cycle. Quantities in reservoirs are in units of 10^{15} mol C, and fluxes are in 10^{15} mol C yr^{-1}.

This chapter is arranged as follows. First the carbon cycle is put in a geological perspective. The earth's fossil fuel reserves accumulated as a result of biological activity over many millions of years, and it will take on the order of a million years for the man-made perturbation to completely disappear. Next, the ocean carbonate equilibria are defined so that we can understand (a) why the oceans contain so much more carbon than the atmosphere and (b) what limits their capacity to take up additional carbon. Attempted explanations of "recent" natural variations of the atmospheric CO_2 concentration, as recorded in air bubbles in ice cores during the last few hundred thousand years, are all based on the impact on the oceanic carbonate system of changes in ocean circulation, sea level, and biological processes. Evidence is presented that the present increase of CO_2 is, beyond any reasonable doubt, due to man's burning of fuels for energy, and elements of a prognosis for CO_2 are given if the world proceeds to burn all fossil fuel reserves.

THE CARBON CYCLE ON
GEOLOGICAL TIMESCALES

The atmosphere, biosphere, and oceans combined contain only an insignificant amount of carbon compared to that in limestone and silicate rocks. The first three reservoirs we will call the mobile reservoirs. An approximate residence time for carbon in the mobile reservoirs can be obtained by dividing the total amount by the total flux, it is of the order of 100,000 years. Likewise, the residence time in the rock reservoirs is about 150 million years, the timescale of the erosion-sedimentation cycle. On a timescale longer than a million years, the amount of carbon in the mobile reservoirs is governed by the requirement that the input of eroded continental materials equals the rate of sedimentation. Carbonic acid, H_2CO_3, formed from H_2O and CO_2, is the dominant global weathering acid, and the rates of chemical weathering are controlled, to a first order of approximation, by the partial pressure of CO_2 in the atmosphere.

Biological organisms produce CO_2 concentrations in soils that are orders of magnitude higher than in the atmosphere. Via this mechanism, soils enhance the rate of weathering, but their presence is not necessary for chemical weathering reactions to take place (Holland, 1978). In the oceans the removal rates of cations and carbon are to a large extent controlled by the concentration of carbonate (CO_3^{2-}) ions, with higher concentrations leading to greater removal rates. Presently, $CaCO_3$ minerals are produced by shell-forming organisms in surface ocean waters in much greater quantity than the rate of input of carbon to the oceans. Much of this $CaCO_3$ falls toward the bottom, together with organic debris. The solubility of $CaCO_3$ increases with pressure and with lower temperature. Deep waters are also more acidic, due to the oxidation of organic material. The concentration of carbonate ions is suppressed at higher acidity levels (see below). As a result, deep waters are corrosive to $CaCO_3$, so that only a relatively small fraction of the mineral present in surface environments survives in deep ocean sediments.

The weathering and burial of carbonate minerals ($CaCO_3$, $CaMg(CO_3)_2$) does not affect the total amount of carbon in the mobile reservoirs on the million year timescale. The overall reaction is

$$CO_2 + CaCO_3 + H_2O \longleftrightarrow Ca^{2+} + 2HCO_3^- \tag{12.1}$$

Dissolution of one $CaCO_3$ removes one carbon from the rock reservoirs and adds it to the oceans and atmosphere, but it is removed again upon formation of $CaCO_3$. The amounts added and removed in this way have to be about equal in a million years because the calcium cycle is approximately balanced (see below). Transformations that are effective in regulating the total carbon content of the mobile reservoirs are (Berner, 1990)

$$CaCO_3 + SiO_2 \longleftrightarrow CO_2 + CaSiO_3 \tag{12.2}$$

and

$$CO_2 + H_2O \longleftrightarrow CH_2O + O_2 \tag{12.3}$$

Reaction (12.2) represents the thermal decarbonation at high temperature and depth of $CaCO_3$ as a result of magmatic and metamorphic processes, where $CaSiO_3$ stands for a generic calcium silicate mineral. (Magnesium can undergo a similar transformation.) The CO_2 thus produced is released to the atmosphere by volcanic processes. It is used up again in the reverse process during the weathering of the calcium silicate once the rock has come to the surface. In reaction (12.3), CH_2O represents a generic carbohydrate (organic material formed as a result of photosynthesis) that escapes oxidation and gets buried in sediments. Called kerogen, it is a component of most sedimentary rocks. It is oxidized after the rocks have returned to the surface. The carbon in fossil fuel deposits could be considered part of this reservoir, and it is reduced further than carbohydrate, but it forms only a tiny fraction of the total amount of kerogen in sedimentary rocks. Processes (12.2) and (12.3) greatly influence the amount of carbon in the mobile reservoirs because there can be enormous time delays between production and consumption of CO_2, and the storage capacity in sedimentary and metamorphic rocks is very large.

Berner and Lasaga (1989) have simulated the atmospheric partial pressure of CO_2 over the last 600 million years in a simple model which treats the geochemical cycle of carbon only—not that of other elements. An important driving force is plate tectonics, which sets the rate of CO_2 outgassing caused by decarbonation. High atmospheric CO_2 levels are assumed to stimulate plant growth, which in turn leads to greater rates of rock weathering. The ratio between the rates of sedimentation of calcite and organic matter is constrained by observations of $^{13}C/^{12}C$ in limestone rocks. This is possible because newly formed kerogen has a lower $^{13}C/^{12}C$ ratio than calcite formed during the same time period. A change in the relative rates of burial will produce a shift in the $^{13}C/^{12}C$ ratio of the mobile reservoirs. Berner and Lasaga's result is that atmospheric CO_2 has changed by more than an order of magnitude over the last 600 million years, with the present as a period of low CO_2.

Although the omission of the geochemical cycles of other elements may not greatly influence the partial pressure of CO_2 in the atmosphere, it does influence the overall budget of carbon. Broecker (1971) assumed that on geological timescales the atmospheric partial pressure of CO_2 (pCO_2) was controlled by the need for weathering reactions to balance the oceanic removal rate of cations. The atmospheric partial pressure of CO_2 in turn determined the concentration of dissolved CO_2 through the Henry's Law solubility constant. Furthermore, Broecker's model also assumes that the concentration of CO_3^{2-} ions in seawater controls the removal of carbon from the ocean such that this removal balances the carbon inputs.

Before we can determine the total amount in the ocean, we need another constraint because C can also be present in the form of bicarbonate (HCO_3^-) ion. Carbonate and bicarbonate ions in seawater provide part of the charge balance. The carbonate alkalinity, defined as $2[CO_3^{2-}] + [HCO_3^-]$, equals $[Na^+] + [K^+] + 2[Ca^{2+}] + 2[Mg^{2+}] - [Cl^-] - 2[SO_4^{2-}] - [Br^-] \pm$ minor species. Therefore, the proportioning of carbon between CO_3^{2-} and HCO_3^- is determined by the excess of cations over anions other than (bi)carbonate, and in this way by the geochemical cycles of other elements.

Once $[HCO_3^-]$ is constrained by this requirement of charge balance, the ocean's pH is also fixed (see next section) as well as the total amount of carbon. The residence time (total amount in reservoir divided by the removal rate) of C in the oceans is short (see Figure 12.1) compared to that of other major elements in seawater, which have residence times of more than a million years. It follows that these elements are so well mixed in the oceans that their ratios are constant, and one variable is sufficient to describe their concentrations. This variable is the salinity, which is a measure of the total salt content of the water expressed as g/(kg seawater). Important processes determining the concentrations of the various salts are continental weathering, new sea floor formation, hydrothermal venting, sedimentation, ion exchange in newly formed sediments and mid-ocean ridge basalts, and evaporation from semienclosed basins.

DEFINITION OF THE OCEANIC CARBONATE SYSTEM

Carbon dioxide in the atmosphere is not in thermodynamic equilibrium with the bulk of the oceans. In order to calculate what that equilibrium partial pressure might be, we will first have to introduce the variables that define the oceanic carbonate system. The thermodynamic equilibria we need are:

$$CO_2(gas) \longleftrightarrow CO_2(dissolved) \qquad [CO_2(aq)] = K_0 pCO_2 \qquad (12.4)$$

$$[CO_2(aq)] + H_2O \longleftrightarrow [H^+] + [HCO_3^-] \qquad K_1 = [H^+][HCO_3^-]/[CO_2(aq)] \qquad (12.5)$$

$$[HCO_3^-] \longleftrightarrow [H^+] + [CO_3^{2-}] \qquad K_2 = [H^+][CO_3^{2-}]/[HCO_3^-] \qquad (12.6)$$

In writing these equations we have adopted the so-called hydrate convention, which does not distinguish between gaseous dissolved carbon dioxide, $CO_2(aq)$, and its hydrated form carbonic acid, H_2CO_3 (Stumm and Morgan, 1981). K_0 is the Henry's Law solubility, and K_1 and K_2 are the first and second dissociation constants of carbonic acid. As defined here they are not proper thermodynamic constants written in terms of activities, but they are practical constants defined in terms of stoichiometric quantities and a particular pH scale. The reason for the use of practical constants is that true thermodynamic activities are difficult to determine in a medium as complex as seawater (e.g., Dickson, 1984). As practical constants, the recently revised values of K_0, K_1, and K_2 (Millero, 1995) are given as a function of temperature and salinity in Table 12.1. At 25°C and a salinity of 35 g/(kg seawater) the pK values (p$K = -\log K$) are p$K_1 = 5.847$ and p$K_2 = 8.916$. The thermodynamic activities of many species in seawater are quite different from their stoichiometric quantities because of considerable complex formation. An example would be HSO_4^-, formed from H^+ and SO_4^{2-}, which ties up part of the "free" hydrogen ions. The stoichiometric quantities are the sum of the free ions and the ions tied up in complexes. Any problems of chemical speciation connected with complex formation are implicitly incorporated into the practical constants themselves because they have been measured (and tabulated) for the particular mix of seawater.

In (12.4)–(12.6) we have five unknowns and three equations. The system is completely determined if we either specify two of those five variables or add two more constraints. We choose the latter, by fixing the total amount of inorganic carbon, $[CO_2(aq)] + [CO_3^{2-}] + [HCO_3^-]$, commonly written as DIC (dissolved inorganic carbon) or as ΣC, and the titration alkalinity, written as TA. Both can be routinely measured. TA is defined by

$$TA = [HCO_3^-] + 2[CO_3^{2-}] + [B(OH)_4^-]$$

$$+ [NO_3^-] + [OH^-] - [H^+] \pm \text{minor species} \qquad (12.7)$$

On a timescale of a thousand years, both ΣC and TA can change only very little in the oceans, and therefore it is useful to study the system with the assumption that the alkalinity and total inorganic carbon are constant. The carbonate alkalinity (CA), defined

TABLE 12.1
Parameters for the Carbonate System in the World's Oceans

The thermodynamic constants given here are as summarized recently by Millero (1995), who incorporated the results of new measurements performed during the last few years. The reference by Millero also presents thermodynamic constants for the dissociation of phosphoric acid, ammonium, silicic acid, and hydrogen sulfide, which have been neglected here. All constants are applicable to concentration units of mol kg^{-1} seawater. T is in degrees Kelvin, t is in degrees centigrade, S is salinity in permil, or g/kg seawater, log is the logarithm base 10, $pK = -\log(K)$, ln is the natural logarithm (base e).

Solubility of CO_2:
$$\ln K_0 = -60.2409 + 9345.17/T + 23.3585*\ln(T/100) + S[0.023517 - 0.00023656T + 0.0047036(T/100)^2]$$

Dissociation constants of carbonic acid:
$$\ln K_1 = 2.18867 - 2275.0360/T - 1.468591*\ln(T) + (-0.138681 - 9.33291/T)S^{0.5} + 0.0726483S - 0.0057493S^{1.5}$$
$$\ln K_2 = -0.84226 - 3741.1288/T - 1.437139*\ln(T) + (-0.128417 - 24.41239/T)S^{0.5} + 0.1195308S - 0.0091284S^{1.5}$$

Dissociation constant of boric acid:
$$\ln K_B = (-8966.90 - 2890.51S^{0.5} - 77.942S + 1.726S^{1.5} - 0.0993S^2)/T + (148.0248 + 137.194S^{0.5} + 1.62247S) + (-24.4344 - 25.085S^{0.5} - 0.2474S)\ln(T) + 0.053105S^{0.5}T$$

Dissociation of water (in the seawater pH$_{SWS}$ scale, consistent with the other constants):
$$\ln K_W = 148.9802 - 13847.26/T - 23.6521*\ln(T) + [-5.977 + 118.67/T + 1.0495*\ln(T)]S^{0.5} - 0.01615*S$$

There are two crystalline forms of $CaCO_3$ in the oceans, aragonite and calcite. Their stoichiometric (again, in terms of total concentrations, not thermodynamic activities) solubility products as a function of temperature and salinity are given by:

$$pK_{sp}(arag) = 171.945 + 0.077993T - 2903.293/T - 71.595 \log T + S^{0.5}(0.068393 - 0.0017276T - 88.135/T) + 0.10018S - 0.0059415S^{1.5}$$
$$pK_{sp}(calc) = 171.9065 + 0.077993T - 2839.319/T - 71.595 \log T + S^{0.5}(0.77712 - 0.0028426T - 178.34/T) + 0.07711S - 0.0041249S^{1.5}$$

As a check, these expressions give at 298.15 K, $S = 35$, and $P = 1$ bar:

$K_0 = 0.028392$	$pK_1 = 5.8468$
$pK_2 = 8.9156$	$pK_B = 8.5819$
$pK_{sp}(arag) = 6.188$	$pK_{sp}(calc) = 6.369$
$pK_W = 13.2107$	

The pressure dependence can be expressed as
$$\ln(K_i^P/K_i^\circ) = -(\Delta V_i/RT)(P - 1) + (0.5\Delta k_i/RT)(P - 1)^2,$$

in which P is in bar, $R = 83.147$ bar cm^3 mol^{-1} K^{-1}, K_i° is the thermodynamic constant at 1 bar, and ΔV_i (cm^3 mol^{-1}) and Δk_i (cm^3 mol^{-1} bar^{-1}) are the changes in molar volume and molar compressibility of the dissociation and dissolution reactions. They are:

K_1: $\Delta V_1 = -25.50 + 0.1271t$	$10^3\Delta k_1 = -3.08 + 0.0877t$
K_2: $\Delta V_2 = -15.82 - 0.0219t$	$10^3\Delta k_2 = 1.13 - 0.1475t$
K_B: $\Delta V_B = -29.48 - 0.1622t + 0.002608t^2$	$10^3\Delta k_B = -2.84$
K_W: $\Delta V_W = -25.60 + 0.2324t - 0.0036246t^2$	$10^3\Delta k_W = -5.13 + 0.0794t$
$K_{sp}(arag)$: $\Delta V_a = -35.0 + 0.5304t$	$10^3\Delta k_a = -11.76 + 0.3692t$
$K_{sp}(calc)$: $\Delta V_c = -48.76 + 0.5304t$	$10^3\Delta k_c = -11.76 + 0.3692t$

above as $2[CO_3^{2-}] + [HCO_3^-]$, makes up the majority of the titration alkalinity, but CA itself is not directly measurable. At the pH of seawater the contributions from $[OH^-]$ and $[H^+]$ are very minor, so that we only need to consider the dissociation of boric acid (a weak acid) and specify $[NO_3^-]$. The borate equilibrium is given by

$$B(OH)_3 + H_2O \leftrightarrow H^+ + B(OH)_4^-, \qquad K_b = [H^+][B(OH)_4^-]/[B(OH)_3] \qquad (12.8)$$

At 25°C in seawater of salinity 35 we have $pK_b = 8.582$ (see Table 12.1), which is fairly close to the second dissociation constant of carbonic acid. Since the residence time of boron in the oceans is very long, its total concentration, $B(OH)_3 + B(OH)_4^-$, is proportional to the salinity. In other words, it is very thoroughly mixed with the other salts that have long residence times in the oceans. The introduction of boron introduced two more variables, $B(OH)_3$ and $B(OH)_4^-$, but also two constraints, namely Eq. (12.8) and the total amount of boron, given by

$$\Sigma B = 1.179 \times 10^{-5} S \text{ mol/kg} \qquad (12.9)$$

where S is the salinity expressed as g/kg seawater. The inclusion of boron is necessary to derive the partitioning between HCO_3^- and CO_3^{2-} from the measured values of the titration alkalinity. Nitrate ion, NO_3^-, is an important biological nutrient in seawater and is very low in most surface waters (see Chapter 18, this volume). Its concentration in deep waters is typically about 50 μmol/kg due to the oxidation of biological material. To determine the carbonate system completely, we take for average deep water (Takahashi et al., 1981) $\Sigma C = 2288$ μmol/kg, TA = 2393 μequiv./kg, and salinity $(S) = 34.78$.

Now we are finally ready to estimate the atmospheric partial pressure of CO_2 in equilibrium with deep ocean water. The equations describing the carbonate system can best be solved by first solving for $[H^+]$ in an iterative manner, which can be done in the following way:

$$\Sigma C = (1 + K_1/[H^+] + K_1 K_2/[H^+]^2)[CO_2(aq)] \qquad (12.10)$$

$$CA = (K_1/[H^+] + 2K_1 K_2/[H^+]^2)[CO_2(aq)] \qquad (12.11)$$

Dividing Eq. (12.10) by Eq. (12.11) and rearranging gives a quadratic equation for $[H^+]$,

$$(CA)[H^+]^2 + K_1(CA - \Sigma C)[H^+] + K_1 K_2(CA - 2\Sigma C) = 0 \qquad (12.12)$$

and CA is either specified, or found from TA by

$$CA = TA - \{K_b/(K_b + [H^+])\}\Sigma B - [NO_3^-]$$

$$+ [H^+] - [OH^-] \pm \text{ minor species} \qquad (12.13)$$

No great errors are incurred if the contributions of H^+ and OH^- to the alkalinity are neglected, but they can be included by making use of the ion product of water in seawater (Table 12.1). The values of all constants mentioned in this section are given in Table 12.1, so that the reader will be able to reproduce the calculations that follow

It we bring average deep seawater of the above composition to the surface and warm it to 16°C (the observed average temperature of surface waters), the partial pressure (fugacity, strictly speaking) would be 938 μatm, and the pH would be 7.73. For simplicity, we have assumed the contribution of all minor species including $[NO_3^-]$, $[H^+]$, $[OH^-]$ to the alkalinity to be zero. The ocean surface would establish a partial pressure of 938 μatm in air saturated with water vapor at the temperature of 16°C, so that the CO_2 mole fraction in dry air would be equal to 955 μatm. The actual CO_2 mixing ratio in the atmosphere during the recent Holocene period was about 280 parts per million (ppm = 280 μatm at sea level). The large difference is caused by the activity of biological organisms near the ocean surface. Carbon is used up during photosynthesis, and nutrients (mainly phosphate and nitrate) are incorporated into the organic matter formed. In the process, ΣC in surface water is decreased by about 13%, and the nutrients remaining at the surface are reduced almost to zero.

Incidentally, if we had included the presence of deep water nitrate (NO_3^-) in our above calculation of alkalinity, the equilibrium partial pressure of atmospheric CO_2 would have been 40% higher yet than 938 μatm. Due to biological processes the global average values for surface waters are (Takahashi et al., 1981) TA = 2311 μequiv./kg, ΣC = 2002 μmol/kg, both normalized to the average ocean salinity of 34.78 At a temperature of 16°C the atmospheric partial pressure in equilibrium with this surface ocean is 278 μatm, and the pH of the water is 8.18, again assuming the contributions of nitrate, H^+, and OH^- to the alkalinity to be zero. The speciation in this surface water is $[CO_2(aq)]$ = 10 μmol/kg, $[HCO_3^-]$ = 1769 μmol/kg, and $[CO_3^{2-}]$ = 223 μmol/kg. It is clear why the oceans contain so much more carbon than the atmosphere. The atmosphere is in equilibrium with $CO_2(aq)$, but ΣC in this seawater is about 200 times larger than $[CO_2(aq)]$.

With a total mass of 5.14×10^{21} g and a preindustrial CO_2 mixing ratio of 280 ppm, the atmosphere holds $280 \times 10^{-6} \times 5.14 \times 10^{21}/29 = 50 \times 10^{15}$ mol C. If the oceans contained only dissolved CO_2 in equilibrium at the surface with an atmosphere of 280 μatm, and no carbonate or bicarbonate, their total C content could be roughly estimated as follows. The solubility at 16°C is 0.036 mol/kg-atm. The average ocean depth is 4 km, and the area covered by the oceans is 350×10^6 km^2. The total amount of CO_2 dissolved in water would then be about 14×10^{15} mol C, less than one-third of the atmospheric amount. However, when bicarbonate and carbonate are included, the upper 70 meters of the oceans alone hold about the same amount of carbon as the atmosphere.

We can now comprehend an important feature of the carbonate system—namely, why an increase in atmospheric CO_2 does not result in the same proportional increase in ocean carbon at equilibrium. If atmospheric pCO_2 is forced to increase by 10%, $CO_2(aq)$ in the surface layer of the ocean will also go up by 10% if equilibrium with the atmosphere is maintained. What happens to CO_3^{2-} and HCO_3^- ? Dividing Eq. (12.5) by Eq. (12.6) we obtain

$$K_1/K_2 = [HCO_3^-]^2/([CO_2(aq)][CO_3^{2-}]) \tag{12.14}$$

Equation (12.14) is always obeyed. Since $CO_2(aq)$ is only a minor species and most of the carbon is present as bicarbonate, we will assume, to a first order of approximation, that the latter does not change. Then, from Eq. (12.14), $[CO_3^{2-}]$ will have to decrease by 10% when $[CO_2(aq)]$ increases by that much. What happens is that most of the invading CO_2 reacts with CO_3^{2-} to form $2HCO_3^-$. HCO_3^- then finds itself increased by a relative amount $223 \times 2 \times 10\%/1769 = 2.5\%$, not too inconsistent with our initial assumption. Equation (12.14) represents the thermodynamic equilibrium equation of the reversible reaction

$$CO_2(aq) + CO_3^{2-} \longleftrightarrow 2HCO_3^- \tag{12.15}$$

In our approximate treatment the total amount of carbon has changed from the original 2002 μmol/kg to $11 + (1769 + 44) + (223 - 22) = 2025$, adding dissolved CO_2, bicarbonate, and carbonate in that order. Thus, for a fractional increase of pCO_2 of 10%, ΣC has increased by only $23/2002 = 1.1\%$. A more precise calculation, using the constants of Table 12.1, gives 11, 1800, and 210 μmol/kg, respectively, for $[CO_2(aq)]$, $[HCO_3^-]$, and $[CO_3^{2-}]$, summing to $\Sigma C = 2021$. In the more precise calculation the increase of HCO_3^- is more than exactly twice the decrease of CO_3^{2-} because not only reaction (12.15), but also the following reaction with borate, has taken place, in competition with reaction (15):

$$CO_2(aq) + B(OH)_4^- \longleftrightarrow HCO_3^- + B(OH)_3 \qquad (12.16)$$

The detailed equilibrium calculations indeed show that $B(OH)_4^-$ has decreased by the amount required. The increase in total carbon is still close to 1%.

The largest contribution to the oceanic capacity for absorbing CO_2 gas is due to reactions (12.15) and (12.16). On average, the sum of the concentrations of CO_3^{2-} and $B(OH)_4^-$ present in today's oceans, at an average temperature of 4°C, is about 110 μmol/kg. With an ocean area of 350×10^6 km^2 and an average depth of 4 km, this translates into an absorbing capacity for CO_2 of about 150×10^{15} mol C. This reasoning is not correct, however. The capacity of surface waters is greater than that of deep water because of biological activity, and the atmosphere "sees" only the surface waters. The sum of CO_3^{2-} and $B(OH)_4^-$ in surface waters, at an average temperature of 16°C, is approximately 300 μmol/kg (see Figure 12.2). Provided that biological activity keeps the chemistry of surface waters different from that of deep waters, this would lead to a total ocean absorbing capacity of about 420×10^{15} mol C.

It does not matter that at greater depths the capacity would then be exceeded in the event that all fossil fuel reserves are burned. That will only result in very high concentrations of $CO_2(aq)$ and high pCO_2 in deep waters, but that would be invisible to the atmosphere. The 420×10^{15} mol C is small compared to the total amount of inorganic carbon in the oceans, which is about 3100×10^{15} mol C. It is comparable to estimates of the total amount of recoverable fossil fuels, which is 400×10^{15} mol C, give or take a factor of two, and not including shales and tar sands (Perry and Landsberg, 1977). Once the ocean's absorbing capacity has been exceeded, any further CO_2 additions to the atmosphere will remain as gaseous and dissolved CO_2, and the atmosphere will have to retain 75% of the excess addition.

The relationship between the fractional changes of pCO_2 and total carbon is often called the buffer factor or Revelle factor, after Roger Revelle, who played a seminal role in the first modern measurements of the global oceanic and atmospheric carbon system. The Revelle factor (R), defined by

$$\Delta pCO_2/pCO_2 = R\Delta\Sigma C/\Sigma C$$

depends on ΣC and temperature and is about equal to 10 for the present global average surface ocean. In the previous paragraph we have seen that the ocean's ability to absorb additional carbon depends mostly on the concentrations of CO_3^{2-} and $B(OH)_4^-$ in surface waters. These amounts will decrease as more and more CO_2, generated by the burning of fossil fuels, is absorbed. As a result the Revelle factor which is approximately equal to the concentration of total inorganic carbon divided by the concentration of carbonate ions, will increase as CO_3^{2-} gets converted, and it gradually becomes harder for the oceans to keep absorbing CO_2. This is shown in Figure 12.2, in

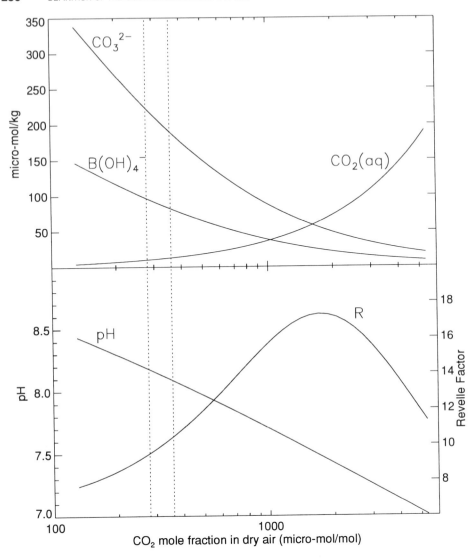

Figure 12.2. The response of surface ocean waters at 16°C to increasing atmospheric concentrations of CO_2, which is plotted on a logarithmic scale. The dissolved $CO_2(aq)$ increase is directly proportional to pCO_2 (Henry's Law). The decreases of CO_3^{2-} and $B(OH)_4^-$ have been calculated with the constants of Table 12.1. In the lower panel, pH is referred to the left axis, and the Revelle factor, R, (see main text) is referred to the right axis. The vertical dotted lines indicate the preindustrial (280 ppm) and the 1995 level of CO_2 (360 ppm).

which $[CO_3^{2-}]$, $[B(OH)_4^-]$, $[CO_2(aq)]$, R, and pH of average surface water are plotted as a function of pCO_2. The ocean reaches equilibrium with the atmosphere after only a relatively small amount has been absorbed, namely a fractional ΣC increase of $1/R$ times the atmospheric increase. At the high pCO_2 values expected when mankind continues to burn fossil fuels, most of the added CO_2 will eventually remain as dissolved CO_2. There will not be enough CO_3^{2-} and $B(OH)_4^-$ left in the oceans to react with. R starts decreasing again at very high pCO_2 values because $CO_2(aq)$, which is proportional to pCO_2, then becomes an increasingly important component of ΣC.

NATURAL VARIATIONS ON THE
TIMESCALE OF ICE AGES

Significant changes in the atmospheric mixing ratios of trace gases have occurred over the last glacial-interglacial cycle (Figure 12.3). During the last decade, convincing evidence of these changes has been obtained from samples of air trapped in ice in Greenland and Antarctica. For a recent review, see Raynaud et al. (1993). Over this geologically short time span there is no requirement that the inputs and outputs from the mobile reservoirs are in balance, and $p\mathrm{CO_2}$ and $[\mathrm{CO_3}^{2-}]$ are free to vary. Furthermore, significant rearrangements between the mobile reservoirs of atmosphere, oceans, and

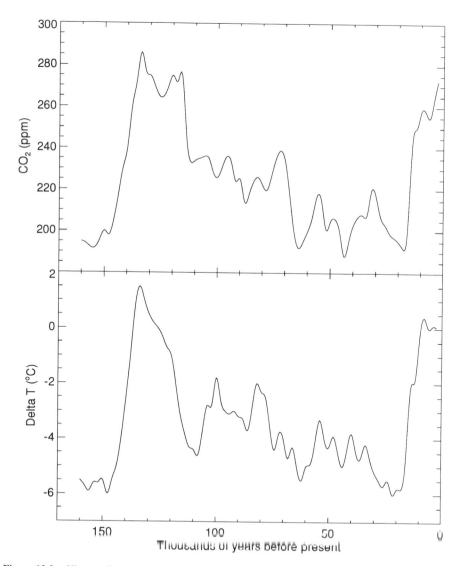

Figure 12.3.   History of temperature and atmospheric CO_2 as recorded in the Vostok ice core, Antarctica. T is plotted as the difference from modern temperatures. (Data from Raynaud et al., 1993).

terrestrial biosphere could have had an impact on the pCO_2 level that has been recorded in the ice cores.

The fact that organisms keep the surface values of the carbonate system different from the bulk of the oceans provides a potentially powerful mechanism to control pCO_2 of the atmosphere on the timescale of thousands of years. The concentrations of phosphate and nitrate, which are the nutrients for the primary photosynthetic producers (phytoplankton), are close to zero in the surface layer over most of the tropical and temperate zones of the oceans. This surface layer is fairly shallow, about 70 m or so, and therefore comprises only a few percent of all ocean water. Deep waters receive a steady rain of organic debris from the sunlit surface layer. The organic material slowly decomposes, increasing the nutrient levels at depth and adding to the total pool of inorganic carbon, ΣC. Only at high latitudes in both hemispheres are the nutrient concentrations at the sea surface significantly different from zero. The high latitudes are also the regions where deep water can outcrop to the surface and where most of the new deep water formation occurs.

The quantities of nutrients in inorganic form contained in this newly formed deep water are of crucial importance to the carbon balance. Chemical oceanographers call these "preformed" nutrients, as opposed to the deep water nutrients that are derived later from the decomposition of the falling organic debris. If all the nutrients of newly formed deep water are tied up in organic material ("zero preformed nutrients"), the gradual carbon increase resulting from organic decomposition will start from approximate equilibrium with the atmosphere and will continue until the average nutrient content of deep water has been reached. The carbon increase of new deep water with an intermediate initial (or preformed) nutrient concentration also starts from approximate equilibrium with the atmosphere, but will be less before the nutrient level of average deep water has been reached.

Thus, with zero preformed nutrients the atmosphere will be further removed from equilibrium with deep water than with nonzero preformed nutrients. The atmosphere is a minor carbon reservoir compared to the deep seas, so we can assume that the carbon content of the deep seas remains constant. Therefore, atmospheric CO_2 has to be lower with fuller biological nutrient utilization in surface waters. This can be formulated in yet another way: What matters for the atmospheric concentration is whether the oceanic nutrients reach the depths while taking carbon with them (as organic debris) or not (as preformed nutrients).

The mechanism outlined in the previous paragraph has often been called the "biological pump" because inorganic carbon and nutrients are continuously removed from the surface layer, falling to the depths due to gravity. Some proposed explanations for the difference in atmospheric CO_2 during an ice age ($\sim$200 ppm) and an interglacial period ($\sim$280 ppm) are based on hypotheses about glacial–interglacial differences in the effectiveness of the biological pump and different nutrient distributions in the oceans. The oceans must have been involved because they are the largest of the mobile carbon reservoirs, and the time scale of ice ages is tens of thousands of years. There is ample time for the atmosphere and oceans to arrive at some form of steady state or equilibrium. In many of these hypotheses, special attention has been given to high latitudes because we know that in today's oceans those are the places in which nutrients are not close to zero at the surface.

Several mechanisms have been postulated as potential causes of shifts in preformed nutrients. Knox and McElroy (1984) explored in a simple ocean model how the efficiency of nutrient utilization, driven by secular changes in the insolation at high latitudes, could alter the CO_2 content of the atmosphere. Sarmiento and Toggweiler (1984) considered both changes in high latitude productivity and in the rate of thermohaline overturning in the oceans. A simultaneous increase in productivity and a decrease in the rate of overturning would reinforce each other because organisms are given more time to decrease the surface nutrient concentrations. Siegenthaler and Wenk (1984) stress the importance of the rate of mixing of deep water with cold, high-latitude surface waters. An attractive feature of all the above mechanisms is that they could alter atmospheric CO_2 at a fast enough rate to explain the observed rapid variations. The ice core record suggests that changes of the atmospheric CO_2 content took place at sustained rates of up to 5×10^{15} mol C per century. Faster increases may have occurred for short periods, but are not visible at the resolution of the ice record. Another common feature of these hypotheses is that they predict very low oxygen concentrations for deep water during glacial times, which should have led to widespread anoxia. There is no evidence in the sediments that this actually occurred, casting considerable doubt on the likelihood of the above mechanisms.

Mangini and collaborators proposed that rapid outgassing of CO_2 from the oceans could have been triggered by an intensification of deep water formation at the beginning of a warm period. This would bring O_2 to the oxygen-poor deep waters causing the oxidation of soluble $Mn^{2+}CO_3^{2-}$ to insoluble $Mn^{4+}O_2$, which then deposits in the sediments and releases HCO_3^- and H^+. As a result, the CO_3^{2-} concentration would drop and the partial pressure of $CO_2(aq)$ would rise. Some evidence for such occurrences has been found in deep sea sediment cores and Mn crusts (Mangini et al., 1991).

Another hypothesis not involving the biological pump was presented by Opdyke and Walker (1992). They noted that most of the present $CaCO_3$ deposition takes place on coral reefs and in shallow waters generally and that it is occurring at a rate ($\sim 2.0 \times 10^{13}$ mol/yr) higher than its long-term average during the Pleistocene period ($\sim 0.8 \times 10^{13}$ mol/yr). Their hypothesis is that during interglacial warm periods sea level is higher, so that carbonate precipitation can occur on the continental shelves. During glacial intervals sea level is low, and the carbonate deposits are exposed and eroded. The average deposition over a full glacial cycle can then be close to the long-term Pleistocene average. The net reaction for this process has already been given in Eq. (12.1). The driving force is sea level, and therefore climate, with atmospheric CO_2 responding. The hypothesis would imply an (asymmetric) sawtooth-like history of atmospheric CO_2 due to short interglacials separated by much longer glacial intervals. A look at the ice core record of CO_2 faintly suggests that such a pattern might indeed be present.

Recently, Archer and Maier-Reimer (1994) proposed yet another intriguing hypothesis for the cause of the glacial-interglacial CO_2 changes. We have seen at the beginning of this chapter that only a small fraction of the calcite reaching the sediments survives dissolution. The oceans are self-regulating with respect to the calcite flux. All deep waters are corrosive to calcite. If the calcium content of the oceans rises, due to increased erosion on the land for instance, the carbonate ion content will rise and the CO_2 partial pressure will drop (same as the two paragraphs above). The increased carbonate ion content implies that the deep waters are less corrosive to calcite or that the portion of the ocean floor bathed in waters that dissolve calcite decreases. Therefore, more of the calcite reaching the ocean floor will be preserved, increasing the calcium output until balance with the higher inputs of calcium is reestablished.

What Archer and Maier-Reimer observed is that the oxidation of organic carbon in ocean sediments promotes the dissolution of calcite, due to the addition of the respired CO_2 to the sedimentary pore water. They employed a three-dimensional ocean model in steady state for the calcium fluxes. In their model the total preservation of calcite is regulated by two contributions, namely, "regular" dissolution in undersatured waters and enhanced dissolution through respiratory CO_2 in sediments rich in organic material. Therefore, if the ratio of organic carbon to calcite reaching the sediments increases, the steady-state contribution to calcite weathering of the "regular" process has to decrease, implying a higher carbonate ion concentration and a lower partial pressure of CO_2.

Any hypothesis trying to explain the glacial–interglacial CO_2 changes leads to a number of other consequences that should have left their signature in the sedimentary record. Examples are the $^{13}C/^{12}C$ ratio, the cadmium content as a proxy record of nutrient concentrations, the types of organisms buried, and the preservation record of $CaCO_3$. The precise and detailed time history of the CO_2 change in the ice cores relative to the timing of the climate change is also an important piece of the evidence. The evidence is still confusing at this time, and no completely satisfactory explanation has yet been advanced.

THE PRESENT

There is no doubt that the current increase in atmospheric CO_2 is mainly due to the burning of fossil fuels. During the early 1990s, 0.5×10^{15} mol was injected into the atmosphere annually (Marland and Boden, 1991). This is an order of magnitude larger than any imbalance between the mobile C reservoirs inferred from the ice core record, although it must be kept in mind that the slow process of air capture in the ice has produced a smoothing of the record (Raynaud et al., 1993). Also, the present level of atmospheric CO_2, about 360 parts per million in 1995, is much higher than at any time during the last several hundred thousand years, and it appeared suddenly during this century, coincident with the rapid rise of fossil fuel consumption.

The isotopic evidence is also consistent with fossil fuels. $^{14}C/C$ ratios were lowered from the 19th century until 1950 when atmospheric testing of nuclear bombs injected a large amount of ^{14}C. Fossil fuels contain no ^{14}C, in contrast to other conceivable sources of atmospheric CO_2 such as biomass or ocean carbon. $^{13}C/^{12}C$ ratios are similarly being lowered, with the trend continuing to the present, which proves that the source of carbon has to be depleted in ^{13}C. This would have to be either fossil fuels or contemporary organic carbon. The photosynthetic process discriminates slightly against the assimilation of ^{13}C relative to ^{12}C. As a result, organic matter has a lower $^{13}C/^{12}C$ ratio than atmospheric CO_2. Fossil fuels have the same low isotopic $^{13}C/^{12}C$ signature, having been derived from organic matter. (The same applies to kerogen, which plays an important role in the carbon balance on long geological timescales.) It appears likely from the detailed recent time history of CO_2 and $^{13}C/^{12}C$ that during the second half of the 19th century and the first half of the 20th century there was appreciable loss of carbon, up to 0.1×10^{15} mol/yr, from terrestrial ecosystems and soils to the atmosphere (Enting and Pearman, 1987), whereas at present the observations indicate that the biosphere has to act globally as a net sink of C (Francey et al., 1995; Ciais et al., 1995).

Further evidence for the present (im)balance of sources and sinks of the global carbon cycle comes from the spatial distribution of CO_2 in the atmosphere. Its mixing ratio is higher in the Northern Hemisphere than in the Southern Hemisphere. The atmosphere is stirred quite vigorously by winds, and the average transport time from one hemisphere to the other is about one year. The higher concentration in the Northern Hemisphere is being actively maintained by a distribution of sources and sinks; the Northern Hemisphere has stronger sources than sinks and the Southern Hemisphere has stronger sinks. Ninety percent of fossil fuel combustion takes place in the Northern Hemisphere.

However, when a numerical model of the atmosphere with properly calibrated transport is used to calculate the north–south gradient of CO_2 resulting solely from the burning of fossil fuels, the calculated gradient is much larger than what is observed. Therefore, there must be sinks of CO_2 in the Northern Hemisphere that partially compensate for the fossil fuel source. Based on observations of the north–south gradient of the $^{13}C/^{12}C$ ratio, Keeling et al. (1989) concluded that the sink had to be mostly in the northern oceans. Tans et al. (1990) studied observations of the difference of the partial pressure of CO_2 in the ocean surface and the atmosphere, which is the thermodynamic driving force of any net flux between those reservoirs. They concluded that the difference was insufficient to sustain a large northern oceanic sink. Therefore, they hypothesized that there has to be substantial CO_2 uptake in forests at temperate latitudes.

It has been demonstrated in many different experiments under controlled conditions that increased CO_2 is generally beneficial to plant growth (see, for instance, Allen et al., 1987). The increased productivity may result in more standing biomass above and below ground and can also lead to increasing amounts of organic matter in soils. It is, however, very difficult to measure this directly because of the tremendous heterogeneity of ecosystems and microclimates. Furthermore, there are other anthropogenic influences that could either counteract or enhance any effects of increased CO_2 in different areas. Examples are increased concentrations of O_3, acid rain, fertilization by low amounts of nitrate deposition, and so on. From an experimental viewpoint the effect on whole ecosystems over a decade or longer is unknown (Mooney et al., 1991). Additional, longer-term processes such as species competition would have to be taken into account. An assessment of net carbon uptake by forest ecosystems based on recent national and regional studies of carbon inventories, carbon fluxes, and land-use statistics concluded that deforestation at low latitudes is still causing emissions of $0.13(\pm 0.04) \times 10^{15}$ mol CO_2 yr^{-1} into the atmosphere, but that carbon storage in temperate forests is increasing by $0.06(\pm 0.02) \times 10^{15}$ mol C yr^{-1} (Dixon et al., 1994). However, this last study did not consider any potential effects of CO_2 fertilization.

FUTURE SCENARIOS FOR CO_2

The odds are very strong that the combustion of fossil fuels will continue to overwhelm other imbalances in the carbon cycle well into the next century. In 1990 the consumption of energy by 1.2 billion people in the developed world was 9×10^{12} W, while the other 4.1 billion people together used 4.5×10^{12} W. Suppose that the developed world manages to lower, through increased conservation and efficiency, its per capita consumption of energy every year by 2% so that in 2025 it would be half of the 1990 number. Then that part of the world would still consume 5.3×10^{12} W in the year 2025 if the population increased to 1.4 billion. In the same time the number of people in the less developed countries might have grown to 6.9 billion, assuming a growth rate of 1.5% per year. If we also assume that their economic well-being improves and that, as a result, they will buy more energy at an increase rate of 2% per year, their total consumption would be 15×10^{12} W in 2025. Global energy consumption would then have increased by 50% compared to 1990, while per capita consumption in the developed countries would still be almost twice that in less developed countries.

In other words, there is ample room for substantial further increases in energy use, and it would be highly desirable as well, in view of the existing great inequities in the use of global natural resources. With current technology, most of the increased energy demand will be met by the burning of coal, oil, and natural gas. Even if an alternative energy source would be economically competitive at this moment, it would still take several decades of investment and replacement of energy infrastucture before fossil fuel burning would cease to be our major source of energy.

In discussing future CO_2 scenarios we will initially neglect the role of the terrestrial biosphere. Without performing any detailed model calculations, a few general observations can then be made, both about the transient effects during the next few centuries and about the long-term prospects. First the latter. The mixing time of the deep oceans is of the order of a thousand years. On that timescale the amount of CO_2 in the atmosphere will come to chemical equilibrium with the oceans (mediated by the biological pump; see the section on the ice ages), and even more slowly with the sediments underlying them.

The atmospheric level would then be determined by four factors: the total net amount added to the ocean–atmosphere system, the ocean's neutralizing capacity as determined by the concentrations of dissolved carbonate and borate ions, the effectiveness of the biological pump, and the amount of $CaCO_3$ in ocean bottom sediments that will be dissolved. For the long term it does not matter much whether it takes one century or five centuries to burn the fossil fuels—only the total amount burned is important. This result was already obtained by Keeling and Bacastow (1977) in an early study of the impact of industrial gas emissions on the atmosphere.

If we neglect the dissolution of sedimentary calcite on the ocean floor for the moment, a fraction of the CO_2 emitted would remain in the atmosphere "indefinitely," determined by equilibrium with the carbonate system in the oceans. At this time approximately 5% of the global fossil fuel resources have been consumed, and a fraction of about 15% will remain permanently in the atmosphere if there is no change in ocean calcium. This long-term airborne fraction would gradually increase from 15% to 75%, at which point the ocean's neutralizing capacity will have become exhausted. Current observations of the atmospheric CO_2 increase show that about half of the industrial emissions remain in the atmosphere. This number is higher than the long-term airborne fraction because the time is too short to reach equilibrium with deeper ocean waters, and the terrestrial biosphere also plays a role.

However, there will likely be an adjustment of the ocean's calcium content. Recall that the surface waters are supersaturated with respect to $CaCO_3$ mineral, but that the deep waters are not. We have seen that an increase in pCO_2 will decrease CO_3^{2-}. Therefore, a greater portion of the seafloor than at present will be bathed in waters that are undersaturated. The fraction of the sea floor that was previously overlain with wa-

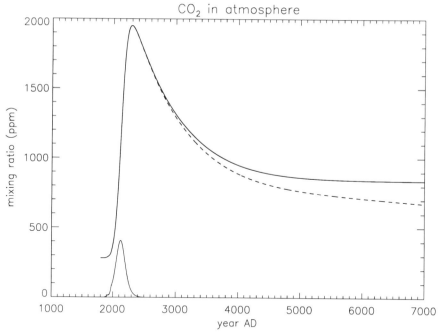

Figure 12.4. Long-term projection of atmospheric CO_2 with a five-box model of the carbon cycle similar to Keeling and Bacastow (1977). The five reservoirs are the atmosphere, ocean surface waters, the deep oceans, living biomass, and soil carbon. It is assumed that all recoverable fossil fuel reserves are burned (400×10^{15} mol C) in a temporal pattern as indicated (not to scale) on the bottom axis, with an initial increase rate of 2% per year during the first decades of the 21st century. The solid curve results if it is assumed that the terrestrial biosphere does not absorb or lose carbon, that the biological pump (see main text) remains as today, and that there is no calcite dissolution of deep ocean sediments. For the dashed curve, calcite dissolution does take place until the present product of $[CO_3^{2-}]$ and $[Ca^{2+}]$ in deep water is restored, with a time constant of 1000 years.

ters supersaturated with respect to $CaCO_3$, but not any longer, becomes susceptible to calcite dissolution. This will have the effect of lowering the partial pressure of CO_2 [see reaction (12.1)]. There are some uncertainties regarding the kinetics of this process in real-life bottom mud.

A possible time course of these events is sketched in Figure 12.4. The solid line assumes the burning of "all" fossil fuel reserves and no calcite dissolution, and the ocean–atmosphere system reaches a new equilibrium with "permanently" enhanced levels of CO_2. If we assume, on the other hand, that calcite dissolution is affected and we restore the calcium balance between inputs and outputs on a timescale of 1000 years (by requiring the product of $[Ca^{2+}]$ and $[CO_3^{2-}]$ in deep waters to be restored to present values on that time scale), we obtain the dashed curve. The atmospheric response to calcite dissolution is slow because there are several delays: The enhanced CO_2 has to reach deeper waters first, the deep water calcium balance has to adjust, and then the deep water signal has to reach the atmosphere again.

Thus, we can expect that the CO_2 generated from the burning of fossil fuels will—slowly, over thousands of years—cause a greater amount of Ca to be mobilized in the oceans, which will lower the pCO_2 of the atmosphere. A new quasi-steady state develops, in which the supply of calcium to the oceans equals the loss and in which the amounts of mobile carbon and mobile calcium are above their preanthropogenic levels. Left to itself, the atmosphere–ocean system will ultimately return to preindustrial conditions through silicate weathering (reaction (12.2)) and through the burial of carbon in reduced form, as in reaction (12.3), and as it was initially in fossil fuel deposits. This, however, will take millions of years.

The calcite dissolution process has no influence on the CO_2 transient expected during the coming centuries. If all the fossil fuel reserves are burned over the next few centuries, the atmospheric CO_2 concentration will go through a transient which could reach maximum levels of five to ten times the preindustrial concentration. How rapidly the transient will rise, and how high the maximum will be, depends sensitively on the rate of fossil fuel consumption relative to the rate of deep sea mixing, as well as on what happens with land plants and soils.

A possible course of events is shown in Figure 12.5. The solid line is the same as in Figure 12.4, whereas the dotted line projects the atmospheric concentrations if the same total amount of fossil fuels is burned more slowly. If the CO_2 transient goes near the high end of the possible range, the CO_3^{2-} concentrations of surface waters will drop low enough that they may become undersaturated with respect to $CaCO_3$ (Figure 12.2). For calcite this will happen at $pCO_2 \approx 2650$ μatm; and for aragonite, which is

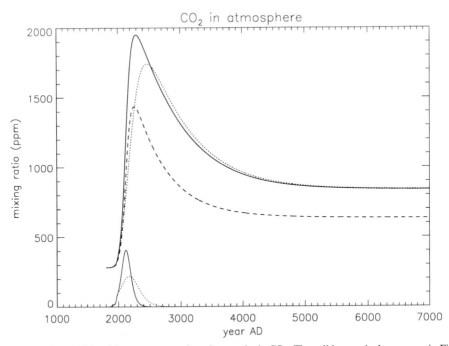

Figure 12.5. Additional long-term scenarios of atmospheric CO_2. The solid curve is the same as in Figure 12.4. All curves assume no change in the effectiveness of the biological pump and no dissolution of deep sea calcite sediments. The dotted curve is different from the solid curve only in the initial rate of increase of fossil fuel burning, 1% per year during the first half of the 21st century (as indicated on the bottom axis), with the total amount burned as in Figure 12.4, 400×10^{15} mol C. For the dashed curve the fossil fuel consumption pattern is the same as for the solid curve, but the global productivity of terrestrial ecosystems is assumed to respond to increased CO_2 in a way similar to Michaelis–Menten kinetics (Allen et al., 1987). Respiration remains proportional to total biomass. These assumptions lead to continued storage of carbon in biomass above today's amounts as long as atmospheric CO_2 remains higher than today.

the other form of crystalline $CaCO_3$ but about twice as soluble, it will occur at $pCO_2 \approx$ 1360 μatm. (The relative abundance of aragonite is only a few percent, however). This could mean a serious threat to the existence of many shell-forming organisms, but it would also put a cap on the maximum atmospheric CO_2 increase.

Thus far the terrestrial biosphere has been left out of our discussion of the future course of anthropogenic CO_2 because it represents one of the more uncertain facets of the problem. Gain or loss of carbon in terrestrial ecosystems will, respectively, decrease or increase the net amount of CO_2 added to the atmosphere–ocean system as described in the previous paragraphs. The total amount of carbon stored in standing biomass has been estimated as 70×10^{15} mol and organic C in soils as 120×10^{15} mol (Schlesinger, 1991). If these amounts could be doubled as a result of fertilization by increased CO_2, or possibly due to human manipulation, a sizable fraction of all the fossil fuel CO_2 would be absorbed. A scenario such as this is illustrated as the dashed line in Figure 12.5.

This possibility was first seriously debated by Revelle and Munk (1977). Stimulation of C uptake by the biosphere has an immediate effect on the CO_2 transient we are in, as well as potentially on the long-term increase of atmospheric CO_2. As discussed above, the atmospheric and oceanic evidence strongly suggests that the biosphere currently absorbs CO_2. Thus it is clear that the biosphere exerts an important influence on today's CO_2 levels, and is likely to continue to do so in the future. An important difference between carbon storage in the deep oceans versus trees and soils is that the latter form of storage remains close to the atmosphere and would appear to be vulnerable to climate change and to continuing human intervention.

The doubling mentioned above is entirely speculative. No credible predictions have been made, but ecologists tend to expect the opposite for the near future, namely that the adjustment of ecosystems to climate change and other perturbations will cause a loss of carbon stored, at least temporarily, which would exacerbate the atmospheric CO_2 problem (Woodwell et al., 1994).

Finally, another crucial uncertainty with regard to the future course of atmospheric CO_2 needs mention. The scenarios pictured in Figures 12.4 and 12.5 assume that the oceanic biological pump keeps working as it does today. If the effectiveness of the pump in lowering the surface values of nutrients and ΣC decreases, the atmospheric CO_2 level will go up further. The effectiveness of the pump depends on the interplay of primary photosynthesis/grazing with ocean circulation/mixing, all of which are likely to be affected by global climate change. The effectiveness of the biological pump could also increase, but it is close to its maximum efficiency (zero nutrient levels in the surface ocean) at the present time.

CONCLUSION

The limitations on the uptake of industrially produced CO_2 are seen clearly when the earth system is considered in a geological perspective. Industrial carbon essentially does not leave the ocean–atmosphere biosphere system for many thousands of years. How much of the extra carbon resides as CO_2 in the atmosphere depends on several processes that govern the partitioning between these three reservoirs. Important are the chemical equilibria in the oceans, how the terrestrial biosphere responds to direct human perturbations and to global environmental change, and the oceanic biological pump. Of these, only ocean carbon chemistry is well understood. The biological pump is likely to undergo change, but it is currently not apparent that we would have a measure of control over it.

The only obvious controls we have over the level of atmospheric CO_2 are the rate and the total amount of fossil fuel combustion. The deliberate manipulation of terrestrial biological processes could be an effective control process if we knew how to muster it. Our current understanding of long-term biological sources and sinks of carbon is grossly insufficient for that purpose, however. Therefore, it is extremely likely that the earth's climate will be under enhanced greenhouse forcing for thousands of years as a result of the utilization of fossil fuels as our main source of energy. There will also be a more transient period, starting during the era of fossil fuel burning and continuing for several centuries afterwards, during which the enhanced greenhouse forcing due to CO_2 will be roughly twice as strong as during the long aftermath.

References

Allen, L. H., Jr., Boote, K. J., Jones, J. W., Jones, P. H., Valle, R. R., Acock, B., Rogers, H. H., and Dahlman, R. C., 1987, Response of vegetation to rising carbon dioxide: photosynthesis, biomass, and seed yield of soybean. *Global Biogeochem. Cycles*, 1, 1.

Archer, D., and Maier-Reimer, E., 1994, Effect of sedimentary calcite preservation on atmospheric CO_2 concentration. *Nature*, 367, 260–263.

Berner, R. A., 1990, Atmospheric carbon dioxide levels over phanerozoic time. *Science*, 249, 1382.

Berner, R. A., and Lasaga, A. C., 1989, Modeling the geochemical carbon cycle. *Sci. Am.*, 260, 74.

Broecker, W. S., 1971, A kinetic model for the chemical composition of seawater. *Quaternary Res.*, 1, 188.

Ciais, P., Tans, P. P., Trolier, M., White, J. W. C., and Francey, R. J., 1995, A large northern hemisphere terrestrial CO_2 sink indicated by the $^{13}C/^{12}C$ ratio of atmospheric CO_2. *Science*, 269, 1098–1102.

Dickson, A. G., 1984, pH scales and proton-transfer reactions in saline media such as seawater. *Geochim. Cosmochim. Acta*, 48, 2299–2308.

Dixon, R. K., Brown, S., Houghton, R. A., Solomon, A. M., Trexler, M. C., and Wisniewski, J., 1994, Carbon pools and flux of global forest ecosystems. *Science*, 263, 185.

Enting, I. G., and Pearman, G. I., 1987, Description of a one-dimensional carbon cycle model calibrated using techniques of constrained inversion. *Tellus*, 39B, 459.

Francey, R. J., Tans, P. P., Allison, C. E., Enting, I. G., White, J. W. C., and Trolier, M., 1995, Changes in oceanic and terrestrial carbon uptake since 1982. *Nature*, 373, 326–330.

Holland, H. D., 1978, *The Chemistry of the Atmosphere and Oceans*, John Wiley & Sons, New York, pp. 103–106.

Keeling, C. D., and Bacastow, R. B., 1977, Impact of industrial gases on climate. In: *Energy and Climate, Studies in Geophysics*, National Academy of Sciences, Washington, D.C., p. 72.

Keeling, C. D., Piper, S. C., and Heimann, M., 1989, A three-dimensional model of atmospheric CO_2 transport based on observed winds: 4. Mean annual gradients and interannual variations, in: *Aspects of Climate Variability in the Pacific and the Western*

Americas, Peterson, D. H., ed., Geophysical Monograph 55, Amer. Geophys. Union, Washington, DC, 305.

Knox, F., and McElroy, M. B., 1984, Changes in atmospheric CO_2: influence of marine biota at high latitude. *J. Geophys. Res.*, 89, 4629.

Mangini, A., Eisenhauer, A., and Walter, P., 1991, A spike of CO_2 in the atmosphere at glacial–interglacial boundaries induced by rapid deposition of manganese in the oceans. *Tellus*, 43B, 97.

Marland, G., and Boden, T., 1991, CO_2 emissions—modern record. In: T. A. Boden, R. J. Sepanski, and F. W. Stoss, eds., *Trends '91: A Compendium of Data on Global Change*, Publ. no. ORNL/CDIAC-46, Oak Ridge National Laboratory, 386.

Millero, F. J., 1995, Thermodynamics of the carbon dioxide system in the oceans. *Geochim. Cosmochim. Acta*, 59, 661–677.

Mooney, H. A., Drake, B. G., Luxmoore, R. J., Oechel, W. C., and Pitelka, L. F., 1991, Predicting ecosystem responses to elevated CO_2 concentrations. *BioScience*, 41, 96.

Opdyke, B. N., and Walker, J. C. G., 1992, Return of the coral reef hypothesis: Basin to shelf partitioning of $CaCO_3$ and its effect on atmospheric CO_2. *Geology*, 20, 733.

Perry, H., and Landsberg, H. H., 1977, Projected world energy consumption, In: *Energy and Climate, Studies in Geophysics*, National Academy of Sciences, Washington, D.C., p. 35.

Raynaud, D., Jouzel, J., Barnola, J. M., Chappellaz, J., Delmas, R. J., and Lorius, C., 1993, The ice record of greenhouse gases. *Science*, 259, 926.

Revelle, R., and Munk, W., 1977, The carbon dioxide cycle and the biosphere, in: *Energy and Climate, Studies in Geophysics*, National Academy of Sciences, Washington, D.C., p. 140.

Sarmiento, J. L., and Toggweiler, J. R., 1984, A new model for the role of the oceans in determining atmospheric pCO_2, *Nature*, 308, 621

Schlesinger, W. H., 1991, *Biogeochemistry. An Analysis of Global Change*, Academic Press, New York.

Siegenthaler, U., and Wenk, T., 1984, Rapid atmospheric CO_2 variations and ocean circulation. *Nature*, 308, 624.

Stumm, W., and Morgan, J. J., 1981, *Aquatic Chemistry*, 2nd ed., John Wiley & Sons, New York, p. 128.

Takahashi, T., Broecker, W. S., and Bainbridge, A. E., 1981, The alkalinity and total carbon dioxide concentration in the world oceans. In: Bolin, B., ed., *Carbon Cycle Modelling*, SCOPE 16, John Wiley & Sons, New York, p. 271.

Tans, P. P., Fung, I. Y., and Takahashi, T., 1990, Observational constraints on the global atmospheric CO_2 budget. *Science*, 247, 1431.

Woodwell, G. M., and Mackenzie, F. T., eds., 1995, *Biotic Feedbacks in the Global Climate System; Will the Warming Feed the Warming?* Oxford University Press, New York. 416 pp.

Stratospheric Chemistry— Perspectives in Environmental Chemistry

WILLIAM H. BRUNE

ABSTRACT. Ozone is the most significant and most abundant reactive chemical in the stratosphere. Because ozone absorbs solar ultraviolet light, decreases in ozone allow more solar ultraviolet light to reach Earth's surface, causing potential ecological and biological harm. Twenty years of observations reveal decreases in stratospheric ozone; concerted studies of the chemistry implicate human-made chlorofluorocarbons and other chlorine- and bromine-containing compounds as being at least partly responsible. When nitrogen- and hydrogen-containing compounds, of mostly natural origin, and chlorine- and bromine-containing compounds, of mostly human origin, enter the stratosphere in the tropics, they are destroyed by solar ultraviolet light and subsequent chemistry. The resulting chemicals react to form free radicals, acids, nitrates, and other compounds as the air moves slowly poleward. A balance is established between the ozone-destroying free radicals and other less-reactive chemicals by gas-phase and heterogeneous reactions. The rates of these reactions in turn depend upon the pressure, temperature, and amount and types of small particles at each latitude and altitude. The free radicals attack ozone in catalytic chemical cycles by which ozone is destroyed while the free radicals are not. However, the ozone abundance is determined not only by this chemical loss; also important are ozone production from the photodestruction of molecular oxygen and the air motions that transport and mix ozone from different regions. Rapid ozone loss in the lower stratosphere above Antarctica, called the ozone hole, has appeared since the late 1970s. Observations confirm that chlorine and bromine chemistry, in which most of the available chlorine and bromine is converted to free radicals, causes the rapid ozone loss. These observations, along with laboratory and computer modeling studies, thus establish the link between human-made chlorofluorcarbons and bromine compounds and stratospheric ozone loss.

INTRODUCTION

Stratospheric chemistry became environmental chemistry in the early 1970s when scientists studied the potential stratospheric effects of supersonic aircraft (Crutzen, 1970; Johnston, 1971). They realized that human activity could affect the chemistry of this cold, remote region 10–50 km above the Earth. Of greatest concern was the destruction of stratospheric ozone, Earth's protective shield against solar ultraviolet light. This concern sparked a flurry of activity that in 1987 led to an international treaty, the Montreal Protocol, for controlling the production and use of human-made chlorofluorcarbons (CFCs) that affect stratospheric ozone.

Studies of stratospheric chemistry began when Hartley (1881) first proposed ozone's presence in the upper atmosphere. A description of ozone chemistry came later when Chapman proposed the reaction sequence, now called the Chapman mechanism (Chapman, 1930):

$$O_2 + h\nu \ (\lambda < 242 \ nm) \rightarrow O + O \tag{13.1}$$

$$O + O_2 + M \ (M = N_2, O_2) \rightarrow O_3 + M \tag{13.2}$$

$$O_3 + h\nu \rightarrow O + O_2 \tag{13.3}$$

$$O + O_3 \rightarrow O_2 + O_2 \tag{13.4}$$

The results from this simple model were later found to differ from the observed ozone in two ways. First, the calculated average total ozone column is more than twice as large as measured (Brewer and Wilson, 1968). The total ozone column is the amount of ozone per unit area of the Earth's surface integrated radially from the surface to space. This difference indicates a problem with the chemistry. Second, the model predicts that ozone concentrations should be largest in the tropics, where the ozone production is greatest, whereas observations have shown that the ozone amount is greatest at high latitudes (Duetsch, 1971). This second difference indicates a problem with ozone transport.

A way to resolve the problem of excess calculated ozone was found in the 1950s, when Hampson (1965) and Bates and Nicolet (1950) proposed that the reactive hydrogen species, hydroxyl (OH) and hydroperoxyl (HO$_2$), form a cycle that catalytically destroys ozone. A second cycle involving the reactive nitrogen species, nitrogen dioxide (NO$_2$) and nitric oxide (NO), was proposed two decades later (Hampson, 1966; Crutzen, 1970; Johnston, 1971). A few years later, cycles involving reactive chlorine (Stolarski and Cicerone, 1974; Molina and Rowland, 1974) and bromine (Wofsy et al., 1975) were proposed. Several other cycles have been found as laboratory studies and atmospheric measurements have uncovered new reactions and chemistry. Adding this chemistry to the Chapman mechanism has greatly improved the agreement between the calculated and observed ozone concentrations.

The differences between the modeled and observed ozone distribution were largely resolved by considering stratospheric transport. Brewer (1949) suggested that the dryness of the stratosphere resulted from air entering the stratosphere in the tropics. Only in the tropics are the temperatures at the bottom of the stratosphere low enough to "freeze-dry" the air to its observed dryness as it enters the stratosphere. Dobson (1956) argued that air entering the stratosphere in the tropics and moving toward high latitudes would create the observed high ozone concentrations at high latitudes. Once at high latitudes, this air descends back into the troposphere, completing the cycle in several years.

This general picture of meridional transport from tropics to high latitudes does not describe the actual paths taken by molecules entering the stratosphere. These paths involve rapid, possibly wobbly, circulation around the globe (weeks), rising in the tropics (months), and transport downward toward the poles (months to years). Groups of neighboring molecules are often called parcels. Parcels of air do not stay intact long but, instead, are mixed with other air parcels by eddies, and they lose their identity in a week or so. This mixing occurs on quasi-horizontal surfaces that slope toward the poles. The mixing is quasi-horizontal because horizontal transport is much faster than vertical transport. The surfaces slope toward the poles because forces caused by atmospheric waves act as a suction pump that pulls air upward and poleward from the tropics and pushes air downward at middle-to-high latitudes (Holton et al., 1995). This quasi-horizontal mixing is rapid in the middle latitudes (McIntyre and Palmer, 1983; Plumb and Ko, 1992), but mixing with the tropics and the wintertime polar region is impeded (McIntyre, 1989). The transport of air into, through, and out of the stratosphere has a profound influence on the chemistry.

A whole new dimension was added to studies of stratospheric chemistry in 1985, when observations of rapid springtime ozone loss over Antarctica were first reported by members of the British Antarctic Survey (Farman et al., 1985). Quick analyses showed that the known chemical cycles could not be responsible. New chemical mechanisms were proposed that involved chlorine chemistry (Solomon et al., 1986; McElroy et al., 1986; Molina and Molina, 1987). Most surprising was the discovery that stratospheric particles composed of water vapor and nitric acid, called polar stratospheric clouds (PSCs), act as sites to produce this halogen-dominated chemistry (Toon et al., 1986; Tolbert et al., 1987; Leu, 1988; Molina et al., 1987). The reactions of gases on particles, called heterogeneous chemistry, are now known to be important not just for the polar regions but also for the entire lower stratosphere.

The focus of much of the research in stratospheric chemistry over the last 30 years has been on ozone and the possibility of human influences on it. This chapter is a primer on stratospheric ozone chemistry. First is a description of stratospheric structure and ozone climatology. Second is a brief tutorial on chemical concepts used frequently in atmospheric chemistry. Third is a description of stratospheric ozone chemistry in both the tropics and middle latitudes and in the wintertime polar regions.

THE STRATOSPHERE

Structure

The stratosphere extends from the tropopause, a temperature minimum near 15 km in the tropics and 8 km at high latitudes, to the stratopause, a temperature maximum, at about 50 km (Figure 13.1). Temperatures at the tropopause are generally 190–215 K, while temperatures at the stratopause are 240–280 K. In the wintertime polar stratosphere, temperatures can drop to 180 K. Although stratospheric temperatures generally increase everywhere with height, the temperature values depend upon the location and the season, particularly in the lower and upper stratosphere.

The average vertical profile of pressure, p, comes from the competition between gravity, which pulls molecules toward Earth's surface, and the molecular kinetic energy, which keeps the molecules moving. Expressed mathematically, this competition results in the Law of Atmospheres: $p = p_{surface}\exp(-mgz/kT)$, where $p_{surface}$ is the surface pressure (1013 hPa), m is the average molecular weight of the molecules that compose air (4.6×10^{-26} kg molecule^{-1}), g is the acceleration due to gravity (9.8 m s^{-1}), z is the height, k is the Boltzmann constant (1.38×10^{-23} J K^{-1}), and T is the tem-

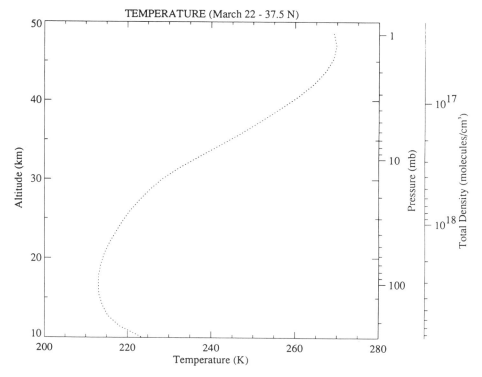

Figure 13.1. The vertical structure of temperature, pressure, and molecular number density for the stratosphere. The smoothed, typical profile is for equinox, 22 March at 37.5°N latitude. The tropopause is defined as the lowest altitude for which the temperature changes by less than 2°C per km for two kilometers above. For pressure, 1 mb = 1 hPa. (Reprinted from DeMore et al., 1994.)

perature. On average, the factor kT/mg, which is called the scale height H, is 7 ± 1 km. Thus, the atmospheric pressure falls off exponentially with height by a factor of e^{-1} (2.7) every 7 km (Figure 13.1). The resulting stratospheric pressure ranges from 120 hPa at 15 km to 0.8 hPa at 50 km. The atmospheric number density (molecules cm^{-3}, or cm^{-3}) is related to pressure and temperature by the ideal gas law, $n = P/kT$. It varies from $4.0 \times 10^{18} \ cm^{-3}$ at 15 km to $2.1 \times 10^{16} \ cm^{-3}$ near 50 km. Because the number density affects individual reactions differently, the importance of individual reactions changes from the lower to upper stratosphere.

Stratospheric temperatures increase with height because stratospheric ozone and, to a lesser extent, molecular oxygen absorb ultraviolet sunlight and convert some of the energy into molecular kinetic energy, or heat. The stratospheric temperature structure gives the stratosphere its stability. To understand this stability, consider a small parcel of air that is forced slightly upward but does not mix with the surrounding air. As it rises, the air in the parcel expands as the parcel's pressure decreases; if no heat is added, it cools. If after it has cooled the air in the parcel is less dense than its surroundings, it will continue to rise. However, if the air parcel is still more dense than the surrounding air, it will sink back down to its original position. The increasing temperature with height in the stratosphere ensures that rising air parcels will be more dense that their surroundings and will sink back down, thus creating the stability of

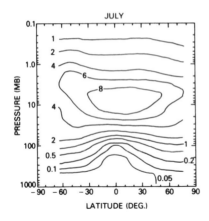

Figure 13.2. The observed altitude distribution of ozone mixing ratio as a function of latitude. Ozone mixing ratios are in ppmv (parts per million volume = 10^{-6} fraction of ozone molecules to air molecules). Observations are for July. The O_3 mixing ratio peaks in the ozone production region in the tropics; mainly transport and chemical loss determine the ozone mixing ratios away from the production region. (Reprinted from McPeters et al., 1984.)

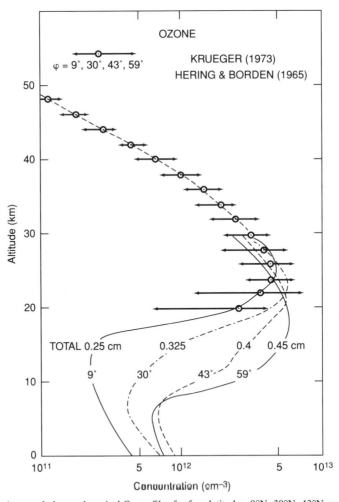

Figure 13.3. Averaged observed vertical O_3 profiles for four latitudes: 9°N, 30°N, 43°N, and 59°N. Concentrations peak lower at higher altitudes. Peaks are broader; and the total O_3 column abundance is greater at higher latitudes (0.45 cm = 450 DU). (Adapted from the U.S. Standard Atmosphere, 1976.)

TOTAL OZONE (DOBSON)

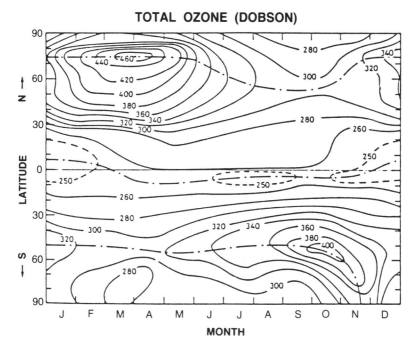

Figure 13.4. The observed total O_3 column abundance as a function of latitude and season. Total ozone column abundance is defined as the vertical integral of ozone concentration from the surface to space. 300 Dobson units (DU) = 3 mm of pure O_3 at 1013 hPa and 273 K or 8.07×10^{18} molecules cm^{-2}. Observations are from the LIMS instrument on Nimbus 6 from 1978 to 1979. (Adapted from London, 1980.)

the stratosphere. Air does move higher in the stratosphere, but must do so by absorbing energy from radiation or atmospheric waves or by mixing with the air above it.

Ozone Climatology and Observed Ozone Change

A combination of production, loss, and transport produces the global distribution of ozone volume mixing ratio as seen in Figure 13.2 (1 ppmv = 10^{-6} fraction of O_3 molecules in air). The vertical profile indicates that the ozone peak is at 7 hPa (35 km) in the tropics. As air is transported away from the ozone production region in the tropics to higher latitudes, ozone loss begins to dominate ozone production. Outside of the tropics, ozone follows mixing surfaces that tend to slope downward toward the poles. Because the concentration of air molecules (N_2, O_2, and minor constituents) is greater at lower altitudes, the downward slope of the mixing surfaces causes the ozone concentration to increase toward the poles, even though the mixing ratio decreases slightly. The total ozone column abundance, which is the sum of all ozone moleclues directly overhead, also increases toward the poles as a result (Figure 13.3).

The observed total ozone column abundance is largest in the springtime in both hemispheres (Figure 13.4). This maximum results from active descent of strato-

Total Ozone Trends

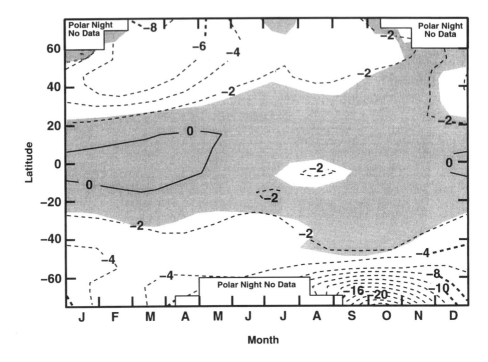

Figure 13.5. Observed total O_3 column trends as a function of latitude and season. The observations are from the Total Ozone Mapping Spectrometer (TOMS) on the Nimbus 7 satellite between 1979 and 1990. The trend in the shaded region is statistically insignificant. (Reprinted from Stolarski et al., 1991.)

spheric air, driven by propagation of eddies from vigorous springtime weather in the troposphere. Because the Northern Hemisphere has a more asymmetric distribution of land mass in the polar region, eddies are more intense in the Northern Hemisphere—hence the greater the downward pull and the greater the ozone column.

The abundance and distribution ozone have changed during the last 20 years. This change is largest in the Southern Hemisphere in austral spring and is called the Antarctic ozone hole. However, the downward ozone trend persists throughout the year at the middle and high latitudes (Figure 13.5). Ozone appears to remain unchanged in the tropics. The altitude of greatest change is in the lower stratosphere, below 20 km at both middle latitudes (Figure 13.6) and the polar regions (Figure 13.7). These changes are linked to changes in trace gas concentrations and chemistry.

COMMONLY USED CONCEPTS FOR STRATOSPHERIC CHEMISTRY

Some concepts are used frequently in atmospheric chemistry. These include photolysis, photochemical lifetimes or time constants, steady-state analyses, and catalytic cycles. The interaction of visible and ultraviolet sunlight with molecules to break them apart is called photolysis, where X, Y, and Z are chemical species and $h\nu$ represents light. A photolysis reaction is written as

$$X + h\nu \rightarrow Y + Z \tag{13.5}$$

The photolysis rate coefficient, J (units $= s^{-1}$), is a first-order rate coefficient for the photolysis of a particular chemical species and the products of that photolysis. The photolysis rate coefficient J is

$$J = \int F(\lambda)T(\lambda)\sigma(\lambda)\phi(\lambda)\,d\lambda \tag{13.6}$$

where $F(\lambda)$ is the solar flux at the top of the atmosphere per unit wavelength (photons $cm^{-2}\,s^{-1}\,nm^{-1}$), $T(\lambda)$ is the transmission of sunlight through the atmosphere at wavelength λ, $\sigma(\lambda)$ is the absorption cross section of the molecule at wavelength λ (cm^2 molecule^{-1}), and $\phi(\lambda)$ is the quantum yield for the formation of a specified product. Photolysis rate coefficients for molecules involved in stratospheric chemistry range from $0.3\ s^{-1}$ to less than $10^{-9}\ s^{-1}$.

For example, the photolysis of N_2O leads to the production of N_2 and $O(^1D)$, where $O(^1D)$ is an excited state of atomic oxygen. The absorption cross section is roughly $10^{-19}\ cm^{-2}$ from 170 to 200 nm and falls off rapidly at longer wavelengths (Figure 13.8). The solar flux times the transmission in the spectral region is roughly 3×10^{10}

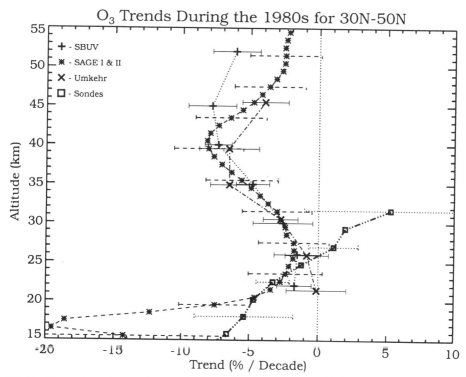

Figure 13.6. Observed trends in the O_3 vertical distribution for 30°N–50°N during the 1980s. SBUV and SAGE I&II are satellite instruments; Umkehr is a ground-based spectroscopic measurement; and Sondes are small balloon-borne instruments. Ozonesonde and Umkehr trends are derived by Miller et al. (1995). Error bars are 95% confidence limits. The negative trends are largest near 40 km and below 25 km. (Reprinted from WMO, 1994.)

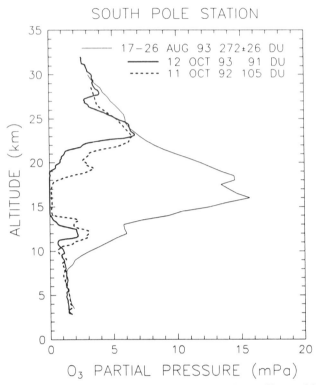

Figure 13.7. Ozonesonde observations of the vertical O_3 concentration profile over McMurdo Station, Antarctica in 1993. 1 nb = 10^{-3} hPa = 4×10^{12} cm^{-3} (approximate). The change from August to October shows the rapid ozone destruction over Antarctica. A much smaller, but significant, change is observed in the Arctic. (Reprinted from Hofmann et al., 1994.)

photons cm^{-2} s^{-1} nm^{-1} at 30 km, and the quantum yield is 1. Thus, the photolysis rate constant is roughly $10^{-19} \times (3 \times 10^{10}) \times 30 \times 1 = 10^{-7}$ s^{-1}. This rough estimate is fairly close to the actual value for 35°N in summer.

The lifetime, or time constant, for a trace gas is the time required for its concentration to decay to $1/e$ of its original value. In the absence of air transport into or out of an air parcel, the change in the concentration of a species X with time is proportional to the production rate minus the loss rate:

$$\frac{d[X]}{dt} = P - L([X]) \tag{13.7}$$

where the loss rate, L (units = molecules cm^{-3}s^{-1}), is almost always a function of [X]. With few exceptions, L is proportional to [X], so that we can define $L([X]) = L'[X]$, where L' has units of s^{-1}. If the production stops, then $P = 0$ and this rate equation reduces to

$$\frac{d[X]}{dt} = -L'[X] \tag{13.8}$$

The solution to this simple, first-order differential equation is

$$[X] = [X]_0 \exp\{-L'(t - t_0)\}$$

where $[X]_0$ and t_0 are the initial values. Because $[X]$ decreases by $1/e$ when $(t - t_0) = 1/L'$, $1/L'$ is called the lifetime and is often designated as τ.

The lifetime or time constant is the reciprocal of the first-order rate coefficient for a photochemical process. For photolysis, $\tau = J^{-1}$. For unimolecular, bimolecular, and termolecular chemical reactions, τ is $(k^I)^{-1}$, $(k^{II} [Y])^{-1}$, and $(k^{III} [M] [Y])^{-1}$ respectively, where Y and M are reactants for bimolecular and termolecular reactions. For simple heterogeneous reactions, the first-order rate coefficient is given by the equation

$$\frac{d[X]}{dt} = - \frac{\gamma A v}{4} [X] \qquad (13.9)$$

where γ is the reactivity coefficient, A is the surface area density (cm^2 cm^{-3} or mm^2 cm^{-3}), and v is the molecular thermal velocity, which is typically $2-5 \times 10^4$ cm s^{-1}. For heterogeneous reactions, τ is $\{\gamma A v/4\}^{-1}$. The relative importance of different processes can be determined if the first-order rate coefficients are known.

If two or more reaction pathways are occurring simultaneously, then the total lifetime, τ_{total}, is given by the expression

$$\tau_{total} = \left(\sum_i 1/\tau_i\right)^{-1} \qquad (13.10)$$

where τ_i are the time constants for the individual processes. As expected, the process with the fastest rate has the greatest effect on the lifetime.

If two or more processes are occurring in sequence, then the total lifetime is given by the expression

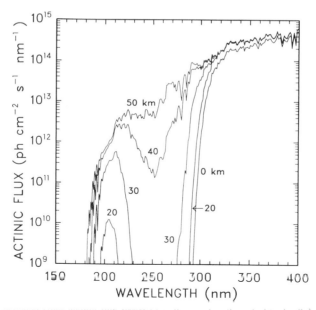

Figure 13.8. Calculated solar actinic flux variation with wavelength and altitude. Solar actinic flux is strongly absorbed by ozone in the Herzberg continuum near 250 nm and by molecular oxygen in the Schumann–Runge bands below 180 nm. At wavelengths above 320 nm the actinic flux is virtually independent of altitude. (Reprinted from DeMore et al., 1994.)

$$\tau_{\text{total}} = \sum_i \tau^i \tag{13.11}$$

The slowest step in a sequence is called the rate-limiting step. The concept of a rate-limiting step is used to find the rate at which ozone is destroyed.

For the Chapman mechanism, the lifetimes are derived from the rate equations for [O] and [O$_3$], which come from the chemical equations (13.1)–(13.4):

$$\frac{d[\text{O}]}{dt} = 2J_{\text{O}_2} + J_{\text{O}_3}[\text{O}_3] - k_{\text{O}+\text{O}_2+\text{M}}[\text{M}][\text{O}_2][\text{O}] - k_{\text{O}+\text{O}_3}[\text{O}_3][\text{O}] \tag{13.12}$$

and

$$\frac{d[\text{O}_3]}{dt} = k_{\text{O}+\text{O}_2+\text{M}}[\text{M}][\text{O}_2][\text{O}] - J_{\text{O}_3}[\text{O}_3] - k_{\text{O}+\text{O}_3}[\text{O}_3][\text{O}] \tag{13.13}$$

The rate equation for O$_x$, the sum of O and O$_3$, comes from summing Eq. (13.12) and (13.13):

$$\frac{d[\text{O}_x]}{dt} = 2J_{\text{O}_2} - 2k_{\text{O}+\text{O}_3}[\text{O}_3][\text{O}] \tag{13.14}$$

The lifetimes of O and O$_3$ are less than an hour because O and O$_3$ are rapidly interchanged by reactions (13.2) and (13.3). However, the lifetime of O$_x$ is weeks to years below 30 km because reactions (13.1) and (13.4), the production and destruction of O$_x$, take weeks to years.

A third concept for stratospheric chemistry is steady state. In steady state, the production rate and loss rate are equal; the concentration of the species does not change in time. If several such species are simultaneously in steady state, then the set of equations can be solved to establish a fixed relationship among them.

The steady-state concept can be applied to the Chapman mechanism. Because the amount of sunlight is roughly constant near midday, [O] and [O$_3$], the concentrations of O and O$_3$ will be roughly constant.

In the steady state,

$$\frac{d[\text{O}]}{dt} = \frac{d[\text{O}_3]}{dt} = \frac{d[\text{O}_x]}{dt} = 0 \tag{13.15}$$

The subtraction of (13.13) and (13.14) from (13.12) leads to the steady-state relationship between [O] and [O$_3$]:

$$\frac{[\text{O}]}{[\text{O}_3]} = \frac{J_{\text{O}_3}}{k_{\text{M}+\text{O}_2+\text{O}}[\text{M}][\text{O}_2]} \tag{13.16}$$

This relationship shows that O exists only during the day, where $J_{\text{O}_3} \neq 0$. Because the reaction rate $k_{\text{M}+\text{O}_2+\text{O}}[\text{M}][\text{O}_2] \gg J_{\text{O}_3}$, [O] is 10^3 to 10^6 times less than [O$_3$] in the stratosphere.

A fourth useful concept is catalytic cycles that destroy ozone. The reaction of O with O$_3$ (13.4) remakes the O$_2$ chemical bond that was broken by photolysis. Other cycles also make the O$_2$ chemical bond. Suppose we have a species X that reacts rapidly with O$_3$ and a species XO that reacts rapidly with O. Then a catalytic cycle that destroys O$_x$ is

$$X + O_3 \rightarrow XO + O_2$$
$$\underline{XO + O \rightarrow X + O_2} \qquad (13.17)$$
$$\text{net:} \quad O_3 + O \rightarrow 2O_2$$

X and XO are not destroyed in these reactions, but simply cycle one into the other. For each cycle, however, $O + O_3 \rightarrow 2O_2$. This cycle is catalytic in the destruction of O_x and thus O_3.

These reactions are generally fast enough to be considered to be in steady state. Usually, the rate-limiting step is the XO + O reaction. The ozone loss rate due to this catalytic cycle is equal to $2k_{XO+O}$ [XO][O] (Johnston and Podolske, 1978). Catalytic cycles that recombine two ozone molecules into three oxygen molecules also catalytically destroy ozone.

Many chemical cycles can be formed. Some of these will neither produce nor destroy ozone and are called null cycles. Other cycles do not involve ozone at all but switch members of chemical families from one form to another. These cycles are dominant processes in stratospheric chemistry and contribute indirectly to ozone loss.

STRATOSPHERIC CHEMICAL SPECIES

Chemical species for each chemical family have different functions in stratospheric chemistry (Figure 13.9, Table 13.1). The source species are generally those chemicals that live long enough in the troposphere to survive transport to the stratosphere. Once they reach the stratosphere, sources species are destroyed, either directly by absorption of solar ultraviolet light or by chemical reactions that are initiated by solar UV. Some products of the photochemical destruction of the source species are either reac-

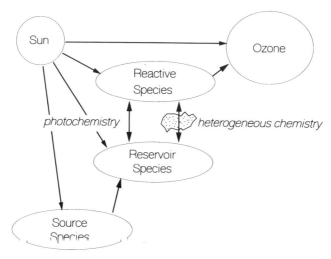

Figure 13.9. The roles of trace gases in the stratosphere. The chemistry is initiated and activated by the sun. Source species are converted into reactive and reservoir species. Reactive species destroy ozone, but are closely coupled to reservoir species by both gas-phase and heterogeneous chemistry.

TABLE 13.1
Source, Reservoir, and Reactive Species for the Chemical Families

Species	Family				
	Oxygen	Hydrogen	Nitrogen	Chlorine	Bromine
Source	O_2	H_2O CH_4	N_2O	CH_3Cl CFCs CH_3CCl_3 CCl_4 HCFCs	CH_3Br Halons
Reservoir	$O_x = O + O_3$	H_2O_2 HNO_3 HO_2NO_2 $HOCl$ $HOBr$	HNO_3 $ClONO_2$ N_2O_5 HO_2NO_2	HCl $ClONO_2$ $HOCl$ $OClO$ $BrCl$	HBr $BrONO_2$ $BrCl$ $HOBr$
Reactive	O O_3	OH HO_2	NO NO_2 NO_3	Cl ClO Cl_2O_2	Br BrO

tive or reservoir species. In this chapter, "reactive" indicates free radicals and other chemical species that are photolyzed into free radicals within minutes. "Reservoir" indicates species such as acids and nitrates that are exchanged with free radicals by reactions or photolysis, but generally over a period of hours to months. All reactive and reservoir species are "trace" species and have volume mixing ratios of less than 20 parts per billion by volume (ppbv $= 10^{-9}$) in air.

Another important trace component is aqueous particles. They are primarily sulfate aerosols throughout the lower stratosphere below 25 km, called the Junge layer, but become polar stratospheric clouds (PSCs) in the cold, wintertime polar regions. The surface area density (cm^2 per cm^3 of air) and the composition of the particles affect stratospheric heterogeneous chemistry through the first-order rate constant given in Eq. (13.9). Sporadic volcanic eruptions inject large amounts of sulfur into the stratosphere, so that the surface area varies in the range of $(0.5–20) \times 10^{-8}$ cm^2 cm^{-3}. In addition, condensible chemicals such as H_2O and HNO_3 deposit on the aerosol at low temperatures, swelling them. Third, the composition affects the uptake and reaction efficiency, which varies from $<10^{-5}$ to almost 1. The chemical effects of particles appear to be limited to below 25 km, but they are profound.

The predominant exchange among source, reservoir, and reactive species is either by photolysis or chemical reactions or by heterogeneous chemistry. Only sunlight and reactive species can directly affect the ozone amount. However, reservoir species have an indirect influence because photochemistry and heterogeneous chemistry determine the balance between the reactive and reservoir species and thus amounts of the reactive species.

Because most source gases enter the stratosphere through the tropics and leave at high latitudes, the types of chemical species and reactions are the same throughout the stratosphere. However, which chemical species and reactions are most important differs from the tropics to the wintertime polar region. The tropics, the main entry points for gases into the stratosphere, are regions of extensive photochemical production, where an abundance of sunlight and reactive species begin to convert source gases into reservoir and reactive species. The wintertime polar region in the lower stratosphere is heavily influenced by the heterogeneous chemistry that occurs on the cold aqueous particles there, and ozone loss dominates production (Chapter 15, this volume). The middle latitudes are a region where the photochemical production and loss rates are more in balance than in the other two regions. These three regions are connected by transport, but semipermeable barriers appear to prevent rapid mixing from one region to the other.

CHEMISTRY OF THE TROPICS AND MIDDLE LATITUDES

Source Gases

The source gases for all of the chemical families originate at Earth's surface, even molecular oxygen (Table 13.1). Some gases are not in this table. Sulfur gases, such as SO_2 and OCS, also enter the stratosphere in significant amounts, but they usually end up as sulfate aerosol within a few months. CO_2 also enters the stratosphere, but its main influence on stratospheric chemistry comes from its absorption and emission of infrared radiation that can alter stratospheric temperatures. Noticeably absent from this list of sources are nonmethane hydrocarbons and soluble gases that dominate tropospheric chemistry. These chemicals do not survive the oxidation and precipitation in the troposphere to enter in the stratosphere in amounts capable of affecting the chemistry. Stratospheric chemistry originates from only a few chemicals.

The destruction of most source gases occurs as they are transported up into the tropical stratosphere. Source gases develop vertical profiles with greatest mixing ratios near the tropopause. As these gases are transported away from the tropics, their destruction decreases.

Oxygen species

Photolysis of O_2 is the only source of O_3 and O, O_x, in the stratosphere (Figure 13.10). Local sources of NO_x in the lower stratosphere can act as O_x sources by the photochemical smog reactions, but these are a smaller source than the photolysis of O_2. Photolysis of O_3 at wavelengths less than 320 nm produces O atoms in an excited state, $O(^1D)$. These excited state atoms usually collide with O_2 and return to being ground-state atoms, $O(^3P)$. Some, however, react with other molecules.

Hydrogen species

Hydrogen has two main sources, H_2O and CH_4 (Figure 13.11). The $O(^1D)$ produced by O_3 photolysis reacts with H_2O to produce two OH molecules. Methane is oxidized by OH and undergoes an oxidation sequence that leads to CO_2 and H_2O. This sequence creates about two water molecules from each fully oxidized CH_4 molecule. The result is that the sum of mixing ratios of $2 \times CH_4 + H_2O$ is approximately constant at 6–7 ppmv throughout the stratosphere.

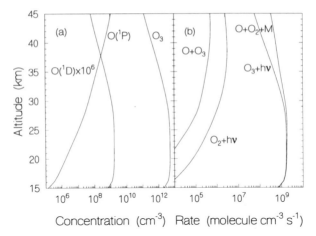

Figure 13.10. The vertical diurnally averaged distributions of (a) the concentrations (molecules cm^{-3}) and (b) the reaction rates (molecules cm^{-3} s^{-1}) for the oxygen chemical family. Calculations are for June at 38°N by the AER 2-D model. O(^{1}D) is oxygen in the first excited electronic state; O(^{3}P) is atomic oxygen in the ground state. (Adapted from D. Weisenstien and J. Rodriguez, private communication.)

Nitrogen species

Reactive and reservoir nitrogen together, called NO$_y$, has three sources in the stratosphere (Figure 13.12) (WMO, 1994): N$_2$O reacting with O(^{1}D) to form NO by the reaction N$_2$O + O(^{1}D) → 2NO (65%); solar proton events and galactic cosmic rays producing NO (10%); and lightning in the equatorial upper troposphere producing NO that is transported into the stratosphere (25%). These estimates are highly uncertain, especially the estimate for the lightning source. Nitrous oxide is the largest source of reservoir and reactive nitrogen. The largest sink of stratospheric N$_2$O is photolysis, N$_2$O + $h\nu$ → N$_2$ + O. Only 7% of the N$_2$O reacts with O(^{1}D) to form NO (Fahey et

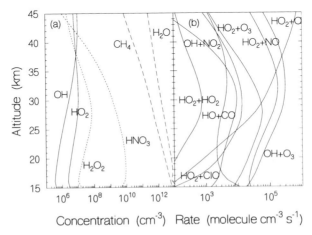

Figure 13.11. The vertical diurnally averaged distributions of (a) the concentrations (molecules cm^{-3}) and (b) the reaction rates (molecules cm^{-3} s^{-1}) for the hydrogen chemical family. Calculations are for June at 38°N by the AER 2-D model. For concentrations, long dashes, short dashes, and solid lines indicate source, reservoir, and reactive species, respectively. For reaction rates, the reactants indicate the reaction. (Adapted from D. Weisenstien and J. Rodriguez, private communication.)

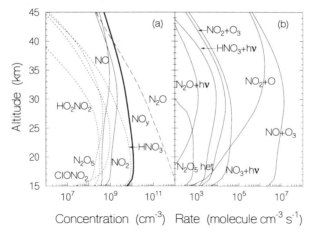

Figure 13.12. The vertical diurnally averaged distributions of (a) the concentrations (molecules cm^{-3}) and (b) the reaction rates (molecules cm^{-3} s^{-1}) for the nitrogen chemical family. Calculations are for June at 38°N by the AER 2-D model. For concentrations, long dashes, short dashes, and solid lines indicate source, reservoir, and reactive species, respectively. For reaction rates, the reactants indicate the reaction. (Adapted from D. Weisenstien and J. Rodriguez, private communication.)

al., 1990). In addition, NO can be destroyed in the upper tropical stratosphere by the photolysis of NO to produce N and O, followed by the reaction N + NO → N$_2$ + O. This sink of NO may result in a loss of 20% of the total stratospheric NO$_y$ as it leaves the tropics for middle latitudes.

Chlorine species

The main source gases for the 3.7 ppbv of stratospheric chlorine in 1995 are the long-lived anthropogenic chlorofluorocarbons (CFCs) methyl chloroform (CH$_3$CCl$_3$)

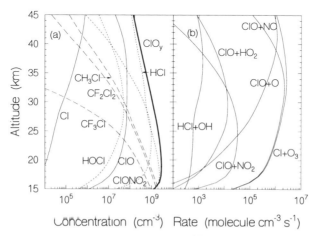

Figure 13.13. The vertical diurnally averaged distributions of (a) the concentrations (molecules cm^{-3}) and (b) the reaction rates (molecules cm^{-3} s^{-1}) for the chlorine chemical family. Calculations are for June at 38°N by the AER 2-D model. For concentrations, long dashes, short dashes, and solid lines indicate source, reservoir, and reactive species, respectively. For the reaction rates, the reactants indicate the reaction. (Adapted from D. Weisenstien and J. Rodriguez, private communication.)

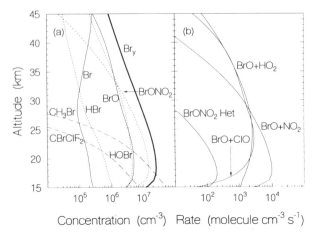

Figure 13.14. The vertical diurnally averaged distributions of (a) the concentrations (molecules cm^{-3}) and (b) the reaction rates (molecules cm^{-3} s^{-1}) for the bromine chemical family. Calculations are for June at 38°N by the AER 2-D model. For concentrations, long dashes, short dashes, and solid lines indicate source, reservoir, and reactive species, respectively. For the reaction rates, the reactants indicate the reaction. (Adapted from D. Weisenstien and J. Rodriguez, private communication.)

and carbon tetrachloride, for a total of 2.7 ppbv, the largely natural methyl chloride CH$_3$Cl at 0.6 ppbv, and increasingly the hydrochlorofluorocarbons (HCFCs) designed to replace the CFCs (Figure 13.13). The most important CFCs are CFC-11 (CCl$_3$F), CFC-12 (CCl$_2$F$_2$), and CFC-113 (CCl$_2$FCClF$_2$). Other small chlorine sources such as volcanos and solid fuel rockets contribute less than 1% of the total stratospheric chlorine burden (WMO, 1994). The tropospheric mixing ratios of the CFCs, carbon tetrachloride, and methyl chloroform are peaking, thanks to the Montreal Protocol, an international treaty for controlling substances that destroy stratospheric ozone.

The only significant sink for CFCs is photochemical destruction in the stratosphere. Most stratospheric destruction is by photolysis. Methyl chloride is predominantly destroyed by OH in the troposphere, but the main stratospheric sink is photolysis. The HCFCs are also mostly lost to tropospheric OH, but once they are in the stratosphere, they are mainly destroyed by photolysis. Measurements confirm that chlorine species are stripped of all their chlorine atoms. These chlorine atoms are incorporated into inorganic chlorine reservoir and reactive species, collectively called Cl$_y$.

Bromine species

The major sources of 20 pptv of stratospheric bromine are methyl bromide (CH$_3$Br) at 12 pptv, halon 1211 (CBrClF$_2$) at 2.5 pptv, and halon 1301 (CBrF$_3$) at 2.0 pptv (Figure 13.14). Because a large fraction of methyl bromide is anthropogenic, the total anthropogenic bromine contribution to the stratosphere is about 40–50%. The bromine atoms incorporated into stratospheric bromine reservoir and reactive species are collectively called Br$_y$.

Other halogen species

The major source of stratospheric fluorine is the CFCs. The fluorine atoms, once released from the CFCs by photochemistry, rapidly are incorporated into HF, which does not participate in stratospheric chemistry. Source molecules such as SF_6, C_2F_6, and CF_4, which are being used more widely, are very stable and have lifetimes of thousands of years. Fluorine plays essentially no known role in stratospheric chemistry.

Long-lived species, such as the source gases N_2O, CH_4, CFCs, and the sums of reservoir and reactive species NO_y, Cl_y, and Br_y, tend to be well-mixed along the quasi-horizontal mixing surfaces. Because the time constant for the destruction (or production) of these long-lived species is faster than vertical mixing but slower than horizontal mixing, long-lived chemical species develop vertical gradients and compact, simple relationships to one another. These compact relationships have been observed and are used to determine the amount of one long-lived tracer, be it source gas or inorganic product, from others. Some observationally determined relationships for NO_y versus N_2O in 1989 and Cl_y versus N_2O in 1992 are given by the equations

$$NO_y \text{ (ppbv)} = 0.082 \, (266 - N_2O \text{ (ppbv)}) \qquad \text{(Fahey et al., 1990)} \qquad (13.18)$$

$$Cl_y \text{ (ppbv)} = 2.79 + 4.1 \times 10^{-3} \, N_2O \text{ (ppbv)} - 4.0 \times 10^{-5} \, \{N_2O \text{ (ppbv)}\}^2$$

$$\text{(Woodbridge et al., 1995)} \qquad (13.19)$$

These relationships apparently do not hold in the tropics, where most of the source destruction is occurring. From observations of N_2O mixing ratios (Figure 13.15), the mixing ratios of other long-lived source gases, NO_y, Cl_y, and Br_y, can be found. Once NO_y, Cl_y, and Br_y are known, the partitioning into reservior and reactive species is determined by photochemistry alone, without regard to transport.

The Partitioning Among Reservoir and Reactive Gases

Photochemistry and heterogeneous chemistry influence the partitioning of the chemical families (NO_y, Cl_y, Br_y) into reservoir and reactive species. Thus they determine the amount of reactive species and are controlling factors in ozone loss. These processes depend upon temperature, pressure, amount of sunlight, and aqueous particle surface area and composition. They vary for different seasons, latitudes, altitudes, and aerosol loadings due to volcanic eruptions (Figures 13.10–13.14).

Reactive species are present mostly during sunlight (Figure 13.16). Often, diurnally averaged concentrations and reaction rates are used for model calculations of

Figure 13.15. The calculated distribution of N_2O as a function of altitude and latitude for March, from the AER 2-D model. The contours are in units of ppbv. (Reprinted from Plumb and Ko, 1992.)

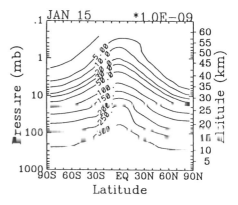

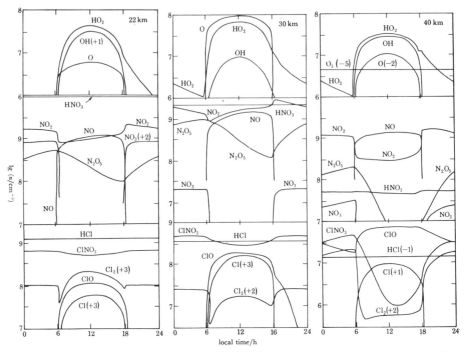

Figure 13.16. Calculated diurnal variation of chemical species in the oxygen, hydrogen, nitrogen, and chlorine chemical families for 30 km and 40 km altitude at 30°N latitude and 0° solar declination. The vertical scale is log(concentration), in units of log(cm^{-3}); the horizontal axis is local time in hours. A number in parentheses, e.g. (+1), indicates that the concentration of that species should be multiplied by 10 to that power, e.g. 10^{+1}. Logan, J. A., et al., 1978, Atmospheric chemistry: a response to human influence, *Phil. Trans. Roy. Soc.*, 290, 187–234. Reprinted by permission of The Royal Society.

many years of atmospheric chemistry. Generally, such calculations correctly represent the chemistry, but they do not correctly represent diurnal variations. The lifetimes of chemical species can be found by dividing the concentration of the species by the reaction rates, such as in Figures 13.10–13.14. However, caution must be used when finding the lifetimes of diurnally varying species with diurnally averaged model results.

Hydrogen family

The reactive species in the hydrogen family, the hydroxyl radical OH and the hydroperoxyl radical HO_2, together are called HO_x. The reservoir species are H_2O_2, HNO_3, HO_2NO_2, HOCl, and HOBr. Hydrogen peroxide is formed by the reaction

$$HO_2 + HO_2 \rightarrow H_2O_2 + O_2 \quad \text{(hours)} \quad (13.20)$$

and is destroyed by photolysis to yield OH:

$$H_2O_2 + h\nu \rightarrow 2\ OH \quad \text{(hours)} \quad (13.21)$$

The other reservoir species result from reactions with the other chemical families. Of these, the most important is HNO_3, which augments H_2O as a source of HO_x in the lower stratosphere.

The main interactions with the halogen families are the reactions:

$$HO_2 + ClO \rightarrow HOCl + O_2 \qquad \text{(minutes–hours)} \qquad (13.22)$$

$$HO_2 + BrO \rightarrow HOBr + O_2 \qquad \text{(minutes)} \qquad (13.23)$$

$$OH + HCl \rightarrow H_2O + Cl \qquad \text{(days–weeks)} \qquad (13.24)$$

$$OH + HBr \rightarrow H_2O + Br \qquad \text{(day)} \qquad (13.25)$$

The halogen reservoirs, HOCl and HOBr, are short-lived. The reactions of OH with HCl and HBr are not important losses of OH, but are important reactions for the chlorine and bromine families.

The exchange between OH and HO2 occurs mainly by the reactions

$$HO_2 + NO \rightarrow OH + NO_2 \qquad \text{(minutes)} \qquad (13.26)$$

$$OH + O_3 \rightarrow HO_2 + O_2 \qquad \text{(minutes)} \qquad (13.27)$$

$$HO_2 + O_3 \rightarrow OH + 2\,O_2 \qquad \text{(minutes–hour)} \qquad (13.28)$$

$$OH + O \rightarrow H + O_2 \qquad \text{(seconds–hour)} \qquad (13.29)$$

$$HO_2 + O \rightarrow OH + O_2 \qquad \text{(seconds–hour)} \qquad (13.30)$$

$$H + O_2 + M \rightarrow HO_2 + M \qquad \text{(subseconds)} \qquad (13.31)$$

These fast reactions establish a steady-state relationship between HO2 and OH in a few minutes:

$$\frac{[HO_2]}{[OH]} = \frac{k_{OH+O_3}[O_3] + k_{OH+CO}[CO] + k_{OH+O}[O]}{k_{HO_2+NO}[NO] + k_{HO_2+O_3}[O_3] + k_{HO_2+ClO}[ClO] + k_{HO_2+O}[O]} \qquad (13.32)$$

Throughout most of the stratosphere, the predominant conversion from HO$_2$ to OH is the reaction with NO. This exchange is hundreds of times more rapid than the conversion of HO$_x$ into its reservoir species. The ratio [HO$_2$]/[OH] is approximately 1 in the upper stratosphere, but is larger than 10 in the lower stratosphere.

Nitrogen family

The nitrogen family is represented by the sum NO$_y$ = HNO$_3$ + ClONO$_2$ + BrONO$_2$ + HO$_2$NO$_2$ + NO$_2$ + NO + NO$_3$ + 2N$_2$O$_5$ + HONO + N + aerosol nitrate. This family is sometimes called odd-nitrogen. The stratospheric NO$_y$ mixing ratio is typically 4–12 ppbv, with the higher values in middle latitudes. The reactive species NO and NO$_2$ together are called NO$_x$. The NO$_x$/NO$_y$ ratio indicates the ability of NO$_y$ to influence ozone destruction.

The gas-phase reactions that partition the family into reservoirs and radicals and their approximate time constants for converting nitrogen are

$$HNO_3 + h\nu \rightarrow OH + NO_2 \qquad \text{(days–weeks)} \qquad (13.33)$$

$$OH + NO_2 + M \rightarrow HNO_3 + M \qquad \text{(hours–weeks)} \qquad (13.34)$$

$$HNO_3 + OH \rightarrow NO_3 + H_2O \qquad \text{(weeks)} \qquad (13.35)$$

$$ClONO_2 + h\nu \rightarrow Cl + NO_3 \qquad \text{(hours)} \qquad (13.36)$$

$$\text{ClO} + \text{NO}_2 + \text{M} \rightarrow \text{ClONO}_2 + \text{M} \qquad \text{(hours)} \qquad (13.37)$$

$$\text{BrONO}_2 + h\nu \rightarrow \text{Br} + \text{NO}_3 \qquad \text{(minutes)} \qquad (13.38)$$

$$\text{BrO} + \text{NO}_2 + \text{M} \rightarrow \text{BrONO}_2 + \text{M} \qquad \text{(hours)} \qquad (13.39)$$

$$\text{NO}_2 + \text{O}_3 \rightarrow \text{NO}_3 + \text{O}_2 \qquad \text{(day(s))} \qquad (13.40)$$

$$\text{NO}_3 + h\nu \rightarrow \text{NO}_2 + \text{O} \qquad \text{(seconds)} \qquad (13.41)$$

$$\text{NO}_2 + \text{NO}_3 + \text{M} \rightarrow \text{N}_2\text{O}_5 + \text{M} \qquad \text{(hours)} \qquad (13.42)$$

$$\text{N}_2\text{O}_5 + h\nu \rightarrow \text{NO}_2 + \text{NO}_3 \qquad \text{(hours)} \qquad (13.43)$$

$$\text{NO} + \text{O}_3 \rightarrow \text{NO}_2 + \text{O}_2 \qquad \text{(seconds)} \qquad (13.44)$$

$$\text{NO}_2 + \text{O} \rightarrow \text{NO} + \text{O}_2 \qquad \text{(minutes–hours)} \qquad (13.45)$$

In competition with the gas-phase reactions are the heterogeneous reactions on sulfate aerosol in the lower stratosphere. The most important of these is the hydrolysis of N_2O_5:

$$\text{N}_2\text{O}_5 \text{ (gas)} + \text{H}_2\text{O} \text{ (liquid)} \rightarrow 2\text{HNO}_3 \text{ (gas)} \qquad \text{(hours to days)} \qquad (13.46)$$

Other heterogeneous reactions are the hydrolysis of ClONO_2 and BrONO_2:

$$\text{ClONO}_2 \text{ (gas)} + \text{H}_2\text{O} \text{ (liquid)} \rightarrow \text{HOCl(liquid)}$$
$$+ \text{HNO}_3\text{(liquid) (days to weeks)} \qquad (13.47)$$

$$\text{BrONO}_2 \text{ (gas)} + \text{H}_2\text{O} \text{ (liquid)} \rightarrow \text{HOBr(liquid)}$$
$$+ \text{HNO}_3\text{(liquid) (hours to days)} \qquad (13.48)$$

where the shorter time constants are for volcanic aerosol clouds and the longer ones are for background aerosol amounts. N_2O_5 and BrONO_2 hydrolysis are almost temperature-independent, whereas ClONO_2 hydrolysis competes with gas-phase chemistry only when $T < 205$ K (Hanson and Ravishanka, 1991). Thus, while ClONO_2 hydrolysis is important mainly in cold regions, N_2O_5 hydrolysis and BrONO_2 hydrolysis affect chemistry throughout the lower stratosphere.

In regions where heterogeneous chemistry is important, the interaction between gas-phase and heterogeneous chemistry results in a "saturation" of N_2O_5 hydrolysis, in which the addition of aerosol surface area does not significantly change the balance between reservoir and reactive NO_y, as represented by the NO_x/NO_y ratio. The gas-phase reactions that determine the N_2O_5 concentration are reactions (13.40)–(13.43). In the lower stratosphere, particularly in the high latitudes in winter, the hydrolysis of N_2O_5 is as fast as the gas-phase chemistry. The net result is that for a large range in aerosol loadings, the effect of this N_2O_5 hydrolysis on the NO_x/NO_y partitioning is approximately constant (Fahey et al., 1993).

Chlorine family

The chlorine family is represented by the sum $\text{Cl}_y = \text{HCl} + \text{ClONO}_2 + \text{HOCl} + \text{ClO} + 2\text{Cl}_2\text{O}_2 + \text{OClO} + \text{BrCl} + \text{Cl}$ (Figure 13.13). In the middle latitudes, the total amount of Cl_y is 2 to 3.5 ppbv. Throughout much of the stratosphere at middle latitudes, more than 80% of the 3.7 ppbv of chlorine is in the reservoir species HCl,

$ClONO_2$, and HOCl. In the lower stratosphere, the reactive species ClO concentration is only 5–30 pptv, at most a few percent of Cl_y.

For some conditions, HCl is the dominant reservoir; for others, $ClONO_2$ is. The HCl reservoir is created by the reaction of Cl with CH_4 and CH_2O and is destroyed by the gas-phase reaction with OH. Its lifetime is generally weeks to months. The $ClONO_2$ reservoir is created by the reaction of ClO with NO_2 and is destroyed mainly by photolysis. Its lifetime is generally weeks.

HOCl is only a few percent of the total chlorine reservoir because it is so rapidly lost by photolysis. Its lifetime is only a few hours. Which reservoir is dominant determines the amount of reactive chlorine that will be present during the day.

The reservoir species OClO and BrCl result from a reaction between the bromine and chlorine families:

$$ClO + BrO \rightarrow Br + OClO \qquad \text{(minutes–hours)} \qquad (13.49a)$$

$$\rightarrow BrCl + O_2 \qquad (13.49b)$$

$$\rightarrow Br + ClOO \qquad (13.49c)$$

These are not important chlorine reservoir species in the tropics and middle latitudes. However, while pathway a is part of a null cycle, pathways b and c of this reaction are part of a catalytic cycle that destroys ozone throughout the lower stratosphere.

During the night, all the chlorine is in the reservoir species. During the day, $ClONO_2$ is photolyzed and a balance is maintained which is defined by the steady-state relationship:

$$\frac{[ClO]}{[ClONO_2]} = \frac{J_{ClONO_2}}{k_{M+ClO+NO_2}[M][NO_2]} \qquad (13.50)$$

The photolysis time constant is a few hours, and the recombination is tens of minutes. As a result, $ClONO_2$ is the larger of the two species. In the upper stratosphere, where termolecular reactions are slower and photolysis is faster, a greater fraction is present as ClO. For a given amount of $ClONO_2$, the amount of ClO during the day is inversely dependent upon $[NO_2]$. Because NO_2 is reduced by N_2O_5 hydrolysis, more ClO will be present during the day in the presence of volcanic aerosols or in the lower stratosphere at high latitudes.

The exchange between ClO and Cl is mainly by the reactions

$$ClO + NO \rightarrow Cl + NO_2 \qquad \text{(minute)} \qquad (13.51)$$

$$Cl + O_3 \rightarrow ClO + O_2 \qquad \text{(subsecond)} \qquad (13.52)$$

$$ClO + O \rightarrow Cl + O_2 \qquad \text{(minutes)} \qquad (13.53)$$

This exchange is in steady-state for all day except sunrise and sunset, and the [Cl]/[ClO] ratio is

$$\frac{[Cl]}{[ClO]} = \frac{k_{ClO+NO}[NO] + k_{ClO+O}[O]}{k_{Cl+O_3}[O_3]} \qquad (13.54)$$

Below 35 km, the reaction between ClO and NO is at least 10 times faster than the reaction of ClO with O, so that the amount of chlorine is approximately proportional to the amount of reactive nitrogen. In the upper stratosphere, the ClO reaction with O becomes important, and with decreasing $[O_3]$ the ratio $[Cl]/[ClO]$ increases to approximately 0.01 above 40 km. When $[Cl]$ is larger, more HCl is created because the reaction rate of $Cl + CH_4 \rightarrow HCl + CH_3$ increases.

Bromine family

The bromine family is represented by the sum $Br_y = HBr + BrONO_2 + HOBr + BrO + BrCl + Br$ (Figure 13.14). Its abundance is 10–20 pptv throughout the stratosphere (Schauffler et al., 1993). The bromine family has the same reactions as the chlorine family. However, differences between the two halogens make a large difference in the partitioning between the reservoir and radical species. First of all, the photolysis rate coefficient for $BrONO_2$ is about 50 times larger than that for $ClONO_2$ in the lower stratosphere. Second, the reaction of OH with HBr is about 20 times faster than the reaction of OH with HCl. As a result, the amount of bromine that is in the form of BrO during the day is roughly 50% of Br_y. HBr is at most about 10% of Br_y. The large BrO/Br_y ratio makes bromine competitive with chlorine for ozone destruction in the lower stratosphere despite the large differences between the Br_y and Cl_y abundances.

Significance of the Interactions Among the Chemical Families

The nitrogen family controls the hydrogen and halogen families for most stratospheric conditions. This control results from the abundance of NO_y being greater than that of HO_x, ClO_y, or Br_y. The direct control of chlorine and bromine is by the formation of $ClONO_2$ and $BrONO_2$. For HO_x, this control is exerted mainly through the formation of nitric acid [reaction (13.34)]. Because NO_x controls HO_x, and HO_x in part controls the amount of HCl and HBr, NO_x also indirectly controls even the amount of HCl and HBr. This control is strongest in the lower stratosphere.

Heterogeneous chemistry converts nitrogen species from reactive to reservoir species and from reservoir to long-lived reservoir species, while it converts halogen species from reservoir species to reactive species. The rough differences for three chemistries—gasphase, in the presence of sulfate aerosol, and in the presence of polar stratospheric clouds—show these trends for the nitrogen and chlorine families (Figure 13.17). Reactive hydrogen, like reactive halogens, tend to be suppressed when NO_x is greater.

Ozone Loss in Middle Latitudes and the Tropics

Ozone loss at middle latitudes and the tropics occurs by catalytic cycles involving reactive species, as presented above. Hence, an important catalytic cycle for the upper stratosphere is

$$HO_2 + O \rightarrow OH + O_2$$
$$OH + O_3 \rightarrow HO_2 + O_2 \qquad (13.55)$$
$$\overline{\text{net:}\quad O + O_3 \rightarrow 2 O_2}$$

Similar cycles exist for NO_2, ClO, and BrO. Additional cycles important for the lower stratosphere involve reactions with ozone only:

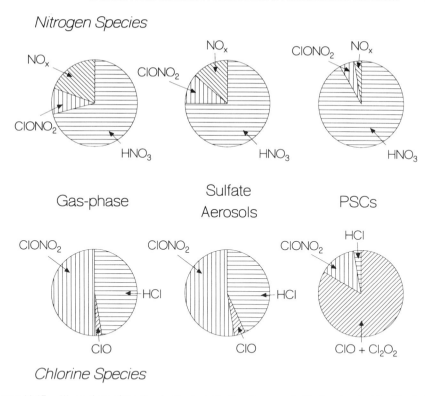

Figure 13.17. Illustrations of the chemical budgets for the nitrogen and chlorine chemical families for three different conditions: gas-phase chemsitry only, heterogeneous chemistry on sulfate aerosols, and heterogeneous chemistry on PSCs. The exact partitioning is very dependent upon season, latitude, and trajectory of the air. Some more minor chemical species are not shown.

$$HO_2 + O_3 \rightarrow OH + 2\,O_2$$
$$OH + O_3 \rightarrow HO_2 + O_2 \qquad (13.56)$$
$$\overline{\text{net: } 2O_3 \rightarrow 3O_2}$$

and

$$NO + O_3 \rightarrow NO_2 + O_2$$
$$NO_2 + O_3 \rightarrow NO_3 + O_2 \qquad (13.57)$$
$$NO_3 + h\nu \rightarrow NO + O_2$$
$$\overline{\text{net: } 2O_3 \rightarrow 3O_2}$$

These cycles involve only one chemical family acting on O_x. However, other catalytic cycles involve the reactions of reactive species from more than one chemical family. A large number of these catalytic cycles exist. Some of the most important are the halogen–halogen and the halogen–hydrogen cycles:

$$ClO + BrO \rightarrow Br + Cl + O_2$$
$$Br + O_3 \rightarrow BrO + O_2$$
$$Cl + O_3 \rightarrow ClO + O_2 \qquad (13.58)$$
$$\overline{\text{net: } O_3 + O_3 \rightarrow 3O_2}$$

and

$$HO_2 + ClO \rightarrow HOCl + O_2$$
$$HOCl + h\nu \rightarrow OH + Cl$$
$$Cl + O_3 \rightarrow ClO + O_2 \qquad (13.59)$$
$$OH + O \rightarrow HO_2 + O_2$$

$$\text{net:} \quad O_3 + O_3 \rightarrow 3O_2$$

The importance of these various cycles will depend upon the location, season, and altitude (Figure 13.18). The hydrogen and halogen catalysis of destruction of ozone are greater than nitrogen catalysis below 20 km and near 40 km; nitrogen catalysis is largest in between. In the winter and spring at high latitudes, hydrogen and halogen catalysis dominate up to 23 km. Interestingly, the ozone trends in Figure 13.6 are greatest exactly where the chlorine and bromine increase have the greatest influence on ozone: below 20 km and near 40 km.

When NO_x changes, either by an increase in the total NO_y or by a shift in the partitioning between reservoir and reactive species within the nitrogen family, the close chemical coupling with the hydrogen and halogen chemical families causes a shift in their partitioning as well. Generally, an increase in NO_x results in a decrease in both HO_x and ClO_x. As a result, smaller amounts of hydrogen and halogen reactive species destroy less ozone, while nitrogen reactive species destroy more (Figure 13.19). At low $[NO_x]$, ozone is destroyed primarily by halogens with a large contribution from hydrogen reactive species. As $[NO_x]$ increases, the more reactive halogen and hydrogen species are controlled by the increasing $[NO_x]$, so that the overall ozone loss rate decreases, and a minimum in the ozone loss rate occurs.

Even in the case of low aerosol loading, the current atmosphere is near the minimum in ozone loss in the lower stratosphere. Higher in the stratosphere, where NO_x dominates ozone loss, any changes in NO_x translate into an almost comparable fractional change in the ozone loss.

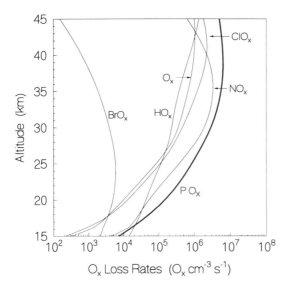

Figure 13.18. Vertical profiles of ozone production and destruction rates for June in the middle latitudes. All reactions involving bromine are counted as part of the bromine destruction rates; all reactions involving chlorine, except those with bromine, are counted as part of the chlorine destruction rates. Ozone is destroyed mostly by reactive hydrogen and halogens in the lower and upper stratosphere and mostly by reactive nitrogen in the middle stratosphere. P O_x is the production rate of O_x, $2J_{O_2}[O_2]$. (Adapted from D. Weisenstein and J. Rodriguez, private communication.)

Figure 13.19. Ozone destruction rates in the lower stratosphere at middle latitudes as a function of NO_x. At low NO_x, O_3 destruction is dominated by the faster hydrogen and halogen catalytic cycles. At high NO_x, hydrogen and halogen reactive species are converted to reservoir species and nitrogen catalytic cycles doiminate ozone loss. (Adapted from Wennberg et al., 1994.)

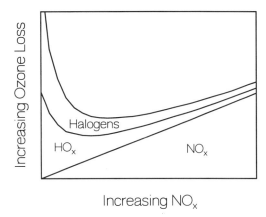

Can these catalytic cycles explain the ozone losses observed in the middle latitudes? Not entirely. Current models can simulate the observed summertime losses but calculate only about one-half of the observed wintertime loss in the Northern Hemisphere (WMO, 1994). Chemical processes from within or near the wintertime Arctic polar vortex may cause the additional ozone loss if polar air mixes sufficiently into the middle latitudes. Another possibility is some additional halogen chemistry that is missing from the models. Both possibilities are being aggressively studied.

WINTERTIME POLAR REGIONS

Observations, laboratory studies, and modeling studies have firmly established that chlorine and bromine chemistry cause the observed rapid ozone loss over Antarctica each October (WMO, 1994). They also show that chlorine and bromine chemistry cause significant ozone loss of about 10–20% of the ozone column in the Arctic each February. Although the photochemistry of the wintertime polar regions appears to be unique, in reality it represents the extremely low NO_x case in Figure 13.19. A different set of reactions becomes most important because of the meteorological conditions of the wintertime polar regions.

The chemistry responsible for the Antarctic ozone hole begins when the sun retreats to the Northern Hemisphere in April. A circumpolar jet in the middle stratosphere picks up strength and the temperatures poleward of the jet begin to fall. The air in this region has spent more than 3 years in the stratosphere, and much of the source gases has been converted to reservoir and reactive species. Air cools and descends for the next six months, and air inside the vortex, while shed by the vortex into middle latitudes, remains generally isolated from the air from middle latitudes.

As the temperatures continue to fall below 205 K, the sulfate aerosols swell with water vapor and nitric acid. These aerosols are larger and the hydrolysis of N_2O_5, $ClONO_2$, and $BrONO_2$ are accelerated. As the temperature continues to fall below 200 K, more HCl is incorporated into the aerosol, and heterogeneous reactions involving HCl become more important, particularly the reactions

$$HCl + ClONO_2 \rightarrow Cl_2 + HNO_3 \qquad \text{(hours)} \qquad (13.60)$$

$$HCl + HOCl \rightarrow Cl_2 + H_2O \qquad \text{(hours)} \qquad (13.61)$$

These reactions become even faster, with time constants of about an hour, at lower temperatures just below 195 K, when polar stratospheric clouds (PSCs), made of frozen water and nitric acid, form. If the temperature reaches the water vapor frost point, near 185 K for the lower stratosphere and 4.5 ppmv of water vapor, then the PSCs can grow to a few microns in size, large enough to settle out of the stratosphere within a few days, taking the HNO_3 with them.

Heterogeneous chemistry on cold aqueous particles, particularly PSCs, initiates the chemical sequence that leads to the observed rapid ozone loss (Figure 13.20). By this process, nitrogen species are shifted into HNO_3, which is bound up into PSCs as long as the temperature remains below 195 K, and chlorine and bromine species are converted from the HCl and $ClONO_2$ reservoir forms into Cl_2 and BrCl. Because Cl_2 and BrCl are quickly photolyzed in weak, visible sunlight, the resulting Cl and Br atoms react with O_3 within milliseconds to form ClO and BrO. Normally, ClO and BrO would react with NO_2 to form $ClONO_2$ and $BrONO_2$. However, because NO_x is shifted into HNO_3 by heterogeneous chemistry, ClO and BrO become the dominant species in their respective chemical families (Figure 13.17).

Under these circumstances, a new catalytic cycle becomes the most important for ozone loss:

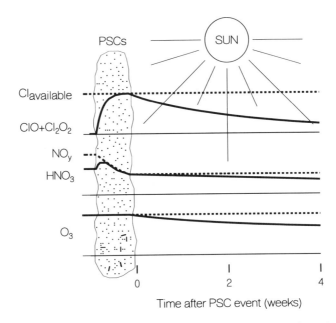

Figure 13.20. A diagram showing the effect of PSCs and sunlight on concentrations of trace gases and ozone. Dotted lines indicate available chlorine, reactive nitrogen, and initial O_3; solid lines indicate reactive chlorine (ClO and Cl_2O_2), HNO_3, and O_3. Reprinted with permission from Brune, W. H., et al., The potential for ozone depletion in the Arctic polar stratosphere.

$$ClO + ClO + M \rightarrow Cl_2O_2 + M$$
$$Cl_2O_2 + h\nu(\lambda < 250 \text{ nm}) \rightarrow Cl + ClOO$$
$$ClOO + M \rightarrow Cl + O_2 + M \qquad (13.62)$$
$$2(Cl + O_3 \rightarrow ClO + O_2)$$

$$\text{net:} \quad 2O_3 \rightarrow 3O_2$$

In the cold polar regions, the rate-limiting step in this catalytic cycle is the formation of Cl_2O_2 during daylight. During the day, ClO mixing ratios can approach 1 to 1.5 ppbv. BrO becomes an even larger fraction of Br_y; and the catalytic cycle (13.58), with the reaction $ClO + BrO \rightarrow Br + ClOO$ (and $BrCl + O_2$), becomes the second most important destruction mechanism. Other cycles contribute, but these are the main two.

The ozone destruction rate by these cycles is written as

$$\frac{d[O_3]}{dt} = -2k_{ClO+ClO}[M][ClO]^2 f_{(photolysis)} - 2k_{ClO+BrO}[ClO][BrO] \qquad (13.63)$$

At temperatures above 215 K, the thermal decomposition of Cl_2O_2 becomes important, so that the ozone destruction rate is modified by the fraction of Cl_2O_2 that is photolyzed compared to the total that are destroyed by photolysis and thermal decomposition, $f_{(photolysis)}$. In the lower stratosphere, where $[M] = 2 \times 10^{18} \text{ cm}^{-3}$, $\chi_{ClO} = 1$ ppbv, and $\chi_{BrO} = 7$ pptv, the loss rate of ozone can approach 1–3% per day in sunlit parts of the vortex. Thus, the total removal of ozone from the vortex can occur in about 50 days.

Evidence that ClO and BrO cause ozone loss was observed by instruments on the NASA ER-2 aircraft during the Airborne Antarctic Ozone Expedition to Punta Arenas, Chile in August and September 1987 (Figure 13.21). The edge of the polar vortex on this date was at approximately 67°S. In August, the ozone inside the vortex shows no significant loss, even though the ClO mixing ratio is large. However, a month later, the ozone mixing ratio inside the vortex has decreased to a third of the outside value. Calculations using the observed ClO, BrO, and O_3 from 12 flights in 1987, combined with knowledge of the amount of descent of the air and the rate constants, show that the calculated change in ozone agrees with the observed change in ozone to within the uncertainty of the calculation (Anderson et al., 1989; Solomon 1990). This agreement has been found for the 15–20% ozone loss in the Arctic as well.

The PSCs do not constantly exist and usually do not fill the vortex, especially in the Northern Hemisphere. When the air warms above 195 K, the PSCs evaporate and release the HNO_3 back into the gas phase. The same sunlight that photolyzes Cl_2O_2 to destroy ozone also photolyzes HNO_3 to form NO_2, which immediately and almost exclusively reacts with ClO to form $ClONO_2$. This reaction reduces the ClO amount and slows the ozone catalysis. Thus, massive ozone loss is possible only if the ClO mixing ratio remains large, which requires that the NO_x mixing ratio remain low.

NO_x concentrations remain lowest in the Antarctic polar vortex. First, the air inside the vortex is relatively isolated from the NO_y-rich air of the middle latitudes. Second, the temperature usually drops below the frost point so that the PSCs become ice covered and a few microns in size. When they settle out, they carry much of the NO_y with them, leaving only a few ppbv behind. Third, because the temperatures remain low through August, PSCs are frequently reformed, thus continually shifting nitrogen from $ClONO_2$ back into HNO_3. Under these conditions, almost complete ozone loss is possible in the volume of air that has been exposed to PSCs. (See Chap-

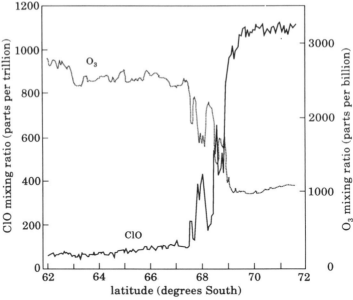

Figure 13.21. Simultaneous meaurements of ClO and O_3 over Antarctica on September 16, 1987 during the Airborne Antarctic Ozone Expedition. The boundary of the ozone-depletion region at 69°S is clearly shown by the rapid increase in ClO mixing ratio and the rapid decrease in O_3 mixing ratio. The rapid fluctuations of anticorrelated ClO and O_3 at the boundary indicate the shedding of air from the vortex. (Adapted from Anderson et al., 1991.)

ter 15, this volume, for a more detailed discussion of heterogeneous chemistry in the stratosphere.)

When the ozone mixing ratio drops to a few hundred ppbv (90% loss), the rate-limiting step in the ozone catalysis sequences shifts toward the reactions of Cl and Br with O_3. The concentrations of Cl and Br begin to build. As this happens, the reaction of Cl with CH_4 shifts more chlorine from Cl into HCl. Because this occurs in October, when the temperatures are generally high enough that no more PSCs are occurring, chlorine is shifted from reactive forms almost exclusively into HCl. The result is an atmosphere in which HCl is most of Cl_y, O_3, and ClO are very low, and photolysis of the remaining HNO_3 creates NO. Chlorine chemistry effectively shuts itself down in a matter of a week. When the polar vortex breaks up in November or December, the ozone-poor polar air mixes with the middle latitude air, but the chlorine is in the form of HCl, limiting further damage at middle latitudes.

The Arctic polar stratosphere is different from the Antarctic polar stratosphere (Brune et al., 1991). First, it does not get as cold, nor is the vortex as stable. As a result, while PSCs composed of nitric acid and water form every year, those large PSCs composed of water ice and that form below the frost point are large enough to settle out of the stratosphere. Much of the NO_y remains in the wintertime Arctic polar stratosphere because it is not removed by the settling of the large, ice-coated PSCs. In addition, PSCs are less frequent and often occur sporadically from February until the vortex break-up in March or April. Although the conversion of chlorine and bromine by PSCs is as complete in the Arctic as in the Antarctic, photolysis of the HNO_3 remaining in the vortex results in NO_x production, which forms $ClONO_2$ (Figure 13.22). Thus, the ozone loss in the

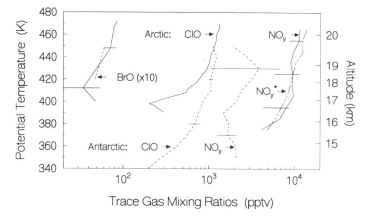

Figure 13.22. Comparison of Antarctic and Artic *in situ* data, taken during the Airborne Antarctic Ozone Expedition in 1987 and the Airborne Arctic Stratospheric Expedition in 1989, respectively. Arctic data are represented by solid lines, and Antarctic data are represented by dashed lines. The dot–dashed line represent NO_y^* mixing ratios for the Arctic, which are about 1000 pptv smaller for the Antarctic. NO_y^* is the mixing ratio of NO_y predicted from the observed N_2O and the NO_y–N_2O relationship. All data are averaged over all flights, and are shown with variability ($[\pm 1\sigma]$), except ClO in the Arctic, which is from a flight on February 10, 1989. Reprinted with permission from Brune, W. H., et al., The potential for ozone depletion in the Arctic polar stratosphere.

Arctic is typically about 15–20% at the affected altitudes, with a column loss of about 10–15%.

Because the ozone is not completely removed before the Arctic polar vortex breaks up in February through April, the chlorine is converted from reactive forms into $ClONO_2$, with conversion into HCl being much slower than in the Antarctic. This air, which has a higher proportion of $ClONO_2$, will have a higher proportion of ClO during the day due to the steady-state relationship between these two species. This difference in the end-product of the wintertime polar chemistry may be part of the reason that the middle latitudes in the Northern Hemisphere experience additional wintertime ozone loss.

The onset of the rapid loss of ozone over Antarctica is related to the increasing levels of stratospheric chlorine and bromine driven by the increase of CFCs and anthropogenic bromine compounds. The increased ozone loss is consistent with the increase in CFCs and anthropogenic bromine chemicals. As the ban on production and use of CFCs and halons continues, the growth of the tropospheric amounts of these compounds has almost ceased. But the long lifetime of CFCs in the atmosphere indicates that the Antarctic ozone hole, along with some ozone loss in the Arctic, is likely to be a common feature for approximately another 50 years, when stratospheric chlorine levels decrease below 2 ppbv.

CONCLUSION

The dominant theme of stratospheric chemistry is the catalytic ozone loss. Key to assessing this loss is the amount of each chemical species in the stratosphere and the degree to which NO_x controls the reactive species from the hydrogen, chlorine, and bromine chemical families. With the inclusion of heterogeneous chemistry, the chemistry of the lower stratosphere below 20 km appears to be fairly well understood from both aircraft and satellite measurements (WMO, 1994). The chemistry of the middle and upper stratosphere also appears to be understood, although more studies are required to resolve remaining questions and inconsistencies.

A worry is that some chemical mechanisms are missing from the understanding. These may not affect the agreement between observations and models under the conditions of the observations, but they may become important if the meteorology or the mixing ratios of trace gases and aerosols change. An example of possible missing chemistry is that of iodine chemistry (Solomon et al., 1994), which would have a considerable affect on the ozone balance of the lower stratosphere through catalytic destruction similar to that of bromine, but more virulent. Another possibility is processes such as photolysis of trace gases in the liquid phase that are not allowed in the gas phase. Yet another possibility is excited-state chemistry. Such possibilities will continue to arise as observations disagree with model results and as new laboratory work uncovers mechanisms previously not seen.

The understanding of stratospheric chemistry has advanced rapidly in the last 20 years. This advancement results directly from the concerted effort of observations, models, and laboratory studies. These efforts will continue to be important as stratospheric chemistry continues to change.

ACKNOWLEDGMENTS

I am grateful to J. Rodriguez and D. Weisenstein at AER, Inc. for providing model calculations of concentrations and reactions rates of chemical species from their well-known two-dimenstional photochemical model.

References

Anderson, J. G., Toohey, D. W., and Brune, W. H., 1991, Free radicals within the Antarctic vortex: the role of CFCs in Antarctic ozone loss. *Science*, 251, 39–46.

Anderson, J. G., Brune, W. H., Lloyd, S. A., Toohey, D. W., Sander, S. P., Starr, W. L., Loewenstein, M., and Podolske, J. R., 1989, Kinetics of O_3 destruction by ClO and BrO within the Antarctic vortex: an analysis based on in situ ER-2 data. *J. Geophys. Res.*, 94, 11480–11520.

Bates, D. R., and Nicolet, M., 1950, The photochemistry of water vapor. *J. Geophys. Res.*, 55, 301.

Brewer, A. W., 1949, Evidence for a world circulation provided by the measurements of helium and water vapor distribution in the stratosphere, *Q. J. R. Meteor. Soc.*, 75, 351–363.

Brewer, A. W., and Wilson, A. W., 1968, The regions of formation of atmospheric ozone, *Q. J. R. Meteor. Soc.*, 94, 249–265.

Brune, W. H., Anderson, J. G., Toohey, D. W., Fahey, D. W., Kawa, S. R., Jones, R. L., McKenna, D. S., and Poole, L. R., 1991, The potential for ozone depletion in the Arctic polar stratosphere. *Science*, 252, 1260–1266.

Chapman, S., 1930, A theory of upper atmospheric ozone, *Mem. R. Meteor. Soc.*, 3, 103.

Crutzen P. J., 1970, The influence of nitrogen oxides on the atmospheric ozone content, *Q. J. R. Meteor. Soc.*, 96, 320.

DeMore, W. B., Sander, S. P., Golden, D. M., Hampson, R. F., Kurylo, M. J., Howard, C. J., Ravishankara, A. R., Kolb, C. E., Molina, M. J., 1994, *Chemical Kinetics and Photochemistry Data for Use in Stratospheric Modeling*, Evaluation number 11, JPL Publications 94–26.

Dobson, G. M. B., 1956, Origin and distribution of polyatomic molecules in the atmosphere. *Proc. R. Soc. London*, A236, 187–193.

Duetsch, H. U., 1971, Photochemistry of atmospheric ozone. *Adv. Geophys.*, 15, 219.

Fahey, D. W., Solomon, S., Kawa, S. R., Loewenstein, M., Podolske, J. R., Strahan, S. E., Chan, K. R., 1990, A diagnostic for denitrification in the winter polar stratospheres. *Nature*, 345, 698–702.

Fahey, D. W., et al., 1993, *In situ* measurements constraining the role of sulfate aerosols in midlatitude ozone depletion. *Nature*, 363, 509–514.

Farman, J. C., Gardiner, B. G., and Shanklin, J. D., 1985, Large losses of total ozone in Atnatarctica reveal seasonal ClO_x/NO_x interaction. *Nature*, 315, 207–210.

Hampson, J., 1965. In: *Les Problems Meteorologiques de la Stratosphere et de la Mesosphere*, Presses Universitaires de France, Paris, p. 393.

Hampson, J., 1966, *Technical Note 1738*, Canadian Armament Research and Development Establishment, Valcartier, Quebec.

Hanson, D. R., and Ravishankara, A. R., 1991, The reaction probabilities of $ClONO_2$ and N_2O_5 on 40 to 75 percent sulfuric acid solutions. *J. Geophys. Res.*, 96, 17307–17314.

Hartley, W. N., 1881, On the absorption of solar rays by atmospheric ozone. *J. Chem Soc.*, 39, 111.

Hering, W. S., and Borden, T. R., 1965, *Ozone Sonde Observations over North America*, Vol. 3, Rep AFCRL-64-30(3), Air Force Cambridge Res. Labs, Bedford, MA, 1965.

Hofmann, D. J., Oltmans, S. J., Lathrop, J. A., Harris, J. M., and Voemel, H., 1994, Record low ozone at the South Pole in the spring of 1993. *Geophys. Res. Lett.*, 21, 421–424.

Holton, J. R., Haynes, P. E., McIntyre, M. E., Douglass, A. R., Rood, R. B., and Pfister, L., 1995, Stratosphere-troposphere exchange. *Rev. Geophys.*, 33, 403–439.

Johnston H. S., 1971, Reduction of stratospheric ozone by nitrogen oxide catalysts from supersonic transport exhaust. *Science*, 173, 517.

Johnston, H. S., and Podolske, J., 1978, Interpretations of stratospheric photochemistry. *Rev. Geophys. Space Phys.*, 16, 491–519.

Krueger, A. J., 1973, The mean ozone distributions from several series of rocket soundings to 52 km at latitudes from 58°S to 64°N. *Pure Appl. Geophys.*, 106–108, 1271.

Leu, M.-T., 1988, Laboratory studies of sticking coefficients and heterogeneous reactions important in the Antarctic stratosphere. *Geophys. Res. Lett.*, 15, 17.

Logan, J. A., Prather, M. J., Wofsy, S. C., and McElroy, M. B., 1978, Atmospheric chemistry: a response to human influence, *Philos. Trans. R. Soc.*, 290, 187–234.

London, J., 1980, Radiative energy sources and sink in the stratosphere and mesosphere. In: M. Nicolet and A. C. Aikin, eds., *Proceeding of the NATO Advanced Study Institute on Atmospheric Ozone: Its Variation and Human Influences*, U.S. Department of Transportation, Washington, D.C., Report No. FAA–EE–80-20, pp. 31–34.

McElroy, M. B., Salawitch, R. J., Wofsy, S. C., and Logan, J. A., 1986, Reduction of Antarctic ozone due to synergetic interactions of chlorine and bromine. *Nature*, 321, 759–762.

McIntyre, M. E., and Palmer, T. N., 1983, Breaking planetary waves in the stratosphere. *Nature*, 305, 593–600.

McIntyre, M. E., 1989, On the antarctic ozone hole. *J. Atmos. Terr. Phys.*, 51, 29–43.

McPeters, R. D., Heath, D. F., and Bhartia, P. K., 1984, Average ozone profiles for 1979 from the NIMBUS 7 SBUV instrument. *J. Geophys. Res.*, 89, 5199–5214.

Miller, A. J., Bishop, L., DeLuisi, J. J., Nagatani, R. M., Kerr, J., Maerr, C. L., Wuebbles, D., Reinsel, G. C., and Tiao, G. C., 1995, Comparisons of the observed ozone trends in the stratosphere though examination of Umkehr and balloon ozonesonde data. *J. Geophys. Res.*, 100, 11209.

Molina, M. J., and Rowland, F. S., 1974, Stratospheric sink for chlorofluoromethanes: chlorine atom-catalyzed destruction of ozone. *Nature*, 249, 810–814.

Molina, M. J., Tso, T. L., Molina, L. T., and Wang, F. C. Y., 1987, Antarctic stratospheric chemistry of chlorine nitrate, hydrogen chloride, and ice: release of active chlorine. *Science*, 238, 1253.

Molina, L. T., and M. J., Molina, 1987, Production of Cl_2O_2 from the self-reaction of the ClO radical. *J. Phys. Chem.*, 91, 433–436.

Plumb, R. A., and Ko, M. K. W., 1992, Interrelationships between mixing ratios of long-lived stratospheric constituents. *J. Geophys. Res.*, 97, 10145–10156.

Schauffler, S. M., Heidt, L. E., Pollock, W. H., Gilpin, T. M., Vedder, J. F., Solomon, S., Lueb, R. A., and Atlas, E. L., 1993, Measurements of halogenated organic compounds near the tropical tropopause. *Geophys. Res. Lett.*, 20, 2567–2570.

Solomon, S., 1990, Progress towards a quantitative understanding of Antarctic ozone depletion. *Nature*, 347, 347–354.

Solomon, S., Garcia, R. R., Rowland, F. S., and Wuebbles, D. J., 1986, On the depletion of Antarctic ozone. *Nature*, 321, 755–758.

Solomon, S., Garcia, R. R., and Ravishankara, A. R., 1994, On the role of iodine in ozone depletion. *J. Geophys. Res.*, 99, 20491–20499.

Stolarski, R. S., and Cicerone, R. J., 1974, Stratospheric chlorine: a possible sink for ozone. *Can. J. Chem.*, 52, 1610.

Stolarski, R. S., Bloomfield, P., and McPeters, R. D., 1991, Total ozone trends deduced from NIMBUS 7 TOMS data. *Geophys. Res. Lett.*, 18, 1015–1018.

Tolbert, M. A., Rossi, M. J., Malhotra, R., and Golden, D. M., 1987, Reaction of chlorine nitrate with hydrogen chloride and water at Antarctic stratospheric temperatures. *Science*, 238, 1258.

Toon, O. B., Hamill, P., Turco, R. P., and Pinto, J., 1986, Condensation of HNO_3 and HCl in the winter polar stratosphere. *Geophys. Res. Lett.*, 13, 1284.

U.S. Standard Atmosphere, 1976, U.S. Government Printing Office, Washington, D.C.

Wennberg, P. O., et al., 1994, Removal of stratospheric O_3 by radicals: *in situ* measurements on OH, HO_2, NO, NO_2, ClO, and BrO. *Science*, 266, 398–404.

WMO, 1994, Scientific Assessment of Ozone Depletion, *Global Ozone Research and Monitoring Project—Report No. 37*, World Meteorological Organization, Geneva, 1995.

Wofsy, S. C., McElroy, M. B., and Yung, Y. L., 1975, The chemistry of atmospheric bromine. *Geophys. Res. Lett.*, 2, 215–218.

Woodbridge, E. L., et al., 1995, Estimates of total organic and inorganic chlorine in the lower stratosphere from insitu and flask measurements during AASEII. *J. Geophys. Res.*,

Laboratory Studies of Heterogeneous Chemistry in the Stratosphere

ANN M. MIDDLEBROOK AND
MARGARET A. TOLBERT

ABSTRACT. In the past decade, scientists have begun to study heterogeneous (gas–surface) chemical reactions of atmospheric relevance. In this chapter, we explain how heterogeneous chemistry plays an important role in the formation mechanism of the Antarctic ozone hole. Although chemical destruction of stratospheric ozone by chlorine atoms occurs in the gas phase, the appearance of the ozone hole cannot be explained by gas-phase chemistry alone. This is because chlorine usually exists in long-lived reservoir molecules, hydrochloric acid and chlorine nitrate, rather than as isolated chlorine atoms. However, laboratory experiments have shown that these reservoir molecules react on model stratospheric aerosol surfaces to produce photochemically active molecules. When the heterogeneous reaction products are photolyzed, they release the chlorine atoms capable of gas-phase ozone destruction. Since inclusion of these heterogeneous chemical reactions in stratospheric models, there has been much better agreement between calculated and observed ozone, both in the polar regions and globally.

INTRODUCTION

Although most atmospheric reactions occur purely in the gas phase, it has recently become apparent that heterogeneous (gas–surface) reactions can have a major impact on the chemistry of the atmosphere. Heterogeneous reactions can occur on either solid or liquid particles. For solids, the reactions are typically confined to the surface or near-surface region and may depend strongly on the microscopic surface properties of the particle. These reactions are typically characterized by a reaction probability, γ, defined as the fraction of collisions that lead to reaction. For liquid particles, the chemistry may depend not only on the surface characteristics, but also the bulk properties such as solubility, diffusion, and condensed phase reaction rates. These parameters will be discussed in more detail below.

Heterogeneous reactions on solid and liquid particles are now recognized to occur in both the troposphere and the stratosphere. However, in the present review, we will focus on heterogeneous chemistry that occurs in the stratosphere, and the consequences of that chemistry on global ozone levels. The stratosphere is characterized by increasing temperature with altitude. This increase in temperature is due to light absorption by stratospheric ozone. Typical temperature and ozone profiles of the atmosphere are shown in Figure 14.1. In the troposphere, that part of the atmosphere closest to Earth, the temperature decreases with increasing altitude up to the tropopause. For water vapor to reach the stratosphere, it must either rise through the "cold trap" of the tropopause or be formed in the stratosphere due to the oxidation of methane. Due to these constraints, the stratosphere is fairly dry with very few clouds.

Because little water vapor reaches the stratosphere, atmospheric scientists have only recently focused attention on the formation and heterogeneous reactivity of stratospheric clouds. The temperatures must be very cold (<189 K) for water–ice particles to condense. However, above Antarctica and occasionally above the Arctic Ocean, stratospheric temperatures do become cold enough in the winter to form ice clouds. These ice particles are called Type II polar stratospheric clouds (PSCs) because they are found predominantly over the poles.

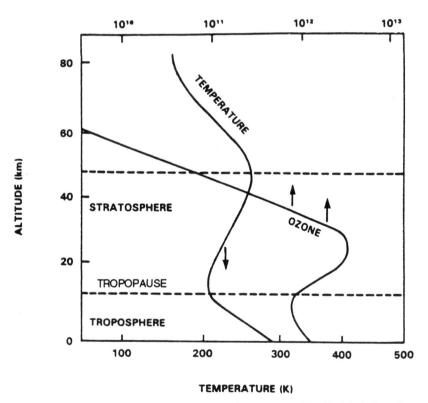

Figure 14.1.   Variation of temperature (bottom axis) and ozone (top axis) with altitude from the surface of the earth. In the troposphere, temperature decreases with altitude. In the stratosphere, temperature increases with altitude. The tropopause is located at the minimum temperature. Ozone concentrations maximize in the stratosphere. The ozone concentration is shown on a log scale in units of molecules per cubic centimeters (cm^{-3}). (Reprinted from Watson et al., 1986, with permission.)

Other cloud particles containing nitric acid (HNO_3) and water (H_2O) may form at warmer temperatures. These HNO_3/H_2O particles are called Type I PSCs because they form at higher temperatures (<195 K) than pure ice clouds. Due to the warmer temperatures necessary for Type I PSC formation, these clouds are more prevalent than Type II PSCs, and are found over both poles during the winter.

In addition to PSCs, there are very different stratospheric aerosols found globally, the so-called stratospheric sulfate aerosol (SSA) layer. At normal stratospheric temperatures (205–240 K), the SSAs are believed to be composed of concentrated solutions (60–80 wt. %) of H_2SO_4 in H_2O. Thus, three main classes of particles exist in the stratosphere that may participate in heterogeneous reactions: liquid, sulfuric acid droplets (global), HNO_3/H_2O particles (Antarctic and Arctic winters), and ice crystals (Antarctic winters). The stratospheric aerosols available for heterogeneous chemistry are summarized in Table 14.1. We will find that each of these surfaces provides a unique environment for heterogeneous chemistry. Thus, the relevant chemistry depends strongly on what type of stratospheric aerosol is present.

Gas-Phase Ozone Reactions

To appreciate the significance of heterogeneous chemistry on the ozone layer, we must first consider the gas-phase (homogeneous) reactions that control the abundance of ozone. Stratospheric ozone is produced by the photolysis of oxygen molecules, followed by the combination of one oxygen atom with an oxygen molecule:

$$O_2 + h\nu \rightarrow O + O \qquad \lambda < 242 \text{ nm} \qquad (14.1)$$

$$\underline{O + O_2 + M \rightarrow O_3 + M} \qquad\qquad\qquad (14.2)$$

$$\text{net:} \quad 2O_2 \rightarrow O + O_3$$

Ozone can be destroyed by photolysis and by reaction with an oxygen atom:

$$O_3 + h\nu \rightarrow O_2 + O \qquad\qquad\qquad (14.3)$$

$$\underline{O + O_3 \rightarrow 2O_2} \qquad\qquad\qquad (14.4)$$

$$\text{net:} \quad 2O_3 \rightarrow 3O_2$$

Chapman (1930) first proposed reactions (14.1) through (14.4) to explain the global distribution of stratospheric ozone. However, the steady-state ozone concentration predicted by the Chapman mechanism is too high to explain the current observations of stratospheric ozone. This is because while reaction (14.1) is the only net source of odd

TABLE 14.1
Stratospheric Particulate

Temperature	Composition	Location
<109 K	Ice (Type II PSCs)	Antarctic winter/spring
<195 K	HNO_3/H_2O (Type I PSCs)	Arctic and Antarctic winters
205 K < T < 240 K	60 % < H_2SO_4 < 80 %(SSAs)	global stratosphere

oxygen ($O_x = O + O_3$), there are many ways in addition to reaction (14.4) by which odd oxygen can be destroyed. For example, ozone is catalytically destroyed in a cycle involving the oxides of nitrogen ($NO_x = NO + NO_2$)(Crutzen, 1970):

$$NO + O_3 \rightarrow NO_2 + O_2 \tag{14.5}$$

$$O + NO_2 \rightarrow NO + O_2 \tag{14.6}$$

$$\text{net:} \quad O + O_3 \rightarrow 2O_2 \tag{14.7}$$

Nitrogen oxides occur naturally in the stratosphere from the oxidation of nitrous oxide, N_2O, released in the troposphere by bacteria. N_2O molecules do not photolyze or react in the troposphere, and thus are able to reach the stratosphere. Once there, N_2O reacts with an electronically excited oxygen atom, $O[^1D]$, to form NO. Nitrogen oxides may also be injected directly into the stratosphere by fuel exhaust from high-flying aircraft as first recognized by Johnston (1971).

Other trace species such as ClO_x, BrO_x, and HO_x, can also lead to catalytic ozone destruction. Molina and Rowland (1974) first proposed that chlorine from chlorofluorocarbons (CFCs) could cause catalytic ozone loss:

$$Cl + O_3 \rightarrow ClO + O_2 \tag{14.8}$$

$$O + ClO \rightarrow Cl + O_2 \tag{14.9}$$

$$\text{net:} \quad O + O_3 \rightarrow 2O_2 \tag{14.10}$$

The CFCs do not react in the troposphere and slowly rise above the ozone layer where they photolyze to release chlorine atoms for reactions (14.8) and (14.9).

When gas phase catalytic ozone destruction cycles involving NO_x, ClO_x, BrO_x, and HO_x were included in photochemical atmospheric models, global ozone levels were predicted to drop slowly over decades. However, in the early 1980s, scientists who were monitoring ozone concentrations began to observe abrupt and massive decreases above Antarctica in the early spring (Farman et al., 1985). Figure 14.2 shows a similar, but more recent, ozone "hole" (Hofmann et al., 1994). The ozone profile during the dark, polar winter (23 Aug. 1993 data) resembles the profile shown in Figure 14.1, with ozone concentrations peaking at roughly 17 km. However, only 7 weeks later (Oct. 12, 1993), all measurable ozone between 14 and 19 km was destroyed. The integrated amount of ozone in this profile was 91 Dobson units (DU), the lowest column amount ever measured! For comparison, the lowest column amount in 1992 was 105 DU on Oct. 11, 1992. By the beginning of the polar summer, ozone concentrations returned to normal levels. By studying the ozone concentration record, the ozone hole was found to be deeper and last longer every other year until 1989, 1990, and 1991, when the holes were comparable. In 1992 and 1993, the area covered by the ozone hole was the largest on record with significant ozone loss occurring over 10% of the Southern Hemisphere (WMO, 1994). These recent observations of dramatic ozone loss during the Antarctic spring cannot be explained by a purely gas-phase mechanism.

Since the models failed to predict the occurrence of dramatic ozone loss over Antarctica, it was clear that something was missing. Field measurements in the spring of 1986 in Antarctica suggested that destruction of ozone by catalytic reactions involving chlorine atoms were responsible for the ozone hole (Solomon et al., 1987; de Zafra et al., 1987). This was confirmed by *in situ* measurements during a 1987 field mission into the ozone hole where low ozone levels were found to be anticorrelated

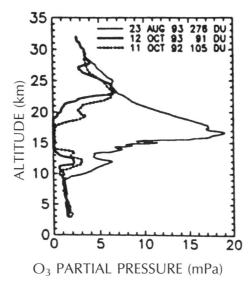

Figure 14.2. Ozone profile measured in 1992 and 1993 at the South Pole Station, Antarctica. On August 23, the ozone concentration profile appears normal. On October 12, ozone concentrations between 14 and 19 km were below the instrument's detection limit, illustrating vertical location of the ozone hole. For comparison, the ozone profile on October 11, 1992 is also shown. (Reprinted from Hofmann et al., 1994, with permission.)

with high ClO levels (Anderson et al., 1989). However, in the winter and early spring above Antarctica, sunlight is unavailable to create oxygen atoms, needed for ozone loss via reaction (14.9). Thus, another mechanism was needed to recycle the ClO back to Cl atoms. Molina and Molina (1987) proposed that the ClO reacts with itself to form a dimer, which in turn photolyzes into Cl atoms. However, if NO_2 is available, then ClO would react with it to form $ClONO_2$:

$$ClO + NO_2 + M \rightarrow ClONO_2 + M \qquad (14.11)$$

This process is often called "chlorine deactivation" because it changes the chlorine from an ozone-destroying species, ClO, to a reservoir form, $ClONO_2$. Therefore, high levels of catalytically active chlorine (ClO and Cl) can only be maintained if levels of NO_2 are very low.

HETEROGENEOUS CHEMISTRY ON POLAR STRATOSPHERIC CLOUDS (PSCs)

Solomon et al. (1986) discovered the missing piece of the ozone hole puzzle by suggesting that heterogeneous reactions could be occurring on the surfaces of polar stratospheric clouds (PSCs). PSCs are observed in the polar winter above Antarctica and remain prevalent during the early spring when the ozone hole forms. Heterogeneous chemistry on these PSCs can affect both the ClO_x and NO_x budgets.

The heterogeneous reactions that are now thought to be the most important for polar ozone loss are

$$ClONO_2 + H_2O \rightarrow HOCl + HNO_3 \qquad (14.12)$$

$$N_2O_5 + H_2O \rightarrow 2HNO_3 \qquad (14.13)$$

$$ClONO_2 + HCl \rightarrow Cl_2 + HNO_3 \qquad (14.14)$$

$$N_2O_5 + HCl \rightarrow ClNO_2 + HNO_3 \qquad (14.15)$$

$$HOCl + HCl \rightarrow Cl_2 + H_2O \qquad (14.16)$$

The role of reactions (14.12) through (14.16) is to repartition chlorine and nitrogen species during the Antarctic winter, thereby setting the stage for massive ozone destruction in the spring. For example, reactions (14.12) and (14.14) through (14.16) on PSCs convert the stable chlorine reservoir species ($ClONO_2$ and HCl) into more active forms of chlorine ($HOCl$, Cl_2, and $ClNO_2$), a process referred to as "chlorine activation." Photolysis of the active chlorine molecules at polar sunrise forms chlorine capable of catalytic ozone destruction. In addition, formation of HNO_3 in the above reactions leads to loss of nitrogen oxides from the polar winter stratosphere. Sedimentation of HNO_3 associated with PSCs can cause permanent removal of nitrogen species from the stratosphere, so-called "denitrification." If denitrification does not occur, the HNO_3 can photolyze to produce NO_2 which then can react with ClO to reform $ClONO_2$ via reaction (14.11) (chlorine deactivation) and stop the ClO/O_3 cycle. This overall sequence of events is shown schematically in Figure 14.3.

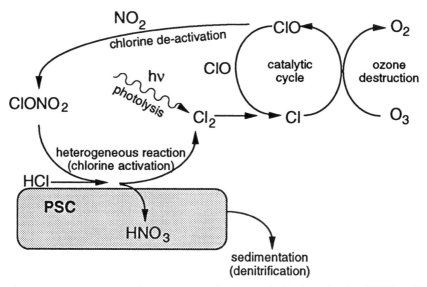

Figure 14.3. Schematic representation of PSC processing. Reservoir chlorine molecules, $ClONO_2$ and HCl, react on PSC surfaces (heterogeneous reactions) to form HNO_3 and Cl_2 (chlorine activation). Sunlight causes the Cl_2 to photodissociate into Cl atoms, which catalytically destroy O_3. This catalytic cycle is sustained when ClO levels are high and NO_2 levels remain low. If the PSC particles containing HNO_3 fall out of the stratosphere, then nitrogen species are permanently removed (denitrification) and the NO_2 concentrations remain low. Otherwise, ClO will react with NO_2 to reform $ClONO_2$ (chlorine deactivation).

It can thus be seen that PSCs are important for polar ozone destruction in two ways: They activate the chlorine species via heterogeneous reactions and they cause denitrification via sedimentation. To include heterogeneous chemical reactions in atmospheric models of the ozone hole, it is necessary to know the rates of the relevant reactions. Recent laboratory studies have begun to make measurements of the rates, thus providing a solid foundation for our understanding of polar ozone chemistry.

Experiments Designed to Study Heterogeneous Reactions

Early experiments on the reactivity of $ClONO_2$ and N_2O_5 indicated that reactions (14.12) through (14.16) occurred very slowly in the gas phase. However, after the discovery of the ozone hole, several research groups quickly designed experiments to investigate these reactions on cold surfaces representing PSCs. The experimental quantity used to characterize the heterogeneous reaction rates is the reaction probability, γ, defined as the fraction of reactant collisions with the surface that leads to reaction:

$$\gamma = \frac{\text{\# molecules reacting (molecules/cm}^2\text{-s)}}{\text{\# gas–surface collisions (collisions/cm}^2\text{-s)}}$$

In several laboratory experiments, the number of molecules reacting (or forming) is determined by measurements of gaseous species using mass spectrometric or other techniques. Some experiments also measure the number of reacting molecules by observing the growth of condensed phase species. The number of molecules lost per second is then divided by the sample surface area. The number of gas–surface collisions depends on the trace gas pressure and is obtained from gas-kinetic theory as $\langle v \rangle [N/V]/4$, where $\langle v \rangle$ is the temperature-dependent average molecular velocity (cm/s), and $[N/V]$ is the gas number density (molecules/cm^3). The rate of nonreactive gas uptake by the surface (still denoted γ) is measured in a·similar manner, but is usually then referred to as the "net uptake efficiency."

Two principal types of reaction chambers have been employed to measure γ: a low-temperature flow tube operated at a pressure of ~ 0.2–3 torr and a low-temperature Knudsen cell operated at $\sim 10^{-3}$ torr. An example of the Knudsen cell flow reactor is shown in Figure 14.4a. This chamber consists of two sections, separated by a large valve. The PSC surface is located in the lower chamber. The reactant molecules are introduced into the upper section of the chamber and monitored with the mass spectrometer through a small calibrated aperture. The valve is then opened and the reactant molecules are exposed to the cold PSC surface in the lower section of the chamber. If the reactant molecules are taken up by the cold surface due to reaction or adsorption, the mass spectrometer signal decreases as indicated in Figure 14.4b. The reaction probability for this type of experiment is measured by:

$$\gamma = \frac{A_a}{A_s} \cdot \left(\frac{I_o - I}{I} \right)$$

where A_a is the area of the aperture to the mass spectrometer, A_s is the area of the cold surface, I_o is the mass spectrometer signal before exposure to the cold surface, and I is the mass spectrometer signal during exposure. If the reaction products are released into the gas phase, they can be monitored with the mass spectrometer simultaneously with the reactants. If the products are retained in the cold surface, the surface can sub-

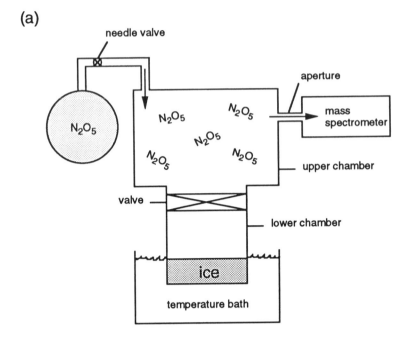

(a)

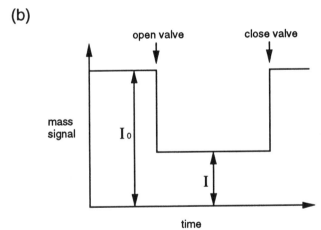

(b)

Figure 14.4. Schematic diagram of (a) a two-chamber Knudsen cell flow reactor and (b) the mass spectrometer signal. The two chambers are isolated by a large valve. In the lower chamber, the sample substrate (i.e. ice) is maintained at constant temperature. With the valve closed, reactant molecules (i.e., N_2O_5) flow into the upper chamber and through an aperture into a mass spectrometer. When the valve is opened, reactant molecules are exposed to the sample surface. A decrease in the mass spectrometer signal indicates loss of the reactant species to the surface, as shown in part (b).

sequently be heated and the product gases are measured with the mass spectrometer during evaporation. One of the major uncertainties in much of the laboratory work is the determination of the surface area. Most often, the geometrical surface area of the sample is used. However, Keyser et al. (1991) found that their PSC surfaces were highly porous, and they adjusted their surface areas accordingly.

Results on ice surfaces representative of Type II PSCs

Several research groups have studied the reaction of chlorine nitrate ($ClONO_2$) and nitrogen pentoxide (N_2O_5) on ice surfaces:

$$ClONO_2 + H_2O \rightarrow HOCl + HNO_3 \qquad (14.12)$$

$$N_2O_5 + H_2O \rightarrow 2HNO_3 \qquad (14.13)$$

For chlorine nitrate hydrolysis [reaction (14.12)] on ice surfaces, γ has been measured to be between 0.01 and 0.3 (Molina et al., 1987; Tolbert et al., 1987; Leu, 1988a; Hanson and Ravishankara, 1991a, 1992; Abbatt and Molina, 1992a). The lower value of gamma has since been determined to be a lower limit due to surface saturation for experiments using relatively high pressures of $ClONO_2$. Hence, the reaction probability of 0.3 for $ClONO_2$ hydrolysis on ice at low pressures is the most accurate measurement (Hanson and Ravishankara, 1991a). In contrast, the reaction probability for the hydrolysis of N_2O_5, reaction (14.13), on ice surfaces is slightly smaller than for $ClONO_2$ hydrolysis, about 0.03 (Leu, 1988b; Tolbert et al., 1988a; Quinlan et al., 1990; Hanson and Ravishankara, 1991a). A summary of recent laboratory measurements of reaction probabilities for reactions (14.12) through (14.16) on model PSCs is given in Table 14.2.

The time constants for reactions (14.12) and (14.13) occurring in the gas phase are very long, about 10 months. In contrast, the timescale of the Antarctic ozone hole is only 6–7 weeks. By including heterogeneous reactions on PSCs, the lifetimes of these species are dramatically reduced. The lifetimes of $ClONO_2$ or N_2O_5 with respect to heterogeneous loss can be expressed as

$$\tau \, (s) = \frac{4}{\langle v \rangle \, A \gamma} \qquad (14.17)$$

where $\langle v \rangle$ is the average molecular velocity of $ClONO_2$ or N_2O_5 molecules (cm/s), A is the aerosol surface area density (cm^2/cm^3), and γ is the reaction probability. For example, assuming a surface area density of 10^{-7} cm^2/cm^3 for Type II PSCs (Turco et al., 1989), Eq. (14.17) can be used to estimate the lifetime of $ClONO_2$ and N_2O_5 in the stratosphere. For $\gamma = 0.1, 0.01$, and 0.001, the lifetimes calculated using Eq. (14.17) are 6 hours, 2 days, and 23 days, respectively. Since the reaction probabilities for $ClONO_2$ and N_2O_5 on ice are >0.01, the presence of Type II PSCs will dramatically lower the lifetimes of these molecules. Thus, the timescale of heterogeneous chemistry is consistent with the ozone hole observations.

Results on nitric acid/ice surfaces representative of Type I PSCs

The results described above were for reaction on pure ice surfaces. However, ice particles (Type II PSCs) form only at very low temperatures (<189 K) and are thus found predominantly in the Antarctic winter and spring stratosphere and occasionally in the Arctic. Nitric acid/ice particles (Type I PSCs) are more prevalent during winter in both polar stratospheres because they form at warmer temperatures (<195 K). Although the exact composition and phase of Type I PSCs are not known, most studies have assumed that they are composed of crystalline nitric acid trihydrate (NAT) because it is the most thermodynamically stable phase of HNO_3/H_2O under stratospheric conditions (Hanson and Mauersberger, 1988). More recent work suggests that some

Type I PSCs are supercooled ternary solutions of $HNO_3/H_2SO_4/H_2O$ (Carslaw et al., 1994; Tabazadeh et al., 1994).

The reactions of chlorine nitrate with water and hydrochloric acid (HCl) have recently been studied on NAT surfaces representative of Type I PSCs (Moore et al., 1990; Leu et al., 1991; Hanson and Ravishankara, 1991a, 1992; Abbatt and Molina, 1992a):

$$ClONO_2 + H_2O \rightarrow HOCl + HNO_3 \tag{14.12}$$

$$ClONO_2 + HCl \rightarrow Cl_2 + HNO_3 \tag{14.14}$$

The reaction probabilities of $ClONO_2$ on HNO_3/H_2O surfaces indicate that the reaction of $ClONO_2$ with HCl on the surface is much more effective than reaction without HCl (see Table 14.2). For example, reaction probabilities of $\gamma = 0.001$ to 0.02 and $\gamma = 0.06$ to 1.0 have been measured for reactions (14.12) and (14.14), respectively. For both of these reactions, recent studies show that γ is a strong function of available water and that it increases as the relative humidity increases (Abbatt and Molina, 1992a; Hanson and Ravishankara, 1993; Tabazadeh and Turco, 1993). In addition, recent work has shown that reactions (14.12) (14.14), and (14.16) are all very efficient on the supercooled ternary solutions of $HNO_3/H_2SO_4/H_2O$ that may be the composition of Type I PSCs (Zhang et al., 1994, 1995; Ravishankara and Hanson, 1996). Therefore, both Type I and II PSCs are highly effective in converting chlorine from the reservoir species ($ClONO_2$ and HCl) to photochemically active chlorine (Cl_2, HOCl, and $ClNO_2$).

TABLE 14.2
Reaction Probabilities on PSC Surfaces

Reaction	Reference	Ice	NAT
$ClONO_2 + H_2O$ (14.12)			
	Molina et al. (1987)	0.02[a]	
	Tolbert et al. (1987)	0.009[a]	
	Leu (1988a)	0.06[a]	
	Hanson and Ravishankara (1991a)	0.3	0.006
	Abbatt and Molina (1992a)	>0.02	0.001–0.002[b]
	Moore et al. (1990)		0.02[b]
	Leu et al. (1991)		0.001[b]
$N_2O_5 + H_2O$ (14.13)			
	Leu (1988b)	0.028[a]	
	Tolbert et al. (1988a)	>0.001[a]	
	Quinlan et al. (1990)	0.03[a]	0.015[a]
	Hanson and Ravishankara (1991a)	0.024	0.0006
$ClONO_2 + HCl$ (14.14)			
	Molina et al. (1987)	0.05–0.1	
	Leu (1988a)	0.06–0.3	
	Hanson and Ravishankara (1991a)	0.3	0.3
	Abbatt and Molina (1992a)	>0.2	>0.2[b]
	Moore et al. (1990)		0.06–1.0
	Leu et al. (1991)		0.1
$N_2O_5 + HCl$ (14.15)			
	Leu (1988b)	0.05	
	Tolbert et al. (1988a)	>0.003	
	Hanson and Ravishankara (1991a)	0.024	0.0032
$HOCl + HCl$ (14.16)			
	Abbatt and Molina (1992b)	0.16–0.24	0.17[b]
	Hanson and Ravishankara (1992)	>0.3	0.1

[a]Values that may reflect surface saturation and/or melting.
[b]Value on "water-rich" NAT. Values on "pure" NAT are considerably lower.

To summarize, the three requirements for massive ozone loss in the stratosphere appear to be (1) chlorine in the atmosphere, (2) cloud surfaces for heterogeneous chemistry, and (3) sunlight. The minimum polar stratospheric temperatures over the last 10 years indicate that the south polar region is always cold enough for PSCs, whereas the north polar region is sometimes cold enough for PSCs. If stratospheric temperatures decrease, as may be possible with tropospheric greenhouse warming, more PSCs could form and possibly more polar ozone loss could occur. In fact, during the 1994–1995 Arctic winter, temperatures in the stratosphere were unusually cold and large-scale ozone depletion was observed (Manney et al., 1996; Wirth and Renger, 1996).

HETEROGENEOUS CHEMISTRY ON STRATOSPHERIC SULFATE AEROSOLS (SSAs)

It has recently become apparent that stratospheric ozone loss is not confined to the cold polar regions. Data from the Total Ozone Mapping Spectrometer (TOMS) instrument on the Nimbus 7 satellite from 1978 to 1990 show an overall decrease of up to 0.8% per year in ozone at mid-latitudes (Stolarski et al., 1991) (see Figure 14.5). In addition, other TOMS data show that the mid-latitude ozone concentrations reached their lowest levels on record in 1992 and 1993 after the eruption of Mt. Pinatubo (Gleason et al., 1993). Since polar stratospheric clouds are not found globally, it may be possible that heterogeneous reactions on global stratospheric sulfate aerosols (SSAs) could be contributing to the observed decreasing ozone levels at mid-latitudes.

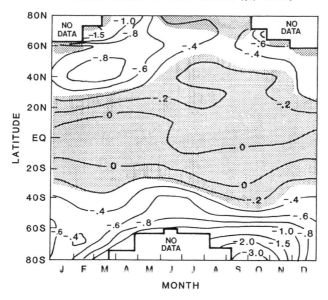

Figure 14.5. Global trends in total ozone from 1978 to 1990. Percent change per year was deduced from TOMS satellite data as a function of latitude and time of the year. The decrease in polar ozone during the Austral spring (September–November) and decreases in ozone at northern mid-latitudes (up to 0.8% per year) are observed. (Reprinted from Stolarski et al., 1991, with permission.)

Global SSAs are believed to be composed of concentrated solutions of sulfuric acid in water. The ambient temperature and water partial pressure dictate the concentration of the sulfuric acid droplets. As the temperature decreases, the aerosols absorb more water to maintain equilibrium. For stratospheric temperatures of 205–240 K, the droplet concentrations are between 60 and 80 wt % H_2SO_4.

The background stratospheric sulfate layer has a surface area density near 10^{-8} cm^2/cm^3 (an order of magnitude lower than PSCs) and is thought to be formed by the oxidation of (natural) OCS (Crutzen, 1976). However, the sulfuric acid particle loading of the stratosphere can increase by up to two orders of magnitude following a major volcanic eruption. For example, the eruption of El Chichón in 1982 injected SO_2 into the stratosphere, which rapidly oxidized to H_2SO_4 and became globally distributed. Ozone loss was detected at mid-latitudes after the El Chichón eruption, possibly due to heterogeneous chemistry on SSAs (Hofmann and Solomon, 1989). In 1991, the eruption of Mt. Pinatubo injected 2–3 times as much SO_2 into the stratosphere as El Chichón. The increase in aerosol surface area due to the Mt. Pinatubo eruption and higher levels of chlorine in the stratosphere may be responsible for the record low ozone concentrations observed recently by TOMS (Gleason et al., 1993).

Results on sulfuric acid surfaces representative of SSAs

Heterogeneous chemistry on sulfuric acid particles could directly produce active chlorine in reactions analogous to those that occur on PSCs:

$$ClONO_2 + H_2O \rightarrow HOCl + HNO_3 \tag{14.12}$$

$$ClONO_2 + HCl \rightarrow Cl_2 + HNO_3 \tag{14.14}$$

Several research groups have performed laboratory experiments to investigate the possibility of reaction (14.12) occurring on sulfuric acid droplets (Tolbert et al., 1988b; Hanson and Ravishankara, 1991b; Golden et al., 1994). Figure 14.6 shows that the reaction probability increases dramatically as the acid content decreases. For example, $\gamma = 0.0002$ for 75 wt % sulfuric acid but increases to $\gamma = 0.07$ for 40 wt % sulfuric acid at 223 K. Therefore, the $ClONO_2$ hydrolysis reaction becomes more important at colder stratospheric temperatures when the SSAs absorb more H_2O.

For reaction (14.14) on sulfuric acid surfaces to cause significant chlorine activation, HCl must be available for reaction in the sulfuric acid droplets. However, laboratory experiments have shown that HCl is soluble only in dilute H_2SO_4 solutions (Hanson and Ravishankara, 1991b; Watson et al., 1990; D. R. Worsnop, personal communication, 1992; Williams and Golden, 1993; Zhang et al., 1993). Thus, significant HCl uptake by H_2SO_4 aerosols is only expected at low temperatures.

Recent work has combined measurements of HCl solubility, diffusion, and reactivity to assess the importance of reaction (14.14) in the stratosphere (Hanson et al., 1994). These authors found that indeed reaction (14.14) could contribute to chlorine activation at cold temperatures. In addition, this reaction becomes increasingly important in chlorine activation for high stratospheric sulfuric acid loading such as that present after a major volcanic eruption.

Sulfuric acid aerosols can also promote heterogeneous reactions involving the oxides of nitrogen:

$$NO_2 + NO_3 + M \longleftrightarrow N_2O_5 + M \qquad \text{(gas-phase equilibrium)} \qquad (14.18)$$

$$N_2O_5 + H_2O \rightarrow 2HNO_3 \qquad \text{(heterogeneous)} \qquad (14.13)$$

The net effect of reactions (14.18) plus (14.13) is to convert NO_2 into HNO_3. This indirectly causes an increase in ClO because less NO_2 is available to form $ClONO_2$ via reaction (14.11). Thus, ClO levels can increase due to heterogeneous reactions involving only nitrogen species. Since both ClO_x and NO_x are important for ozone depletion, atmospheric models must be used to assess the net impact of reactions (14.18) and (14.13).

The laboratory measurements in Figure 14.7 show that the reaction probability for N_2O_5 with H_2O, reaction (14.13), on sulfuric acid solutions is quite high, near 0.1 (Mozurkewich and Calvert, 1988; Hanson and Ravishankara, 1991b; Van Doren et al., 1991; Fried et al., 1994; Golden et al., 1994). In contrast to the chlorine nitrate hydrolysis reaction, the N_2O_5 reaction on sulfuric acid surfaces appears to be essentially independent of sulfuric acid concentration and temperature and should occur readily under all stratospheric conditions. Because the conversion of N_2O_5 to HNO_3 has such a large reaction probability, this reaction could be important in lowering NO_2 and increasing ClO globally.

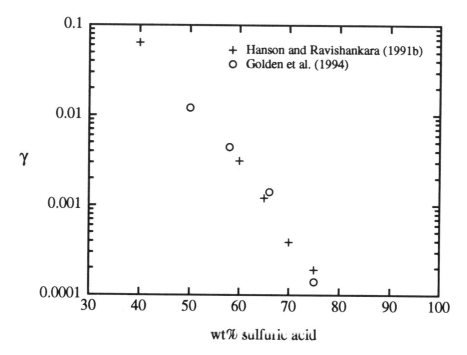

Figure 14.6. $ClONO_2$ reacting with H_2O on sulfuric acid surfaces. The log of the reaction probability, γ, is plotted against weight percent sulfuric acid. As the concentration of the acid decreases, the reaction probability increases.

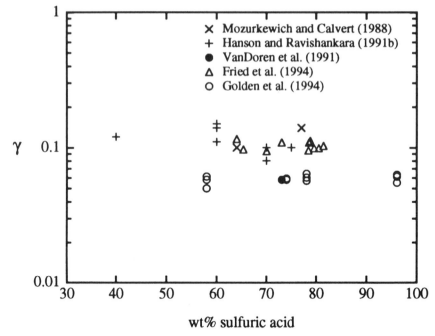

Figure 14.7. N$_2$O$_5$ reacting with H$_2$O on sulfuric acid surfaces. The log of the reaction probability, γ, is plotted versus the weight percent sulfuric acid. The reaction probability appears to be independent of acid concentration.

INCLUDING HETEROGENEOUS CHEMISTRY IN OZONE MODELS

Since heterogeneous chemistry affects ozone concentrations in the stratosphere, this new chemistry must be included in photochemical models. Brasseur et al. (1990; see also Chapter 16, this volume) have modeled the potential impact of heterogeneous chemistry on stratospheric ozone (see Figure 14.8). For the model results shown in Figure 14.8, the change in ozone between 1960 and 2010 was calculated assuming a 95% phaseout of CFCs by the year 2000. By using only gas-phase reactions, their model predicts changes of +0.6 to −0.8% over the 50-year period (see Figure 14.8a). As discussed previously, homogeneous chemistry alone is not able to predict the Antarctic ozone hole. However, when heterogeneous reactions on PSCs are included (Figure 14.8b), the ozone hole is evident in their calculations.

The observed decrease in the northern, mid-latitude ozone levels was not predicted by the model including gas-phase chemistry and heterogeneous chemistry on only PSCs. A closer approximation of the observed decrease could be simulated if heterogeneous chemistry on sulfuric acid aerosols, specifically reaction (14.13), was included in the model (see Figure 14.8c). Indeed, their model with PSC and sulfuric acid surface reactions projects loss of ozone at mid-latitudes on the order of current observations. However, the magnitude of the modeled ozone loss is still not as large as the observed ozone loss.

Detailed models must include heterogeneous chemistry to predict how the atmosphere will respond to future natural and anthropogenic events. For example, a fleet of

Figure 14.8. Model predictions of future ozone change. Parts a–c show the calculated total change in ozone for the period of 1960 to 2010 as a function of latitude and time of the year. The three parts a–c show the results of (a), homogeneous chemistry alone (b), homogeneous chemistry with heterogeneous reactions on PSCs and (c) homogeneous chemistry with heterogeneous reactions on both PSCs and SSAs. Observed decreases in ozone (Antarctic ozone hole and global trends) are only predicted when model includes heterogeneous chemistry on both PSCs and SSAs. Copyright 1990 Macmillan Magazines Limited.

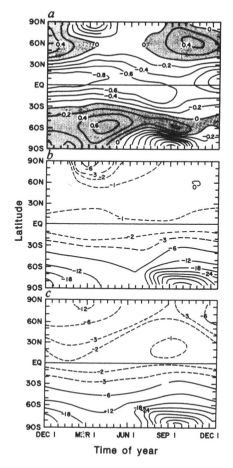

high-speed civil transports (HSCTs) is currently under consideration. The HSCTs will directly inject NO_x into the stratosphere, which could cause ozone destruction via the NO_x cycle, reactions (14.5) and (14.6). However, when heterogeneous chemistry is included in predicting the effects of HSCTs, NO_x is depleted by reaction (14.18) and (14.13), shutting off ozone destruction by the NO_x cycle but increasing the importance of the ClO_x and HO_x cycles. Thus, complete photochemical models including the temperature-dependent heterogeneous reaction rates are needed to predict the effect of HSCTs on stratospheric ozone. Such modeling efforts are currently underway by several research groups (Ko and Weisenstein, 1993).

One final difficulty in predicting ozone concentrations in stratospheric models is determining when PSC formation occurs. Although PSCs are thought to condense on the existing SSAs, the precise mechanism and formation temperature are not known. If the SSAs remain liquid at temperatures below 195 K, they most likely form supercooled ternary solutions (STS) composed of H_2SO_4, HNO_3, and H_2O (Carslaw et al., 1994; Tabazadeh et al., 1994). The reactivity of STS surfaces is very similar to that of supercooled sulfuric acid surfaces (Zhang et al., 1994, 1995; Ravishankara and Hanson, 1996). However, when the STS particles or SSAs crystallize, their reactivity switches to that of solid HNO_3/H_2O surfaces (Hanson and Ravishankara, 1993). Because the temperatures where the STS particles or SSAs freeze are not well established, the conditions for change in reactivity cannot be accurately incorporated in models. Furthermore, the amount of polar ozone loss appears to be strongly coupled to the extent of denitrification (removal of NO_x through particle sedimentation). Denitrification is a function of particle size, which in turn is affected by PSC composition. Therefore, determining the phase and composition of these particles is critical for understanding and predicting their impact on ozone.

ACKNOWLEDGMENTS

We gratefully acknowledge support from NASA (SASS-94-091) and the National Science Foundation (ATM-9321582). In addition, AMM received a NASA Global Change Research Graduate Fellowship (NGT-30132) and a DOE Distinguished Postdoctoral Fellowship in Global Change Research, and MAT was supported as a National Science Foundation Young Investigator and a Camille Dreyfus Teacher Scholar.

References

Abbatt, J. P. D., and Molina, M. J., 1992a, Heterogeneous interactions of $ClONO_2$ and HCl on nitric acid trihydrate at 202 K. *J. Phys. Chem.*, 96, 7674–7679.

Abbatt, J. P. D., and Molina, M. J., 1992b, The heterogeneous reaction of HOCl + HCl → Cl_2 + H_2O on ice and nitric acid trihydrate: reaction probabilities and stratospheric implications. *Geophys. Res. Lett.*, 19, 461–464.

Anderson, J. G., Brune, W. H., and Proffitt, M. H., 1989, Ozone destruction by chlorine radicals within the Antarctic vortex: the spatial and temporal evolution of ClO-O_3 anticorrelation based on *in situ* ER-2 data. *J. Geophys. Res.*, 94, 11,465–11,479.

Brasseur, G. P., Granier, C., and Walters, S., 1990, Future changes in stratospheric ozone and the role of heterogeneous chemistry. *Nature*, 348, 626–628.

Carslaw, K. S., Luo, B. P., Clegg, S. L., Peter, T., Brimblecombe, P., and Crutzen, P. J., 1994, Stratospheric aerosol growth and HNO_3 gas phase depletion from coupled HNO_3 and water uptake by liquid particles. *Geophys. Res. Lett.*, 21, 2479–2482.

Chapman, S., 1930, A theory of upper-atmosphere ozone. *Mem. R. Meteorol. Soc.*, 3, 103–125.

Crutzen, P. J., 1970, The influence of nitrogen oxides on the atmospheric ozone content, *Q. J. R. Meteorol. Soc.*, 96, 320–325.

Crutzen, P. J., 1976, The possible importance of CSO for the sulfate layer of the stratosphere. *Geophys. Res. Lett.*, 3, 73–76.

de Zafra, R. L., Jaramillo, M., Parrish, A., Solomon, P., Connor, B., and Barrett, J., 1987, High concentrations of chlorine monoxide at low altitudes in the Antarctic spring stratosphere: diurnal variation. *Nature*, 328, 408–411.

Farman, J. C., Gardiner, B. G., and Shanklin, J. D., 1985, Large losses of total ozone in Antarctica reveal seasonal ClOx/NOx interaction. *Nature*, 315, 207–210.

Fried, A., Henry, B. E., Calvert, J. G., and Mozurkewich, M., 1994, The reaction probability of N_2O_5 with sulfuric acid aerosols at stratospheric temperatures and compositions. *J. Geophys. Res.*, 99, 3517–3532.

Gleason, J. F., Bhartia, P. K., Herman, J. R., McPeters, R., Newman, P., Stolarski, R. S., Flynn, L., Labow, G., Larko, D., Seftor, C., Wellemeyer, C., Komhyr, W. D., Miller, A. J., and Planet, W., 1993, Record low global ozone in 1992. *Science*, 260, 523–526.

Golden, D. M., Manion, J. A., Reihs, C. M., and Tolbert, M. A., 1994, Heterogeneous chemistry on global stratospheric particulate: Reaction of $ClONO_2$ and N_2O_5 on sulfuric acid surfaces. In: J. G. Calvert, ed., *Chemrawn VII: Chemistry of the Atmosphere: The Impact of Global Change*, Blackwell Scientific Publications, Boston, pp. 39–52.

Hanson, D., and Mauersberger, K., 1988, Laboratory studies of the nitric acid trihydrate: implications for the south polar stratosphere. *Geophys. Res. Lett.*, 15, 855–858.

Hanson, D. R., and Ravishankara, A. R., 1991a, The reaction probabilities of $ClONO_2$ and N_2O_5 on polar stratospheric cloud materials. *J. Geophys. Res.*, 96, 5081–5090.

Hanson, D. R., and Ravishankara, A. R., 1991b, The reaction probabilities of $ClONO_2$ and N_2O_5 on 40 to 75% sulfuric acid solutions. *J. Geophys. Res.*, 96, 17,307–17,314.

Hanson, D. R., and Ravishankara, A. R., 1992, Investigation of the reactive and nonreactive processes involving $ClONO_2$ and HCl on water and nitric acid doped ice. *J. Phys. Chem.*, 96, 2682–2691.

Hanson, D. R., and Ravishankara, A. R., 1993, The reaction of $ClONO_2$ with HCl on NAT, NAD, and frozen sulfuric acid and hydrolysis of N_2O_5 and $ClONO_2$ on frozen sulfuric acid. *J. Geophys. Res.*, 98, 22,931–22,936.

Hanson, D. R., Ravishankara, A. R., and Solomon, S., 1994, Heterogeneous reactions in sulfuric acid aerosols: a framework for model calculations. *J. Geophys. Res.*, 99, 3615–3629.

Hofmann, D. J., and Solomon, S., 1989, Ozone destruction through heterogeneous chemistry following the eruption of El Chichón. *J. Geophys. Res.*, 94, 5029–5041.

Hofmann, D. J., Oltmans, S. J., Lathrop, J. A., Harris, J. M., and Voemel, H., 1994, Record low ozone at the South Pole in the spring of 1993. *Geophys. Res. Lett.*, 21, 421–424.

Johnston, H. S., 1971, Reduction of stratospheric ozone by nitrogen oxide catalysts from supersonic transport exhaust. *Science*, 173, 517–522.

Keyser, L. F., Moore, S. B., and Leu, M.-T., 1991, Surface reaction and pore diffusion in flow-tube reactors. *J. Phys Chem.*, 95, 5496–5502.

Ko, M. K. W., and Weisenstein, D. K., 1993, Ozone response to aircraft emissions: Sensitivity to heterogeneous reactions. In: R. S. Stolarski and H. L. Wesoky, eds., *The Atmospheric Effects of Stratospheric Aircraft: A Second Program Report*, NASA Reference Publication 1293, pp. 83–113.

Leu, M.-T., 1988a, Laboratory studies of sticking coefficients and heterogeneous reactions important in the Antarctic stratosphere. *Geophys. Res. Lett.*, 15, 17–20.

Leu, M.- T., 1988b, Heterogeneous reactions of N_2O_5 with H_2O and HCl on ice surfaces: implications for Antarctic ozone depletion. *Geophys. Res. Lett.*, 15, 851–854.

Leu, M.- T., Moore, S. B., and Keyser, L. F., 1991, Heterogeneous reactions of chlorine nitrate and hydrogen chloride on Type I polar stratospheric clouds. *J. Phys. Chem.*, 95, 7763–7771.

Manney, G. L., Froidevaux, L., Waters, J. W., Santee, M. L., Read, W. G., Flower, D. A., Jarnot, R. F., and Zurek, R. W., 1996, Arctic ozone depletion observed by UARS MLS during the 1994–95 winter. *Geophys. Res. Lett.*, 23, 85–88.

Molina, M. J., and Rowland, F. S., 1974, Stratospheric sink for chlorofluoromethanes: chlorine atom-catalyzed destruction of ozone, *Nature*, 249, 810–812.

Molina, L. T., and Molina, M. J., 1987, Production of Cl_2O_2 from the self-reaction of the ClO radical. *J. Phys. Chem.*, 91, 433–436.

Molina, M. J., Tso, T.-L., Molina, L. T., and Wang, F. C.-Y., 1987, Antarctic stratospheric chemistry of chlorine nitrate, hydrogen chloride and ice: release of active chlorine. *Science*, 238, 1253–1257.

Moore, S. B., Keyser, L. F., Leu, M.-T., Turco, R. P., and Smith, R. H., 1990, Heterogeneous reactions on nitric acid trihydrate. *Nature*, 345, 333–335.

Mozurkewich, M., and Calvert, J. G., 1988, Reaction probability of N_2O_5 on aqueous aerosols. *J. Geophys. Res.*, 93, 15,889–15,896.

Quinlan, M. A., Reihs, C. M., Golden, D. M., and Tolbert, M. A., 1990, Heterogeneous reactions on model polar stratospheric cloud surfaces: reaction of N_2O_5 on ice and nitric acid trihydrate. *J. Phys. Chem.*, 94, 3255–3260.

Ravishankara, A. R., and Hanson, D. R., 1996, Differences in the reactivity of Type I polar stratospheric clouds depending on their phase. *J. Geophys. Res.*, 101, 3885–3890.

Solomon, S., Garcia, R. R., Rowland, F. S., and Wuebbles, D. J., 1986, On the depletion of Antarctic ozone. *Nature*, 321, 755–758.

Solomon, S., Mount, G. H., Sanders, R. W., Schmeltekopf, A. L., 1987, Visible spectroscopy at McMurdo Station, Antarctica 2. Observations of OClO. *J. Geophys. Res.*, 92, 8329–8338.

Stolarski, R. S., Bloomfield, P., McPeters, R. D., and Herman, J. R., 1991, Total ozone trends deduced from Nimbus 7 TOMS data. *Geophys. Res. Lett.*, 18, 1015–1018.

Tabazadeh, A., and Turco, R. P., 1993, A model for heterogeneous chemical processes on the surfaces of ice and nitric acid trihydrate particles. *J. Geophys. Res.*, 98, 12,727–12,740.

Tabazadeh, A., Turco, R. P., Drdla, K., Jacobson, M. Z., and Toon, O. B., 1994, A study of Type I polar stratospheric cloud formation. *Geophys. Res. Lett.*, 21, 1619–1622.

Tolbert, M. A., Rossi, M. J., Malhotra, R., and Golden, D. M., 1987, Reaction of chlorine nitrate with hydrogen chloride and water at Antarctic stratospheric temperatures. *Science*, 238, 1258–1260.

Tolbert, M. A., Rossi, M. J., and Golden, D. M., 1988a, Antarctic ozone depletion chemistry: reactions of N_2O_5 with H_2O and HCl on ice surfaces. *Science*, 240, 1018–1021.

Tolbert, M. A., Rossi, M. J., and Golden, D. M., 1988b, Heterogeneous interactions of chlorine nitrate, hydrogen chloride, and nitric acid with sulfuric acid surfaces at stratospheric temperatures. *Geophys. Res. Lett.*, 15, 847–850.

Turco, R. P., Toon, O. B., and Hamill, P., 1989, Heterogeneous physicochemistry of the polar ozone hole. *J. Geophys. Res.*, 94, 16,493–16,510.

Van Doren, J. M., Watson, L. R., Davidovits, P., Worsnop, D. R., Zahniser, M. S., and Kolb, C. E., 1991, Uptake of N_2O_5 and HNO_3 by aqueous sulfuric acid droplets. *J. Phys. Chem.*, 95, 1684–1689.

Watson, L. R., Van Doren, J. M., Davidovits, P., Worsnop, D. R., Zahniser, M. S., and Kolb, C. E., 1990, Uptake of HCl molecules by aqueous sulfuric acid droplets as a function of acid concentration. *J. Geophys. Res.*, 95, 5631–5638.

Watson, R. T., Geller, M. A., Stolarski, R. S., and Hampson, R. F., 1986, Present state of knowledge of the upper atmosphere: an assessment report—processes that control ozone and other climatically important tract gases. NASA Reference Publication 1162, p. 20.

Williams, L. R., and Golden, D. M., 1993, Solubility of HCl in sulfuric acid at stratospheric temperatures. *Geophys. Res. Lett.*, 20, 2227–2230.

Wirth, M. and Renger, W., 1996, Evidence of large scale ozone depletion within the arctic polar vortex 94/95 based on airborne LIDAR measurements. *Geophys. Res. Lett.*, 23, 813–816.

World Meteorological Organization (WMO) Global Ozone Research and Monitoring Project, 1994, *Scientific Assessment of Ozone Depletion: 1994*, 37, p. 1.47.

Zhang, R., Wooldridge, P. J., and Molina, M. J., 1993, Vapor pressure measurements for the $H_2SO_4/HNO_3/H_2O$ and $H_2SO_4/HCl/H_2O$ systems: incorporation of stratospheric acids into background sulfate aerosols. *J. Phys. Chem.*, 97, 8541–8548.

Zhang, R., Leu, M.-T., and Keyser, L. F., 1994, Heterogeneous reactions of $ClONO_2$, HCl, and HOCl on liquid sulfuric acid surfaces. *J. Phys. Chem.*, 98, 13,563–13,574.

Zhang, R., Leu, M.-T., and Keyser, L. F., 1995, Hydrolysis of N_2O_5 and $ClONO_2$ on the $H_2SO_4/HNO_3/H_2O$ ternary solutions under stratospheric conditions. *Geophys. Res. Lett.*, 22, 1493–1496.

The Stratospheric Aerosol and Its Impact on Stratospheric Chemistry

JAMES CHARLES WILSON

ABSTRACT. A persistent mist of fine particles is found in the lower stratosphere. Away from the winter pole, these particles consist primarily of sulfuric acid and water. Explosive volcanic eruptions often provide the sulfur. Between eruptions, anthropogenic and natural sources of long-lived sulfur species appear sufficient to maintain the sulfate layer, but they have not been quantified. Near the winter pole, polar stratospheric clouds form from nitric acid and water. The details of the composition and phase of the nitric acid–water clouds remain to be discovered, but they are known to form at temperatures above the frost point. Below the frost point, clouds are predominately water ice. The size, composition and dynamics of the aerosol and cloud particles help determine their lifetimes and chemical impact.

The stratospheric aerosol has a very significant impact on the chemistry of the stratosphere. In the polar winter, reactions on cloud particles free the chlorine in hydrochloric acid and chlorine nitrate. When the polar air is exposed to sunlight, this chlorine participates in rapid, catalytic destruction of ozone. In mid-latitudes, the hydrolysis of N_2O_5 on the surface of sulfate particles leads to the reduction of NO_x and the increase of ozone-destroying chlorine and hydrogen radicals. The performance of models describing the chemistry of the lower stratosphere is dramatically improved when reactions on surfaces of aerosol and cloud particles are included. The new models have been tested against coordinated measurements of many of the key species made simultaneously from aircraft.

BACKGROUND

The term "aerosol" refers to a mixture of solid or liquid particles dispersed in a gas. In still air, particles will settle out of the gas due to gravity. However, the atmosphere is sufficiently well stirred so that small particles remain suspended for long times. Away from the winter pole, stratospheric aerosol particles consist primarily of supercooled solutions of sulfuric acid and water. In the cold air over the winter pole, polar stratospheric clouds (PSCs) can form when supersaturated species condense. Two main types of PSCs are recognized. Type I PSCs consist mainly of nitric acid and water which have probably condensed on a nucleus of sulfuric acid. Type II PSCs consist mainly of water ice. The diameters of these particles range from about 10 nm (in the case of newly formed sulfate aerosol) to hundreds of microns (in the case of large ice crystals).

It has been recently learned that chemical reactions occurring on stratospheric aerosol and cloud particles play central roles in ozone loss. The rates at which these reactions occur often depend upon the composition of the particles, and their net effect often increases as the amount of aerosol surface area increases. This chapter treats the characteristics and sources of the stratospheric aerosol, the reactions that involve the particles, and the effects of the reactions on the chemistry of the stratosphere. Some basic facts about the atmosphere provide the context for this discussion, and they are recalled in this section.

The Stratosphere and the Troposphere

The troposphere and stratosphere are the two lowest layers in the atmosphere. The troposphere extends upward from the ground and is characterized by strong vertical mixing. Tropospheric temperatures generally decrease with altitude. The temperature becomes nearly constant with altitude at altitudes near 18 km in the tropics and 9 km at the winter pole. This region is referred to as the tropopause. Above it lies the stratosphere, where vertical mixing is much weaker and temperature generally increases with altitude.

Atmospheric circulation takes air into the stratosphere from below in the tropical regions. The subsequent motions are upward and poleward. Air descends at high latitudes, where it cools radiatively. It leaves the stratosphere at mid-latitudes and at the winter pole where descent is strongest. Since air enters the stratosphere through the cold tropical tropopause, much of the water is removed and the stratosphere is quite dry compared to the troposphere. Chemicals entering the stratosphere from the troposphere must survive long enough to be transported to the tropical tropopause. Many species emitted in the troposphere are oxidized and condense or removed in precipitation before reaching the stratosphere. Chloroflurocarbons (CFCs) have very long lifetimes in the troposphere. They survive transport to the tropical tropopause, where they enter the stratosphere and play a critical role in ozone chemistry.

The Antarctic ozone hole occurs in the vortex which forms in the winter. As the polar air cools and descends, air from lower latitudes flows toward the pole. These motions establish a jet stream at a latitude near 70° and an altitude near 20 km. This high-speed jet defines the equatorward edge of the polar vortex. Transport across the jet is weak and the air in the vortex is somewhat isolated from the rest of the stratosphere while the vortex persists. The polar vortex breaks up in the spring when warming occurs. In the Northern Hemisphere, a warmer, less symmetrical and shorter-lived vortex forms.

Stratospheric ozone

The stratospheric ozone layer forms as a result of photolysis of molecular oxygen by solar radiation having wavelengths less than 242 nm. The resulting oxygen atoms rapidly combine with oxygen molecules to form ozone. Ozone itself absorbs ultraviolet (UV) radiation and shields the lower atmosphere and earth's surface from these energetic photons. This shielding protects the biosphere from UV radiation and is primarily responsible for the thermal structure of the stratosphere. Therefore, ozone formation and destruction are of considerable interest. Ozone formation occurs continuously in the upper stratosphere and at tropical latitudes. It is transported throughout the stratosphere. Ozone is destroyed continuously at higher latitudes and lower altitudes in cycles involving oxides of nitrogen, odd-hydrogen radicals (OH and HO_2), and reactive chlorine. The steady-state abundance of ozone is established by the balance between the formation and destruction (see Chapter 13, this volume).

Impact of reactions on particles

Reactions occurring on the surface of polar stratospheric cloud particles dramatically shift the population balance among the chlorine compounds. In the absence of these reactions, most chlorine is found in inactive compounds which do not participate in ozone destruction. In the presence of these heterogeneous reactions, much of the chlorine in the polar vortex is in active or ozone-destroying forms. The presence of the PSCs and the effect of these surface reactions make it possible to understand why the ozone hole occurs over the pole and in the spring.

Reactions on sulfate aerosol particles in mid-latitudes affect the partitioning of the nitrogen atoms amoung a family of compounds and radicals. The abundance of HNO_3 is increased at the expense of NO and NO_2. This shift increases the amount of active chlorine and the amount of HO_2 relative to that which would be present if only gas-phase chemistry were important. Thus, these heterogeneous reactions tend to decrease the ozone loss due to oxides of nitrogen and increase the loss due to chlorine and hydrogen radicals.

CHARACTERISTICS AND ORIGINS OF THE STRATOSPHERIC AEROSOL

The Sulfate Aerosol Layer

Measurements made by the U.S. National Aeronautics and Space Administration (NASA) SAGE II instrument from a satellite show that the earth is shrouded by a thin mist of particles which extends about 15 km above the tropopause. The aerosol extinction coefficient (km^{-1}) for light having wavelength of 1 μm is plotted in Figure 15.1. In light to moderate aerosol loading, the extinction coefficient is the inverse of the distance that the beam must travel in order to be attenuated by $1/e$. The data in Figure 15.1a were acquired prior to the eruption of Mt. Pinatubo and represents what is thought to be a nonvolcanic stratosphere.

It is likely that the most common stratospheric particles consist of sulfuric acid and water in equilibrium with the water vapor. At normal stratospheric temperatures they are thought to be supercooled, liquid solutions. A number of indirect arguments concerning the state of the aerosol at colder temperatures have been advanced (Dye et al., 1992), but direct determination of the state has not been done. This is of particular interest since these particles probably serve as the nuclei for the formation of polar stratospheric clouds and their effectiveness as nuclei may depend upon their state.

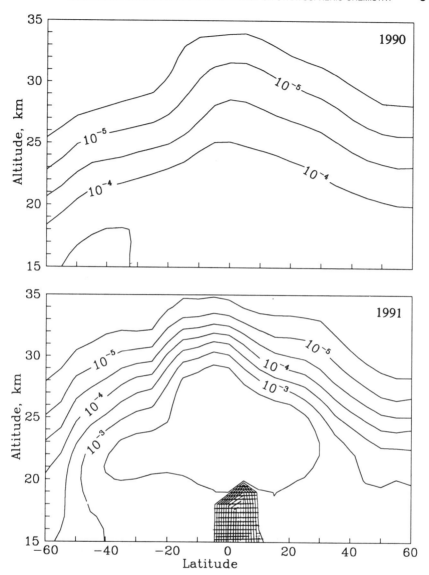

Figure 15.1.   Aerosol extinction coefficient (km^{-1}) measured at wavelength of 1 μm with the NASA SAGE II instrument. These satellite measurements show amount of attenuation of direct solar radiation caused by aerosol. In light to moderate aerosol loadings, the extinction coefficient equals the inverse of the distance that the beam must travel to be attenuated by $1/e$. Measurements were made from mid-September to mid-October 1990 (a) and from mid-September to mid-October 1991 (b). Crosshatched area indicates that extinction exceeded the instrument range. (Reprinted from Poole et al., 1991.)

Measurements of the sulfuric acid mass fraction have been reported in the literature (Rosen, 1971; Hofmann and Solomon, 1989) and are in agreement with compositions derived from thermodynamic extrapolations of laboratory data (Steele and Hamill, 1981). The measurements were made by exposing stratospheric particles to increasing temperatures until they could no longer be detected by an optical counter capable of measuring particles as small as 0.3 μm in diameter. The cause of the particles disappearance was assumed to be boiling, and the composition of the particles was determined from the boiling points of sulfuric acid and water solutions. However, heat and mass transfer rates in the gas phase and in the particles have not been investigated in these cases. It is possible that water escapes preferentially as the particles are heated (Wilson et al., 1992) and that the particles disappeared as the result of evaporation and not boiling. Thus the thermodynamic extrapolations providing the acid and water mass fractions of these particles may require additional confirming studies.

Size distributions and dynamics of the aerosol

The impact of particles often depends upon their size. Size distributions of stratospheric particles have been measured by collecting samples with impactor samplers and returning them to the laboratory for electron microscopic analysis (Farlow et al., 1979; Gras and Laby, 1981; Woods and Chuan, 1983). *In situ* measurements have been made with balloon-borne optical aerosol spectrometers (Hofmann et al., 1975). Stratospheric measurements from NASA ER-2 aircraft have involved the recently developed the focused cavity aerosol spectrometer (FCAS; Jonsson et al., 1995), the forward scattering spectrometer probe (FSSP; Baumgardner et al., 1992), and condensation nucleus counter (CNC; Wilson et al., 1983). This array of instruments covers the diameter range from 0.02 to 20 μm. The aerosol spectrometers measure the light scattered by individual particles as they are passed through a laser beam and determine the size in the diameter range from about 80 nm to 20 μm. The CNC subjects particles to a saturated vapor and allows them to grow to a size which permits optical detection and counting. The CNC provides a measure of total concentration. The number of particles in the 20- to 80-nm-diameter range is determined by subtraction.

A size distribution measured with these instruments (Figure 15.2) at 19.2 km altitude nearly 3 months after the eruption of Mt. Pinatubo illustrates many features of the stratospheric aerosol and processes that affect it. The number distribution (Figure 15.2a) extends over several decades in size and concentration. The distribution function, $dN/d(\log(Dp))$, is plotted on the y axis. dN is the number of particles-cm^{-3} in the diameter interval Dp to $Dp + dDp$, where dDp is an infinitesimal increment in diameter, Dp.

In the linear-log form, the area under the distribution curve between any two diameters is proportional to the concentration of the aerosol property present in the diameter interval. The number distribution (Figure 15.2b) is dominated by particles in the 0.02-μm to 0.1-μm-diameter range. The area under the distribution curve is much greater in the diameter range from 0.02 to 0.07 μm than it is for any other diameter range. These new particles were formed in the three months since the eruption of Mt. Pinatubo by condensing H_2SO_4 and water. The acid vapor was produced in the gas phase by the oxidation of SO_2 injected by the eruption. In order to form new particles, the supersaturation of H_2SO_4 must be quite high. The rapid oxidation of the injected SO_2 provides the needed supersaturation (McKeen et al., 1984). The mode of new, small particles is referred to as the nuclei mode (Whitby, 1978). Since particles collide with one another in proportion to the square of their concentration and mobility, these small, highly mobile particles suffer rapid coagulation. Thus this mode of new particles decays over a period of a few months.

Typically, much more of the H_2SO_4 vapor condenses on preexisting aerosol surface than goes into new particle formation. All particles grow as a result of this con-

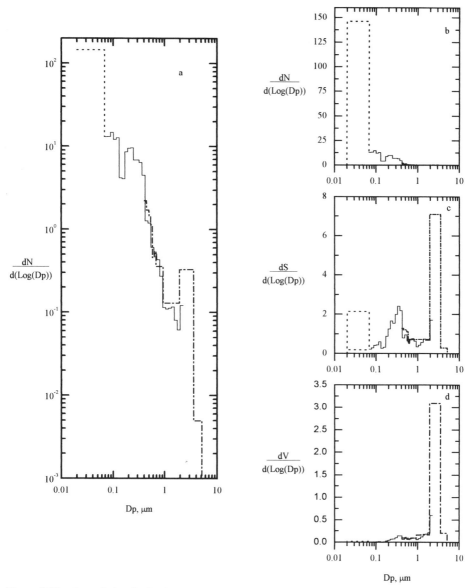

Figure 15.2. Aerosol size distribution measured at an altitude of 19 km, at 31°N two months after the eruption of Pinatubo. Solid line shows FCAS data; the dot–dashed line is FSSP data, and the dotted line is determined from the difference between the CNC and the FCAS. See text for explanation. (Adapted from Wilson et al., 1993.)

densation. Coagulation and condensational growth will eventually move many of the nuclei mode particles into the mode seen between 0.1 and 1.0 μm. This mode, called the accumulation mode, dominates the aerosol surface and volume in periods between eruptions and is seen most clearly in the surface distribution (Figure 15.2c). The surface distribution is determined by multiplying the number distribution by πD_p^2. Thus the larger particles, though fewer in number, contribute significantly to the total surface area concentration of the aerosol. Since size information was not acquired for particles in the 0.02- to 0.08-μm-diameter range, only the minimum and maximum possible values of surface in this size range are indicated in the plot.

The volume distribution, Figure 15.2d, is determined by multiplying the number distribution by $\pi Dp^3/6$ and is dominated by a third population of particles found at diameters greater than 1 μm, even though these particles are orders of magnitude less numerous than the smaller ones. This mode contains particles which are large enough to have significant fall velocities. Some of them consist of crustal materials injected at higher altitudes during the eruption. In the months between the eruption and measurement, those particles fell and were coated with condensing sulfuric acid. Others have grown from accumulation mode particles at higher altitudes and have fallen to the altitude at which they were measured. These processes have been described quantitatively in models of the stratospheric aerosol (Turco, 1982) which produce many of the observed features of these distributions.

The nonvolcanic sulfate aerosol layer

The stratospheric aerosol layer persists between volcanic injections and appears to be sustained by tropospheric sources of sulfur gases. OCS is emitted naturally and an-

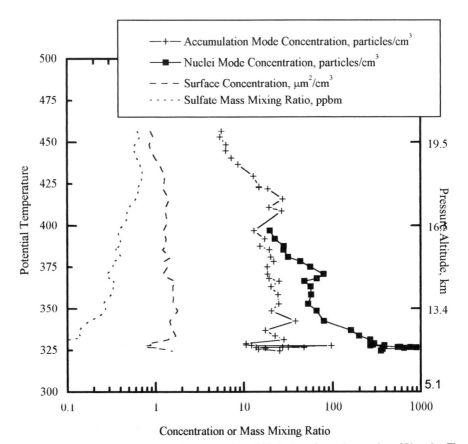

Figure 15.3. Profiles of aerosol properties measured at mid-latitudes prior to the eruption of Pinatubo. The nuclei mode is not seen above 16.5 km in this profile. (Adapted from Wilson et al., 1993.)

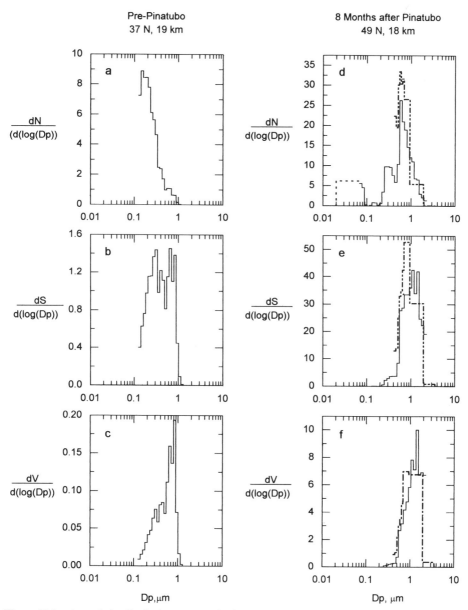

Figure 15.4.   Aerosol size distributions measured prior to (a–c) and after (d–f) the eruption of Pinatubo at similar altitudes and mid- latitudes. (Adapted from Wilson et al., 1993.)

thropogenically and is robust enough to survive tropospheric chemical processes and be transported to the tropical tropopause (Crutzen, 1976). Although most SO_2 emitted at ground level is removed by chemical processes in the troposphere, some may make it to the stratosphere. Direct injection of SO_2 by aircraft may also be important in sustaining or increasing the stratospheric aerosol loading between eruptions (Hofmann, 1990b).

Some of the sulfur gases reaching the vicinity of the tropical tropopause are converted to H_2SO_4 vapor which co-condenses with water vapor to form new particles in this cold, relatively humid region. The upwelling air moves these new particles to higher altitudes in the stratosphere along with the remaining precursor gases. In the process of this transport, the number of small particles is reduced by coagulation (Brock et al., 1995) and the remaining sulfur gases are oxidized and that vapor condenses, causing all particles to grow. If there is sufficient aerosol surface present, the H_2SO_4 vapor will rapidly diffuse to the existing surface and prevent the supersaturation from reaching the values required to form new particles.

The number concentration of particles decreases with altitude and the mass mixing ratio increases. Horizontal transport and mixing causes these features to be seen at mid-latitudes as well (Figure 15.3a). A nuclei mode is seen near the tropopause but disappears with altitude. At altitudes near 20 km, the accumulation mode dominates the number, surface, and volume distributions (Figure 15.4a). The mass mixing ratio is seen to increase with height, and the surface is nearly constant (Figure 15.3a).

Production of new particles has also been observed in the descending air in the polar vortices (Hofmann et al., 1988; Hofmann, 1990a; Wilson et al., 1990). It is possible that these particles result from the recondensation of sulfuric acid vapor which evaporated from particles as the circulation carried them to high, warmer altitudes before descending into the vortex. This source of new particles appears to be much less important than the tropical tropopause.

The volcanic sulfate aerosol

Explosive volcanic eruptions deposit gas and crustal material in the stratosphere and greatly increase the stratospheric aerosol loading. A recent time history of volcanic impact is displayed in Figure 15.5. The data were acquired with an aerosol laser radar

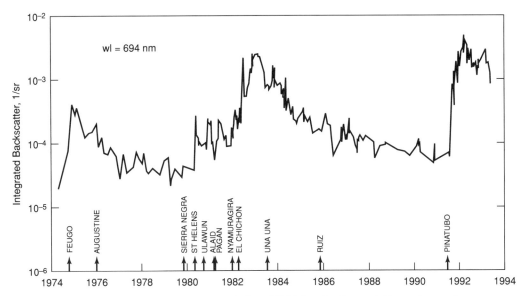

Figure 15.5.   Aerosol lidar measurements made from NASA Langley Research Center. Laser pulses are shot upward and the light scattered back is recorded. The total light scattered back is plotted as a function of time. The large enhancements in light scattering are due to the aerosol resulting from volcanic eruptions. The rather slow decay of the injected volcanic material is also evident. (Adapted from Poole et al., 1991.)

(lidar) operated at the NASA Langley Research Center (LaRC). The eruptions of El Chichon in 1982 and Pinatubo in 1991 had significant impact on the stratosphere. It is believed that eruptions of this size occur a few times per century. The large effect that Pinatubo had on the extinction observed by SAGE II is seen by comparing the measurements of Fall 1991 (Figure 15.1b) with those of Fall 1990 (Figure 15.1a). Extinction is roughly proportional to aerosol surface area concentration. It is clear that aerosol surface area was dramatically greater following the eruption.

The evolution of the aerosol properties in the 18- to 20-km altitude range following the eruption of Mt. Pinatubo in June 1991 is shown in Figure 15.6 (Jonsson et al., 1996). New particles formed soon after the eruption decay by coagulation, and the highest concentrations are seen to disappear within months after the eruption. Aerosol surface concentration grew by factors of 20 following the eruption as injected SO_2 was converted to H_2SO_4 and condensed on preexisting particles. The volume grew by factors of 50 and decayed over a period of a few years. The impact of the coagulation and condensation of vapor is seen in size distributions measured 8 months after the eruption at 18-km altitude (Figure 4b). At that altitude in a nonvolcanic stratosphere, the particles are typically 0.1–0.2 μm in diameter (Figure 15.4a). However, at the height of Pinatubo's impact, the particles were several times larger. These large particles account for the added surface and volume concentrations and have significant sedimentation rates so they fall to lower altitudes in the atmosphere.

The removal of particles from lower levels in the stratosphere is accomplished through atmospheric motions as well as sedimentation (Jonsson et al., 1996). The impact of these changes on stratospheric chemistry are discussed below.

Polar Stratospheric Clouds (PSCs)

Clouds have been observed in the polar stratosphere for over a century (Stanford and Davis, 1974). Satellite and lidar observations of PSCs (McCormick et al., 1981) prompted a number of artful thermodynamic extrapolations predicting the composition of the clouds (Toon et al., 1986; Crutzen and Arnold, 1986; McElroy et al., 1986; Solomon et al., 1986). The association of the PSCs with ozone depletion in the Antarctic led to the proposal that they were perturbing the nitrogen and chlorine reservoirs in the region.

Type I PSCs

Laboratory studies (see Chapter 14, this volume) have determined the phase diagrams for several candidate PSC substances. These include nitric acid trihydrate (NAT), nitric acid dihydrate (NAD), and the ternary solution of nitric acid, sulfuric acid and water. Figure 15.7 shows the phase diagram of the nitric acid trihydrate system including isotherms (Hanson and Mauersberger, 1988). If the ambient water vapor pressure and nitric acid vapor pressure fall in the trihydrate region, then nitric acid trihydrate (NAT) can exist in equilibrium with the gases at the temperature whose isotherm passes through the point defined by the vapor pressures. The equilibrium point defines the saturation temperature for the vapors with respect to NAT. At higher temperatures, the vapors are not saturated with respect to NAT and any NAT would evaporate. At colder temperatures, the vapors would be supersaturated with respect to NAT.

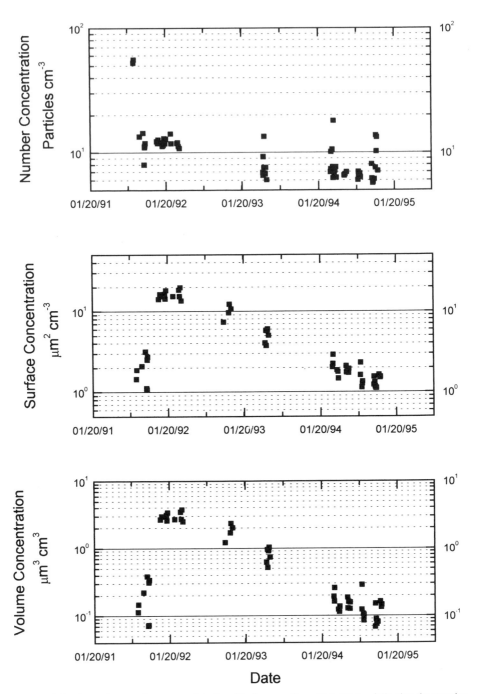

Figure 15.6. Aerosol properties measured in the altitude range from 18 to 20 km following the eruption of Mt. Pinatubo in the Northern Hemisphere. (Adapted from Jonsson et al., 1996.)

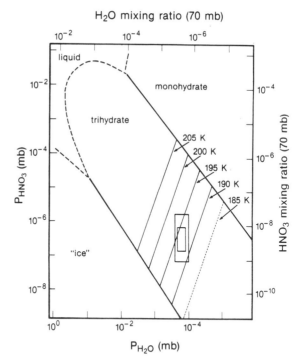

Figure 15.7. Phase diagram of nitric acid trihydrate as determined by Hanson and Mauersberger (1988). Mixing ratios corresponding to stratospheric conditions have been plotted on the top and right. (Reprinted from Fahey et al., 1989a. Copyright by the American Geophysical Union.)

The larger box includes the range of observed mixing ratios of water and nitric acid throughout the Airborne Antarctic Ozone Experiment (AAOE) of 1987 (Fahey et al., 1989a). The smaller box includes the range of mixing ratios observed coincident with PSCs during a particular flight. The observed abundances of nitric acid and water imply that NAT could have existed in equilibrium at temperatures in the approximate range from 189 to 197 K in the air encountered in AAOE. Temperatures this low occur frequently in the polar winter.

Type I clouds contain nitric acid (Pueschel et al., 1989; Gandrud et al., 1989; Pueschel et al., 1990). However, the composition or crystal structure of Type I PSCs has not been determined since sampling from aircraft involves heating (Wilson et al., 1992) and no one has yet devised a sampling technique which leaves the composition and phase intact long enough to do the analysis.

Reactive nitrogen, NO_y, is defined as the following sum:

$$NO_y = NO_2 + NO + HNO_3 + ClONO_2 + 2N_2O_5 + \cdots$$

Simultaneous measurements of reactive nitrogen, water vapor, temperature, pressure, and aerosol made from an aircraft have been interpreted to make a strong case for the presence of NAT in an Antarctic PSC (Fahey et al., 1989a). Figure 15.8e shows the ratio of measured NO_y to the saturation vapor pressure of HNO_3 over NAT, $PS(HNO_3)$, where $PS(HNO_3)$ was determined from the relationships shown in Figure 15.7. Figures 15.8a-8d show the fraction of particles appearing in each of several size ranges. Since this fraction is determined from independent measurements, each with finite errors, it is occasionally larger than one.

Prior to entering the cloud, nearly all the particles were smaller than 0.81 μm and the ratio $NO_y/PS(HNO_3)$ was less than one. In the absence of clouds, this ratio approximates the saturation ratio of HNO_3 since HNO_3 is the main constituent of NO_y in this case. In the presence of particles containing NO_y, this ratio increases dramati-

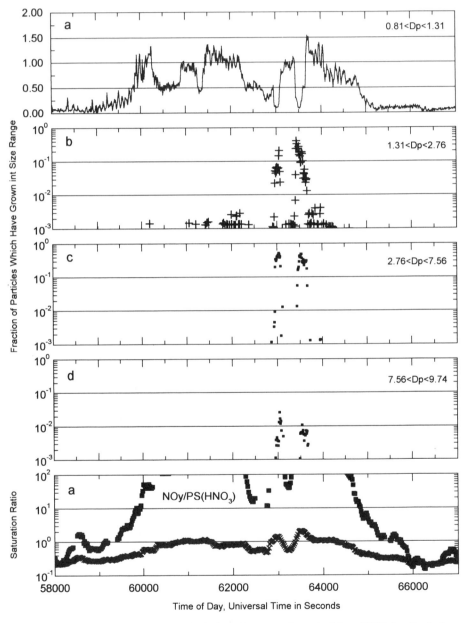

Figure 15.8. Fraction of particles appearing in four size ranges determined from FSSP size distributions (the numerator) and the CNC (the denominator). Measurement errors permit the ratio to exceed 1 in (a). The saturation of water vapor over ice and $NO_y/PS(HNO_3)$ are also shown (e). Horizontal axis is time in GMT seconds. The measurement was made while flying through a PSC over Antarctica in 1987. (Adapted from Wilson et al., 1989.)

cally since the NO_y on the particles is amplified in the signal due to the inlet used on the NO_y instrument (Fahey et al., 1989a). The appearance of particles in the 0.81- to 1.31-μm range is associated with the dramatic increases in NO_y/PS(HNO_3). Inside the cloud, many of the particles grew to diameters in the 0.81- to 1.31-μm range. These sizes are consistent with the condensation of the available NO_y and associated water onto the available number of sulfate particles. The appearance of the PSC particles was consistent with the equilibrium thermodynamics of NAT.

In a number of other observations, significant amounts of cloud material did not appear even though the inferred nitric acid vapor pressures, the ambient temperatures, and the measured water pressures would have permitted NAT to exist in equilibrium with the gases (Schlager et al., 1990; Kawa et al., 1992a; Dye et al., 1992). Saturation of HNO_3 over NAT reaching 10 was frequently reached before significant development of clouds occurred. At these elevated saturations, particles containing nitrogen species were formed at temperatures still above the frost point. Some of these observations seem consistent with the behavior of a ternary solution of water, nitric acid, and sulfuric acid (Carslaw et al., 1994; Tabazadeh et al., 1994), but no single composition explains all the observations and the exact composition of Type I PSCs is still unknown.

Type II PSCs

The region of the curve in Figure 15.7 labeled "ice" refers to water ice with some nitric acid in it. The line separating the ice and NAT regions is very close to the line defining the frost point over pure water ice. Therefore, as a parcel is cooled in the polar winter, NAT would be expected to form at temperatures warmer than the frost point. If temperatures reached the frost point, the formation of ice particles is expected. These are referred to as Type II PSCs. They are typically larger than Type I PSCs because there is usually much more water available than nitric acid.

Around 63,000 GMTs and 63,500 GMTs in Figure 15.8, water vapor had become saturated with respect to ice, as is shown by the H_2O saturation curve, and large particles appeared. Although the composition of the particles was not directly determined by measurement, the inference that they are ice is hard to avoid. They contained too much mass to be HNO_3, and captured crystals had shapes appropriate for ice (Goodman et al., 1989).

In a number of instances, large particles, presumed to be ice, have been observed in air whose temperatures were above the frost point (Gandrud et al., 1990; Wofsy et al., 1990; Deshler et al., 1994). The particles would be expected to sublimate under these conditions. Some of these observations suggest that the particles had formed at higher altitudes in colder air and had settled into warmer air. Physical mechanisms to slow the sublimation have been suggested. Slowing the sublimation would increase their lifetime and permit them to settle a greater distance in air which is too warm for the ice to exist in equilibrium.

Denitrification and dehydration by PSCs

Air parcels from which nitric acid has been removed by some cloud process are described as being denitrified (Fahey et al., 1989b). Denitrification is more thorough in the Antarctic than in the Arctic because the Antarctic winter is colder and longer than the arctic winter (Kawa et al., 1990). HNO_3 is photolyzed or reacts with OH to produce NO_2. Because NO_2 is a sink for ClO (ClO + $NO_2 \rightarrow ClONO_2$), reducing levels of NO_2 permits ClO levels to remain elevated. Since ClO participates in the catalytic destruction of ozone and $ClONO_2$ does not, removal of HNO_3 permits the catalytic destruction of ozone to continue until more NO_2 is mixed into the air parcel by the breakup of the vortex. Dehumidified air parcels have also been observed in the stratosphere (Kelly et al., 1989).

It is widely agreed that the removal of HNO_3 occurs through the gravitational sedimentation of particles carrying HNO_3 and/or H_2O. However, the mechanism by which this occurs is not understood. Type I PSC particles often have diameters near 1 μm, and their sedimentation is rather slow. Ice particles can be much larger, but their concentration tends to be much smaller. Studies of denitrified air suggests that a few of the largest sulfate particles were removed in the process of denitrification (Wilson et al., 1992). So it is necessary to find a mechanism which places most of the removed material on a few large particles which form on the largest sulfate nuclei. There are a number of candidate mechanisms for achieving denitrification (Toon et al., 1990; Wofsy et al., 1990).

HETEROGENEOUS CHEMISTRY OF THE STRATOSPHERE

In heterogeneous reactions, one or more of the reactants originates in the gas phase and the reaction occurs on or in the particle. The products can end up anywhere. If the reaction occurs on the surface of the particle and all portions of that surface can host the reaction, then the number of reactions which occur in a fixed volume of air and time interval will be proportional to the amount of aerosol surface present, the number of collisions between the rate-limiting reactant and the surface, and the fraction of collisions which lead to a reaction. This fraction is referred to as the reaction probability.

Heterogeneous Reactions Occurring on PSCs

Stratospheric levels of chlorine are approximately 3–4 ppbv, and most of that chlorine entered the stratosphere as chlorofluorocarbons (CFCs). After organics have been destroyed by photolysis, the released chlorine is mostly found in two inorganic forms, HCl and $ClONO_2$. As is made clear in other chapters in this volume, these species do not participate directly in the catalytic destruction of ozone and are referred to as inactive chlorine species. Active chlorine-containing species such as Cl, ClO, OClO, and OClClO are participants in catalytic cycles that destroy ozone and are usually present at 100 pptv or less. In late polar winter, ClO reaches mixing ratios in excess of 1 ppbv (Anderson et al., 1989; Kawa et al., 1992a) in the polar vortex. This dramatic increase in active chlorine is due mainly to heterogeneous reactions occurring on PSCs and listed in Table 15.1 (Fahey, 1994).

Reactions which are labeled "Fast" in Table 15.1 have reaction probabilities between a few hundredths and a few tenths. Reactions labeled "$f(RH)$" have reaction probabilities which are a function of relative humidity, RH. Rates labeled "$f(wt \% H_2SO_4)$" have reaction probabilities that depend on the weight percent of sulfuric acid in the su-

TABLE 15.1
Heterogeneous reactions of importance in the stratosphere[a]

Reaction	Ice (Type II PSCs)	HNO$_3$ Hydrates (Type I PSCs)	Supercooled Sulfate	Frozen Sulfate	Reaction Number
ClONO$_2$ + HCl → Cl$_2$ + HNO$_3$	Fast: f(RH)	Fast: f(RH)	f(wt % H$_2$SO$_4$)	Fast: f(RH)	1
HOCl + HCl → Cl$_2$ + H$_2$O	Fast: f(RH)	Fast: f(RH)	f(wt % H$_2$SO$_4$)	Fast: f(RH)	2
ClONO$_2$ + H$_2$O → HOCl + HNO$_3$	Fast	Slow	f(wt % H$_2$SO$_4$)	Slow	3
N$_2$O$_5$ + H$_2$O → 2 HNO$_3$	Fast	Slow	Fast	Slow	4

[a]Adapted from Fahey (1994).

percooled particle. In stratospheric air, lower temperatures usually result in increased relative humidity, decreased weight percent of sulfuric acid, and faster reactions due to a larger fraction of effective collisions

In polar winter, as temperatures drop below 210 K, nitric acid dissolves in the dilute sulfuric acid and water solutions and particle growth continues. Reactions 1, 2, and 3 will convert inactive chlorine into Cl$_2$ and HOCl, which are readily photolyzed into active chlorine in sunlight. These reactions produce more product as the reaction probability increases with increasing RH, and as more aerosol surface area becomes available due to particle growth. In a nonvolcanic stratosphere, the condensation of all the available nitric acid will lead to a nearly 5-fold increase particle diameter and 20-fold increase in surface.

Observations of the impact of heterogeneous PSC chemistry

Figure 15.9 shows the classic observation of the difference in ClO mixing ratios inside and outside the Antarctic polar vortex on September 16, 1987. The air in the polar vortex had been processed in PSCs throughout the Antarctic winter and much of the chlorine had been converted to active forms. Air outside vortex, north of about 69°S, had not been processed. The result was an enhancement in ClO mixing ratio by orders of magnitude. Since 1 ppbv of active chlorine will remove ozone at the rate of a few percent per day, the significant destruction of ozone and the anti-correlation of ozone and ClO shown in Figure 15.8 required time to develop (Anderson et al., 1989). The ozone loss demonstrates the ultimate impact on ozone in the polar vortex of the repartitioning of the chlorine species by heterogeneous reactions.

In the Arctic, ClO was observed while flying through a PSC. The aerosol volume and ClO are plotted in Figure 15.10 for flights in the early morning (Figure 15.10a) and afternoon (Figure 15.10b). Large aerosol volumes are observed in the cloud (Poole et al., 1990). The ClO levels were larger in the cloud than out of the cloud and they were larger in the afternoon traverse through the cloud than in the morning. This is consistent with the occurrence of Reaction 1 on the PSC particles. The morning tra-

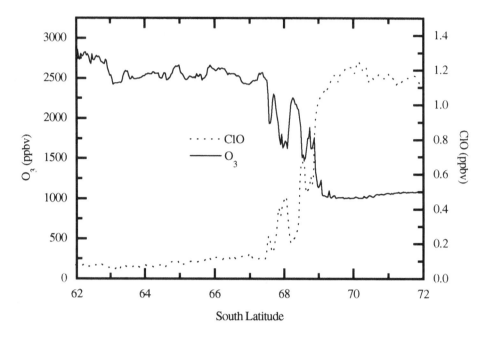

Figure 15.9. Measurement of ClO and O_3 on September 16, 1987 as the NASA ER-2 aircraft flew into the Antarctic polar vortex. Note the negative correlation between ClO and O_3 south of 67.5°S. This correlation illustrates the impact of the activation of chlorine by reactions on PSCs and the effect of the chlorine on ozone. (Adapted from Anderson et al., 1989.)

verse occurred before the air had seen much sunlight. Therefore, the Cl_2 which had been formed by Reaction 1 had not yet been photolyzed, and the ClO levels were lower than those during the afternoon pass. Model calculations of the expected ClO levels duplicated most of the main features of the observations (Jones et al., 1990). These calculations took into account the history of the air parcels' temperature and exposure to sunlight.

The impact of heterogeneous reactions is seen in measurements of ClO and HCl in the polar vortices. Calculations of back trajectories of air parcels permit lowest temperatures seen by air parcels in the previous weeks to be estimated. Elevated values of ClO near local noon and depressed values of HCl serve as markers for air which has experienced heterogeneous processing. Diminished HCl (Webster et al., 1993) and elevated ClO (Toohey et al., 1993) occurred primarily in parcels whose back trajectories showed temperatures sufficiently low to cause PSC formation. Parcels having back trajectory temperatures always above the temperature required for PSC formation equilibrium were much less likely to have elevated ClO or reduced HCl.

Heterogeneous Reactions on Sulfate

Reactions 1, 2, and 3 also proceed rapidly on sulfate particles as temperatures approach those required for PSC formation. Following large volcanic injections, surface concentrations of sulfate aerosol approach those provided by PSCs. Thus reactions on

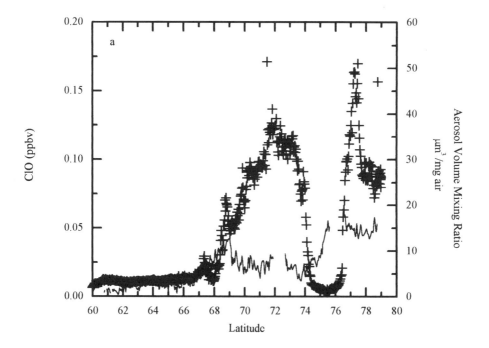

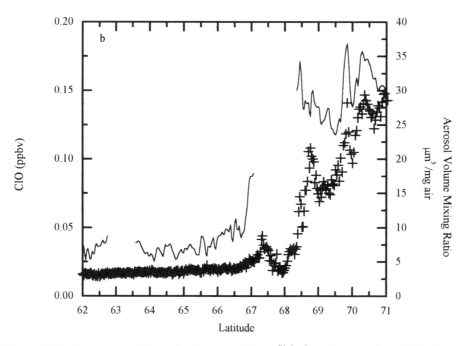

Figure 10.10. Measurement of aerosol volume and ClO on a flight through a polar stratospheric cloud in the morning (a) and the afternoon (b). Elevated aerosol volumes indicate the presence of a PSC. ClO is more elevated after the PSC has seen some sunlight. (Adapted from Poole et al., 1990, and Jones et al., 1990.)

361

sulfate may result in elevated levels of active chlorine prior to PSC formation and following large eruptions.

Partitioning of Reactive Nitrogen Species

Reactive nitrogen appears as NO, NO_2, HNO_3, $ClONO_2$, N_2O_5, and so on. Moving nitrogen from one species to another is referred to as repartitioning and affects chlorine, hydrogen radicals, and ozone as well as the nitrogen species. The following gas phase reactions convert NO and NO_2 to N_2O_5:

$$NO + O_3 \rightarrow NO_2 + O_2$$

$$NO_2 + O_3 \rightarrow NO_3 + O_2$$

$$NO_3 + NO_2 \rightarrow N_2O_5$$

Since N_2O_5 is photolyzed rapidly in the daytime but persists at night, N_2O_5 serves as a nighttime reservoir for NO_2 and NO. Reaction 4 (Table 15.1) converts the N_2O_5 to HNO_3. Since HNO_3 has a longer photolysis lifetime than does N_2O_5, the effect of Reaction 4 is to move reactive nitrogen from NO and NO_2 to HNO_3 (Fahey et al., 1993).

Since NO_2 serves as a sink for ClO, Reaction 4 has the effect of shifting the population of chlorine toward ClO from $ClONO_2$. By decreasing NO, Reaction 4 also tends to increase HO_2 (Wennberg et al., 1994). Reducing the amount of NO tends to reduce ozone destruction in catalytic cycles involving NO. Reducing ClO tends to reduce ozone destruction in catalytic cycles involving chlorine and bromine. Increasing HO_2 tends to increase the destruction of ozone by catalytic cycles involving hydrogen radicals (Wennberg et al., 1994). The effect of these changes is summarized in Figure 15.11, where NO_x is defined as NO + NO_2.

Observations of the effects of the N_2O_5 + H_2O reaction

Following the eruption of Mt. Pinatubo, the spatial variability of aerosol surface concentration was large at mid-latitudes in the fall. Evidence for the occurrence of Reaction 4 was gathered by comparing air parcels which contained varying amounts of aerosol surface (Fahey et al., 1993; Wilson et al., 1993). The ratios of NO_x/NO_y and ClO/Cl_y were found to vary with aerosol surface when other important parameters were controlled. Cl_y is the total amount of inorganic chlorine and can be determined from the amount of N_2O in the air parcel (Kawa et al., 1992b). As aerosol surface increased from background levels to five times background levels, NO_x/NO_y decreased by about 30% and ClO/Cl_y increased by about 50% (Figure 15.12). Reaction 4 would be expected to increase the destruction of ozone by chlorine and to decrease the destruction of ozone by NO and NO_2 (Rodriguez et al., 1991; Granier and Brasseur, 1992). However, the saturation of the effect seen in Figure 15.12 seems to preclude this reaction from causing catastrophic loss of ozone.

Incorporation of heterogeneous reactions into models describing the chemistry of the mid-latitude, lower stratosphere has dramatically improved their performance. Simultaneous measurements of key members of the chlorine, nitrogen, odd-hydrogen families provide rigorous tests of the models. Measurements of the actual radiation

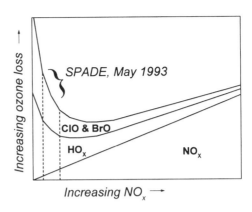

Figure 15.11. Ozone loss in mid-latitudes by HO_x, ClO and BrO, and NO_x as a function of the amount of NO_x present. Reaction 4 occurs readily in the mid-latitudes and had the effect of reducing NO_x. Aerosol surface areas were elevated relative to a nonvolcanic stratosphere when these measurements were made in the 1993 Stratospheric Photochemistry, Aerosol and Dynamics Exeperiment (SPADE). The volcanic aerosol accelerated ozone loss in this instance. (Adapted from Wennberg et al., 1994.)

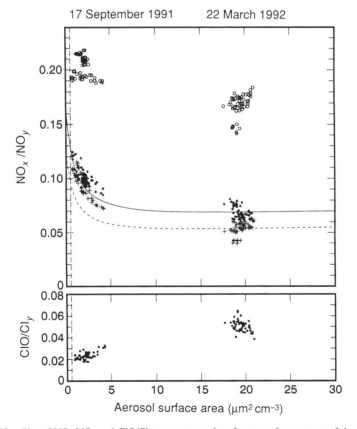

Figure 15.12. Plot of NO_x/NO_y and ClO/Cl_y versus aerosol surface area for two sets of data acquired at similar conditions of latitude and solar zenith angle. ClO, NO, NO_y, and aerosol surface area were measured. Cl_y was determined from measurements of N_2O and NO_x was determined from NO, ClO, and O_3 measurements. The solid circles indicate the values determined from the measurements. The + symbols indicate the result of models which take into account the heterogeneous reactions, and the open circles represent the result of models which consider only gas phase reactions. The solid line is the model calculation initialized on the September 17 observations, and the dashed line is the model calculation initialized on the March 22 observations. (From Fahey et al., 1993. Reprinted with permission from *Nature*, Vol. 363, p. 509. Copyright 1993, Macmillan Magazines Ltd.)

field improve the characterization of photochemistry, and measurements of aerosol properties permit quantification of heterogeneous processes. Resulting agreement between measurements and models confirms the role that heterogeneous chemistry plays in the chemistry of the stratosphere and shows that we are approaching a quantitative understanding of these effects (Kawa et al., 1993; Salawitch et al., 1994a,b).

Explaining the downward trend in mid-latitude ozone over the last two decades presents a very interesting problem since the detailed features of the trend do not match the increases in chlorine perfectly. The satellite record permits reconstruction of the aerosol surface area concentration over the time period, and a recent investigation has shown that many of the features of the downward trend can be understood if the increase in chlorine and the variations in aerosol surface due to volcanoes are included in the calculations of ozone (Solomon et al., 1996).

CONCLUSION

In the last decade, our understanding of heterogeneous chemistry has progressed dramatically. It is clear that we cannot explain ozone loss in the polar regions or abundance of ClO, NO, NO_2, and HO_2 at mid-latitudes if the effects of reactions on particles are not considered. In addition, incorporating the laboratory rates of heterogeneous reactions into models often leads to accurate quantitative descriptions of the chemistry of the lower stratosphere. However, many questions remain to be answered concerning PSCs, sulfate aerosol, and the heterogeneous reactions occurring on them.

Aircraft are expected to perturb the concentrations of water vapor and NO_y in the stratosphere (Stolarski et al., 1995). Temperatures may change in response to the accumulation of greenhouse gases. The impact of these changes cannot be predicted until the phase and composition of PSCs and their formation kinetics are determined. Similarly, the extent of denitrification may change, but those changes are also unpredictable until the mechanisms of denitrification are understood.

The nonvolcanic sources of the stratospheric sulfate layer have not been quantified nor is the state of the sulfate aerosol certain. Both issues cloud efforts to predict the future impact of these particles.

The list of important heterogeneous reactions is not long. As more reactions are studied, it may lengthen.

References

Anderson, J. G., Brune, W. H., and Proffitt, M. H.,1989, Ozone destruction by chlorine radicals within the Antarctic vortex: the spatial and temporal evolution of ClO-O_3 anticorrelation based on ER-2 data. *J. Geophys. Res.*, 94, 11,465.

Baumgardner, D., Dye, J. E., Gandrud, B. W., and Knollenberg, R. G.,1992, Interpretation of measurements made by the Forward Scattering Spectrometer Probe (FSSP-300) during the airborne Arctic stratospheric expedition. *J. Geophys. Res.*, 97, 8035.

Brock, C. A., Hamil, P., Wilson, J. C., Jonsson, H. H., and Chan K. R., 1995, Particle formation in the upper tropical troposphere: a source of nuclei for the stratospheric aerosol. *Science*, 270, 1650–1653.

Carslaw, K. S., Luo, B. P., Clegg, S. L., Peter, T., Brimblecombe, P., and Crutzen, P. J., 1994, Stratospheric aerosol growth and HNO_3 gas phase depletion from coupled HNO_3 and water uptake by liquid particles. *Geophys. Res. Lett.*, 21, 2479–2482.

Crutzen, P. J., 1976, The possible importance of CSO for the sulfate layer of the stratosphere. *Geophys. Res. Lett.*, 3, 73.

Crutzen, P. J., and Arnold, F., 1986, Nitric acid cloud formation in the cold Antarctic stratosphere: a major cause for the springtime "ozone hole". *Nature*, 324, 651.

Deshler, T., Peter, T., Muller, R., and Crutzen, P., 1994, The lifetime of leewave-induced ice particles in the Arctic stratosphere. I. Balloonborne observations. *Geophys. Res. Lett.*, 21, 1327–1330.

Dye, J. E., Baumgardner, D., Gandrud, B. W., Kawa, S. R., Kelly, K. K., Loewenstein, M., Ferry, G. V., Chan, K. R., and Gary, B. L.,1992, Particle size distributions in arctic polar stratospheric clouds, growth and freezing of sulfuric acid droplets and implications for cloud formation. *J. Geophys. Res.*, 97, 8015.

Fahey, D. W., Kelly, K. K., Ferry, G. V., Poole, L. R., Wilson, J. C., Murphy, D. M., Loewenstein, M., and Chan, K. R., 1989a, *In situ* measurements of total reactive nitrogen, total water, and aerosol in a polar stratospheric cloud in the Antarctic. *J. Geophys. Res.*, 94, 11,299.

Fahey, D. W., Murphy, D. M., Kelly, K. K., Ko, M. K. W., Proffitt, M. H., Eubank, C. S., Ferry, G. V., Loewenstein, M., and Chan, K. R., 1989b, Measurements of nitric oxide and total reactive nitrogen in the Antarctic stratosphere: observations and chemical implications. *J. Geophys. Res.*, 94, 16,665.

Fahey, D. W., Kawa, S. R., Woodbridge, E. L., Tin, P., Wilson, J. C., Jonsson, H. H., Dye, J. E., Baumgardner, D., Borrmann, S., Toohey, D. W., Avallone, L. M., Proffitt, M. H., Margitan, J., Loewenstein, M., Podolske, J. R., Salawitch, R. J., Wofsy, S. C., Ko, M. K. W., Anderson, D. E., Schoeberl, M. R., and Chan, K. R., 1993, *In situ* measurements constraining the role of sulphate aerosols in mid-latitude ozone depletion. *Nature*, 363, 509.

Fahey, D. W., 1994, Polar ozone. In: D. L. Albritton, R. T. Watson, and P. J. Aucamp, eds., *Scientific Assessment of Ozone Depletion*. World Meteorological Organization Report 37, Geneva, Chapter 3.

Farlow, N. H., Ferry, G. V., Lem, H. Y., and Hayes, D. M., 1979, Latitudinal variations of stratospheric aerosol. *J. Geophys. Res.*, 84, 733.

Gandrud, B. W., Sperry, P. D., Sanford, L., Kelly, K. K., Ferry, G. V., and Chan, K. R.,1989, Filter measurement results from the Airborne Antarctic Ozone Experiment. *J. Geophys. Res.*, 94, 11,285.

Gandrud, B. W., Dye, J. E., Baumgardner, D., Ferry, G. V., Loewenstein, M., Chan, K. R., Sanford, L., Gary, B., and Kelly, K., 1990, The January 30, 1989 Arctic polar stratospheric (PSC) event: evidence for a mechanism of dehydration. *Geophys. Res. Lett.*, 17, 457.

Goodman, J., Toon, O. B., Pueschel, R. F., Snetsinger, K. G., and Verma, S., 1989, Antarctic stratospheric ice crystals. *J. Geophys. Res.*, 94, 16,449.

Granier, C., and Brasseur, G., 1992, Impact of heterogeneous chemistry on model predictions of ozone changes. *J. Geophys. Res.*, 97, 18015.

Gras L. J., and Laby J. E., 1981, Southern hemisphere stratospheric aerosol measurements J. Size distributions 1974–1979. *J. Geophys. Res.*, 86, 9767.

Hanson, D., and Mauersberger, K., 1988, Laboratory studies of the nitric acid trihydrate: implications for the south polar stratosphere. *Geophys. Res. Lett.*, 15, 855.

Hofmann, D. J., 1990a, Measurements of the condensation nuclei profile to 31 km in the Arctic in January, 1989 and comparisons with Antarctic measurements. *Geophys. Res. Lett.*, 17, 357.

Hofmann, D. J., 1990b, Increase in the stratospheric background sulfuric acid aerosol in the past 10 years. *Science*, 248, 996.

Hofmann, D. J., and Solomon, S., 1989, Ozone destruction through heterogeneous chemistry following eruption of El Chichon. *J. Geophys. Res.*, 94, 5029.

Hofmann, D. J., Rosen, J. M., Pepin, T. J., and Pinnick, R. G., 1975, Stratospheric aerosol measurements. I. Time variations at northern mid-latitudes. *J. Atmos. Sci.*, 32, 1446.

Hofmann, D. J., Rosen, J. M., and Harder, J. W., 1988, Aerosol measurements in the Winter/Spring Antarctic stratosphere, 1. Correlative measurements with ozone. *J. Geophys. Res.*, 93, 665.

Jones, R. L., Solomon, S., McKenna, D. S., Poole, L. R., Brune, W. H., Toohey, D. W., Anderson, J. G., and Fahey, D. W., 1990, The polar stratospheric cloud event of January 24, Part 2, Photochemistry. *Geophys. Res. Lett.*, 17, 541.

Jonsson, H. H., Wilson, J. C., Brock, C. A., Knollenberg, R. G., Newton, R., Dye, J. E., Baumgardner, D., Borrmann, S., Ferry, G. V., Pueschel, R., Woods, D. C., and Pitts, M. C., 1995, Performance of a focused cavity aerosol spectrometer for measurements in the stratosphere of particle size in the 0.06–2.0 μm diameter range. *J. Ocean. Atmos. Technol.*, 12, 115–129.

Jonsson, H. H., Wilson, J. C., Brock, C. A., Dye, J. E., Ferry, G. V., and Chan, K. R., 1996, Evolution of the stratospheric aerosol in the northern hemisphere following the June 1991 volcanic eruption of Mount Pinatubo: role of tropospheric–stratospheric exchange and transport. *J. Geophys. Res.*, 101, 1553–1570.

Kawa, S. R., Fahey, D. W., Anderson, L. C., Loewenstein, M., and Chan, K. R., 1990, Measurements of the total reactive nitrogen during the Airborne Arctic Stratospheric Expedition. *Geophys. Res. Lett.*, 17, 485.

Kawa, S. R., Fahey, D. W., Heidt, L. E., Pollock, W. H., Solomon, S., Anderson, D. E., Loewenstein, M., Proffitt, M. H., Margitan, J. J., and Chan, K. R., 1992a, Photochemical partitioning of the reactive nitrogen and chlorine reservoirs in the high-latitude stratosphere. *J. Geophys. Res.*, 97, 7905.

Kawa, S. R., Fahey, D. W., Kelly, K. K., Dye, J. E., Baumgardner, D., Gandrud, B. W., Loewenstein, M., Ferry, G. V., and Chan, K. R., 1992b, The Arctic polar stratospheric cloud aerosol: aircraft measurements of reactive nitrogen, total water, and particles. *J. Geophys. Res.*, 97, 7925.

Kawa, S. R., Fahey, D. W., Wilson, J. C., Schoeberl, M. R., Douglass, A. R., Stolarski, R. S., Woodbridge, E. L., Jonsson, H., Lait, L. R., Newman, P. A., Profitt, M. H., Anderson, D. E., Loewenstein , M., Chan, K. R., Webster, C. R., May, R. D., Kelly, K. K., 1993, Interpretation of NO_x/NO_y observations from AASE-II using a model of chemistry along trajectories. *Geophys. Res. Lett.*, 20, 2507–2550.

Kelly, K. K., Tuck, A. F., Murphy, D. M., Proffitt, M. H., Fahey, D. W., Jones, R. L., McKenna, D. S., Lowenstein, M., Podolske, J. R., Strahan, S. E., Ferry, G. V., Chan, K. R., Vedder, J. F., Gregory, G. L., Hypes, W. D., McCormick, M. P., Browell, E, V,, and Heidt, L, F,, 1989, Dehydration in the lower Antarctic stratosphere during late winter and early spring, 1987. *J. Geophys. Res.*, 94, 11,317.

McElroy, M. B., Salawitch, R. J., and Wofsy, S. C., 1986, Antarctic O_3: chemical mechanisms for the spring decrease. *Geophys. Res. Lett.*, 13, 1296.

McCormick, M. P., Chu, W. P., Grams, G. W., Hamill, P., Herman, B. M., McMaster, L. R., Pepin, T. J., Russell, P. B., Steele, H. M., and Swissler, T. J., 1981, High-Latitude stratospheric aerosols measured by SAM II satellite system in 1978–1979. *Science*, 214, 328.

McKeen, S. A., Liu, S. C., and Kiang, C. S., 1984, On the chemistry of stratospheric SO_2 from volcanic eruptions. *J. Geophys. Res.*, 89, 4873–4881.

Poole, L. R., Solomon, S., Gandrud, B. W., Powell, K. A., Dye, J. E., Jones, R. L., and McKenna, D. S., 1990, The polar stratospheric cloud event of January 24, 1989, Part 1, Microphysics. *Geophys. Res. Lett.*, 17, 537.

Poole, L. R., Kurylo, M. J., Jones, R. L., and Wahner, A., 1991, Heterogeneous processes: laboratory, field and modeling studies. In: D. L. Albritton and R. T. Watson, eds., *Scientific Assessment of Ozone Depletion*, World Meteorological Organization Report 25, Geneva.

Pueschel, R. F., Snetsinger, K. G., Goodman, J. K., Toon, O. B., Ferry, G. V., Oberbeck, V. R., Livingston, J. M., Verma, S., Fong, W., Starr, W. L., and Chan, K. R., 1989, Condensed nitrate, sulfate, and chloride in Antarctic stratospheric aerosols. *J. Geophys. Res.*, 94, 11,271.

Pueschel, R. F., Snetsinger, K. G., Hamill, P., Goodman, J. K., and McCormick, M. P., 1990, Nitric acid in polar stratospheric clouds: similar temperature of nitric acid condensation and cloud formation. *Geophys. Res. Lett.*, 17, 429.

Rodriguez, J. M., Ko, M. K. W., and Sze, N. D., 1991, Role of heterogeneous conversion if N_2O_5 on sulphate aerosols in global ozone losses. *Nature*, 352, 134.

Rosen, J. M., 1971, The boiling point of stratospheric aerosols. *J. Appl. Meteorol.*, 10, 1044.

Salawitch, R. J., et al., 1994a, The distribution of hydrogen, nitrogen, and chlorine radicals in the lower stratosphere: implications for changes in O_3 due to emission of NO_x from supersonic aircraft. *Geophys. Res. Lett.*, 21, 2547–2550.

Salawitch, R. J., et al., 1994b, The diurnal variation of hydrogen, nitrogen, and chlorine radicals: implications for the heterogeneous production of HNO_2. *Geophys. Res. Lett.*, 21, 2550–2554.

Schlager, H. F., Arnold, F., Hofmann, D., and Deshler, T., 1990, Balloon observations of nitric acid aerosol formation in the Arctic stratosphere. *Geophys. Res. Lett.*, 17, 1275.

Solomon, S., Garcia, R. R., Rowland, F. S., and Wuebbles, D. J., 1986, On the depletion of Antarctic ozone. *Nature*, 321, 755.

Solomon, S., Portmann, R. W., Garcia, R. R., Thomason, L. W., Poole, L. R., and McCormick, M. P., 1996, The role of aerosol variations in anthropogenic ozone depletion at northern midlatitudes. *J. Geophys. Res.*, 101(3), 6713–6727.

Stanford, J. L., and Davis, J. S., 1974, A century of stratospheric cloud reports: 1870–1972. *Bull. Am. Meteorol. Soc.*, 55, 213.

Steele, H. M., and Hamill P., 1981, Effects of temperature and humidity on the growth and optical properties of sulphuric acid–water droplets in the stratosphere. *J. Aerosol Sci.*, 12, 517.

Stolarski, R. S., et al., 1995, *1995 scientific assessment of the atmospheric effects of stratospheric aircraft, NASA Reference Publication 1381*, National Aeronautics and Space Administration, Washington, D. C.

Tabazadeh, A., Turco, R. P., Didla, K., and Jacobson, M. Z., 1994, A study of Type I polar stratospheric cloud formation. *Geophys. Res. Lett.*, 21, 1619–1622.

Toohey, D. W., Avallone, L. M., Lait, L. R., Schoeberl, M. R., Newman, P. A., Fahey, D. W.,

Woodbridge, E. L., and Anderson, J. G., 1993, The seasonal evolution of reactive chlorine in the northern hemisphere stratosphere. *Science*, 261,1134.

Toon, O. B., Hamill, P., Turco, R. P., and Pinto, J., 1986, Condensation of HNO_3 and HCl in the winter polar stratospheres. *Geophys. Res. Lett.*, 13, 1284.

Toon, O. B., Turco, R. P., and Hamill, P., 1990, Denitrification mechanisms in the polar stratospheres. *Geophys. Res. Lett.*, 17, 445.

Turco, R. P., 1982, Models of stratospheric aerosols and dust. In: Whitten, R. C., ed., *The Stratospheric Aerosol Layer*, Springer-Verlag, New York, Chapter 4.

Webster, C. R., May, R. D., Toohey, D. W., Avallone, L. M., Anderson, J. G., Newman, P., Lait, L., Schoeberl, M. R., Elkins, J. W., and Chan, K. R., 1993, Chlorine chemistry on polar stratospheric cloud particles in the Arctic winter. *Science*, 261, 1130.

Wennberg, P. O., Cohen, R. C., Stimpfle, R. M., Koplow, J. P., Anderson, J. G., Salawitch, R. J., Fahey, D. W., Woodbridge, E. L., Keim, E. R., Gao, R. S., Webster, C. R., May, R. D., Toohey, D. W., Avalone, L. M., Proffitt, M. H., Loewenstein, M., Podolske, J. R., Chan, K. R., and Wofsy, S. C., 1994, Removal of stratospheric O_3 by radicals: *in situ* measurements of OH, HO_2, NO, NO_2, ClO, and BrO. *Science*, 226, 398–404.

Whitby, K. T., 1978, The physical characteristics of sulfur aerosols, *Atmos. Environ.*, 12, 135.

Wilson J. C., Hyun, J. H., and Blackshear, E. D., 1983, The function and response of an improved stratospheric condensation nucleus counter. *J. Geophys. Res.*, 88, 6781.

Wilson, J.C., Loewenstein, M., Fahey, D. W, Gary, B., Smith, S. D., Kelly, K. K.,,. Ferry, G. V, and Chan, K. R., 1989, Observations of condensation nuclei in the airborne antarctic ozone experiment: implications for new particle formation and stratospheric cloud formation. *J. Geophys. Res.*, 94, 16437–16448.

Wilson, J. C., Stolzenburg, M. R., Clark, W. E., Loewenstein, M., Ferry, G. V., and Chan, K. R., 1990, Measurements of condensation nuclei in the Airborne Arctic Stratospheric Expedition: observation of particle production in the polar vortex. *Geophys. Res. Lett*, 17, 361.

Wilson J. C., Stolzenburg, M. R., Clark, W. E., Loewenstein, M., Ferry, G. V., Chan, K. R., and Kelly, K. K., 1992, Stratospheric sulfate aerosol in and near the northern hemisphere polar vortex: the morphology of the sulfate layer, multimodal size distributions and the effect of denitrification. *J. Geophys. Res.*, 97, 7997.

Wilson, J. C., Jonsson, H. H., Brock, C. A., Toohey, D. W., Avallone, L. M., Baumgardener, D., Dye, J. E., Poole, L. M., Woods, D. C. , DeCoursey, R. J., Osborn, M., Pitts, M. C., Kelly, K. K., Chan, K. R., Ferry, G. V., Loewenstein, M., Podolske, J. R., and Weaver, A., 1993, *In situ* observations of aerosol and ClO after the eruption of Mt. Pinatubo: effects of reactions on sulfate aerosol. *Science*, 261,1140.

Wofsy, S. C., Salawitch, R. J., Yatteau, J. H., McElroy, M. B., Gandrud, B. W., Dye, J. E., and Baumgardner, D., 1990, Condensation of HNO_3 on falling ice particles: mechanism for denitrification of the polar stratosphere. *Geophys. Res. Lett.*, 17, 449.

Woods, D. C., and Chuan, R. L., 1983, Size specific composition of aerosols in the El Chichon volcanic cloud. *Geophys. Res. Lett.*, 10, 1041.

Chemical-Transport Models
of the Atmosphere

GUY P. BRASSEUR, FRANCK LEFÈVRE,
AND ANNE K. SMITH

ABSTRACT. This chapter describes the methodology commonly used to develop global chemical-transport models of the atmosphere. An introduction to numerical techniques used to represent advective, diffusive, and convective transport of trace gases as well as chemical transformations is presented. The concepts described in this chapter are illustrated through the presentation of a three-dimensional chemical-transport model of the stratosphere which simulates at different spatial resolutions the behavior of 40 chemically active species, taking into account heterogeneous chemical conversion mechanisms (polar stratospheric clouds, sulfate aerosols) in addition to 99 gas-phase chemical and photochemical reactions.

Nature is written in mathematical language.

GALILEO GALILEI

INTRODUCTION

Numerical models are now commonly used to simulate dynamical and chemical processes in the atmosphere. These models can be regarded as mathematical representations of fundamental physical laws in a given spatial domain; they require adequate parameterizations of unresolved processes and specified constraints at the boundaries of this domain. Their purpose is not only to describe individual physical, chemical, and perhaps biological processes, but also to identify key variables and quantify important interactions in a system which is generally nonlinear. Ultimately, models will be used to predict the evolution of the state of the atmosphere—for example, changes in the chemical composition in response to natural or anthropogenic perturbations, changes in global or regional climate conditions, and so on. The development of a model is necessarily an iterative process, involving the design of codes, the validation of results against observations, the refinement of model components, the completion of sensitivity tests, and finally the interpretation of the model output. In most cases, models are unable to simulate the full complexity of reality, but they can reveal the consequences of plausible assumptions.

Historically, atmospheric models were first developed to improve the largely subjective methods used for weather forecasting. The philosophy behind the new approach was directly influenced by the success of Newton's theory on the behavior of physical systems, which, as mentioned by Pierre Simon de Laplace in 1812, stated that complete knowledge of the masses, positions, and velocities of particles at any single instant enables precise calculation of any past and future events. At the beginning of the 20th century, the Norwegian meteorologist V. Bjerknes (1904) suggested that the state of the atmosphere could be forecast from an accurate knowledge of initial conditions and a detailed understanding of the fundamental laws governing the development of one state of the atmosphere from another. The methodology for obtaining numerical solutions of the basic equations for atmospheric motions was presented for the first time in a book published in 1922 by the British scientist Lewis Fry Richardson. When first applied to a specific meteorological situation in Europe, however, the method failed to produce an accurate forecast. It was only in the late 1940s, as one of the first electronic computers was being developed at Princeton University, that John von Neumann and Jule Charney were able to develop a rather successful method for weather prediction on the basis of simplified hydrodynamical equations. Intensive work in the last 40–50 years has led to significant improvements in meteorological models and to the development of more comprehensive general circulation and climate models.

During the same period of time, our understanding of the chemical processes in the atmosphere has greatly improved and the role of chemical constituents in the radiative balance of the atmosphere (and hence in the climate system) has been recognized as very important. Chemical models describing the production and fate of these constituents have also been developed and used for pollution studies at the urban, regional, and even the global scale. Similar models, generally coupled to transport schemes, are now being developed to investigate the fundamental questions related to tropospheric and stratospheric chemistry and are being regarded as potential components of comprehensive earth system models. A hierarchy of models has been developed, and the choice of a specific model is dictated by the scientific questions to be addressed. For example, chemistry-intensive models take into account a relatively large number (several hundred or more) of chemical and photochemical reactions between constituents, but generally ignore or crudely parameterize the effects of atmospheric dynamics. At the opposite extreme, transport-intensive models simulate the effects of advection, convection, and other transport processes more accurately, but include only a small number of constituents and ignore or greatly simplify the role of chemical reactions.

Because of the large computer power required, comprehensive three-dimensional (3-D) chemical transport models with detailed chemical schemes and advanced transport formulations are not yet available, although several attempts to develop such models are underway. Some of these coupled models are intended to study the behavior of ozone and related constituents in the middle atmosphere (e.g., Lefèvre et al., 1994) while others are being developed to understand the fate in the troposphere of chemical constituents and to obtain better estimates of their global budget (e.g., Levy et al., 1982, 1985; Prather et al., 1987; Zimmermann, 1988; Jacob and Prather, 1990; Langner and Rodhe, 1991; Erickson et al., 1991). Regional and continental-scale chemical transport models (e.g., Liu et al., 1984; Chang et al., 1987; Carmichael et al., 1991; McKeen et al., 1991; Roselle et al., 1991; Jacob et al., 1993) have also been developed to simulate the behavior of chemical compounds during a limited period of time (several days) and, for example, to study the formation of high-ozone episodes in industrialized areas.

Because available computer capabilities remain limited, simpler models have also been developed. For example, two-dimensional models, which reproduce the chemical distribution of trace constituents as a function of latitude and altitude, are commonly used for studies of the stratosphere, a region where spatial gradients in the longitudinal direction are generally small. These models provide information on seasonal evolution, meridional transport, and global budgets of trace constituents. The net meridional transport of heat and constituents by three-dimensional waves needs, however, to be expressed by closure relations or included in the formulation of the dynamical equations. The effects of weather patterns on the scale of several hundred kilometers, which are important in the lower stratosphere and in the troposphere, are not well represented in such models. One-dimensional models, which provide vertical distributions of calculated concentrations but ignore the horizontal variations, are even simpler. The one-dimensional formulation, which was extensively used for pioneering studies of atmospheric chemistry, is clearly limited by its elimination of all horizontal exchange, but provides a flexible tool for sensitivity studies in situations where chemistry is not too sensitive to transport processes.

The central element of chemical-transport models is the continuity equation, which expresses mass conservation equivalently in the *flux form*

$$\frac{\partial \rho_i}{\partial t} + \nabla \cdot (\rho_i \mathbf{v}) = S_i \tag{16.1a}$$

or in the *advective form*

$$\frac{\partial f_i}{\partial t} + \mathbf{v} \cdot \nabla f_i = \frac{S_i}{\rho_a} \tag{16.1b}$$

In these expressions, ρ_i is the mass (or number) density of species i, $f_i = \rho_i/\rho_a$ is the mass (or volume) mixing ratio, ρ_a is the air mass (or number) density, $\mathbf{v}$ is the wind velocity vector, and t is time. The source term S_i (expressed in mass or number of particles produced per unit volume and time) accounts for chemical production and destruction of species i. The equivalence between (16.1a) and (16.1b) arises from the fact that the air density ρ_a satisfies the continuity equation

$$\frac{\partial \rho_a}{\partial t} + \nabla \cdot (\rho_a \mathbf{v}) = 0$$

Note that Eq. (16.1b) can also be written

$$\frac{df_i}{dt} = \frac{S_i}{\rho_a} \tag{16.1c}$$

Because the source term for a given constituent is generally a function of the concentration of other constituents, a system of N coupled nonlinear equations (N being the number of species included in the chemical system) needs to be solved with appropriate initial and boundary conditions. These latter constraints account for the influence of the external environment on the system and can be expressed as specified concentrations or fluxes. Boundary conditions are often representative of complex processes at the interface between different components of the earth system (ocean, continental ecosystems, etc.), and their estimation can be a difficult problem.

Because the analytical solution of the conservation Eqs. (16.1) is generally unavailable, numerical techniques have to be invoked to represent the behavior of trace constituents in the atmosphere. In many of them, the differential equations are replaced by finite difference equivalents and the solutions are calculated at discrete times and discrete locations (called grid points). An alternative is to represent the spatial distribution of chemicals in one or more dimensions by a functional fit (such as spherical harmonics) and to solve derivatives analytically. Different classes of methods are commonly used. Methods are said to be *explicit* when the new state of the variable is expressed explicitly in terms of the (known) state at previous times. Conversely, they are said to be *implicit* when the variable at a given time is expressed as a function of the (unknown) state of the system at that particular time.

Modern chemical-transport models generally rely on numerical schemes which are too complex to be described here in any detail. Because these advanced techniques can often be regarded as improvements to elementary methods, this chapter will describe the fundamental concepts and present basic definitions whose knowledge is a prerequisite to the understanding of the most efficient schemes. Examples of results obtained by relatively sophisticated models will, however, be presented. The following section discusses the formulation of transport by advection, diffusion, and convection. In the ensuing section, methods adopted to represent chemical reactions are described. For the sake of illustration, the formulation adopted in a specific three-dimensional, chemical-transport model of the stratosphere is then presented, followed by a discussion of selected results.

TRANSPORT

Transport of atmospheric constituents is driven by air motion and, therefore, can in principle be formulated by a multidimensional advection model. The winds that drive the advective transport must be provided to the chemical transport model either from observations or from calculations performed by general circulation or regional scale models. Observations of the three components of the winds are never available, but partial data (e.g., the two horizontal components) are reported at specific locations (near meteorological stations) and selected time periods. These data can be assimilated in dynamical models and analyzed to produce relatively continuous winds and temperatures consistent with first principles such as mass conservation.

Not all transport processes are captured by advective processes on the numerical spatial grid because the resolution of the models or the data is limited. Small-scale processes, which can contribute significantly to the transport of constituents, must, therefore, be parameterized. Eddy diffusion is frequently used to represent processes whose typical scales are smaller than the spatial resolution of the model and, specifically, to simulate exchanges of mass and heat in the planetary boundary layer. Cloud processes, including deep convection in cumulo-nimbus, are generally poorly resolved in chemical-transport models and are parameterized by a variety of methods.

Advection

Several methods have been developed to represent the advective transport of trace constituents and are reviewed, for example, by Oran and Boris (1987) and by Rood (1987). In spite of the progress made to improve accuracy and computational efficiency, there is no single universal method that can be recommended for all applications.

The desirable properties of transport algorithms have been summarized by Rasch and Williamson (1990) and by Williamson (1992). *Accuracy* is a fundamental, but not the only, requirement. The scheme should also be *stable*, which often imposes a restriction on the time step allowed; it should be *transportive*, which means that perturbations should be advected downwind only. The method should also be *local*; that is, the solution of the advection problem at a given point should neither influence nor be influenced by the field far from that point; and it should be *conservative*, so that no mass (or any other conserved quantity) is lost or gained through the advective process. Another desirable property is that the scheme be *shape-preserving*, so that no new extrema (overshoot, undershoot) are artificially created in the solution. For example, fields which are initially positive everywhere (e.g., densities or mixing ratios of trace constituents) should never become negative in any point of the domain (positive definite solution). Finally, the scheme should be *computationally affordable*, so that it can be applied to realistic problems. Because no scheme possesses all these properties, the choice of a specific transport algorithm results generally from a compromise based on the best judgment of the modeler.

As an introduction to the most advanced techniques, it is useful to first consider elementary algorithms applied to the *one-dimensional* advection equation (flux form)

$$\frac{\partial \psi}{\partial t} + \frac{\partial (F)}{\partial x} = 0 \qquad (16.2)$$

where ψ is the advected quantity (e.g., density), $F = c\psi$ is the flux, c is the velocity (assumed here to be constant in time and space), x and t are the spatial and temporal variables, respectively. In many numerical algorithms, this equation is replaced by a finite difference equivalent. Table 16.1 presents some of the most elementary finite difference formulations. With the exception of the Euler forward scheme, which is unconditionally unstable, the explicit algorithms are stable if the Courant–Friedricks–Lewy (CFL) condition (Courant et al., 1928)

$$\frac{|c|\Delta t}{\Delta x} \le 1 \qquad (16.3)$$

is satisfied. ($\Delta t = t_{n+1} - t_n$ is the time step and $\Delta x = x_{j+1} - x_j$ is the space interval between grid points j and $j + 1$.) This relationship bounds the maximum time step by the ratio of the spatial resolution and the wind speed (or the phase velocity of the fastest propagating wave in the model). Over a time step Δt, the displacement of the "material" included inside a grid box should be limited to a distance less than or equal to the grid size Δx (see Figure 16.1). Thus, as the resolution of the model increases, the time step must decrease proportionally. In many cases this restriction makes explicit schemes impractical when applied on a sphere because Δx on a longitude grid becomes small (tends to zero) near the pole. Alternatives have been proposed to circumvent this difficulty. For example, the CFL criterion does not apply to implicit methods, although

TABLE 16.1
Explicit Conservative Methods for the One-Dimensional Continuity Equation (Flux Form)

Method	Algorithm	Remarks
Euler forward	$\psi_j^{n+1} = \psi_j^n - \dfrac{\Delta t}{2\Delta x}[F_{j+1}^n - F_{j-1}^n]$	Always unstable
Lax	$\psi_j^{n+1} = \dfrac{1}{2}[\psi_{j+1}^n + \psi_{j-1}^n] - \dfrac{\Delta t}{2\Delta x}[F_{j+1}^n - F_{j-1}^n]$	Stable for $\Delta t \leq \dfrac{\Delta x}{|c|}$
Upwind	$\psi_j^{n+1} = \psi_j^n - \dfrac{\Delta t}{\Delta x}[F_j^n - F_{j-1}^n] \qquad$ for $c > 0$ $= \psi_j^n - \dfrac{\Delta t}{\Delta x}[F_{j+1}^n - F_j^n] \qquad$ for $c < 0$	First-order accurate in space and time, positive definite, conservative Stable for $\Delta t \leq \dfrac{\Delta x}{|c|}$
Two step Lax-Wendroff	$\psi_{j+1/2}^{n+1/2} = \dfrac{1}{2}[\psi_{j+1}^n - \psi_j^n] - \dfrac{\Delta t}{2\Delta x}[F_{j+1}^n - F_j^n]$ $\psi_j^{n+1} = \psi_j^n - \dfrac{\Delta t}{\Delta x}[F_{j+1/2}^{n+1/2} - F_{j-1/2}^{n+1/2}]$	Stable for $\Delta t \leq \dfrac{\Delta x}{|c|}$
Leap-frog	$\psi_j^{n+1} = \psi_j^{n-1} - \dfrac{\Delta t}{\Delta x}[F_{j+1}^n - F_{j-1}^n]$	Second-order accurate in time; dispersive; not positive definite

ψ is the advected quantity, c is the velocity assumed to be constant in time and space, $F = c\psi$ is the flux, Δt is the time step, and Δx is the space interval between grid points (assumed to be constant).

large phase errors appear in the solution if (16.3) is violated. Reduced grids, constructed by increasing the zonal space interval when approaching the pole, allows much longer time steps to be used (Kurihara, 1965); these can lead to significant errors (Williamson and Browning, 1973), although accurate and conservative algorithms for two-dimensional transport have recently been proposed (Rasch, 1994). An alternative to reduced

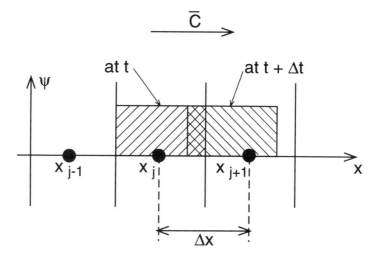

Figure 16.1. Illustration of the displacement of the material initially included in box j during a time step Δt. Stability requires that this displacement be less than Δx.

grids is to adopt differential time steps with uniform grids (Grimmer and Shaw, 1967) or to filter the fields in longitude near the poles and remove the computationally unstable modes at high latitudes. Rose and Brasseur (1989), for example, have used for this purpose a fast Fourier transform technique; diffusion operators and grid point filters are other possible options. These filters are, however, expected to degrade the quality of the solution.

Figure 16.2 measures the success of different schemes relative to the analytical solution of (16.2) when the advected quantity (ψ) is represented as a function of space (x) by a triangle (Rood, 1987). The upwind method (also called donor cell) is diffusive. The leapfrog scheme produces noise which, if not filtered, rapidly dominates the solution. Negative solutions are obtained occasionally in the case of the leapfrog and Lax–Wendroft algorithms.

Figure 16.2. Numerical solution of the one-dimensional advection obtained by different methods (leapfrog (a), Lax–Wendroff (b), and upwind (c)) after 100 time steps and for a Courant number = 0.5, compared to the exact (analytic) solution.

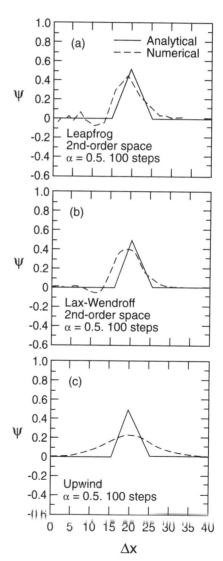

Modern advective schemes have been designed to avoid the excessive diffusion found in low-order methods or the numerical noise (over- and undershoots in regions of strong gradients) common to high-order algorithms. Although these schemes are more complex and in most cases nonlinear, they are attractive since they usually produce stable solutions, maintain steep gradients (shape preserving), and preserve the monotonicity and/or the sign of the solution. Smolarkiewicz (1984), for example, has introduced in the upwind scheme (see Table 16.1) an appropriate antidiffusive velocity which reduces the numerical diffusion characteristic of this algorithm. Other schemes, rather than assuming a constant value of the field inside each grid box, prescribe a spatial variation of the solution between grid points. An example of such techniques is provided by the schemes of Crowley (1968) and Tremback et al. (1987) (named *polynomial fitting*) in which the solution inside a grid box is expressed through a polynomial whose coefficients are determined from information in neighboring boxes. Bott (1989) has presented a positive definite version of the algorithm proposed by Tremback et al. (1987), and Smolarkiewicz (1984) has developed an advection algorithm based on Crowley's second-order scheme, which is conservative and positive definite and produces less diffusion than the upstream method.

The algorithm developed by Prather (1986), which is an extension of the slope scheme of Russell and Lerner (1981), represents the solution inside a grid box by a second-order polynomial; the first-order (slope) and second-order (curvature) moments are transported in addition to the zero-order moment (e.g., average concentration in the cell). The method is therefore characterized by little numerical diffusion; it is one of the most accurate algorithms and is recommended in particular for coarse resolution grids. It is, however, computationally expensive since for each grid point 11 variables need to be stored. A "fixer" can ensure positivity of the solution. A comparison of the

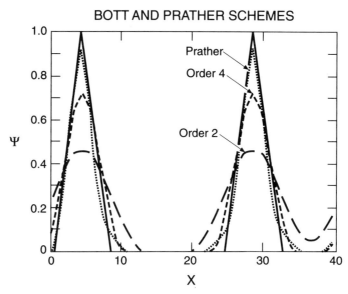

Figure 16.3. Comparison of the positive-definite scheme of Bott for various orders of the polynomial fit and of the Prather scheme (with limiters to ensure positive definiteness) with the analytic solution (solid lines) (Adapted from Müller, 1992).

Figure 16.4. Schematic description of the semi-Lagrangian scheme. The point (x, t) represents the arrival point on the regular mesh, while $(x_0, t - \Delta t)$ represents the departure point. The value of the tracer mixing ratio at the departure point is estimated by interpolating the values of the mixing ratios on the regular grid points at time $t - \Delta t$.

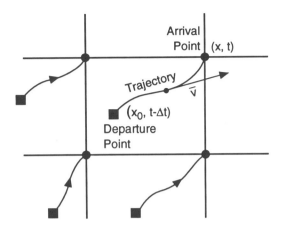

positive definite scheme of Bott (1989) and the method of Prather (1986) is shown in Figure 16.3 (Müller, 1992).

The methods discussed above fall within the category of Eulerian formulations, which means that they solve for transport at fixed spatial locations (either on a grid or using an orthogonal expansion such as spherical harmonics). The limitation associated with the CFL criterion in Eulerian methods can be circumvented by using Lagrangian or semi-Lagrangian formulations. In the Lagrangian scheme (e.g., Walton et al., 1988; Taylor et al., 1991), distinct air parcels, in which the tracers are assumed to be homogeneously mixed, are followed as they are displaced by the winds. Some assumptions are usually made concerning the exchange rate of air between the parcels and the surrounding environment (equivalent to subgrid mixing in Eulerian models). The accuracy of the model depends on the number of air parcels prescribed inside the atmospheric domain. In their global transport model, Taylor et al. (1991) used 100,000 parcels initially distributed uniformly over the globe.

In semi-Lagrangian formulations (Robert, 1981), the solution on prescribed grid points is derived at each time step on the basis of a Lagrangian calculation. In the absence of dissipative processes (e.g., chemical reactions), the mixing ratio at a point $(\mathbf{x}, t)$ on the regular mesh is the same as the value at point $(\mathbf{x}_0, t - \Delta t)$, where $\mathbf{x}_0$ represents the location at time $t - \Delta t$ of the air parcel arriving at $(\mathbf{x}, t)$ (see Figure 16.4). The departure point of the trajectory is calculated from

$$\mathbf{x}_0 = \mathbf{x} + \int_t^{t_0} \mathbf{v}(\mathbf{x},t) \, dt \qquad (16.4)$$

where $\mathbf{v}(\mathbf{x}, t)$ is the wind velocity. $\mathbf{x}_0$ does not in general coincide with a point of the grid, so that the mixing ratio at this point has to be estimated by an interpolation procedure. The success of the method (accuracy, monotonicity, shape preserving, etc.) is greatly dependent on the interpolation scheme used (see Staniforth and Coté, 1991 for a review). The method conserves transported variables with an accuracy only to the truncation error, so that most chemical-transport models (CTMs) adopting the scheme include a "fixer" to correct for potential deviations from exact mass conservation. Semi-Lagrangian advection schemes are now commonly used in general circulation models (to simulate the transport of water vapor) and in global tracer models.

Diffusion

Diffusion terms are often introduced in the transport equations to account for mixing associated with subgrid processes or, in one-dimensional models, to parameterize vertical exchanges associated with complex dynamical processes (closure relations). Even in three-dimensional models, mixing processes in the planetary boundary layers are usually parameterized by diffusion schemes, in which the diffusion coefficient is related to the vertical gradient of the horizontal wind and the static stability. In two-dimensional models of the middle atmosphere, irreversible transport associated with the absorption or breakdown of dynamical waves is often parameterized by eddy diffusivity. Different forms can be adopted for these diffusion terms; for the sake of illustration we consider again the simple one-dimensional case

$$\frac{\partial f}{\partial t} = \frac{\partial}{\partial x}\left(K\frac{\partial f}{\partial x}\right) \tag{16.5}$$

where $K > 0$ represents the diffusion coefficient (assumed here to be constant). The explicit scheme

$$\frac{f_j^{n+1} - f_j^n}{\Delta t} = \frac{K}{(\Delta x)^2}[f_{j+1}^n - 2f_j^n + f_{j-1}^n] \tag{16.6}$$

is stable only if

$$\frac{2K\Delta t}{(\Delta x)^2} \leq 1 \tag{16.7}$$

which puts a limit on the time step (Δt) that can be used. The fully implicit scheme

$$\frac{f_j^{n+1} - f_j^n}{\Delta t} = \frac{K}{(\Delta x)^2}[f_{j+1}^{n+1} - 2f_j^{n+1} + f_{j-1}^{n+1}] \tag{16.8}$$

which is unconditionally stable, is used in most applications. The Crank–Nicholson algorithm, in which the right-hand side of (16.5) is approximated by the average between the explicit and implicit formulations, provides more accurate solutions. In these two latter cases, a linear tridiagonal system must be solved.

Multidimensional problems can be treated by alternating direction methods, in which each spatial direction is considered separately, using, for example, the implicit algorithm. This method is second-order accurate and unconditionally stable.

Convection

Vertical transport produced by convective systems (e.g., cumulonimbus) greatly affects the distribution of chemical species in the troposphere, especially those whose lifetime is of the order of months or less. Convection is episodic and, in most cases, occurs over areas that are substantially smaller than the typical grid cell of a global or even a regional three-dimensional model. It must, therefore, be treated as a subgrid process that can simulate the large-scale effect of relatively isolated, yet very active, transport events.

Motions inside and in the vicinity of convective cloud systems are complex. They include strong updrafts and downdrafts, side entrainment, and detrainment, and they are generally accompanied by strong precipitation. Complex chemical processes (aqueous-phase reactions in cloud droplets, production of NO_x by lightning, etc.) are also occurring. The total mass of air processed by convective systems varies from typically 10^4 kg/hr in growing thunderstorms to 10^6 kg/hr in tropical cloud clusters (see Brost and Heimann, 1991).

Several methods are used to parameterize convective transport in global chemical-transport models and are illustrated in Figure 16.5. (Note that a fraction of the convective motions may be resolved by the advection, especially in high-resolution models.) The simplest of these methods, used in the transport models developed at the Geophysical Fluid Dynamics Laboratory (Princeton, NJ), is based on a diffusive formulation (Mahlman and Moxim, 1978; Levy et al., 1982, 1985); in regions experiencing convection events, the vertical eddy diffusivity (K) is increased by several orders of magnitude. This method tends to mix species in the vertical direction when convection takes place, and it does not explicitly distinguish between ascending and descending motions.

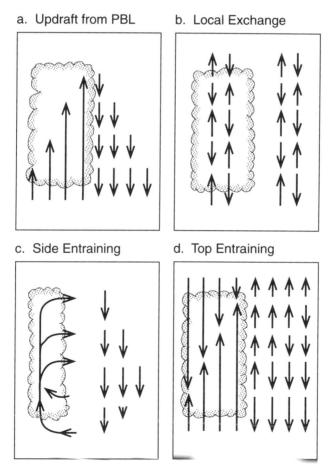

Figure 16.5. Schematic representations of several formulations for convective transport in clouds (Adapted from Brost and Heimann, 1991).

The asymmetry between up and down motions is more easily expressed when a matrix approach is used. In this formulation, the fraction of the air mass in any level z' that is transported to another level z is specified by the transition probability $\Pi(z, z')$, so that the effect of convection on the mixing ratio f is given by

$$\frac{\partial f(z)}{\partial t} = \int \Pi(z, z') f(z') \, dz' \tag{16.9}$$

In the parameterization developed by Costen et al. (1988) and based on an earlier work of Chatfield and Crutzen (1984), the transition probabilities account for (a) ascending motions of air parcels from the planetary boundary layers (PBLs) to all levels of the free troposphere and (b) subsidence from each layer of the model to the adjacent lower layer. The vertically integrated mass of air is unchanged by the scheme. A matrix formulation is used in the GISS/Harvard model (see, e.g., Jacob et al., 1987) and in the IMAGES model (Müller and Brasseur, 1995). Other schemes account for additional processes such as side entrainment and detrainment, and even top entrainment (which is believed to be important in nonprecipitating clouds).

Because of the complexity of the processes involved, the representation of convective transport in chemistry transport models is a major source of uncertainty. The schemes have often been validated by comparing the vertical profiles of radon-222 calculated in the troposphere with available observations. Radon-222 is a decay product of uranium-238 in soils; its emission over land surfaces is relatively well quantified (although it varies greatly with soil conditions and meteorological factors; see, e.g., Turekian et al., 1977; Lambert et al., 1982), and its atmospheric destruction results only from radioactive decay (with a well-established half-life of 3.824 days). A comparison between radon-222 profiles observed at northern mid-latitudes and calculated by Feichter and Crutzen (1990) with and without convective transport is shown in Figure 16.6.

Wet and Dry Deposition

Highly soluble gases (e.g., HNO_3, H_2O_2) are easily incorporated into cloud droplets and removed from the troposphere in rain water. The parameterization of wet removal of chemical constituents in atmospheric models is not straightforward because of the complexity of the processes involved. A further complication arises from the fact that the typical size of clouds is usually smaller than the size of the numerical grid. A simple and commonly used formulation for wet removal is a first-order parameterization in which the rate of change in the gas-phase concentration of soluble species (inside the clouds) is assumed to be proportional to their concentrations

$$\frac{d\rho}{dt} = -\lambda\rho \tag{16.10a}$$

where λ is the "rainout frequency" (expressed in s^{-1}). Giorgi and Chameides (1985) have formulated λ as a function of the rate of removal of liquid water in clouds (W, expressed in molecules cm^{-3} s^{-1}), the liquid water content of the cloud (L, expressed in g/m^3), the effective Henry's Law constant of the chemical constituent (H expressed in M $(atm)^{-1}$, see Table 16.2), and the temperature T (in Kelvin)

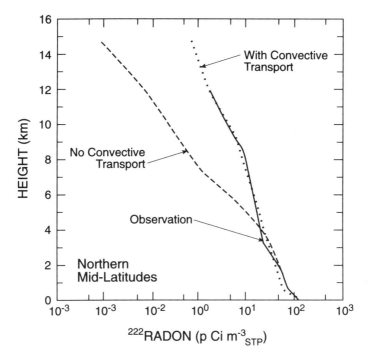

Figure 16.6. Comparison between radon-222 profiles observed at northern mid-latitudes and calculated with and without convective transport. Adapted from Feichter, J., and P. J. Crutzen, 1990, Parameterization of vertical tracer transport due to deep cumulus convection in a globa transport model and its evaluation with ^{222}radon measurements. *Tellus*, 42B, 100–117. Copyright 1990 Munksgaard International Publishers Ltd., Copenhagen, Denmark. Reprinted with permission.

$$\lambda = \frac{W}{55} [AL \times 10^{-9} + (HR'T)^{-1}]^{-1} \qquad (16.10b)$$

where A is Avogadro's number (6.023×10^{-23} molecules per mole) and $R' = 1.36 \times 10^{-22}$ atm cm^3 K^{-1} molecule^{-1}, a constant related to the perfect gas constant (55 converts moles of H_2O to liters of H_2O). Figure 16.7 shows the rainout lifetime $1/\lambda$ calculated as a function of altitude by Giorgi and Chameides (1985) for water vapor re-

TABLE 16.2
Effective Henry's Law or Solubility
Coefficients, H, for Atmospheric
Species of Interest

Species	H (M/atm)[a]
SO_2	4×10^3
HCHO	1.3×10^4
HO_2	3.3×10^4
HCOOH	1.5×10^3
H_2O_2	2×10^5
NH_3	1.1×10^{11}
HNO_3	7×10^{11}

[a]Values are calculated for $T = 290$ °K and pH = 5 (After Chameides, 1984).

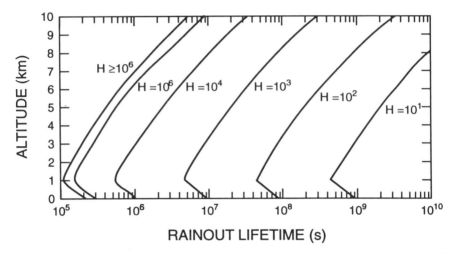

Figure 16.7. Rainout lifetime represented as a function of altitude in the troposphere for gases with different solubilities in water (Henry's Law constants). (After Giorgi and Chameides, 1985.)

moval rates varying from approximately 3×10^{11} molecules $cm^{-3} s^{-1}$ near the surface to 10^{10} molecules $cm^{-3} s^{-1}$ at 9 km altitude, and for a liquid water content of 1 $g m^{-3}$, typical of precipitating clouds. It is apparent from the figure that, for less soluble species, the rainout frequency varies linearly with the Henry Law's constant, but that, for more soluble species ($H > 10^5$ M $(atm)^{-1}$), it is affected only by the physical properties of the clouds. Giorgi and Chameides (1985) have adapted their parameterization to account for the intermittent character of the rain, which leads, on the average, to longer rainout lifetimes.

Dry deposition is treated by introducing at the lower boundary of the model a (descending) flux Φ_D which is proportional to the surface density ρ_S

$$\Phi_D = -v_D \rho_S \tag{16.11a}$$

v_D is the so-called "deposition velocity" which, for a given chemical constituent, is a function of the type of surface and aerodynamic conditions in the surface layer. Table 16.3 provides typical values of deposition velocities (Hauglustaine, 1993).

TABLE 16.3
Typical Deposition Velocities (cm/s) for Several Chemical Constituents Over the Ocean, the Continent and Ice/Snow

Constituents	Ocean	Continent	Ice/Snow
NO	0.003	0.016	0.002
NO_2	0.02	0.1	0.01
NO_3	0.02	0.1	0.01
HNO_3	1	4	0.5
N_2O_5	1	4	0.5
HO_2NO_2	1	4	0.5
O_3	0.07	0.4	0.07
H_2O_2	1	0.5	0.32
CH_3OOH	0.25	0.1	0.01

Over the ocean, the rate at which gas transfer between the ocean and the atmosphere occurs can be expressed by

$$\Phi = -k_w[C_a H^{-1} - C_w] \tag{16.11b}$$

where C_a represents the concentration of the gas in air (expressed for example in atm), C_w its concentration in water (e.g., in moles/L), H the Henry's Law constant, and k_w a kinetic factor which is a function of the wind intensity and turbulence above the surface of the sea.

CHEMISTRY

Formulation of Chemical Processes

The abundances of the chemical constituents in the atmosphere are directly affected by a large number of chemical and photochemical reactions. At the heart of any atmospheric chemical model is a system of nonlinear equations which takes the form

$$\frac{\partial \boldsymbol{\rho}}{\partial t} = \mathbf{S}(\boldsymbol{\rho}, t) \tag{16.12}$$

where t is time, ρ is a vector whose elements $(\rho_1, \rho_2, \rho_i, \ldots, \rho_N)$ denote the density of the N chemical compounds, and the vector $\mathbf{S}(\rho, t)$ is the source term representing

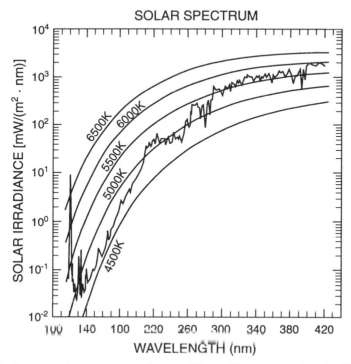

Figure 16.8. Extraterrestrial solar spectrum between 110 and 420 nm measured by the SOLSTICE instrument on board UARS. Values for black-body emissions at different temperatures are also shown (Adapted from London et al., 1993).

the rates of elementary reactions, condensation, evaporation, local injection, and so on, leading to the production (**P**) or loss (**L**) of the chemical constituents. For gas-phase reactions, taking into account zero order (e.g., local injection), first order (e.g., photolysis or thermal decomposition), and second order (e.g., chemical bi- or trimolecular reactions), the source term of species i can be written

$$S_i = a_i + \sum_j b_{i,j}\rho_j + \sum_{j,k} c_{i,j,k}\rho_j\rho_k \qquad (16.13)$$

where coefficients a, b, and c depend on local conditions such as temperature, pressure, and solar intensity. The nature of some of these terms will be discussed in the following paragraphs.

The photolysis frequency J_i of molecule i, an important first-order process, is given by

$$J_i\,(z,\,t) = \int_\lambda \epsilon_i(\lambda)\sigma_i(\lambda)q(z,\,t;\,\lambda)\,d\lambda \qquad (16.14)$$

where $\sigma_i(\lambda)$ represents the absorption cross section of constituent i at wavelength λ, $\epsilon_i(\lambda)$ the quantum efficiency of the photodissociation, and $q\,(z,t;\,\lambda)$ the solar actinic flux at altitude z, time t, and wavelength λ. This flux depends on the extraterrestrial

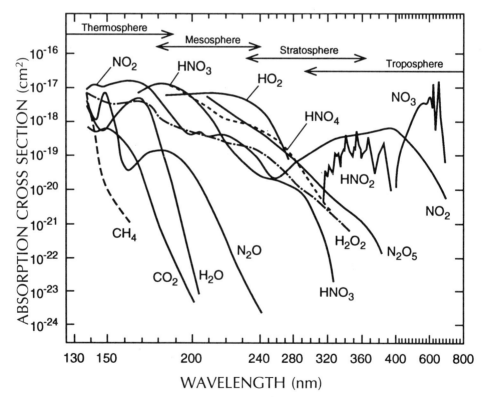

Figure 16.9. Examples of absorption cross sections for selected chemical species. Adapted from Shimazaki, T., 1985, *Minor Constituents in the Middle Atmosphere*. Terra Scientific Publishing Co., Tokyo, Japan. Copyright 1985 Munksgaard International Publishers Ltd., Copenhagen, Denmark. Reprinted with permission.

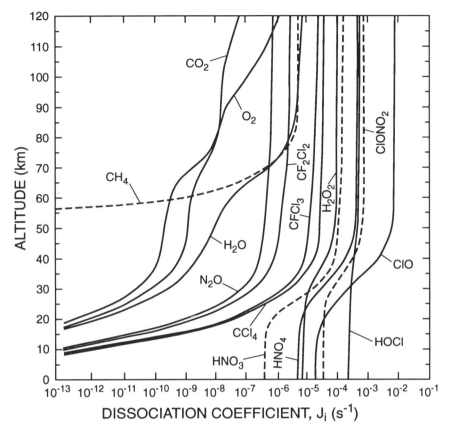

Figure 16.10. Photodissociation frequencies for various molecules represented as a function of altitude for an overhead sun. Adapted from Shimazaki, T., 1985, *Minor Constituents in the Middle Atmosphere.* Terra Scientific Publishing Co., Tokyo, Japan. Copyright 1985 Munksgaard International Publishers Ltd., Copenhagen, Denmark. Reprinted with permission.

solar intensity, its absorption in the atmosphere (mainly by ozone and molecular oxygen), and its scattering by underlying layers and the earth's surface. The accuracy to which these photodissociation rates are calculated depends on the accuracy to which absorption and quantum efficiencies are measured in the laboratory or solar fluxes are measured at the top of the atmosphere by rockets and satellites, and also on the radiative code which is used to calculate the solar penetration through atmospheric layers. An extraterrestrial solar spectrum measured by the SOLSTICE instrument on board the Upper Atmosphere Research Satellite (UARS) (Rottman et al., 1993; London et al., 1993) and covering the 110- to 420-nm range is shown in Figure 16.8. Absorption cross sections for selected species and calculated photolysis frequencies are represented in Figures 16.9 and 16.10, respectively.

Rates (R) associated with chemical reactions between species i and j are expressed by

$$R_{ij} = k_{ij} \rho_i \rho_j \qquad (16.15)$$

where k_{ij}, the temperature and pressure dependent rate constant, must be measured in the laboratory. In the case of bimolecular reactions such as dissociative recombination, the rate constant is usually expressed by

$$k(T) = A \exp \left[-\frac{E}{RT} \right] \tag{16.16}$$

where A is the so-called Arrhenius factor, E the activation energy, R the gas constant, and T the temperature. For trimolecular reactions such as collisionally stabilized recombinations, the rate constant, which is highly dependent on pressure (or air density), is often calculated from the Troe's expression

$$k(M, T) = \frac{k_0(T)[M]}{1 + k_0(T)[M]/k_\infty(T)} 0.6^{\{1+[\log_{10}(k_0(T)[M]/k_\infty(T)]^2\}^{-1}} \tag{16.17}$$

where $k_0(T)$ and $k_\infty(T)$ are low- and high-pressure limits of the rate constant, respectively, and $[M]$ is the air number density.

Finally, collisional (or thermal) dissociations can be regarded as inverse reactions for trimolecular recombinations. Their rate constants, which can be derived from measured equilibrium constants, are highly variable with temperature.

Heterogeneous reactions on the surface of aerosols, water droplets, or ice particles play an important role, especially in the formation of the Antarctic ozone hole and in acid precipitation. The rates of such reactions depend on complex processes such as the adsorption and dilution of reacting gases into particles or droplets, the reactions occurring inside the condensed phase, and the desorption and evaporation of reaction products from the particles or droplets. For example, in the case of chlorine activation on ice particles inside polar stratospheric clouds, the rate of reaction is provided by the collision rate between air molecules and ice crystals multiplied by the probability that this collision leads to a reaction. This reaction probability (accommodation coefficient) is measured in the laboratory.

Numerical Methods

Most of the finite differencing methods used to integrate nonlinear systems of kinetics equations (16.12) are subsets of multistep techniques (Byrne and Hindmarsh, 1975):

$$\rho^{n+1} = \sum_{h=0}^{K} \alpha_h \rho^{n-h} + \Delta t \sum_{h=-1}^{K} \beta_h S(\rho^{n-h}) \tag{16.18}$$

where α_h and β_h are specified constants and K is the order of the method. Equation (16.18) expresses the solution (ρ) at time step $n + 1$ as a function of the solution at previous time steps ($n, n - 1, n - 2, \ldots, n - K$) and the tendencies (sources) at times ($n + 1, n, \ldots, n - K$), with relevant weighting factors (α_h and β_h). The simplest application of (16.18) is provided by the Euler forwards method (all α and β factors equal zero, except $\alpha_o = \beta_o = 1$)

$$\rho^{n+1} = \rho^n + \Delta t S(\rho^n) \tag{16.19}$$

which is rarely practical, since it is only stable for time steps Δt smaller than

$$\Delta t \le \frac{2}{\max_j |\text{Re}(\lambda_j)|} \quad \text{for } 1 \le j \le N \tag{16.20}$$

where λ_j is one of the N eigenvalues of the Jacobian matrix $J = \partial S / \partial \rho$. The implicit

Euler backwards scheme (all α and β factors equal zero, except $\alpha_0 = \beta_{-1} = 1$)

$$\boldsymbol{\rho}^{n+1} = \boldsymbol{\rho}^n + \Delta t S(\boldsymbol{\rho}^{n+1}) \tag{16.21}$$

and the Crank–Nicholson scheme (Crank and Nicholson, 1947), corresponding to $\alpha_0 = 1$ and $\beta_0 = \beta_{-1} = 0.5$

$$\boldsymbol{\rho}^{n+1} = \boldsymbol{\rho}^n + \Delta t [S(\boldsymbol{\rho}^n) + S(\boldsymbol{\rho}^{n+1})]/2 \tag{16.22}$$

are unconditionally stable (stable for any Δt). Equations (16.21) or (16.22) can be solved by linearization and iteration (e.g., the Newton–Ralphson method). The solution can be computationally expensive since matrix systems have to be solved. In many applications, the convergence of the solution once again puts constraints on the adopted time step.

Because the chemical rates and chemical lifetimes of the various species belonging to a chemical system can be different by several orders of magnitude, the ratio between the largest and the smallest eigenvalue of the Jacobian matrix $\partial S/\partial \mathbf{p}$ is often much larger than unity and the system is said to be "stiff." Since most classic numerical methods provide stable and accurate solutions for stiff systems only if very small time steps (of the order of the smallest chemical time constant involved) are used, these methods are rarely practical. Gear (1967, 1971a,b) has developed a multistep method, which is "stiffly stable" and can therefore be applied to nonlinear chemical systems with a relatively large time step. The scheme can be expressed by (16.18) in which all $\beta_h = 0$, except β_{-1}:

$$\boldsymbol{\rho}^{n+1} = \alpha_0 \boldsymbol{\rho}^n + \alpha_1 \boldsymbol{\rho}^{n-1} + \cdots + \alpha_K \boldsymbol{\rho}^{n-K} + \Delta t \beta_{-1} S(\boldsymbol{\rho}^{n+1}) \tag{16.23}$$

The order of the method K and the time step are usually variable and chosen to minimize computational burden for a specified accuracy. The implicit nature of the scheme leads to a nonlinear algebraic system, which is solved by iteration and matrix manipulation. The Gear solver is regarded as one of the most robust and accurate methods available for nonlinear systems, but rapidly becomes computationally expensive and has large memory requirements, thereby rendering it impractical for many applications, particularly multidimensional ones.

Linearization techniques are commonly used to solve nonlinear chemical systems when implicit schemes are adopted. For example, expression (16.21) in the Euler backwards scheme can be solved through a classic Newton–Raphson iteration technique. If the continuity equation (16.12) is linearized and written as

$$\frac{\partial \rho_i}{\partial t} = P_i - L_i \rho_i \tag{16.24}$$

The concentration ρ_i at time $(n + 1)$ in semiexplicit form is given by

$$\rho_i^{n+1} = \frac{\rho_i^n + P_i^n \Delta t}{1 + L_i^n \Delta t} \tag{16.25}$$

or, using the semianalytical solution of Hesstvedt et al. (1978), by

$$\rho_i^{n+1} = (P_i/L_i)^n + [\rho_i^n - (P_i/L_i)^n]\exp(-L_i^n \Delta t) \tag{16.26}$$

Ramarosson et al. (1992), in their semi-implicit symmetric (SIS) method, have linearized the quadratic chemical terms, $k\rho_i\rho_j$, as

$$k\rho_i\rho_j \cong \frac{k}{2}[\rho_i{}^n\rho_j{}^{n+1} + \rho_i{}^{n+1}\rho_j{}^n] \tag{16.27}$$

The integration scheme is 20 times faster than the Gear scheme and has a comparable accuracy for stiff problems. It can, however, produce numerical oscillations for the fast-reacting species, and the positivity of the solution is not guaranteed.

An important criterion for the success of numerical techniques is their ability to conserve exactly the number of atoms during the course of the integration. Although both Euler methods (16.19) and (16.21) and the procedure (16.27) proposed by Ramarosson et al. (1992) are fully conservative, most linearization techniques are not. Conservation can, however, be obtained with the Shimazaki method (16.25) if an iteration over the species is performed. The semianalytical method (16.26) is inherently nonconservative, but an iterative procedure greatly improves the solution.

In order to reduce the stiffness of the chemical systems, strongly coupled chemical (rapidly interchanging) species are often grouped into chemical families whose lifetimes are significantly longer than those of their individual members. Examples of such families are $O_x = O_3 + O$ and $NO_x = NO + NO_2$. Equations for the concentration of a family are obtained by summing up the equations of the individual members. In doing so, terms corresponding to rapid conversion processes disappear and the resulting equations can often be solved by using a simple numerical scheme (e.g., explicit—see Garcia and Solomon, 1983) with relatively large time steps. The partitioning between individual members of the families is performed by assuming photochemical equilibrium conditions for the fast-reacting species or by solving time-dependent kinetics equations constrained by the calculated concentration of the related family. When a linearization procedure is used, a correction to ensure conservation of the number of atoms needs to be introduced. The definition of chemical families is not unique and no general rule can be given to optimize the method.

Multiple Solutions

As a result of their nonlinearity, atmospheric chemical systems admit multiple steady-state solutions. The possibility that more than one of these solutions is stable and physically acceptable (e.g., all concentrations are real and positive numbers) has been examined by different authors. Prather et al. (1979), for example, have shown that two stable chemical states (one with low NO_x and low ClO and the other with high NO_x and low ClO) could characterize the winter high-latitude stratosphere. A study by Yang and Brasseur (1994) has investigated the nonlinear O_x/HO_x system in the mesosphere. Finally, the behavior of chemical systems in the troposphere has been analyzed by White and Dietz (1984), Kasting and Ackerman (1985), and more recently by Stewart (1993). Using a box model with chemical schemes of different complexities, Stewart (1993) was able to find up to three stable, physically acceptable solutions of the continuity equation when the source of NO_x was increased to levels that are somewhat higher than the globally averaged emission of this compound but typical of polluted areas. As shown in Figure 16.11, the different solutions of an O_x–HO_x–NO_x–CH_4 system correspond to ozone and OH concentrations which differ by several orders of magnitude. Changes in the chemical composition of the troposphere, and hence in the ox-

Figure 16.11. Multiple solutions for the concentrations of ozone and OH represented as a function of the NO flux affecting the nonlinear O_x–HO_x–NO_x–CH_4 system (Stewart, 1993). Scales are logarithmic for concentrations expressed in cm^{-3} and fluxes in $cm^{-2} s^{-1}$.

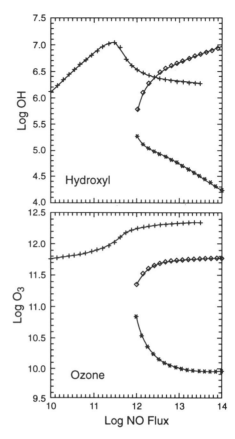

idizing capacity (self-cleaning ability) of the atmosphere in response to human activities, rather than being gradual and reversible, could correspond to abrupt and perhaps irreversible transitions. An important question is to establish which perturbations could indeed cause such abrupt changes, and at what spatial scale.

Uncertainties in Calculated Concentrations

Chemical and photochemical rate constants are never measured without experimental error. Kinetics uncertainties propagate in the model and consequently affect calculated concentrations. When a chemical scheme includes a large number of reactions, the uncertainty on the model results can be estimated using a Monte-Carlo method. The likely range of concentrations is determined from hundreds of runs in which each rate constant or photolysis rate is selected randomly within its experimental uncertainty. Stolarski et al. (1978) have used this methodology to estimate error propagation in a stratospheric model. More recently, Thompson and Stewart (1991) have computed, on the basis of 800 model runs, the uncertainty in the calculated tropospheric composition. Figure 16.12 shows the mean vertical profile of tropospheric ozone calculated in the model of Thompson and Stewart, as well as the standard deviation in the mixing

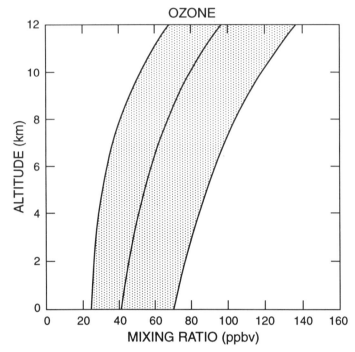

Figure 16.12. Mean vertical profile of tropospheric ozone and standard deviation associated with uncertainties in the rate constants adopted by Thompson and Stewart (1991) for their chemical scheme. Reprinted from *Chemomet. Intell. Lab. Sys.*, 10, Thompson, A. M., and R. W. Steward, How chemical kinetics uncertainties affect concentrations computed in an atmospheric photochemical model, pp. 69–79, 1991, with kind permission of Elsevier Science—NL, Sara Burgerhartstraat 25, 1055 KV Amsterdam, The Netherlands.

ratio. Based on the chemical kinetics data compiled in the JPL/NASA Evaluation of 1987, the uncertainties in the calculated ozone and carbon monoxide concentrations are 17% and 31%, respectively. The corresponding errors are of the order of 20% for NO and NO_2, and 30% for the OH radical.

AN ILLUSTRATION: A THREE-DIMENSIONAL CHEMICAL TRANSPORT MODEL OF THE STRATOSPHERE

Three-dimensional chemical-transport models are being developed to simulate the behavior of trace constituents in the atmosphere, to better understand the complex relations between chemical and dynamical processes, to diagnose observations, and to predict future changes in the atmosphere. In order to illustrate some of the concepts presented earlier, we will now present a CTM developed specifically for the stratosphere. Two versions of the model will be considered: The low-resolution version (10° longitude, 5° latitude and 3 km in altitude) is global and extends from approximately 10 km to 80 km altitude, and the high-resolution version (2° longitude, 2° latitude and 2 km in altitude) is hemispheric and extends from 10 to 32 km. In this latter case the number of grid points at which the concentration of 40 species is calculated at each time step (10 minutes) is thus equal to 182,160.

The species included in the model belong to the oxygen, hydrogen, nitrogen, chlorine, and bromine families. Among them, 26 constituents are explicitly transported; these include long-lived species such as nitrous oxide, methane, the chlorofluorocarbons, and hydrogen chloride, but also more reactive species such as OClO, Cl_2O_2, or NO_3, whose lifetimes are rather long in darkness. The continuity/transport equations are solved for each individual species; families are formed, however, with the most reactive constituents, such that $O_x = O + O_3$, $NO_x = NO + NO_2$, $ClO_x = Cl + ClO$, and $BrO_x = Br + BrO + BrCl$. The partitioning between members of these families is performed by assuming photochemical equilibrium conditions during daytime and by specifying the concentrations of O, NO, Cl, and Br as being equal to zero during nighttime. The concentrations of total nitrogen, total chlorine, and total bromine are also calculated, so that mass conservation can be verified and appropriate corrections on the concentrations of individual species can be made if necessary. The continuity equations for the different species are solved in sequence, using the linearization technique of Shimazaki (1985; see Eq. (16.25)) with four iterations to improve mass conservation (see above).

The model solves 73 gas-phase reactions and the 26 photodissociation reactions. The rate constants are taken from the compilation of DeMore et al. (1992), while the photodissociation frequencies are derived at each time step and every grid point from a four-dimensional look-up table which provides these frequencies as a function of altitude, solar zenith angle, albedo, and the ozone column computed by the model above the grid point.

Five heterogeneous reactions occurring on the surface of particles in polar stratospheric clouds (PSCs) are also included in the model. The occurrence of clouds is predicted in the model as a function of ambient conditions (temperature, and H_2O and HNO_3 partial pressures), using the expressions given by Hanson and Mauersberger (1988) for type I PSCs (nitric acid trihydrate) and by Murray (1967) for type II PSCs (ice). The frequency for collisions between a PSC particle and a molecule is given by (Cadle et al., 1975)

$$\nu = \frac{A}{4}\left[\frac{8\,kT}{\pi m}\right]^{1/2} \qquad (16.28)$$

where k is the Boltzmann's constant, T temperature, m the molecular mass of the molecule, and A the surface area density provided by the PSC particles. The surface area density can easily be calculated if the mean radius of the particles and the density of HNO_3 and H_2O in the condensed phase are known. We have assumed particle radii of 1 micron and 10 microns for type I and type II particles, respectively. The density of nitric acid and water molecules in the condensed phase is given by the difference between the calculated atmospheric density and the density corresponding to saturation. Gravitational sedimentation of cloud particles is assumed to occur with a mean vertical velocity of 15 m/day and 1.5 km/day for type I and type II particles, respectively. The calculated distributions of gas-phase H_2O, HCl, HNO_3, and NO_y are affected by condensation and evaporation processes.

Heterogeneous conversion processes on sulfate aerosol particles at all latitudes in the lower stratosphere are also represented in the model. The reactions involved are

$$N_2O_5 + H_2O(\text{aerosol}) \rightarrow 2HNO_3$$

$$ClONO_2 + H_2O(\text{aerosol}) \rightarrow HNO_3 + HOCl$$

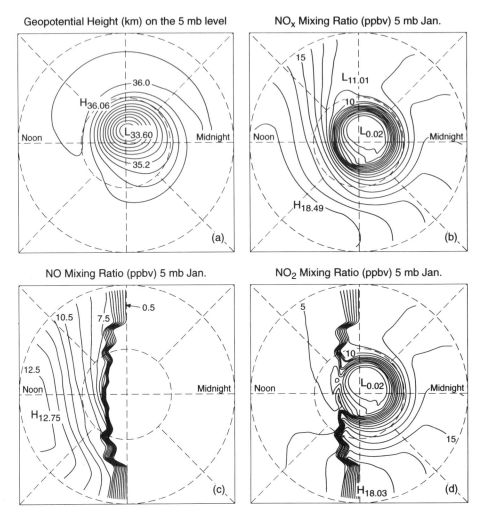

Figure 16.13. Fields provided by a three-dimensional chemical-transport model at the pressure level of 5 mbar: (a) geopotential height (km); (b–h) mixing ratios (ppbv) of NO_x, NO, NO_2, N_2O_5, ClO, $ClONO_2$, and HCl.

The aerosol surface area per unit volume is specified at the initial stage to represent either background conditions or a post-volcanic situation, and it evolves during the model integration as a result of advective transport.

Transport of chemical species is represented by the semi-Lagrangian formulation of Smolarkiewicz and Rasch (1991) [see also Smolarkiewicz and Grell (1992) and Smolarkiewicz and Pudykiewicz (1992)]. In this simulation intended to represent a winter situation, the winds and temperature are obtained by solving the primitive equations (Rose and Brasseur, 1989) with a tropospheric forcing of planetary waves applied at 300 mbar and corresponding to a climatological situation (January). As the wave propagates into the stratosphere, it affects the distribution of trace constituents.

Figure 16.13 shows the geopotential height and the distributions of several chemically active species calculated at the 5-mbar level (approximately 37 km) with the

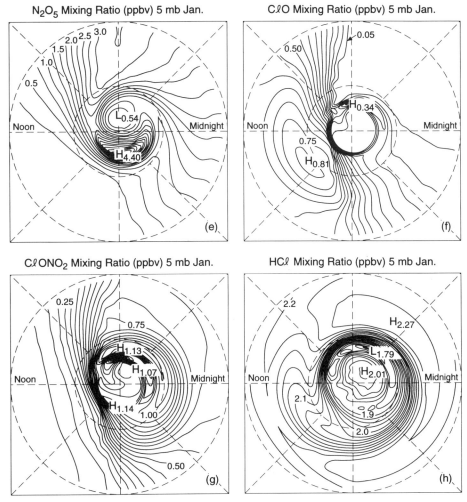

Figure 16.13.   Continued

low-resolution version of the model. The diurnal variation associated with daytime and nighttime chemistry is clearly visible. At sunrise, NO_2 is immediately photolyzed and NO is produced; an equilibrium between NO and NO_2 is achieved almost instantaneously. At sunset, NO, which reacts efficiently with O_3, is rapidly converted back to NO_2. At the same time, NO_3 is formed as a result of the reaction between NO_2 and O_3, but is dissociated in less than a second as soon as the sun returns. The formation during nighttime of N_2O_5 and $ClONO_2$ and their destruction during daytime occur at this altitude with time constants of several hours. These chemical conversions explain the relatively slow variations seen in the concentrations of nitrogen compounds both during daytime and nighttime. For this particular situation the dynamical perturbation (deviation of the geopotential from axisymmetry) is relatively weak, although it is visible at mid- and high latitudes. Under specific circumstances, the perturbation may amplify dramatically and break into much smaller structures. This event, characterized by a warming in certain areas of the polar regions and a weakening (or even a reversal) of the polar jet, is accompanied by strong mixing and enhanced meridional transport of species such as ozone toward the pole. The global distribution of species in the stratosphere might therefore be extremely complex, especially during disturbed periods in winter, and be governed simultaneously by photochemical processes and dynamical transport at a variety of spatial and temporal scales.

The high-resolution version of the model has been used to simulate the behavior of chemically active constituents in the Northern Hemisphere (especially the Arctic lower stratosphere) during winter and to help with the interpretation of data obtained by the NASA ER-2 airplane during the AASE II campaign (December 1991 to March 1992) and by the Upper Atmosphere Research Satellite. For this model, the winds needed to drive the transport are provided every six hours from analysis and mapping of global weather observations performed at the European Centre for Medium Range Weather Forecasts. In the model calculations for early January 1992 (Figure 16.14), polar stratospheric clouds were present and a large fraction of HCl (a reservoir of inorganic chlorine in the lower stratosphere) was converted into reactive ClO, providing the potential for ozone depletion near 20-km altitude. The model also showed the predicted role played by sulfate aerosols resulting from the eruption of Mount Pinatubo (The Philippines, June 1991) and suggested that the exceptionally low ozone column abundances observed over Europe and the North Atlantic during January 1992 were in large part due to the particular meteorological situation that prevailed during this pe-

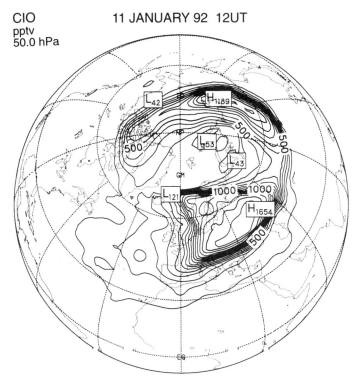

Figure 16.14. ClO distribution calculated on the 50-mbar surface on January 11, 1992 at 12:00 UT, using analyzed winds and temperatures provided by ECMWF (Adapted from Lefèvre et al., 1994).

riod, rather than massive chemical destruction. A detailed analysis of this simulation is provided by Lefèvre et al. (1994).

CONCLUSION

Chemical transport models are used to simulate the distribution of chemically active gases in the atmosphere, to establish their global budgets, and to predict potential changes in the atmospheric composition and climate resulting from human activities. These models include a formulation of the surface emission and deposition of these gases, of their chemical and photochemical conversion in the atmosphere, and of their transport by numerous dynamical mechanisms including advection, convection, and other small-scale processes. With larger computers becoming available and physical parameterizations becoming more accurate, integrated Earth system models in which atmospheric, ocean and ecological processes are coupled will be developed and used to assess potential changes at the global scale. The degree of success of such complex models remains questionable, so that for the time being, simpler and less comprehensive approaches will be used to address specific questions and test scientific hypotheses. It is important to remember that the quality of a model is directly dependent on the quality of the input data. Results need to be validated and compared with observations.

ACKNOWLEDGMENTS

The authors thank Paul Ginoux, Philip J. Rasch, and Stacy Walters for useful comments. The National Center for Atmospheric Research is sponsored by the National Science Foundation.

References

Bjerknes, V., 1904, Das Problem von der Wettervorhersage, betrachted vom Standjunkt der Mechanik und der Physik. *Meteor. Z.*, 21, 1–7.

Bott, A., 1989, A positive definite advection scheme obtained by nonlinear renormalization of the advective fluxes. *Monit. Weather Rev.*, 117, 1006–1015.

Brost, R. A., and Heimann, M., 1991, Parameterization of cloud transport of trace species in global 3-D models. In: H. van Dop and D. G. Steyn, eds., *Air Pollution Modeling and Its Applications*, VIII, Plenum Press, New York. pp. 465–483.

Byrne, G. D., and Hindmarsh, A. C., 1975, A polyalgorithm for the numerical solution of ordinary differential equations. *ACM Trans. Math. Software*, 1, 71–96.

Cadle, R. D., Crutzen, P., and Ehhalt, D., 1975, Heterogeneous chemical reactions in the stratosphere. *J. Geophys. Res.*, 80, 3381–3385.

Carmichael, G. R., Peters, L. K., and Saylor, R. D., 1991, The STEM-III regional scale acid deposition and photochemical oxidant model, I. An overview of model development and applications. *Atmos. Environ.*, 25(A), 2077–2090.

Chameides, W. L., 1984, The photochemistry of a remote marine stratiform cloud, *J. Geophys. Res.*, 89, 4739–4755.

Chang, J. S., Brost, R. A., Isaksen, I. S. A., Madronich, S., Middleton, P., Stockwell, W. R., and Walcek, C. J., 1987, A three-dimensional Eulerian acid deposition model: physical concepts and formulation. *J. Geophys. Res.*, 92, 14,681–14,698.

Chatfield, R. B., and Crutzen, P. J., 1984, Sulfur dioxide in remote oceanic air: cloud transport of reactive precursors. *J. Geophys. Res.*, 89, 7111–7132.

Costen, R. C., Tennille, G. M., and Levine, J. S., 1988, Cloud pumping in a one-dimensional model. *J. Geophys. Res.*, 93, 15,941–15,954.

Courant, R., Friedrichs, K. O., and Lewy, H., 1928, Uber die partiellen Differenzengleichungen der mathematischen Physik. *Math. Ann.*, 100, 67–108.

Crank, J., and Nicholson, P., 1947, A practical method for numerical integration of solutions of partial differential equations of heat-conductive type. *Proc. Cambridge Philos. Soc.*, 43, 50.

Crowley, W. P., 1968, Numerical advection experiments. *Mon. Weather Rev.*, 96, 1–11.

DeMore, W. B., Sander, S. P., Golden, D. M., Hampson, R. F., Kurylo, M. J., Howard, C. J., Ravishankara, A. R., Kolb, C. E., and Molina, M. J., 1992, *Chemical Kinetics and Photochemical Data for Use in Stratospheric Modelling: Evaluation number 10*, Technical Report 92–20, Jet Propulsion Laboratory, Pasadena, CA.

Erickson, D. J., III, Walton, J. J., Ghan, S. J., and Penner, J. E., 1991, Three-dimensional modeling of the global atmospheric sulfur cycle: a first step. *Atmos. Environ.*, 25A, 2513–2520.

Feichter, J., and Crutzen, P. J., 1990, Parameterization of vertical tracer transport due to deep cumulus convection in a global transport model and its evaluation with ^{222}radon measurements. *Tellus*, 42B, 100–117.

Garcia, R. R., and Solomon, S., 1983, A numerical model of the zonally averaged dynamical and chemical structure of the middle atmosphere. *J. Geophys. Res.*, 88, 1379–1400.

Gear, C. W., 1967, The numerical integration of ordinary differential equations. *Math. Comp.*, 21, 146–156.

Gear, C. W., 1971a, The simultaneous numerical solution of differential-algebraic systems, *IEEE Trans. Circuit Theory*, 18, 89–95.

Gear, C. W., 1971b, *Numerical Initial Value Problems in Ordinary Differential Equations*, Prentice-Hall, New York, 253 pp.

Giorgi, F., and Chameides, W. L., 1985, The rainout parameterization in a photochemical model. *J. Geophys. Res.*, 90, 7872–7880.

Grimmer, M., and Shaw, D. B., 1967, Energy-preserving integrations of the primitive equations on the sphere. *Q. J. R. Meteorol. Soc.*, 93, 337–349.

Hanson, D., and K. Mauersberger, 1988, Laboratory studies of the nitric acid trihydrate: implications for the south polar stratosphere, *Geophys. Res. Lett.*, 15, 855–858.

Hauglustaine, D., 1993, *Modeling of the evolution of atmospheric chemical composition and climate: one- and two-dimensional approaches*, Ph.D thesis (in French), University of Liege, Belgium.

Hesstvedt, E., Hov, Ö., and Isaksen, I. S. A., 1978, Quasi steady-state approximations in air pollution modeling: comparison of two numerical schemes for oxidant prediction. *Int. J. Chem. Kinetics*, 10, 971–994.

Jacob, D. J., and Prather, M. J., 1990, Radon-222 as a test of convective transport in a general circulation model. *Tellus*, 42B, 118–134.

Jacob, D. J., Prather, M. J., Wofsy, S. C., and McElroy, M. B., 1987, Atmospheric distribution of ^{85}Kr simulated with a general circulation model, *J. Geophys. Res.*, 92, 6614–6626.

Jacob, D. J., Logan, J. A., Yevich, R. M., Gardner, G. M., Spivakovsky, C. M., Wofsy, S. C., Munger, J. W., Sillman, S., Prather, M. J., Rodgers, M. O., Westberg, H., and Zimmerman, P. R., 1993, Simulation of summertime ozone over North America. *J. Geophys. Res.*, 98, 14,797–14,816.

Kasting, J. F., and Ackerman, T. P., 1985, High atmospheric NO_x levels and multiple photochemical steady states. *J. Atmos. Chem.*, 3, 321–340.

Kurihara, Y., 1965, Numerical integration of the primitive equations on a spherical grid. *Monit. Weather Rev.*, 93, 399–415.

Lambert, G., Polian, G., Sanak, J., Ardouin, B., Buisson, A., Jegou, A., and LeRoulley, J. C., 1982, The cycle of radon and its decay products: An application to the study of tropospheric-stratospheric exchange (in French). *Ann. Geophys.*, 38, 497–531.

Langner, J., and Rodhe, H., 1991, A global three-dimensional model of the tropospheric sulfur cycle. *J. Atmos. Chem.*, 13, 225–263.

Lefèvre, F., Brasseur, G. P., Folkins, I., Smith, A. K., and Simon, P., 1994, The chemistry of the 1991–92 stratospheric winter: three-dimensional model simulations, *J. Geophys. Res.*, 99, 8183–8195.

Levy, H., II, Mahlman, J. D., and Moxim, W. J., 1982, Tropospheric N_2O variability. *J. Geophys. Res.*, 87, 3061–3080.

Levy, H., II, Mahlman, J. D., Moxim, W. J., and Liu, S. C., 1985, Tropospheric ozone: the role of transport. *J. Geophys. Res.*, 90, 3753–3772.

Liu, M.-K., Morris, R. E., and Killus, J. P., 1984, Development of a regional oxidant model and application to the northeastern United States. *Atmos. Environ.*, 18, 1145–1161.

London, J., Rottman, G. J., Woods, T. N., and Wu, F., 1993, Time variations of solar UV irradiance as measured by the SOLSTICE (UARS) instrument. *Geophys. Res. Lett.*, 20, 1315–1318.

Mahlman, J. D., and Moxim, W. J., 1978, Tracer simulation using a global general circulation model: results from a mid-latitude instantaneous source experiment. *J. Atmos. Sci.*, 35, 1340–1374.

McKeen, S. A., Hsie, E.-Y., Trainer, M., Tallamraju, R., and Liu, S. C., 1991, A regional model study of the ozone budget in the eastern United States. *J. Geophys. Res.*, 96, 10,809–10,845.

Müller, R., 1992, Modern finite, volume advection schemes: numerical experiments in a one-dimensional test-bed. *Ber. Bunsenges. Phys. Chem.*, 96, 510–513.

Müller, J. F., and Brasseur, G. P., 1995, IMAGES: a three-dimensional chemical transport model of the global troposphere. *J. Geophys. Res.*, 96, 16,445–16,490.

Murray, F. W., 1967, On the computation of saturation vapour pressure. *J. Appl. Met.*, 6, 203–204.

Oran, E. S., and Boris, J. P., 1987, *Numerical Simulation of Reactive Flow*, Elsevier, New York, 601 pp.

Prather, M. J., McElroy, M. B., Wofsy, S. C., and Logan, J. A., 1979, Stratospheric chemistry: multiple solutions. *Geophys. Res. Lett.*, 6, 163–164.

Prather, M. J., McElroy, M. B., Wofsy, S. C., Russel, G., and Rind, D., 1987, Chemistry of the global troposphere: fluorocarbons as tracers of air motion. *J. Geophys. Res.*, 92, 6579–6613.

Prather, M. J., 1986, Numerical advection by conservation of second-order moments. *J. Geophys. Res.*, 91, 6671–6681.

Ramarosson, R. A., Pirre, M., and Cariolle, D., 1992, A box model for on-line computations of diurnal variations in a 1-D model: potential for application in multidimensional cases. *Ann. Geophys.*, 10, 416–428.

Rasch, P. J., and Williamson, D. L., 1990, Computational aspects of moisture transport in global models of the atmosphere. *Q. J. R. Meteorol. Soc.*, 116, 1071–1090.

Rasch, P. J., 1994, Conservative monotonic two-dimensional transport on a spherical reduced grid. *Monit. Weather Rev.*, 122, 1337–1350.

Robert, A., 1981, A stable numerical integration scheme for the primitive meteorological equations. *Atmos. Oceans*, 19, 35–46.

Rood, R. B., 1987, Numerical advection algorithms and their role in atmospheric transport and chemistry models. *Rev. Geophys.*, 25, 71–100.

Rose, K., and Brasseur, G., 1989, A three-dimensional model of chemically active trace species in the middle atmosphere during disturbed winter conditions. *J. Geophys. Res.*, 94, 16387–16403.

Roselle, S. J., Pierce, T. E., and Schere, K. L., 1991, The sensitivity of regional ozone modeling to biogenic hydrocarbons. *J. Geophys. Res.*, 96, 7371–7394.

Rottman, G. J., Woods, T. N., and Sparn, T. P., 1993, Solar Stellar Irradiance Comparison Experiment I: 1. Instrument design and operation. *J. Geophys. Res.*, 98, 10,667–10,677.

Russell, G. L., and Lerner, J. A., 1981, A new finite-differencing scheme for the tracer transport equation. *J. Appl. Meteorol.*, 20, 1483–1498.

Shimazaki, T., 1985, *Minor Constituents in the Middle Atmosphere*, Terra Scientific Publishers, Tokyo, Japan, 443 pp.

Smolarkiewicz, P. K., 1984, A fully multidimensional positive definite advection transport algorithm with small implicit diffusion. *J. Comput. Phys.*, 54, 325–362.

Smolarkiewicz, P. K., and Rasch, P. J., 1991, Monotone advection on the sphere: an Eulerian versus semi-Lagrangian approach. *J. Atmos. Sci.*, 48, 793–810.

Smolarkiewicz, P. K., and Grell, G. A., 1992, A class of monotone interpolation schemes. *J. Comput. Phys.*, 101, 431–440.

Smolarkiewicz, P. K., and Pudykiewicz, J. A., 1992, A class of semi-Lagrangian approximations for fluids. *J. Atmos. Sci.*, 49, 2082–2096.

Staniforth, A., and Coté, J., 1991, Semi-Lagrangian integration schemes for atmospheric models. A review. *Monit. Weather Rev.*, 119, 2206–2223.

Stewart, R. W., 1993, Multiple steady states in atmospheric chemistry. *J. Geophys. Res.*, 98, 20,601–20,611.

Stolarski, R. S., D. M. Butler, and R. D. Rundel, 1978, Uncertainty propagation in a stratospheric model. 2. Monte Carlo analysis of imprecisions due to reaction rates. *J. Geophys. Res.*, 83, 3074–3078.

Taylor, J. A., Brasseur, G. P., Zimmerman, P. R., and Cicerone, R. J., 1991, A study of sources and sinks of methane and methylchloroform using a global three-dimensional Lagrangian tropospheric tracer transport model. *J. Geophys. Res.*, 96, 3013–3044.

Thompson, A. M., and Stewart, R. W., 1991, How chemical kinetics uncertainties affect concentrations computed in an atmospheric photochemical model. *Chemomet. Intell. Lab. Sys.*, 10, 69–79.

Tremback, C. J., Powell, J., Cotton, W. R., and Pielke, R. A., 1987, The forward-in-time up-stream advection scheme: extension to higher orders. *Monit. Weather Rev.*, 115, 540–555.

Turekian, K. K., Nozaki, Y., and Benninger, L. K., 1977, Geochemistry of atmospheric radon and radon products. *Annu. Rev. Earth Planet. Sci.*, 5, 227–255.

Walton, J. J., McCracken, M. C., and Ghan, S. J., 1988, A global-scale Lagrangian trace species model of transport, transformation, and removal processes. *J. Geophys. Res.*, 93, 8339–8354.

White, W. H., and Dietz, D., 1984, Does the photochemistry of the troposphere admit more than one steady state? *Nature*, 309, 242–244.

Williamson, D. L., and Browning, G. L., 1973, Comparison of grids and difference approximations for numerical weather prediction over a sphere. *J. Appl. Meteorol.*, 12, 264–274.

Williamson, D. L., 1992, Review of numerical approaches for modeling global transport. In: H. van Dop and G. Kallos, eds., *Air Pollution Modeling and Its Application IX*, Plenum Press, New York.

Yang, P., and Brasseur, G. P., 1994, Dynamics of the oxygen–hydrogen system in the mesosphere. Part 1. Photochemical equilibria and catastrophe. *J. Geophys. Res.*, 99, 20,955–20,965.

Zimmermann, P. H., 1988, MOGUNTIA: A handy global tracer model. In: H. van Dop, ed., *Proceedings of the Sixteenth NATO/CCMS International Technical Meeting on Air Pollution Modeling and Its Applications*, Plenum, New York, pp. 593–608.

PART III

APPLIED ENVIRONMENTAL CHEMISTRY

CHAPTER 17

Sample Collection and Handling of Environmental Matrices

MICHAEL J. BARCELONA

ABSTRACT. Representative sampling of contaminated environmental media requires a clear project purpose and appropriate data quality objectives. The chemist, working within an interdisciplinary team, can assist in identifying critical design criteria for monitoring sensitive chemical constituent concentrations. Appropriate sampling and analytical methods exist for volatile organic compounds (e.g., fuels and solvents) which provide the basis for quantitative determinations. In this manner, phase distributions of mobile chemical constituents may be determined at known levels of certainty.

INTRODUCTION

Sampling environmental matrices for chemical analysis often involves a delicate balance. One must balance the costs of obtaining sufficient types and numbers of samples with the need to know what the detailed spatial or temporal distributions of chemical constituents are in specific situations. We often do not have the benefit of preliminary sampling and analysis to guide the design of sampling efforts for these specific situations. We must therefore approach the design of environmental monitoring efforts particularly carefully. *There are no true mean concentration values of environmental chemical constituent populations*; instead, continua of values exist. The identity and heterogeneity of the sample populations in turn may depend on physical, chemical and biological processes. Monitoring designs at a minumum must be developed with the clear purpose to generate representative data of known accuracy and precision (Barcelona, 1988). With the specific project purpose and data quality objectives in mind, it should then be possible to select appropriate sampling and analytical technologies which guard against bias or systematic error. These technologies may then be employed with sufficient replication to establish the precision of the *mean* or *median* of the *sample* population.

The Role of the Chemist

Cost effectiveness of the monitoring design may depend on thorough review of the network design by a team of interdisciplinary scientists and engineers. The initial responsibility of the chemist in large-scale environmental sampling and analysis efforts would usually be to collect samples and generate analytical results of known accuracy and precision which can be linked to the unique attributes of the environmental system under investigation. This chapter seeks to identify promising strategies for environmental sampling which permit us to achieve these objectives. A generally useful strategy to follow is to identify the chemical constituents of concern and to determine if they are present in specific matrices within the environmental system. Then, identify the major sources of *controllable* error or variability involved in sampling, analysis, and environmental conditions (i.e., temporal and spatial trends) so that the results are meaningful to the purposes of the investigation. Sampling and analysis approaches for volatile organic contaminants (e.g., fuels, chlorinated solvents) in difficult mixed matrices (e.g., subsurface solids and ground water) are particularly instructional in this regard. The constituents are prone to negative biases due to losses in the collection and handling of samples and they partition between the dissolved and solid-associated phases. These are difficult matrices since the subsurface environment must be disturbed by drilling, probing, and/or pumping to obtain samples where considerable heterogeneity may be present due to geochemical, microbial, and flow related process. The major challenge is to minimize losses in subsample handling and transfer and ensure the comparability between dissolved and solid concentration estimates.

This treatment of the subject is selective and the reader is directed to a number of recent references for broader coverages of important issues for sampling in general environmental applications (Keith, 1988; Keith et al., 1991; Sanders et al., 1983) and hazardous waste investigations (Gammage and Berven, 1992; Barcelona and Helfrich, 1992; Puls, 1994).

TARGET ANALYTES AND PURPOSES OF INVESTIGATIONS

The purpose of an investigation (i.e., contaminant detection, assessment, or remediation performance monitoring) will usually identify classes or individual compounds or elements of concern. Volatile organic compounds (VOCs) are regulated chemical constituents common to investigations of gas, liquid, or solid media. Some representative VOC compounds are shown in Table 17.1 in halocarbon and aromatic hydrocarbon classes (Gillham and Rao, 1990; Havinga and Cotruvo, 1990). Since these compounds are volatile at ambient temperatures and partition effectively between vapor, liquid, or solid media, they represent a good test for robust sampling and analytical protocols within monitoring designs. By robust, it is meant that the methods perform within acceptable accuracy and precision limits on wide ranges of sample characteristics (i.e., grain size, mineralogy, carbon content, etc.).

Volatile organic compounds also pose dual challenges to environmental quality assessment and protection efforts. The first challenge presents itself through the per-

TABLE 17.1
List of Common VOCs and Selected Physical Properties[a]

	Density (g/cm^3)	Viscosity (cP)	Pressure (mm)	Solubility (μg/L)	EPA[b]
DNAPLs					
Methylene chloride	1.33	0.44	349	20,000	—
Chloroform	1.49	0.56	151	8,200	—
Carbon tetrachloride	1.59	0.97	90	785	0.005
Bromoform	2.89	2.07	5	3,010	—
Bromodichloromethane	1.97	1.71	50	4,500	—
1,2-Dichloroethane	1.26	0.84	61	8,690	0.005
1,1,1-Trichloroethane	1.35	0.84	100	720	0.2
1,1,2-Trichloroethane	1.44	NA[c]	19	4,500	—
1,1,2,2-Tetrachloroethane	1.60	1.76	5	2,900	—
1,1-Dichloroethylene	1.22	0.36	590	400	0.007
trans-1,2-Dichloroethylene	1.26	0.4	326	600	0.1
Trichloroethylene	1.46	0.57	58	1,100	0.005
Tetrachloroethylene	1.63	0.90	14	200	0.005
1,2-Dichloropropane	1.16	NA	42	2,700	0.005
Chlorobenzene	1.11	0.80	12	488	0.1
1,2-Dichlorobenzene	1.31	1.41	1.0	100	0.6
1,3-Dichlorobenzene	1.29	1.08	2.3	123	—
LNAPLs					
Benzene	0.88	0.60	76	820	0.005
Gasoline	0.72–0.78	0.5–0.7	—	150–300	—
Crude Oil	0.8–0.9	3–35	—	5–25	—

[a]Adapted from Gillham and Rao (1990) and Havinga and Cotruvo (1990).

[b]EPA drinking water/ground-water standard or guidelines.

[c]NA, not available.

vasive nature of soil (i.e., surface and subsurface solids) and water contamination in our nation's population and industrial centers (Havinga and Cotruvo, 1990; Westrick, 1990). This aspect of VOC-related problems suggests that the contamination has occurred for some time and that land-use considerations as well as environmental variables must be made part of any meaningful assessment effort. The second challenge posed by VOCs is their persistence in the environment, particularly when they are present with contaminated solids (Mackay et al., 1985). In this regard, our approaches in environmental assessment (i.e., sampling and analysis) and protection (i.e., interpretation and remediation) must be tuned to address both short-term and long-term risk management goals. It is clear that, in many sampling and analytical activities, gaseous or aqueous symptoms of VOC contamination have been assessed rather than the long-term sources.

It is perfectly reasonable to utilize soil gas and ground-water investigations to *estimate* the spatial extent of VOC contaminant *influences* on the subsurface environment. However, in order to identify the VOC *source* distributions (i.e., the long-term problem), it is essential to focus on the free-product and sorbed VOC phases which, through dissolution and desorption, will continue to release mobile contaminants for decades (Gillham and Rao, 1990; Ball et al., 1992). Soil permeability, moisture content, subsurface lithology, aquifer properties, and preferential pathways (e.g., sewer/pipeline excavations, foundations, etc.) for gas and liquid migration should be determined early in the course of assessment activities. These measurements provide the basis for sampling and analytical protocol selection as well as for data interpretation.

The fates and transport properties of individual VOCs differ substantially based on their vapor pressure, viscosity, aqueous solubility, sorptive characteristics, degradability, and density. The listing of compounds in Table 17.1 categorizes VOCs as DNAPLS (dense nonaqueous-phase liquids) and LNAPLS (light nonaqueous-phase liquids) and shows the range of properties they incorporate. In most investigations, there is little information on the original source, mass of the contaminant mixture released, or the time frame over which transport may have occurred. Nonequilibrium transport and degradation conditions are common in the shallow subsurface. It is clear then that, even with detailed source information, the spatial distribution of free-phase contaminants could not be predicted. Sampling and analysis are therefore necessary components of investigative programs which require substantial care in design, execution, and interpretation.

Objectives of specific programs range from determinations of the severity and extent of contamination to the selection of remedial action options for contaminated surface soils, subsurface solids, and ground water. Data-quality objectives during successive phases of an investigation may be expected to change as the dimensions of the problem are better defined. However, adequate spatial coverage of samples and accurate (i.e., unbiased) analytical detail on major regulated contaminants are universal concerns (U.S. Environmental Protection Agency, 1987, 1995). Identification of principal contaminant concentration distributions (particularly free phase occurrences) in soil, aquifer materials, and ground water is desirable despite the fact that many risk assessment and management methodologies do not address free-phase contaminants adequately (Calabrese and Kostecki, 1992).

The design and execution of sampling and analytical programs, nonetheless, should be approached with both short- and long-term contaminant occurrence and fates in mind. Determinations of both major contaminants and potential breakdown products over a wide range of VOC concentrations must also be done within limited sampling holding periods. With respect to VOCs, this will require comprehensive sample coverage, rapid field screening, and laboratory analytical methods. To date, most sampling and analytical methods which minimize sample transfer, handling, and VOC losses have focused on soil–gas and water sample screening. Considerably more attention should be placed on contaminated solids since the bulk of the organic contaminant mass may often reside on the solid and in free phases (Ball et al., 1992; National Research Council, 1990). The need is particularly acute for the field preservation of discrete solid samples after screening cored materials for quantitative laboratory determinations of contaminant–breakdown-product mixtures.

SAMPLING AND ANALYSIS

Much has been written regarding the fundamentals of the design and execution of environmental sampling efforts (Nielsen and Sara, 1992). Cost, statistical rigor, and coping with the uncertainties in relating sample populations to the universe of soil and aquifer environments are real concerns to scientific professionals and the public. The ideal sample population would pinpoint the primary media and time frames of exposures as well as provide data of known quality for risk assessment and management decision making. For VOCs the ideal sampling design must include rapid turn-around from sampling to analysis. In addition, sampling and preservation steps should enable rather than constrain the analytical process. An example would be sealing a subsample of soil or aquifer solids with a solvent mixture into a closed vial immediately on collection. In this instance, the analytical separation step (i.e., extraction) begins at sample collection.

The compromise between the ideal and the practical has most often been wrought on the anvil of expediency. Compromises can be minimized if uncertainties in contaminant distributions (Keith, 1988; Borgman and Quimby, 1988) and environmental characteristics (Hoeksema and Kitanidis, 1985; Urban and Gburek, 1988) are anticipated from the outset, and further sampling and analytical efforts are planned with the basis of initial field analytical results, hydrogeologic data, and geostatistical interpretations.

New Directions in Sampling

The techniques available for field sampling and analysis of gas, water, and solids have grown rapidly (U.S. Environmental Protection Agency, 1988; Robbins et al., 1989; Simmons, 1991), and they may be expected to advance further in the future (Barcelona and Helfrich, 1992). One of the major emerging research directions deals with immediate preservation of VOC samples with the extraction solvent used for more detailed laboratory analysis. Bone (1988) has published an application of this type of method (i.e., methanol preservation of solid samples) which essentially begins the analytical process in the field and affords the possibility of more meaningful VOC analyses on heterogeneous materials. His work included sampling, field preservation, and analyses of 13 chlorinated and nonchlorinated VOCs in soils. For 12 of 13 compounds, the methanol-preserved samples consistently showed several orders of magnitude higher concentrations than bulk samples held for subsequent laboratory purge and trap analyses. These observations have been supported by the work of a number of other groups (University of Wisconsin Extension et al., 1993; Maskarinec and Moody, 1988).

Another example further illustrates both (a) the impact of minimal sample handling and field preservation on determinations of VOCs in cores of aquifer solids and (b) the relative amounts in the solids versus the associated ground water. In this experiment, split-spoon core samples were taken adjacent to the screens of four monitoring wells in a plume at a VOC contamination site. The primary contaminants were trichloroethane (TCA), *cis*-1,2-dichloroethylene (c1,2DCE), and dichloroethane (DCA). Table 17.2 shows the relative concentrations of TCA in the water versus the solids for a representative 1-L aquifer volume containing ~300 mL of H_2O and 1750 g of aquifer solid. There is a marked difference between the samples preserved in the field with ~75% (v/v) methanol versus samples placed in a bulk sampling jar and refrigerated at 4°C prior to analysis. In three out of four cases the bulk jar sample yielded TCA levels below the quantification limit (i.e., ~5 $\mu g/g$). In fact, none of the bulk jar samples showed any evidence of solvent contamination from other components present in the plume.

The majority of potential errors involved in collection, handling, and analysis of VOC samples may be expected to lead to compound losses and low results. One clearly can avoid gross sampling and analysis errors for VOC determinations by field preservations of solids with methanol.

TABLE 17.2
Relative Masses (μg) of TCA in Ground Water and Aquifer Solids for a Representative One-Liter Aquifer Volume

Well	Ground Water	MeOH Preserved (Solid)	Bulk Jar, 4°C (Solid)	% Total TCA in Solid
15	13.5	21.0	—[a]	60
31	85.7	516.0	52	86
32	11.1	30.9	—	74
37	83.7	239	—	74

[a](—) *denotes no detectable levels of TCA in bulk jar sample by static headspace gas chromatography–Hall electrolytic conductivity detection (7-day holding time) (M. J. Barcelona, and D. M. Shaw, 1992, unpublished data).*

It should be noted that national regulatory guidance for the collection and handling of VOC samples recommends bulk jar sampling of cored solids or water with 4°C refrigeration for up to 14 days (U.S. Environmental Protection Agency, 1986; University of Wisconsin Extension et al., 1993). This presents an obvious conflict between the results of recent research (West et al., 1995; Hewitt and Grant, 1995) and recommended practice for environmental sampling. Continuing to apply flawed sampling and preservation methods will have major consequences on the accuracy of measured VOC distributions in the subsurface environment and potentially the remedial actions chosen to clean it up. Several states have recognized this problem and have formulated specific guidance for field preservation steps (Wisconsin Department of Natural Resources, 1993).

New Directions in Analysis

Laboratory analysis techniques exist which can handle difficult matrices adequately (Keith, 1988). The critical considerations here are the difficulties associated with spiking solid samples with standards (e.g., internal or calibration standards) for quality control and the need for minimizing sample handling and transfers prior to VOC analyses (e.g., for purge-and-trap methods). Headspace methods for initial separation, field screening, and lab analysis have promise in this regard.

Several groups have also been working to further integrate sampling and analysis steps which should lead to improved accuracy and precision of VOC determinations. Voice and Kolb (1993) observed superior accuracy and precision of static headspace determinations of a range of VOCs in soil–water slurries as compared to conventional purge-and-trap analyses. Similarly, Hewitt and co-workers (Hewitt et al., 1992; Hewitt and Grant, 1995) have compared headspace–gas chromatography (GC) field screening analysis with laboratory purge-and-trap GC methods for VOCs in soils. They found, in general, that the field preservation and analytical techniques provided quantitative results essentially equivalent to the slower, more expensive lab methods.

ERROR IDENTIFICATION AND CONTROL

The common analytical practice area of environmental chemists frequently places them in the role of "suspects" when outliers or unforeseen results are reported. Traditionally, quality assurance/quality control (QA/QC) programs have been limited to the chain of custody and analysis of samples as received. It should be recognized that the uncertainties associated with environmental sampling designs (i.e., from sampling location in space and time to the selection of sampling devices) far exceed those involved in the analytical process itself.

The need to extend QA/QC measures to the design and operation of monitoring networks is well illustrated clearly by common practice for ground-water quality investigations. In these situations, spatial and temporal variations in contaminant or other chemical constituent distributions may need to be determined at scales of meters to kilometers and months to decades. The role of the environmental chemist should include, at a minimum, the identification of the types and magnitudes of important chemical distributions and the identification of major sources of variabilities in order to minimize errors in sampling and analysis. It may be anticipated that these lines of research will eventually improve our overall estimation of VOC distributions in the subsurface, enabling more extensive spatial sample coverage at reduced cost and effort.

Major concerns in the evaluation of contaminant distributions in the subsurface include the spatial and temporal comparability of VOC concentrations determined in both water and aquifer solid samples. Spatial trend analysis and averaging techniques are quite sensitive to the uncertainty in chemical concentrations determined on discrete samples (Black, 1988; Barth and Mason, 1984). Temporal trend analyses share this sensitivity and also suffer from the use of different sampling techniques, personnel, frequencies and analytical laboratories over the course of an investigation (Keith, 1988). Clearly, we should strive to reduce variability from human and methodologic sources and identify processes which control contaminant distributions, transformations, and fates.

In order to reliably estimate actual spatial or temporal concentrations trends, rather than systematic errors or biases in sampling and analysis, it is necessary to conduct controlled field experiments employing simple well-documented sampling protocols. Temporal trend monitoring of ground-water quality and dissolved geochemical constituents can be done effectively without extraordinary efforts to minimize sampling and analytical errors (Stoline et al., 1993). The use of dedicated bladder pumps, low-flow-rate purging, and uniform water sample handling and analysis techniques reduces these sources of error to less than 10% of observed variability in inorganic parameters such as dissolved oxygen, major cations/anions, or nutrients (Barcelona et al., 1989). The application of this proven ground-water sampling protocol to VOC contamination of a large ($>10/\text{mi}^2$) site in a shallow sand and gravel aquifer has yielded similarly encouraging results (Barcelona et al., 1994).

In this study, a network of more than 40 wells was constructed in an intensive study area of $\sim 4 \text{ mi}^2$ to supplement several synoptic surveys of residential wells and the overall NPL (National Priority List) site investigation. The major objectives of the study were to evaluate levels of spatial and temporal variability of VOCs in the ground water and to intercalibrate selected water and aquifer solid sampling/preservation methods for site characterization. Quarterly sampling of the study wells for 21 months and several sampling intercalibration experiments were conducted focusing on five principal VOC contaminants (i.e., 1,1-DCE, 1,1-dichloroethylene; 1,1-DCA, 1,1 dichloroethane; 1,2c-DCE, 1,2 cis dichloroethylene, 1,1,1 TCA, 1,1,1-trichloroethane, and TCE, trichloroethylene).

Sources of variability were categorized by their respective contributions to the total variance in VOC concentrations for ground-water samples. Variance contributions from analytical (e.g., calibration and QA/QC standards), sampling (e.g., spiked field samples and blanks), and "natural" (e.g., temporal or spatial variations) sources were summed for each VOC contaminant and each well over the study period. The average results of this apportionment of sources of VOC variability for all wells are shown in Table 17.3.

The major source of variability in VOC concentrations in ground water at this site was clearly natural variability. Analytical and sampling errors combined accounted for less than 15% of the total variance within the dataset. Since these sources contribute such a small portion of the total variability, one would not gain much by attempts to further reduce error in sampling or analysis procedures. In this study, normal care was taken in laboratory QA/QC. Though a number of field and lab personnel took part in the sampling and analytical preparations, adherence to the simple written sampling protocol was sufficient to minimize potential errors in sample collection and handling. The protocol has consistency among its principal virtues because it minimizes the potential for operator error, equipment malfunction, and the influence of adverse weather conditions.

The results of this study identify the principal sources of error when minimal QA/QC considerations are extended to the field via a proven sampling protocol. The resolution of meaningful spatial and temporal concentration trends would certainly be possible with this type of study design.

The substitution of a less reproducible and negatively biased sampling device (e.g., a bailer) for the bladder pump in the proven sampling protocol could introduce significant sampling errors and mask real concentration trends. In fact, common practice in ground-water investigations includes the use of bailers for sampling despite consistently low VOC recoveries and poor, operator-dependent precision. More recent regulatory guidance acknowledges the difficulties associated with bailer sampling for VOCs and recommends the use of positive displacement pumps (U.S. Environmental Protection Agency, 1992).

Technological advances will continue to be very important to the development of environmental chemistry as a diverse discipline. We should strive to employ better technology or more rigorous QA/QC procedures where they will achieve the most return for the effort. Clearly, we need to maintain a focus on the sources as well as the symptoms of VOC contamination to extend our understanding of the processes and effects which control their environmental distributions.

TABLE 17.3
Overall Mean, Relative Standard Deviation, and Percentage of Total Variance Attributable to Laboratory or Field (Sampling) Error and Natural Variability (November 1990–September 1992)

			Percent of Total Variability		
VOC	Overall Mean (μg/L)	Relative Standard Deviation (%)	Laboratory (%)	Field (%)	Natural (%)
TCA	119.5	36	1.29	3.26	95.45
TCE	29.8	43	1.95	12.75	85.30
c-1,2-DCE	45.2	32	1.69	4.72	93.59
1,1-DCA	44.3	28	1.02	5.22	93.76
1,1-DCE	16.3	31	3.61	4.15	92.24

ACKNOWLEDGMENTS

The author gratefully acknowledges the contributions of H. A. Wehrmann, D. M. Shaw, G. T. Blinkiewicz, B. Dube, and D. Patt.

DISCLAIMER

Although the information in this document has been funded in part by the U.S. Environmental Protection Agency under Cooperative Agreement No. CR-815681 with the Illinois State Water Survey and No. CR-822922 with the University of Michigan from EPA and the Department of Defense S.E.R.D.P. Program, it has not been subject to Agency review and, therefore, does not necessarily reflect the views of the Agency, and no official endorsement should be inferred. Mention of trade names or commercial products does not constitute endorsement or recommendation for use.

References

Ball, J. W., Curtis G. P., and Roberts, P. V., 1992, Physical–chemical interactions with subsurface solids: the role of mass transfer. In: *Proceedings of Subsurface Restoration Conference—Third International Conference on Ground Water Quality Research*, Dallas, TX, June 21–24, 1992, Rice University, Houston, TX, pp. 8–10.

Barcelona, M. J., 1988, Overview of the sampling process. In: L. H. Keith, ed., *Principles of Environmental Sampling*, ACS Professional Reference Book, American Chemical Society, Washington, D.C., Chapter 1.

Barcelona, M. J., and Helfrich, J. A., 1992, Realistic expectations for ground water investigations in the 1990's. In: D. M. Nielson and M. N. Sara, eds., *Current Practices in Ground-Water and Vadose Zone Investigations*, American Society for Testing and Materials. Philadelphia, 431 pp.

Barcelona, M. J., and Shaw, D. M., 1992, Water Quality Laboratory, Institute for Water Sciences, Western Michigan University, unpublished data.

Barcelona, M. J., Lettenmaier, D. P., and Schock, M. R., 1989, Network design factors for assessing temporal variability in Ground water quality. *Environ. Monit. Assess.*, 12, 149–179.

Barcelona, M. J., Wehrmann, H. A., and Varljen, M. D., 1994, Reproducible well purging procedures and VOC stabilization criteria for ground-water sampling. *Ground Water*, 32, 1, 12–22.

Barth, D. S., and Mason, B. J., 1984, Soil sampling quality assurance and the importance of an exploratory study. In: G. E. Schweitzer and J. A. Santolucito, eds., *Environmental Sampling for Hazardous Wastes*, ACS Symposium Series, No. 267, American Chemical Society, Washington, D.C., Chapter 10, pp. 97–98.

Black, S. C., 1988, Defining Control Sites and Blank Sample Needs, Chapter 7, p. 1077–1117, In: L. H. Keith, ed., *Principles of Environmental Sampling*, ACS Professional Reference Book, American Chemical Society, Washington, D.C., 458 pp.

Bone, T. T., 1988, Preservation techniques for samples of solids, sludges and non aqueous liquids. In: L. H. Keith, ed., *Principles of Environmental Sampling*, ACS Professional Reference Book, American Chemical Society, Washington, D.C., Chapter 29, pp. 409–423.

Borgman, C. E., and Quimby, W. F., 1988, Sampling for tests of hypothesis when data are correlated in space and time. In: L. H. Keith, ed., *Principles of Environmental Sampling*, ACS Professional Reference Book, American Chemical Society, Washington, D.C., Chapter 2.

Calabrese, E. J., and Kostecki, P. T., 1992, *Risk Assessment and Environmental Fate Methodologies*, Lewis Publishers, Chelsea, MI, 150 pp.

Gammage, R. B., and Berven, B. A., 1992, *Hazardous Waste Site Investigations—Towards Better Decisions*, Lewis Publishers, Chelsea, MI, 288 pp.

Gillham, R. W., and Rao, P. S. C., 1990, Transport, distribution and fate of volatile organic compounds in ground water. In: N. M. Ram, R. F. Christman, and K. P. Cantor, eds., *Significance and Treatment of Volatile Organic Compounds in Water Supplies*, Lewis Publishers, Chelsea, MI, Chapter 9, p. 141–181.

Havinga, A., and Cotruvo, J. A., 1990, Statutory and regulatory basis for control of volatile organic chemical in drinking water. In: N. M. Ram, R. F. Christman, and K. P. Cantor, eds., *Significance and Treatment of Volatile Organic Compounds in Water Supplies*, Lewis Publishers, Chelsea, MI, Chapter 1, pp. 3–13.

Hewitt, A. D., and Grant, C. L., 1995, Round robin study of performance evaluation soils vapor fortified with volatile organic compounds. *Environ. Sci. Technol.*, 29, 769–774.

Hewitt, A. D., Mijares, P. H., Leggett, D. C., and Jenkins, T. F., 1992, Comparison of analytical methods for determination of volatile organic compounds in soils. *Environ. Sci. Technol.*, 26, 10, 1932–1938.

Hoeksema, R. J., and Kitanidis, P. K., 1985, Analysis of the Spatial Structure of properties of selected aquifers. *Water Resources Res.*, 21, 4, 563–572.

Keith, L. H., 1988, *Principles of Environmental Sampling*, ACS Professional Reference Book, American Chemical Society, Washington, D.C., 458 pp.

Keith, L. H., Mueller, W., and Smith, D. L., 1991, *Compilation of EPA's Sampling and Analysis Methods*, Lewis Publishers, Chelsea, MI, 803 pp.

Mackay, D. M., Roberts, P. V., and Cherry, J. A., 1985, Transport of organic contaminants in ground water. *Environ. Sci. and Technol.*, 19, 5, 384–392.

Maskarinec, M. P., and Moody, R. L., 1988, Storage and preservation of environmental samples. In: Keith, L. H., ed., *Principles of Environmental Sampling*, ACS Professional Reference Book, American Chemical Society, Washington, D.C., Chapter 9, pp. 145–155.

National Research Council, 1990, *Ground Water and Soil Contamination Remediation Toward Compatible Science, Policy and Public Perception*, Water Science and Technology Board, National Academy Press, Washington, D.C., 261 pp.

Nielsen, D. M., and Sara, M. N., 1992, *Current Practices in Ground Water and Vadose Zone Investigations ASTM 1118*, American Society for Testing and Materials, Philadelphia, 431 pp.

Puls, R. W., 1994, Ground water sampling for metals. In: B. Markert, ed., *Environmental Sampling for Trace Analysis*, VCH, Weinheim, F.R. of Germany, Chapter 14.

Robbins, G. A., Bristol, R. D., and Roe, V. D., 1989, A field screening method for gasoline contamination using a polyethylene bag sampling system. *Ground Water Monit. Rev.*, 9, 3, 87–97.

Sanders, T. G., Ward, R. C., Loftis, J. C., Steele, T. D., Adrian, D. D., and Yevjevich, V., 1983, *Design of Networks for Monitoring Water Quality*, Water Resources Publications, Littleton, CO, 328 pp.

Simmons, M. S., ed., 1991, *Hazardous Waste Measurements*, Lewis Publishers, Chelsea, MI, 315 pp.

Stoline, M. R., Passero, R. N., and Barcelona, M. J., 1993, Statistical trends in ground water monitoring data at a landfill superfund site. *Environ. Monit. Assess.*, 27, 201–219.

University of Wisconsin Extension, U.S. Environmental Protection Agency, Oak Ridge National Laboratory, U.S. Department of Energy, U.S. Toxic and Hazardous Materials Agency, America Petroleum Institute, 1993, *National Symposium on Measuring and Interpreting VOC's in Soils: State of the Art and Research Needs*, January 12–14, Las Vegas, NV.

Urban, J. B., and Gburek, W. J., 1988, A geologic and flow-system based rationale for ground water sampling. In: A. G. Collins and A. I. Johnson, eds., *Ground Water Contamination: Field Methods, ASTM 963*, American Society for Testing and Materials, Philadelphia, pp. 468–481.

U.S. Environmental Protection Agency, 1986, *Test Methods for Evaluating Solid Waste—Physical/Chemical Methods,* SW-846, Office of Solid Waste, Washington, D.C.

U.S. Environmental Protection Agency, 1987, *Data Quality Objectives for Remedial Response Activities Development Process*, U.S. EPA 540/G-87/003, March.

U.S. Environmental Protection Agency, 1988, U.S. Army Toxic and Hazardous Materials Agency, Instrument Society of America, *First International Symposium on Field Screening Methods for Hazardous Waste Site Investigations*, October.

U.S. Environmental Protection Agency, 1992, *RCRA Ground-Water Monitoring: Draft Technical Guidance*, Office of Solid Waste, EPA 1530-R-93-001, Washington, D.C., November.

U.S. Environmental Protection Agency, 1995, *Ground Water Sampling—A Workshop Summary*, U.S. EPA Office of Research and Development, EPA/600/R-94/205, 98 pp.

Voice, T. C., and Kolb, B., 1993, Static and dynamic headspace analysis of volatile organic compounds in soils. *Environ. Sci. Technol.*, 27, 4, 709–713.

West, O. R., Siegrist, R. C., Mitchell, T. J., and Jenkins, R. A., 1995, Measurement error and spatial variability effects on characterization of volitle organics in the subsurface. *Environ. Sci. Technol.*, 29, 647–656.

Westrick, J. J., 1990, National surveys of volatile organic compounds in ground and surface waters. In: N. M. Ram, R. F. Christman, and K. P. Cantor, eds., *Significance and Treatment of Volatile Organic Compounds in Water Supplies*, Lewis Publishers, Chelsea, MI, Chapter 7, p. 103–126.

Wisconsin Department of Natural Resources, 1993, *Leaking Undergorund Storage Tank and Petroleum Analytical and Quality Assurance Guidance*, publication number PUBL-SW-130 93, July 1993.

Metal–Phytoplankton Interactions in Marine Systems

KEITH A. HUNTER, JONATHAN P. KIM,
AND PETER L. CROOT

ABSTRACT. The effects of trace metals on biological organisms in the marine environment have been of considerable research interest over the last two decades. This period of time has seen major improvements in analytical techniques for the measurement and study of trace metals and their subsequent application to a broad spectrum of water systems around the globe. This work has included studies of the geographical distributions of trace metals in different water types in the ocean, studies of the modes of transport and uptake of trace metals, and studies of historical changes in trace metal accumulation. Initial research has highlighted strong similarities in the behaviors of many trace metals and those of well-understood chemical nutrients such as phosphate, nitrate and silicate, known to be essential for phytoplankton growth in the ocean. This suggests that phytoplankton regulate the trace metal composition of seawater for their own benefit. A major achievement in this regard has been the ability to measure, at the sub-nmol/L level, trace metal species that are complexed by naturally occurring organic ligands of biological origin in ocean waters. It is now clear that the free ion activities of many essential and/or toxic trace metals are regulated by highly specific, strongly complexing ligands exuded by marine phytoplankton. This research encourages a new paradigm in which the growth rates and species composition of primary marine organisms are affected by trace metals at concentration levels orders of magnitude lower than is conventionally believed. As a consequence, the capacity of natural waters to assimilate trace metal contaminants may be correspondingly much lower than is currently thought reasonable.

INTRODUCTION

Metallic elements, notably Fe, Cr, Mn, Co, Cu, Ni, Zn, Cd, Hg, and Pb, have been ubiquitous by-products of human activities since the earliest civilized times. Therefore, it is hardly surprising that contamination of the aquatic systems by metallic elements, even in trace amounts, has been of considerable concern for decades. However, it is only over the last 10 or 20 years that scientific methodology has improved to the point that the geochemical behavior of trace metals and how they interact with biological organisms can be adequately studied. Indeed, recent progress has been so great that conventional thinking about the role of trace metals in environmental systems has been largely overturned.

To date, most advances in our understanding of trace metals in aquatic systems are derived from studies of oceanic waters. This is because the open oceans provide a geochemically stable milieu that receives relatively low and constant external inputs of material from coastal waters and the atmosphere. Moreover, the vertical structure of ocean waters is highly stabilized by salinity and temperature-derived density gradients. Thus open-ocean systems can be driven to the limits of their ecological and evolutionary responses to trace metals (Morel and Hudson, 1985). By contrast, freshwater systems (lakes and rivers) are subject to rapid compositional and seasonal changes and exhibit great variety in chemical and biological properties (Meybeck, 1981).

PROBLEMS OF CHEMICAL ANALYSIS

The practical problems involved in the chemical analysis of trace metals in aquatic systems are now widely known (Turekian, 1977; Hunter, 1982; Bruland et al., 1991). The concentrations of most trace metals in aquatic systems are so low in comparison to those prevailing in the materials and flotsam of the average chemistry laboratory (dust, dandruff, paint flakes, chocolate vapor) that without special precautions, a water sample is quickly overwhelmed by external, spurious sources of trace metals. Much of our progress in developing sample collection and analysis methods appropriate to the very low concentrations found in environmental samples can be traced to the pioneering efforts of Clair Patterson at the California Institute of Technology, whose painstaking work on the measurement of Pb and stable Pb isotopes (Patterson and Settle, 1976) has been the inspiration for those who have followed.

Reliable measurements of trace metals in natural waters require scrupulous attention to all steps in the analytical protocol from sample collection to analysis. This includes (a) the preparation of trace-metal–clean-sample containers and (b) rigorously controlled collection and handling of the sample. Also necessary are carefully designed analytical techniques, including the preparation of specially purified reagents, and the strict control throughout of external sources of contamination by use of specialized clothing and clean-room working conditions (Patterson and Settle, 1976; Bruland et al., 1979; Ahlers et al., 1990).

BIOLOGICAL ROLES FOR TRACE METALS

Modern biochemistry makes it clear that a variety of trace metals are essential for biological processes in all organisms, including oceanic phytoplankton. Examples of metals that have key roles in the functioning of metalloprotein enzymes are Zn, Fe, Cu, Co, Mo, and Mn. What has not been clear until recently is whether the availability of any of these elements is ever a factor that limits biological growth under oceanic conditions. The main physical and chemical requirements for photosynthetic activity are well known:

- Sunlight as the basic energy source
- Pigments for the capture and conversion of light energy
- Nitrogen (mainly as nitrate NO_3^-) for synthesis of protein amino acids
- Phosphorus (as phosphate PO_4^{3-}) for ATP, the energy transfer agent
- Trace elements—for example, Zn, Fe, Cu, Co, Mo, and Mn

In addition, oceanic phytoplankton have two further chemical requirements related to the formation of the exoskeleta:

- Silicate $Si(OH)_4$ for formation of the opal exoskeleta of diatoms
- Ca^{2+} and CO_3^{2-} for formation of the $CaCO_3$ exoskeleta of coccolithophores

Over 60 years ago, Redfield (1934) noted that the oceanic distributions of nitrate and phosphate were highly correlated with those of inorganic carbon and dissolved oxygen, the essential precursor and reaction product, respectively, of photosynthetic algal growth. By correlating the concentrations of these substances in different waters, Redfield showed that the variations were related to each other by approximately constant atomic ratios (140 C:140 O_2:20 N:1 P), values remarkably close to the respective atomic ratios of C, N, and P in phytoplankton and the oxygen deficit compared to CO_2 caused by photosynthesis.

This is shown in Figure 18.1, which illustrates data from almost 10,000 analyses for phosphate and nitrate made during the GEOSECS Program, a large international program of research into the chemistry of the oceans conducted during the 1970's. A remarkably constant relationship between these two nutrients is demonstrated by the data.

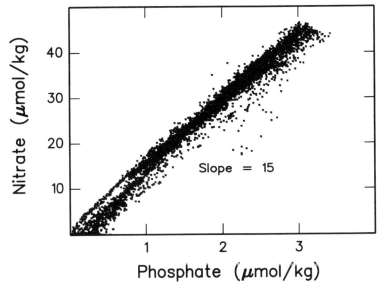

Figure 18.1. Relationship between nitrate and phosphate concentrations in the global ocean. Drawn using data gathered during the GEOSECS Program, all stations (9836 points).

Similarly, Figure 18.2 shows the relationship between total dissolved carbon dioxide SCO_2 and phosphate from the same GEOSECS Program. From relationships of this type, Redfield concluded that the chemical variations in the ocean were a result of photosynthetic uptake of C, N, and P (plus associated O_2 formation) followed by the corresponding oxidative decomposition of the photosynthesized organic matter. This simple conceptual model is now a central paradigm of chemical oceanography and limnology.

The role of carbon and O_2 in Redfield's paradigm is an obvious one. Phosphate and nitrate are also involved because these two substances represent the main chemical form in seawater of the elements P (needed, for example, for biosynthesis of phosphate esters such as those based on adenosine) and N (required for, *inter alia*, biosynthesis of amino acids). Some phytoplankton are capable of fixing N from N_2 gas dissolved in seawater rather than using nitrate. Redfield (1934) also noted that the nutrients phosphate and nitrate both tended to disappear from surface waters almost simultaneously, suggesting that algal growth may be co-limited by supply of N and P. He went on to propose that such a co-limitation was likely to arise if oceanic phytoplankton have, through long-term feedback mechanisms, regulated the supply of nitrogen by denitrification and nitrogen fixation to match that of phosphate.

From these considerations we can build up a picture of the chemical requirements for photosynthesis. Table 18.1 shows a comparison of the molar amounts of CO_2, NO_3^-, Ca^{2+}, and $Si(OH)_4$ relative to that of PO_4^{3-}, needed for photosynthesis with the availability of each chemical substance in seawater. What emerges from this comparison is that to a first approximation, NO_3^-, $Si(OH)_4$, and PO_4^{3-} will all be simultaneously depleted in seawater by phytoplankton growth while, at the same time, large excesses of CO_2 and Ca^{2+} will remain. As a consequence, the three former substances are often termed *limiting nutrients* because their availability in seawater often limits the extent of biological productivity.

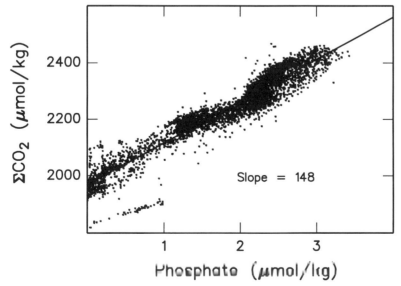

Figure 18.2. Relationship between total dissolved CO_2 and phosphate concentrations in the global ocean. Drawn using data gathered during the GEOSECS Program, all stations (5249 points).

TABLE 18.1
Comparison of the Molar Amounts (relative to PO_4^{3-}) of Different Chemical Substances Needed for Phytoplankton Growth in the Ocean with Their Availability in Seawater

Purpose	CO_2	NO_3^-	PO_4^{3-}	Ca^{2+}	$Si(OH)_4$
Photosynthesis	108	15	1	—	—
Coccolith exoskeleta	40	—	—	40	—
Diatom exoskeleta	—	—	—	—	40
Total	148	15	1	40	40
Availability in seawater	1200	15	1	4000	40
Surplus	88%	0	0	99%	0

At first examination, this picture seems to be in agreement with what is known about plankton productivity in the ocean. Figure 18.3 shows the distribution of the nutrients phosphate and nitrate in surface waters of the Western Atlantic Ocean, along with ΣCO_2, alkalinity, and temperature. These data show that the surface waters of the vast central region of the Atlantic, between latitudes 30°S and 50°N, are almost completely devoid of both phosphate and nitrate. A similar situation prevails for silicate.

The Pacific Ocean shows the same depletion at mid-latitudes except that the equatorial region of the Pacific is affected by the upwelling of deep, nutrient-rich waters (Figure 18.4). The relatively low phytoplankton productivity of the mid-latitude waters of the Atlantic and Pacific Oceans would seem, therefore, to be consistent with the very low concentrations of the nutrients.

What does not fit well with the Redfield paradigm is the corresponding situation in high-latitude waters where high concentrations of the nutrients are observed (Figures 18.3 and 18.4). For example, at latitudes south of 30°S in the cold sub-Antarctic and Antarctic regions, phosphate, nitrate, and silicate concentrations rise to extremely high values that should support prodigious phytoplankton growth. Indeed, it has long been a puzzle to oceanographers why phytoplankton growth has not also depleted these waters in nutrients.

Examination of images of phytoplankton pigment concentrations (an indirect measure of phytoplankton productivity) provided by the Nimbus-7 Coastal Zone Scanner satellite strengthens this puzzling view, but also reveals a possible clue. The open ocean, high nutrient areas of the Southern Ocean have quite low pigment concentrations, indicating that they are low in productivity, just like the nutrient-poor mid-latitude regions. However, high pigment concentrations are observed uniformly near the continents. This distribution strongly implies that another essential component for photosynthesis, one that is supplied from the continents, is necessary for phytoplankton growth. The waters of the Southern Ocean, which are remote from the influence of major continents, therefore appear to be lacking an essential ingredient for phytoplankton growth. Today, we believe that this magic ingredient is one or more of the biologically essential trace metals that are supplied to the ocean by the weathering of continental rocks and soils. It is important, therefore, to consider the behavior of these biologically active trace metals in the ocean.

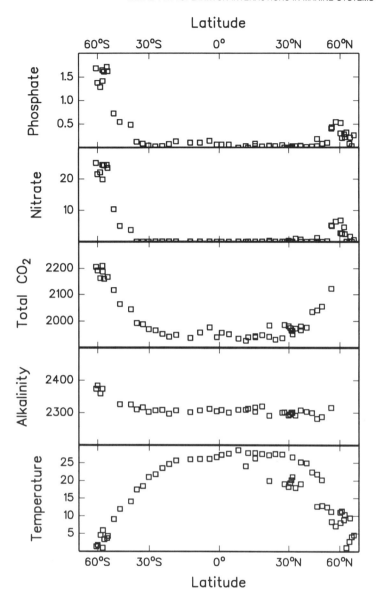

Figure 18.3 Surface water concentrations of nitrate, phosphate, total dissolved CO_2, alkalinity, and temperature in the Atlantic Ocean as a function of latitude. Drawn using data gathered during the GEOSECS Program.

BIOACTIVE TRACE METALS IN THE OCEAN

The bioactive trace metals Fe, Mn, Co, Cu, and Zn, as well as Cd, have figured prominently in the development and use of clean room methods of analysis (Bruland, 1983; Bruland et al., 1991). Consequently, the geographical distributions of these elements have been systematically studied by several research groups in almost all of the major

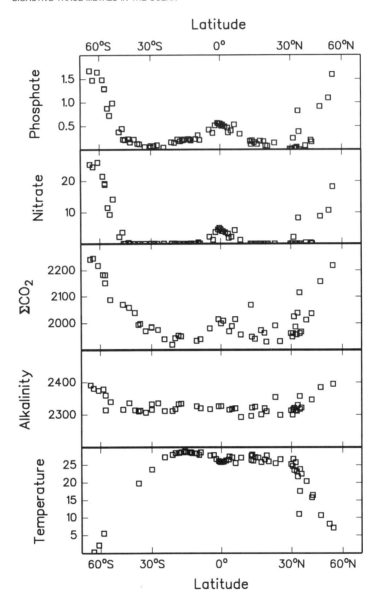

Figure 18.4. Surface water concentrations of nitrate, phosphate, total dissolved CO$_2$, alkalinity, and temperature in the Pacific Ocean as a function of latitude. Drawn using data gathered during the GEOSECS Program.

regions of the global ocean and the results obtained are largely consistent with our understanding of oceanographic processes (Boyle et al., 1976, 1977, 1981; Bruland, 1980; Bruland et al., 1978a,b; Danielsson, 1980; Knauer and Martin, 1981; Bruland and Franks, 1983; Bender et al., 1977; Martin and Gordon, 1988; Martin and Fitzwater, 1988; Hunter and Ho, 1991; Frew and Hunter, 1992).

The first important observation arising from these studies is that trace metal concentrations are generally orders of magnitude lower than was believed in the mid-1970s

(Turekian, 1977; Bruland, 1983; Bruland et al., 1991). Indeed, they are generally well below the practical detection limits of most "conventional" analytical techniques when the blank contribution from spurious contamination is properly accounted for.

A second important observation is that trace metals show vertical concentration profiles that are related to those of well-understood, conventionally measured parameters such as temperature, salinity, dissolved oxygen, dissolved inorganic carbon, or the phytoplankton micronutrients phosphate, nitrate, and silicate (Boyle et al., 1976). This property not only provides a means for assessing the reliability of data, but also provokes interesting questions about the biological role of trace metals (Morel and Hudson, 1985).

The general behavior of bioactive trace metals in the ocean will now be illustrated using data reported by Bruland (1980) for a station at 32°41′N, 144°59′W in the North Pacific. Figure 18.5 shows vertical profiles of salinity, silicate, phosphate, and nitrate, while the corresponding profiles for Cd, Zn, Cu, and Ni are presented in Figure 18.6. It can be noted that the trace metal data show the same smooth trend of increasing concentration with depth in the water column that is exhibited by the conventional parameters in Figure 18.5. More detailed consideration of several trace metals is presented below.

Cadmium

As first noted by Martin et al. (1976) and confirmed by later work, the vertical distribution of cadmium in the ocean closely mimics that of the labile nutrients phosphate and nitrate and is therefore highly interesting because of the special role of the latter nutrients in regulating algal growth in the ocean.

Figure 18.7 compares the correlation of both nitrate and cadmium with phosphate for the results of Bruland (1980) given in Figures 18.5 and 18.6. It is clear that in this case, Cd is linearly related to phosphate as least as well as is the case for nitrate. Similar close correlations between Cd and phosphate (or nitrate) have been reported by

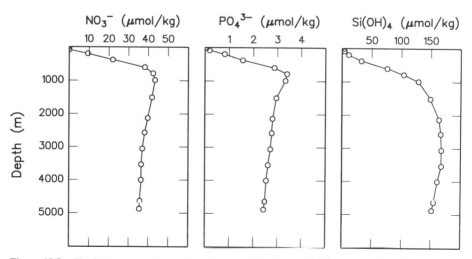

Figure 18.5.   Vertical concentration profiles of nitrate, phosphate, and silicate at a station in the North Pacific (32°41.0′N, 144°59.5′W). Drawn using data gathered by Bruland (1980).

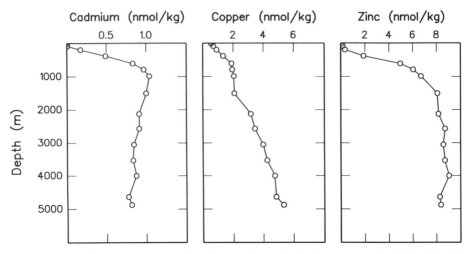

Figure 18.6. Vertical concentration profiles of copper, zinc, and cadmium at a station in the North Pacific (32°41.0′N, 144°59.5′W). Drawn using data gathered by Bruland (1980).

many authors (Bruland et al., 1978a; Boyle et al., 1976; Danielsson, 1980; Bruland and Franks, 1983; Hunter and Ho, 1991; Frew and Hunter, 1992).

However, later work has shown that the slope of the Cd–phosphate relationship is not constant throughout the global ocean, exhibiting marked depletion in sub-Antarctic waters (Nolting et al., 1991; Frew and Hunter, 1992, 1995) and enrichment in some Antarctic waters (Frew, 1995). Significant departures from the Cd–phosphate ratio typical of deep Pacific waters have been noted in surface waters and more shallow shelf waters (Boyle et al., 1981; Hunter and Ho, 1991). Figure 18.8 presents a summary of Cd-phosphate data for sub-surface waters of the global ocean taken from Boyle (1984) and Frew and Hunter (1992). It shows a distinctive "kink" that corresponds to the transition from Atlantic and mid-depth waters to the deep waters of the Pacific and Indian

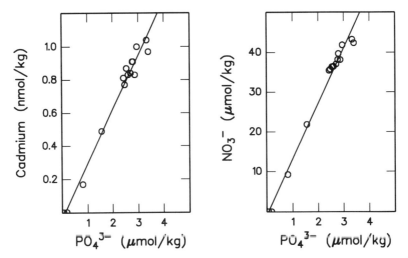

Figure 18.7. Comparison of the relationships of nitrate and cadmium with phosphate for the North Pacific (32°41.0′N, 144°59.5′W) in Figures 18.5 and 18.6. Drawn using data gathered by Bruland (1980).

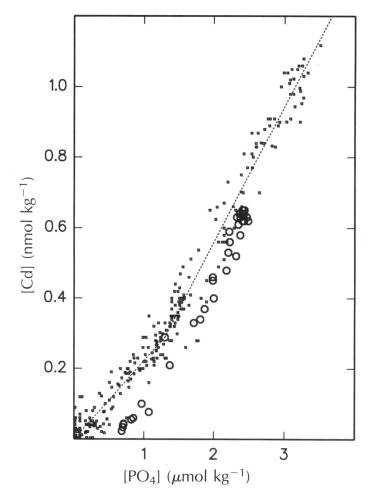

Figure 18.8. The global cadmium–phosphate relationship for subsurface waters. Drawn using data compiled by Boyle (1984). The large open circles represent samples collected in low-Cd sub-Antarctic waters. Reprinted with permission from *Nature* 360: 144–146. Copyright 1992 Macmillan Magazines Limited.

Oceans. This kink is thought to be caused by injection of low-Cd sub-Antarctic waters into the intermediate waters of the global ocean (Frew and Hunter, 1992) and by formation of high-Cd Antarctic Bottom waters near the Antarctic continent (Frew, 1995).

Despite these variations, it is clear that Cd is closely involved in the Redfield model of organic matter formation through photosynthesis and its subsequent remineralization through oxidation deeper in the water column. Exactly why it is involved so intimately is not clearly understood at present. Cadmium has no known or specific biochemical function and is often regarded as exhibiting only toxic properties. Although Cd containing metallothioneins have been reported in many organisms, their formation is generally regarded to be a detoxification mechanisms (Webb, 1979). However, Price and Morel (1990) have shown that Cd^{2+} can substitute for Zn^{2+} in essential growth mechanisms of the diatom *Thalassiosira weissflogii*. The substitution is highly effective, allowing Zn-deficient cells to grow at 90% of their maximum rate when supplied with Cd. Whether Cd is sufficiently abundant in ocean waters relative to Zn for this substitution mechanism to be globally important, however, remains in doubt (*vide infra*).

Walsh and Hunter (1992) showed that the macroalgae *Macrocystis pyrifera* accumulates Cd 40–50 times more efficiently during formation of intracellular polyphosphate bodies (PPB) than in controls without PPB formation. Moreover, both Cd and phosphate levels in the alga were seasonally correlated and Cd levels were maximal when PPB were present in the cells. PPB are ubiquitous phosphate storage products in most organisms and form rapidly during periods of excess phosphate supply (Kulaev, 1979). Walsh and Hunter (1992) argued that if this mechanism also applies to marine microalgae, then it may provide an important link between the oceanic behaviors of Cd and phosphate. In particular, it may explain why Cd appears to be depleted relative to phosphate in high-latitude and upwelling regions of the ocean (Nolting et al., 1991; Frew and Hunter, 1992), because these nutrient-rich waters would provide the best conditions for PPB formation.

Although the biochemical mechanism for Cd uptake by marine algae is not yet known with certainty, the consistent relationship between Cd and phosphate concentrations in subsurface waters of the global ocean (Figure 18.8) has been elegantly exploited for the estimation of historical and paleochemical nutrient levels in the ocean (Boyle and Keigwin, 1982, 1985, 1987; Hester and Boyle, 1982; Boyle, 1984, 1986; Shen and Boyle, 1987). This technique is based on the close similarity in ionic radius of the divalent ions Cd^{2+} and Ca^{2+}, which means that Cd^{2+} substitutes into the crystal lattice of $CaCO_3$ minerals formed by foraminifera, coccoliths, and aragonitic corals.

Zinc

Early reliable measurements of zinc made by Bruland's group (Bruland et al., 1978a; Bruland 1980; Bruland and Franks, 1983) showed a close similarity between the vertical profiles of Zn and the nonlabile nutrient silicate (Figures 18.5 and 18.6). For example, Figure 18.9 shows the correlation between Zn and silicate for the North Pacific station discussed earlier. The close correlation between Zn and silicate implies that Zn is taken up into the mineral parts of phytoplankton such as the $CaCO_3$ or SiO_2 exoskeleta of coccolithophores and diatoms, respectively. These biogenic mineral components of phytoplankton are more refractory than the organic tissue containing the labile nutrients and therefore have a deeper regeneration cycle in the ocean, as illustrated by comparing the vertical profile of silicate with those of phosphate and nitrate in Figure 18.5.

Examination of further sets of data reveals that, in fact, the Zn–Si relationship curves downward slightly. This is shown in Figure 18.10, which contains recent reliable data for different parts of the global ocean. This curvature implies that a fraction of the Zn is remineralized at shallower depths than silicate, perhaps indicating that some of the Zn has been taken up into the organic tissue component of the phytoplankton.

The observation that Zn is closely correlated with silicate does not imply, by itself, that Zn is actually carried by biogenic SiO_2 phases rather than $CaCO_3$ phases because both minerals have rather similar dissolution profiles down the water column.

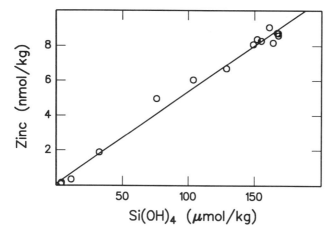

Figure 18.9. Relationship between zinc and silicate concentrations at a station in the North Pacific (32°41.0′N, 144°59.5′W). Drawn using data compiled by Bruland (1980; see also Figures 18.5 and 18.6).

Unlike Cd, Zn is a recognized essential trace element for plankton growth. It is an essential cofactor in several important enzyme systems, including alkaline phosphatase and the carbonic anhydrase needed by diatoms to dehydrate seawater HCO_3^- to provide CO_2 for photosynthesis (Morel et al., 1994). Therefore it would be very curious if diatoms do incorporate Zn into their exoskeleta. Perhaps Zn is essential to the enzyme system that provides the molecular template for SiO_2 formation. Alternatively, the exoskeleton may act as a store of Zn that becomes available for other purposes during periods of Zn shortage.

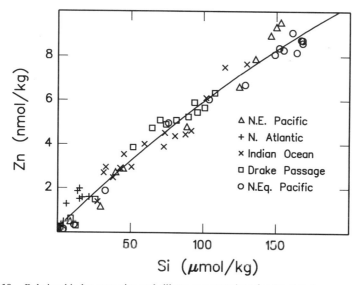

Figure 18.10. Relationship between zinc and silicate concentrations for the global ocean.

Copper

Figure 18.6 shows that this element exhibits mixed behavior relative to the labile and nonlabile nutrients. In these Pacific waters, Cu increases almost linearly with depth from the surface to deep waters, ranging from about 0.5 nmol/L at the surface waters to about 5.0 nmol/L at 5000-m depth. This profile results partly from the superposition of the effects of Cu scavenging by particles in deep waters onto the nutrient-type assimilation and regeneration behavior (Bruland et al., 1991).

In addition, Cu levels in surface water never reach the very low values exhibited by Cd, Zn, and the nutrients. These general trends have been confirmed by other results (e.g., Boyle and Edmond, 1975; Boyle et al., 1977; Bruland and Franks, 1983; Martin et al., 1990c).

High Cu concentrations are found in bottom waters of the east equatorial Pacific (Moore, 1978) and in the Arabian Sea (Saager et al., 1992), indicating the presence of strong benthic Cu inputs (Klinkhammer, 1980). Adsorptive removal of Cu from the water column is most important where particle concentrations are high, such as in thermocline waters underlying productive areas and in benthic nepheloid layers (Moore, 1978).

Nickel

Like Cu, Ni concentrations are always depleted in surface waters relative to the thermocline and deep waters but are rarely smaller than 1 nmol/kg. In general, Ni profiles bear mixed characteristics of regeneration through both labile organic tissue and biogenic mineral phases (Sclater et al., 1976; Bruland, 1980). However, there are considerable regional differences in the extent to which Ni appears coupled to the regeneration cycles of the labile and nonlabile nutrients. In the Arctic and North Pacific Oceans, the effect of deep water regeneration on Ni is very pronounced, producing concentrations in excess of 10 nmol/kg (Bruland, 1980; Martin et al., 1993).

Ni/Si ratios below the thermocline are highest in the Arctic (Martin et al., 1993) and northeast Atlantic Oceans (Yeats and Campbell, 1983). However, much lower Ni/Si ratios are observed in deep waters in the Antarctic Ocean (Westerlund and Ohman, 1991), Indian Ocean (Morley et al., 1993; Saager et al., 1992), and the South Pacific (Tasman Sea; Ho, 1988).

Iron

Iron is perhaps the most bioactive of the trace metals in the ocean, but its chemistry is very complex and poorly understood (Bruland et al., 1991). The main oxidation state is Fe(III), an ion that is strongly hydrolyzed at the pH of seawater, resulting in a high ratio of hydrolyzed products to free Fe^{3+} ions. Turner et al. (1981) and Byrne et al. (1988) separately estimated from thermodynamic data that the ratio of hydrolyzed to free Fe^{3+} is about 10^{12} at pH 8.2, 25°C. Their calculations use thermodynamic data that indicate that $Fe(OH)_3^0$ is the predominant hydrolysis product. By contrast, Hudson and Morel (1990) discounted the importance of $Fe(OH)_3^0$ and calculated a ratio of 10^{10}, with $Fe(OH)_2^+$ being the major hydrolysis product.

Chemical analysis of "dissolved" Fe(III) in seawater uses filtration of the water sample through a membrane filter having a pore size in the range 0.2–0.5 μm as an operational definition of the dissolved state. For most trace metals already considered (Cd, Zn, Cu, and Ni), the proportion of metal found on the filter is very small compared to that passing through, which makes it unlikely that a significant fraction of these elements is present in a colloidal state having a particle size close to that of the filter pores. However, for Fe, a major fraction is clearly particulate, making it highly likely that most, if not all, of the Fe passing through a membrane filter is actually in the form of colloidal particles.

A few studies have been undertaken to determine the solubility equilibrium of iron in seawater by using radiolabeled iron with filtration and dialysis techniques (Byrne and Kester, 1976; Kuma et al., 1992). There is close agreement between the two studies for the thermodynamic solubility product of iron. Kuma et al. (1992) measured $pK_{so} = 37.47 \pm 0.03$ compared to Byrne and Kester's value of $pK_{so} = 37.5 \pm 0.1$. As noted by Kuma et al. (1992), this value is consistent with a solution Fe(III) concentration that is an order of magnitude higher than concentrations of dissolved iron measured at mid-depth in the open ocean using the same pore-size filters. This implies that open ocean waters are not in saturation equilibrium with a metastable colloidal hydrous ferric oxide similar to that used in the laboratory experiments. Wells (1989) suggested that the maximum solubility of iron is 0.1 nmol/L and credited the discrepancy in values to the existence of colloids.

Landing and Bruland (1981) and Gordon et al. (1982) made the first reliable studies of dissolved and particulate Fe in the northeast Pacific Ocean. They found that vertical profiles of dissolved iron had the following features in common: Iron was severely depleted (0.15–0.30 nmol/kg) in surface waters; a concentration maximum (up to 2.6 nmol/kg) occurred at the depth of minimum oxygen concentration; Fe levels appeared to vary little in mid-depth waters (0.5–1.0 nmol/kg). They also reported that total iron profiles followed a similar pattern.

In some regions, surface waters above the thermocline show evidence of atmospheric Fe input (Bruland et al., 1991). According to Duce (1986), more than 95% of the Fe supply to oceanic surface waters comes from the atmosphere.

In the northwest Atlantic, Symes and Kester (1985) found that at open-ocean stations, total Fe was depleted near the surface and increased in the vicinity of the oxygen minimum. A characteristic bottom water maximum was also observed at stations where near-bottom sampling was conducted. They ascribed this feature "partially to resuspension of particulate iron associated with the nepheloid layer." They also found that near the surface, in the vicinity of the chlorophyll maximum, iron was almost totally in particulate form. Dissolved iron increased to be approximately 50% of total iron at the nutrient maximum and approached 100% in the mid-portion of the water column, with particulate iron becoming important again in the bottom waters. Iron levels in all fractions increased near to the coast.

Martin et al. (1990c) measured dissolved iron in the South Drake Passage (Southern Ocean). Levels of dissolved iron were extremely low in the surface waters (0.10 0.16 nmol/kg) and increased gradually down to the bottom waters (0.76 nmol/kg). There was no pronounced oxygen minimum in these waters. Samples taken from the shallow Gerlache Strait showed a high dissolved iron content (4.70–7.40 nmol/kg).

Other studies in the northeast Atlantic (Danielsson et al., 1985), Indian Ocean (Saager et al., 1989), Pacific Ocean (Landing and Bruland, 1987; Martin and Gordon,

1988; Martin et al., 1989), Mediterranean (Sherrell and Boyle, 1988; Saager et al., 1993), Black Sea (Lewis and Landing, 1991), and Antarctic waters (Westerlund and Ohman, 1991; Nolting et al., 1991) have increased the global ocean coverage for iron data and have increased the iron fractions under investigation.

Hong and Kester (1986) measured Fe(II) and total Fe in a transect along the continental shelf off the Peruvian coast. They detected up to 40 nmol/kg of Fe(II) in bottom waters, and the Fe(II) concentration decreased markedly upward in the water column and with distance from shore. The Fe(II) in bottom waters correlated well with nitrite, indicating a common source from the sediment. Elevated levels of Fe(II) near the surface and a diel change were ascribed to photochemical processes.

O'Sullivan et al. (1991) measured subnanomolar concentrations of Fe(II) in the surface waters of the equatorial Pacific. The distribution of Fe(II) varied both spatially and temporally in the upper 100 m. Landing and Westerlund (1988) determined the vertical distribution of Fe(II) in the water column in Framvaren Fjord, Norway. They found iron concentrations increased dramatically below the oxic–anoxic interface at 20 m. Dissolved Fe(II) concentrations in the anoxic bottom waters appeared to be controlled by the solubilities of iron-sulfide phases (mackinawite or greigite).

Wells and Mayer (1991) measured the labile fraction of iron by oxine complexation in Gulf of Maine waters. They found that the labile fraction of iron in seawater varied both temporally and spatially. Particulate iron was generally less labile than dissolved iron, the particulate fraction often contributing substantially to the overall labile iron concentration. By contrast, as much as 75% of the dissolved iron was nonlabile in some cases.

SPECIATION OF TRACE METALS IN THE OCEAN

The trace metal concentration data discussed in the previous section have all been obtained by variations of the same analytical technique. Metal ions have been extracted from seawater by complexing with a strong chelating ligand such as ammonium pyrrholidene dithiocarbamate, or by equivalent methods using ligand functional groups immobilized on a solid phase (Boyle and Edmond, 1975; Bruland et al., 1979; Danielsson et al., 1978). Because the ligands used are very strong complexing agents for the metals determined (i.e., have very high stability constants), almost all complexes formed by metal ions with organic and inorganic ligands naturally present in seawater undergo ligand exchange with the analytical complexing agent. In addition to the high stability of the analytical complexing agent, it is also common to carry out metal ion extraction at a solution pH much lower than the original seawater, further assisting in the release of free metal ion for solvent extraction. Thus these techniques measure the total concentration of metal ion present in solution.

It has long been recognized that the total solution concentration of a metal ion is not particularly valuable in assessing its availability to organisms (Anderson and Morel, 1982), or indeed for a variety of chemical processes such as metal ion–particle reactivity (Davis and Leckie, 1978; Morel, 1983). It now seems clear that the uptake of metal ions by cellular organisms is controlled by the activity of the free metal ion in solution. This leads to important considerations about how metal ions interact with organisms. In the presence of complexing ligands, the free ion activity may be orders of magnitude lower than the total concentration in solution. Moreover, the free ion activity of different waters cannot necessarily be compared with each other using total metal ion concentration data because the concentration and nature of ligands may be different. Therefore, it is important to be able to measure the *speciation* of metal ions— that is, the concentrations (or activities) of all the complexes formed with naturally occurring organic and inorganic ligands.

The inorganic speciation of metal ions is readily calculated from the known concentrations of the main complex-forming anions in seawater (OH^-, Cl^-, SO_4^{2-}, CO_3^{2-}, Br^-, and F^-) and available thermodynamic data for their complexes formed with different metal cations (Turner et al., 1981). By comparison, isolation and direct study of metal–organic complexes in seawater is much more difficult, with the exception of tetrapyrrole complexes such as vitamin B_{12} and the chlorophylls. This is because the concentrations of both metal ion and ligands are extremely low, making attempts to measure speciation highly susceptible to contamination artifacts. In recent years, however, the techniques of clean chemical analysis discussed earlier have been applied to the study of trace metal speciation in seawater, with very satisfactory results.

Several techniques have been used for measuring the concentration of metal–organic complexes in seawater. These may be classified into three types:

- *Competitive techniques* in which the seawater is equilibrated with an synthetic complexing ligand of known concentration and stability constant that closely matches the natural ligands in the seawater. The competing ligand may a solution species that is isolated by solvent extraction (Moffett and Zika, 1987; Moffett et al., 1990), or it may form a metal complex that adsorbs onto a mercury electrode and can be determined by cathodic stripping voltammetry (Van den Berg, 1984). Alternatively, the metal ion may be competitively adsorbed onto a metal oxide surface for later separation (Van den Berg, 1982).

- *Kinetic techniques* in which electrochemical methods are used to distinguish between electrochemically labile forms of the metal and inert organic complexes (Tuschall and Brezonik, 1981; Huizenga and Kester, 1983; Waite and Morel, 1983; Coale and Bruland, 1988, 1990; Bruland, 1989, 1992).

- *Isolation techniques* in which metal–organic complexes are isolated from seawater by adsorption onto suitable liquid chromatography phases and subsequently eluted for analysis (e.g., Mackey, 1983; Sunda and Hanson, 1987).

All three methods yield data for the concentrations of metal–organic complexes in a seawater sample. However, if the uncomplexed free ligands in the sample are titrated with known additions of the metal ion of interest, more useful information on the concentrations and apparent stability constants of the ligand(s) may be obtained (Coale and Bruland, 1988, 1990; Bruland, 1989, 1992).

Speciation of Zinc

This experimental approach is illustrated in Figure 18.11, which shows results obtained by Bruland (1989) for titration of a seawater sample from the Northeast Pacific with Zn(II). The left panel of the figure compares the apparent Zn concentration as found by differential pulse anodic stripping voltammetry (DPASV) at a rotating disk

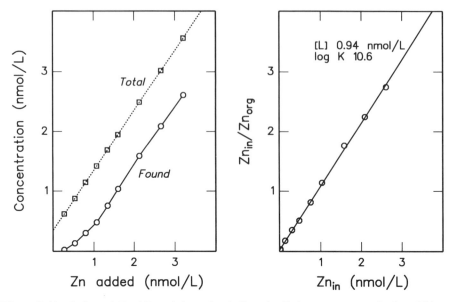

Figure 18.11.   Left panel: Total Zn and electrochemically active Zn in a seawater sample after addition of known aliquots of Zn(II). Right panel: Graphical analysis of the data to determine the ligand concentration and conditional stability constant. Drawn using data reported by Bruland (1989).

thin-film Hg electrode with the total Zn concentration calculated from the amount of Zn(II) added and the concentration present in the original seawater.

It is clear that the apparent Zn concentration is systematically lower than the total concentration indicating the uptake of added Zn(II) by an organic ligand (denoted L, charge omitted):

$$Zn^{2+} + L \longleftrightarrow ZnL$$

Once more than 1.2 nmol/L of Zn(II) has been added, the two plots become linear and parallel, indicating that the ligand has been fully complexed with Zn(II).

The concentration of the ligand and the stability constant of the complex can be calculated from the experimental data using a number of methods, of which the following is a simple approach. The conditional stability constant K for the formation of the Zn–L complex can be written

$$K = \frac{[ZnL]}{[Zn_{in}]([L] - [ZnL])}$$

where [L] is the total concentration of ligand and $[Zn_{in}]$ is the concentration of *inorganic* Zn—that is, all forms of Zn not complexed by the ligand. This equation is easily rearranged to

$$\frac{[Zn_{in}]}{[ZnL]} = \frac{1}{K[L]} + \frac{[Zn_{in}]}{[L]}$$

This shows that a plot of the ratio of the concentrations of inorganic to organic Zn against that of inorganic Zn will be a straight line, having a slope 1/[L] and an intercept 1/K[L], yielding values for both [L] and K. Experimentally, the organic Zn concentration for each titration point in the left panel of Figure 18.11 is given by the DPASV-determined concentration, while the inorganic Zn concentration is calculated from the difference between the latter value and the total Zn concentration. The right panel of Figure 18.11 shows such a plot from which values of [L] = 0.94 nmol/L and log K = 10.6 were obtained.

Figure 18.12 shows the results of such speciation measurements for depths down to 600 m in the Northeast Pacific Ocean reported by Bruland (1989). The left panel shows the concentrations of ligand L, total Zn and inorganic Zn. It is evident that the ligand concentration exceeds that of total Zn throughout the upper 360 m, with the result that only a minor fraction of Zn is present in an inorganic form over this depth range. Below 500 m, the ligand concentration becomes extremely small and exercises little control over Zn speciation.

In the right panel of Figure 18.12 the free Zn^{2+} activity calculated from the measured ligand concentrations and complex stability constants at different depths are compared with the total concentration of Zn. The latter quantity is the zinc measured by the conventional analytical methods such as those described earlier. Note that a logarithmic scale is used in the figure. The results show that the free ion activity of Zn^{2+} is maintained well below 1 pmol/L throughout the upper 200 m, three orders of magnitude lower than the total Zn concentration. The dominance of such a strong ligand in controlling the speciation of Zn over this biologically important depth range is excellent indirect evidence that the ligand has a biological origin and may be produced by one or more organisms for a specific biological purpose.

Table 18.2 summarizes data from recent Zn(II) speciation studies in marine waters. In all cases, a majority of the Zn is found as organic complexes. The last two studies made in coastal and estuarine waters seem to imply less strong Zn complex-

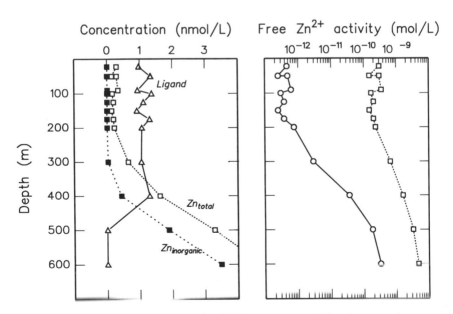

Figure 18.12.   Left panel: Vertical profiles of total Zn, inorganic Zn, and ligand concentrations at a station in the Northeast Pacific. Right panel: Vertical profiles of free Zn^{2+} activity (open circles) and total Zn(II) concentration (open squares). Drawn using data reported by Bruland (1989).

ing. However, it is not certain whether these studies have failed to detect the effects of a strong ligand because of sample contamination with Zn.

Speciation of Copper

Like Zn(II), Cu(II) speciation in the ocean is also dominated by a highly selective strong ligand in the biologically active upper ocean. Figure 18.13 shows the results of DPASV speciation studies conducted by Coale and Bruland (1988) in the Northeast Pacific Ocean. In this case, two Cu-binding ligands were found: a strong ligand L_1 having a conditional formation constant log $K_1 = 11.5$ found only in surface waters (0–200 m depth, maximum concentration 2 nmol/L) and a weaker ligand L_2 having a conditional formation constant log $K_2 = 8.5$ found at higher concentrations (8–10 nmol/L) throughout the water column. This weaker ligand is probably a component of the ubiquitous dissolved organic matter in seawater generated by degradation of plant and animal tissues, and it is similar in nature to the humic acids present in freshwater and soil systems (Chapter 5, this volume).

The left panel of Figure 18.13 shows that the strong ligand L_1 exists only in a narrow depth range near the surface, strongly suggesting that it has a direct biological origin. It exceeds total Cu(II) in concentration in the euphotic zone (upper 200 m), with the result that the free Cu^{2+} activity (right panel of Figure 18.13) is decreased by two orders of magnitude relative to other depths in the water column.

At depths greater than 200 m, the concentration of L_1 decreases to zero and that of uncomplexed, inorganic Cu(II) accordingly increases. However, inorganic Cu(II) remains less than 50% of total dissolved Cu(II) because of the effects of the less strongly binding ligand L_2. As indicated in the right panel of Figure 18.13, the free Cu^{2+} activity in the euphotic zone is kept well below 0.1 pmol/L as a consequence of complexing with the strong ligand.

These findings, which provide compelling evidence of a biological source for strong Cu-complexing ligands, have been confirmed by several more recent studies which find that strong ligand complexing of Cu(II) is greatest at the depth of the bio-

TABLE 18.2
Measurements of Organically Complexed and Free Zn^{2+} in Oceanic, Coastal, and Estuarine Waters

Location	% Organic Zn	$-\log$ $[Zn^{2+}]_{free}$	Reference
NE Pacific	99.9	12.7	Bruland, 1989
NE Pacific	>98	>11	Donat and Bruland, 1990
NE Pacific	>95	>11	Donat and Bruland, 1990
W Atlantic slope	98.7	10.6	Muller and Kester, 1991
Irish Sea	97.8	9.2	Van den Berg, 1985
Scheldt estuary	60–95	7.8–8.7	Van den Berg et al., 1987

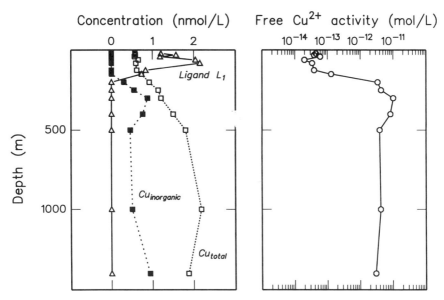

Figure 18.13. Left panel: Vertical profiles of total Cu, inorganic Cu, and strong ligand L_1 concentrations at a station in the Northeast Pacific. Right panel: Vertical profiles of free Cu^{2+} activity (open circles) and total Cu(II) concentration (open squares). Drawn using data reported by Coale and Bruland (1988).

logical productivity maximum and decreases by 1–3 orders of magnitude down the water column. Similar results have also been found elsewhere in the northeast Pacific (Coale and Bruland, 1990) and in the Sargasso Sea (Moffett et al., 1990). The latter authors provided the first direct evidence that a strong Cu-binding ligand is produced by a marine photoautotroph. They carried out laboratory cultures of four marine phytoplankton species (three eucaryotes and one procaryote). The results showed that the cyanobacterial genus *Synechococcus* produced a Cu-binding ligand whose conditional formation constant was identical to the ligand L_1 reported in their study and by Coale and Bruland (1988, 1990). This organism is widespread in the ocean and may therefore have a strong influence on the global speciation of Cu(II) in surface waters. These results are also consistent with early work by McKnight and Morel (1979), who reported that only cyanobacteria seem to exude strong Cu-binding ligands.

The production of a strong Cu-binding ligand by cyanobacteria in the ocean may represent the result of selective evolutionary pressure to detoxify the environment by lowering the Cu^{2+} activity. In support of this, Brand et al. (1986) measured the reproduction rates of 38 marine phytoplankton species in which the Cu^{2+} activity was controlled through the use of a metal ion buffer system. Amongst the different species examined, cyanobacteria were the most sensitive to the toxic effects of Cu while diatoms were the least sensitive. The reproduction rates of the cyanobacteria were diminished at free Cu^{2+} activities greater than 0.4 pmol/L, whereas most of the eucaryotic algae were able to maintain optimal growth rates at free ion activities as high as 4 nmol/L. If these results are compared with the Cu^{2+} activities observed in the right panel of Figure 18.13, it is clear that in the euphotic zone the Cu^{2+} activities are uniformly low enough that they would not limit the reproduction rate of even the most sensitive cyanobacteria.

Speciation of Cadmium

Figure 18.14 shows the results of speciation studies on Cd(II) in the northeast Pacific reported by Bruland (1992). The general findings of this study are very similar to those already discussed for Zn and Cu. A strong Cd-complexing ligand is found in the upper 150 m of the water column having a maximum concentration of 0.1 pmol/L and a conditional formation constant of log $K = 12$. In this case, the ligand exceeds total Cd(II) in concentration down to almost 200 m depth, with the result that the free Cd^{2+} activity in this depth range is as much as three orders of magnitude lower than in the deep ocean. This must also be viewed in the light of the very strong inorganic complexing undergone by this ion, which means that 97% of the inorganic Cd(II) exists as complexes formed with chloride ion.

Comparison of Zinc, Copper, and Cadmium Speciation

Figure 18.15 shows a comparison of the vertical profiles of free ion activity for Zn^{2+}, Cu^{2+}, and Cd^{2+} in the Northeast Pacific taken from the right panels of Figures 18.12–18.14. It shows clearly the influence of strong ligand complexing in diminishing the activities of the two toxic ions (Cd^{2+}, Cu^{2+}) in the euphotic zone. The activity of the biologically essential ion Zn^{2+} is greater than those of the latter two ions by an order of magnitude in the upper 100 m. Although it is tempting to conclude that the activities of both Cu^{2+} and Cd^{2+} are controlled by selective ligands of biological origin as a detoxification mechanism, this is not so clear in the case of Cd which, as already discussed, exhibits strong correlations with the labile nutrients nitrate and phosphate. The reasons for this are presently unknown, but may signal a biological role for Cd.

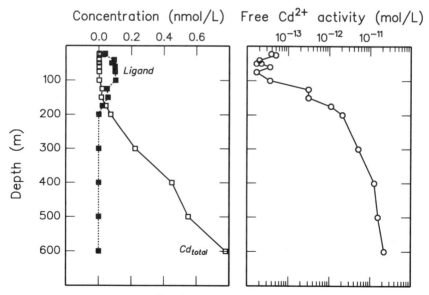

Figure 18.14.   Left panel: Vertical profiles of total Cd and ligand concentrations at a station in the Northeast Pacific. Right panel: Vertical profiles of free Cd^{2+} activity. Drawn using data reported by Bruland (1992).

Figure 18.15. Comparison of the vertical profiles of free ion activity for Cu, Zn, and Cd. Drawn using data from Figures 18.12–18.14.

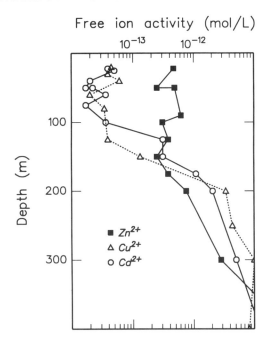

Price and Morel (1990) showed that the growth rates of phytoplankton are reduced at low Zn^{2+} activity (less than 0.2 pmol/L) as a result of insufficient Zn(II) availability. This activity level is comparable to that observed in the Northeast Pacific (Figure 18.15). They also showed that such Zn-deficient cultures could maintain up to 90% of the maximum growth rate when supplied with Cd(II). They suggested that this biochemical substitution of Cd for Zn could be the process that accounts for the pronounced similarity between Cd and the labile nutrients. However, Figure 18.15 shows that the free Cd^{2+} activity is well below that of Zn^{2+}, at least in the Northeast Pacific, ruling out the likelihood of significant substitution of Cd^{2+} for Zn^{2+}. Thus the unusual behavior of Cd remains a mystery to be uncovered by further research.

TRACE METAL INTERACTIONS WITH PHYTOPLANKTON

Laboratory Culture Studies

The knowledge gained from the speciation studies described in the previous section has opened a door to studying trace metal interactions with phytoplankton. Most importantly, it provides information on the free metal ion activities that must be reproduced in laboratory cultures in order to provide culture conditions that realistically model the ocean. In particular, the realization that free metal ion activities are controlled at levels of 1 pmol/L or less by the presence of strong ligands explains the puzzling, but widely known, observation that marine phytoplankton cannot be successfully cultured in the laboratory unless a chelating agent like EDTA is added to the culture medium. We can now see that the role of the chelating agent is to reduce the free ion activities of the metal ions in the medium to the sub-pmol/L range.

The free metal ion activity in a culture medium can be set, at will, to a desired value by adjusting the proportions of complexing agent (EDTA) and total metal ion added, taking account of the known thermodynamic properties of the medium constituents. This has now become a standard approach for the study of metal ion–phytoplankton interactions. Figure 18.16 shows, as reported by Sunda and Huntsman (1992), how the growth rates of four phytoplankton species are affected by free Zn^{2+} activity over ranges typical of the ocean. The growth rates of the two oceanic species (*Thalassiosira oceanica* and *Emiliania huxleyii*) were unaffected by Zn^{2+} activity in the range 1–1000 pmol/L, whereas those of the two neritic species (*Thalassiosira pseudonana* and *Thalassiosira weissflogii*) were significantly reduced at activities below 10 nmol/L. In addition, *T. weissflogii* growth rates were also diminished at Zn^{2+} activities greater than 10 nmol/L, presumably through toxic effects.

Figure 18.17 presents results of the same study for the specific uptake rate of Zn(II) by the phytoplankton, again expressed as a function of the free Zn^{2+} activity. This exhibits an 'S'-shaped curve characteristic of regulated uptake. Indeed, in the large activity range 5–5000 pmol/L (spanning three orders of magnitude), the specific Zn(II) uptake rate changes by less than a factor of 10. This dramatically demonstrates the ability of these phytoplankton to regulate their Zn(II) uptake under conditions of widely varying Zn^{2+} availability. Moreover, the very close similarity in the uptake rates of the four different phytoplankton species is suggestive of a common chemical mechanism. This may be, for example, a Zn(II)-based enzyme system such as alkaline phosphatase which is located on the exterior wall of the cell.

Figure 18.18 shows results from a similar experiment carried out with the same species using Cu(II), as reported by Sunda and Huntsman (1995). In this case, three of the species show a very similar dependence of Cu(II) specific uptake rate on free Cu^{2+} activity, but *T. weissflogii* is quite different.

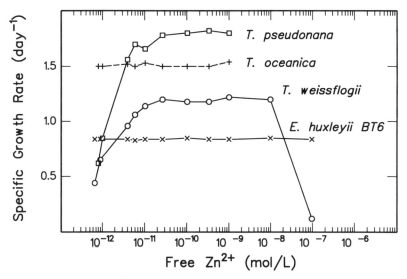

Figure 18.16. Effect of free Zn^{2+} activity on the growth rates of four phytoplankton species. Drawn using data reported by Sunda and Huntsman (1992).

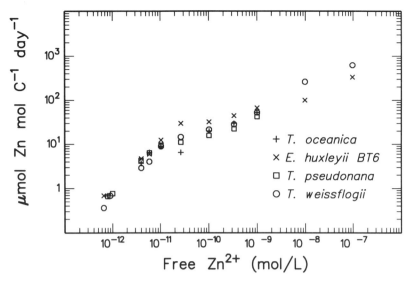

Figure 18.17. Effect of free Zn^{2+} activity on the specific uptake rate of Zn relative to carbon. Drawn using data reported by Sunda and Huntsman (1992).

Finally, Figure 18.19 illustrates the antagonistic interactions of two metal ions, Cu^{2+} and Zn^{2+}, with the neritic diatom *T. pseudonana* (Rueter and Morel, 1981). In this case, Zn is essential for growth of the diatom whereas Cu^{2+} inhibits the effects of Zn^{2+}, possibly by blocking its coordination site in the active enzyme. Thus, in the left panel of Figure 18.19, we see than the uptake rate of silicon by the cells (a measure of their rate of exoskeleton growth) decreases as the free Zn^{2+} concentration is de-

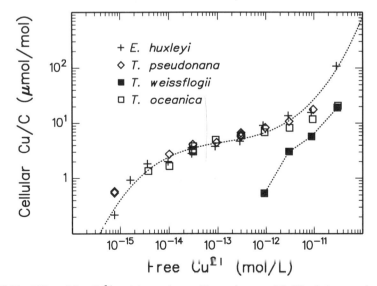

Figure 18.18. Effect of free Cu^{2+} activity on the specific uptake rate of Cu(II) relative to carbon for four phytoplankton species. Drawn using data reported by Sunda and Huntsman (1995).

creased. However, at a pCu of 13 the overall Si uptake rate is much higher than it is at the higher Cu concentration of pCu = 8.9. In the right panel, the same data are presented in terms of the ratio of Zn^{2+} to Cu^{2+} on the ordinate axis. This reveals that the Si uptake rate is enhanced by increasing the free Zn^{2+} ion concentration and diminished by increasing the free Cu^{2+} ion concentration. Interestingly, the Si uptake rates become negative at very low Zn^{2+} availability; that is, the exoskeleton actually dissolves. This may indicate that the diatom is taking advantage of Zn(II) stored in the opal exoskeleton.

Seawater Incubation Studies

Another approach to demonstrating the effects of trace metals on phytoplankton growth rates is to add small quantities of an essential trace metal to a seawater sample and see if this enhances biological activity, compared to suitable controls. This approach is experimentally very difficult because sample collection and handling during incubation must be carried out under the strictest trace metal clean conditions so that natural organic ligands are not swamped by the introduction of metal ion contaminants from reagents and equipment. The application of trace metal clean methods to measuring primary productivity has shown that sampling and handling procedures are important in obtaining accurate and precise results (Fitzwater et al., 1982). Martin et al. (1993) analyzed the contents of primary productivity bottles used by several research groups and found significant zinc, copper, and iron contamination. Zinc is notorious for contamination problems and is suspected to markedly lower primary productivity measurements (Martin et al., 1993). Despite the awareness of trace metal contamination, many research groups that measure phytoplankton productivity still do not use clean techniques for water sampling. Not surprisingly, therefore, the conflicting results gained from "clean" versus "dirty" incubation studies have engendered considerable

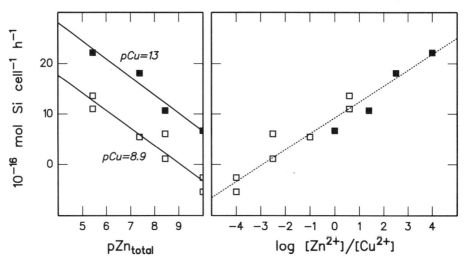

Figure 18.19. Left panel: Effect of free Zn^{2+} and Cu^{2+} activity on the rate of silicon uptake by the diatom *Thalassiosira pseudonana*. Right panel: Si uptake rates as a function of the Zn/Cu activity ratio. Drawn using data reported by Rueter and Morel (1981).

controversy in the literature. This situation is discussed in more detail in the next section with respect to iron.

EFFECT OF IRON ON PHYTOPLANKTON GROWTH

The hypothesis that iron is a limiting factor for phytoplankton growth in regions of the open ocean is not new. For example, Gran (1931, p. 41), based on growth experiments in culture, concluded that the low concentration of iron in seawater probably limited plant growth in areas where it was not replenished by land drainage. Later, Hart (1934) suggested, "Among the ... chemical constituents of sea water ... possibly limiting phytoplankton production, iron may be mentioned ... it may help to explain the observed richness of the neritic plankton ... the land being regarded as a source of iron." It is important to note that trace metal clean techniques did not exist at this time. Consequently, the iron content of seawater was erroneously believed to be range from 50 to 400 nmol/kg (Harvey, 1937a,b).

Traditionally it has been believed that marine and estuarine phytoplankton are nitrogen-limited (Hecky and Kilham, 1988). Other researchers have suggested that phosphorus must ultimately limit marine phytoplankton production (Redfield et al., 1963; Brand, 1991) because of the abundance of N_2 and the potential for N_2 fixation by some procaryotes in tropical ocean regions. Limitation of productivity by silicate is also suspected in the equatorial Pacific and the Southern Ocean (Dugdale and Wilkerson, 1990). Many researchers now believe that co-limitation by macronutrients (N, P, or Si) and trace metal micronutrients may simultaneously control phytoplankton composition and growth in the ocean (Morel and Hudson, 1985).

Early experiments involving the addition of micromolar amounts of iron to Sargasso seawater resulted in increases in productivity (Menzel et al., 1963). However, a similar stimulation of growth was achieved by adding aluminum. Aluminum is not a required nutrient but does form particulate hydroxides in seawater. It is thought that the addition of either metal will lead to the adsorption of toxic metals onto hydroxide surfaces. The resulting drop in trace metal toxicity provides the stimulus for greater production (Sunda, 1989). Addition of copper may also increase productivity by poisoning grazing zooplankton, while phytoplankton are more able to cope with increased copper concentrations (Thomas et al., 1977). Brand et al. (1983) have argued that simultaneous limitation by Zn, Mn, and Fe may be more severe than limitation by any one of the metals alone.

As mentioned earlier, the operational definition of dissolved iron is that fraction which passed through a 0.4-μm membrane filter. Dissolved iron defined this way is often used as a proxy for bioavailable iron, even though it is evident that much particulate iron is also bioavailable and some dissolved iron is not (Wells et al., 1983; Wells, 1989). Some colloidal iron will pass through an 0.4-μm filter, and there is much controversy over whether iron even exists in seawater in a truly dissolved form (Lewin and Chen, 1971).

Hudson and Morel (1990) investigated the uptake of iron in cultures by two coastal phytoplankton species—the coccolithophorid *Pleurochrysis carterae* and the diatom *Thalassiosira weissflogii*. They observed direct uptake from the cell surface using radiolabeled iron in pulse-chase experiments with cells grown under iron limitation. Utilizing a titanium(III)–citrate–EDTA complex as a reductant, they were able to experimentally determine extra- and intracellular iron (Hudson and Morel, 1989). Hudson and Morel found that iron uptake rates were controlled either by the rate of coordination of dissolved iron to surface transport ligands or, when ambient iron is very low, by diffusion to the cell. They suggested that since oceanic phytoplankton were subject to the same strongly size-dependent limit on uptake rates imposed by diffusion, large species would be more likely to experience growth rate limitation under low-iron conditions.

One strategy by which phytoplankton can reduce their iron quotas and adapt to low iron conditions is by substitution of iron-containing proteins such as ferredoxin by functionally similar proteins that do not contain iron (e.g., the redox protein flavodoxin) (Entsch et al., 1983). There is also evidence that some trace metals can replace others in enzymes without significant loss of physiological activity (Price and Morel, 1990). However, no replacement for iron is yet evident.

The methodology for studying trace metal limitation on cultures involves using trace metal ion buffers (some typical examples are EDTA, CDTA, DTPA, and NTA). By manipulating the concentration of either the buffer or the trace metal in question, control of free metal iron activity can be achieved. Use of these buffers has demonstrated that free metal activity, rather than total or chelated metal concentration, controls biological availability (Anderson and Morel, 1982; Brand et al., 1983).

Brand et al. (1983) measured the reproductive rates of 21 species of marine phytoplankton when the free zinc, iron, and manganese ion activities were controlled by EDTA buffer systems. Neritic species have reduced reproductive rates for low iron conditions (<100 nmol/L iron), while oceanic species reproduce at close to the maximal rates even at the lowest iron concentrations (1 nmol/L). Similar findings were reported for other oceanic phytoplankton by Sunda et al. (1991) using similar methodology.

Brand (1991) extended his initial work in a further study of 22 species of marine phytoplankton. He found that among eucaryotic phytoplankton, coastal species have high iron (Fe:P) requirements and become iron-limited in low iron conditions. By contrast, oceanic species adapt their biochemical composition to the low availability of iron in the open ocean and have low iron requirements. However both coastal and oceanic cyanobacteria have high iron requirements, indicating that new production of cyanobacterial biomass may, under the right conditions, be limited by iron availability. Brand (1991) also suggested that the distribution and abundance of cyanobacteria in the ocean is related to iron inputs to the ocean.

Culture experiments utilizing different chemical forms of iron have been undertaken to determine the bioavailability of iron to phytoplankton (Wells et al., 1983; Rich and Morel, 1990; Anderson and Morel, 1982; Wells et al., 1991). Generally, cultures were found to grow better on fresh colloidal ferric hydroxide than on any crystalline iron hydroxide phase (hematite, goethite, or limonite). Indeed, it seems that the more thermodynamically stable the iron hydroxide phase used as an iron source, the lower the resulting growth rate. Addition of iron-chelating agents increased growth rates, probably as a result of ligand-induced dissolution of the oxyhydroxides.

Wells et al. (1991) developed an analytical technique using 8-hydroxyquinoline (oxine) to estimate iron bioavailability. They found a strong relationship between iron lability and the bioavailability of iron to phytoplankton.

Anderson and Morel (1982) found that photoreduction processes also enhanced growth rates in culture experiments. They determined that Fe(III)–EDTA complexes were photodegraded to produce Fe(II) in their cultures, which was reoxidized to form labile Fe(III).

Siderophores are compounds produced by microorganisms for scavenging iron from the environment. They are defined as low-molecular-weight compounds (100–500 daltons) possessing a high affinity for iron(III). Their biosynthesis is regulated by iron levels, and their function is to supply iron to the cell (Neilands, 1981). Siderophores, coordinated to iron, are accumulated by microorganisms by specific translocation mechanisms, and the tightly bound iron is removed for utilization by the cell. Siderophore uptake varies between the different classes of molecule, some enter the cytoplasm, while others apparently donate iron at the cytoplasmic membrane surface. Some siderophores may be secreted in order to deprive competing organisms of iron (e.g., blue-green algae in freshwater lakes; Murphy et al., 1976) and as such will influence the ecology of the environment occupied by the secreting colony of microorganisms. No direct evidence yet exists for siderophores in the open ocean.

Because of its high charge density, Fe(III) is a powerful Lewis acid and forms stable bonds with ligands containing weakly polarizable atoms, typically oxygen. The basic design of siderophores is for bidentate functional units to arrange themselves around a central iron(III) atom to form an intramolecular hexadentate complex (Hider, 1984). Most siderophores bind to iron via either catechol- or hydroxamate-type ligands. Catechol ligands provide greater thermodynamic stability than hydroxamate due to their stronger electrostatic interactions with iron, coupled with resonance stabilization of the complex. However, catechol ligands have the disadvantage of being susceptible to oxidation. Unlike catechol complexes, hydroxamate ligands can form uncharged complexes. This leads to important differences in uptake mechanisms (Hider, 1984). High-spin iron(III) possesses a d^5 configuration which lacks any crystal field stabilization energy, thus rendering the complexes kinetically labile. This lability can be further enhanced if the ligand forms a distorted complex. Siderophores can out-compete other iron complexing ligands and colloidal iron hydroxide for iron due to the formation of a highly kinetically and thermodynamically stable hexadentate complex (Hider, 1984). Reduction of the siderophore complex to form iron(II), which is less kinetically and thermodynamically stable, appears to be the mechanism by which the microorganisms utilize the iron obtained by siderophore scavenging. The siderophore complex itself may be regenerated by this process (Hider, 1984).

Trick et al. (1983a) isolated a hydroxamate-siderophore (prorocentrin) produced by the neritic dinoflagellate *Prorocentrum minimum*. The production of prorocentrin is stimulated by culturing the diatom under iron-deficient conditions. They later extended the study to include other neritic phytoplankton (Trick et al., 1983b). Of five *Prorocentrum* species they studied, prorocentrin is produced by three: *P. mariae-lebouriae*, *P. minimum* and *P. gracile*. Two species, *P. maximum* and *P. micas*, grow poorly under iron-deficient conditions and do not produce siderophores. *Thalassiosira pseudonna* and *Dunaliella teriolecta* also produce extracellular siderophores under iron-depleted conditions, but no siderophore production was detected in *Skeletonema costatum* and *Olithsthodiscus luteus*. For all the siderophore-producing species, production took place 1–2 days after the cessation of growth in the stationary phase and only lasted 1–2 days.

In other studies, strains of open ocean bacteria from the genera *Vibrio*, *Alteromonas*, *Alcaligenes*, *Pseudomonas*, and *Photobacterium* have been surveyed for their ability to produce siderophores (Trick, 1989; Goyne and Carpenter, 1974). In all cases these open-ocean bacteria produce siderophores and their biosynthesis is controlled by iron levels (Trick, 1989). Reid and Butler (1991) isolated and purified two siderophores (designated S2E and S2L) produced by a marine bacterium, *Alteromonas luteviolaceus*. The siderophores isolated were partially characterized by mass spectral analysis, amino acid analysis, qualitative analytical tests, chemical degradation, and nuclear magnetic resonance. A new set of outer-membrane proteins is also produced when the bacterium was grown under iron-limited conditions, consistent with a high-affinity iron uptake system. The isolated siderophores tested positively for the catechol moiety by both NMR and qualitative chemical analysis. Fast atom bombardment mass spectral analysis of the siderophores S2E and S2L established that the exact masses are 945.3961 and 927.3954.

High-Nutrient, Low-Chlorophyll Waters: the "Antarctic Paradox"

As mentioned earlier, the low primary productivity measured in Antarctic waters seems to contradict the persistently high surface concentrations of the major plant nutrients. A large supply of nutrients from depth is assured by continuous upwelling at the circumpolar continental margins. During the Antarctic growth season, phytoplankton standing stock remains low despite the presence of high nutrient concentrations. At no stage do major nutrient (phosphate, nitrate) levels become depleted. Many other large regions of the world's oceans exhibit these so-called high-nutrient, low chlorophyll (HNLC) conditions in their surface waters (Cullen, 1991)—for example, the sub-Arctic and equatorial regions of the Pacific.

This retardation of phytoplankton growth may be dictated by "bottom-up" control. This is where growth is limited by physical factors such as temperature, light conditions, mixed layer depths, and bioavailability of essential chemical species. Standing stocks of phytoplankton may also be limited by processes which control loss rates for phytoplankton ("top-down" control). Such losses may occur by passive sinking or more importantly grazing by zooplankton. In the southern ocean, the high biomasses of krill provide high grazing pressure, which may control phytoplankton populations (Jacques, 1989).

Martin and Gordon (1988) measured dissolved iron levels along the 1600-km inshore–offshore VERTEX transect in the North Pacific. From their data they deduced that advective processes within the ocean could only supply a few percent of the iron required by open ocean phytoplankton. The majority of iron supplied to surface waters must be from atmospheric deposition. They postulated that atmospheric deposition rates to the HNLC sub-Arctic region would not be sufficient to meet phytoplankton demand.

Atmospheric iron deposition and primary productivity in the North Pacific was investigated by Young et al. (1991). Chemical analyses of the atmospheric particles revealed that they were iron-rich (10–15%). They noted that production systematically increased with depth and time. This may have resulted from a continual release of iron from settling particles in the euphotic zone. At all depths, systematic decreases in production followed the initial increase in production, indicating that the growth limitation may have evolved from being iron-limited to nitrogen-limited.

Zhuang et al. (1990) measured the dissolution of atmospheric iron aerosols in open-ocean surface seawater. They found that for all the marine aerosols examined, the dissolved fraction (0.4-μm filter) was found to be 10–17 nmol/kg. The dissolution of atmospheric iron in seawater took place in a few minutes. In later studies, Zhuang et al. (1992) found that Fe(II) contributed 56 $\pm$ 32% of total iron in marine aerosols collected over the central North Pacific. They suggested that Fe(II) is produced by photoreduction of Fe(III) during long-range aerosol transport and proposed that the iron and sulfur cycles in both the atmosphere and the oceans may be closely coupled.

Duce and Tindale (1991) reviewed the atmospheric fluxes of iron to the ocean and commented on the highly episodic nature of dust generation and deposition. In general they concluded wet deposition of iron dominated over dry deposition to the ocean surface.

The Controversy: Does Iron Really Limit Phytoplankton Growth?

Martin et al. (1989) extended their previous work on a further series of VERTEX stations in the Gulf of Alaska. This study included phytoplankton growth experiments in which the effect of adding nanomolar amounts of iron was investigated. They found that iron enhances the growth of phytoplankton from sub-Arctic waters and induces a change in species abundances. Larger diatom cells become more abundant with added iron, compared to higher concentrations of smaller coccolithophorids in the natural, low-iron conditions. They concluded that the input of atmospheric iron to the Gulf of Alaska is sufficient to support a moderate productivity that is not large enough to fully deplete available nutrient levels.

These iron enrichment experiments and their results were soon subject to much scrutiny as controversy raged over Martin's iron limitation hypothesis (Banse, 1990, 1991a,b; Martin et al., 1990a–c; 1991a,b). At issue was the interpretation of growth rates and whether the exclusion of grazing zooplankton from the sample bottles has a significant effect. For example, the absence of grazing pressure may allow phytoplankton to grow unhindered. In addition, the observation that control samples exhibited at least some growth was considered to be significant evidence that other, unknown processes might be occurring in the experiments. Attention also became focused on the rates of nitrate uptake as a means of determining the amount of new production versus regenerated production (Banse, 1991a). Finally, concern was also raised over the degree to which iron was adsorbed to the walls of the polycarbonate culture bottles and how this might be wrongly ascribed to iron uptake by the plankton (Banse, 1991a).

Martin et al. (1993) performed further iron enrichment experiments as part of the JGOFS North Atlantic Bloom Experiment. They found no evidence of iron deficiency. They concluded that the phytoplankton's iron requirements were probably being met by lateral transport from the continental margins. In agreement with this, the North Atlantic does not typically exhibit HNLC conditions.

Martin et al. (1990a,b) investigated the possibility of iron limitation on Antarctic phytoplankton. They found that in the high-productivity neritic waters of the Gerlache strait the dissolved iron levels are high (7.4 nmol/kg), while in the offshore Drake Passage waters the dissolved iron levels are very low (0.16 nmol/kg). Iron enrichment experiments were also performed on water samples from the Ross Sea. Nitrate uptake rates were 2 to 10 times higher after addition of unchelated iron than in the control samples. Total decreases in nitrate were balanced by increases in particulate nitrogen. By contrast, addition of manganese to water samples produced no measurable increase in productivity.

Dugdale and Wilkerson (1990) reanalyzed the data from the iron enrichment experiments in terms of specific uptake rates and found no evidence for phytoplankton growth. They suggested that adding iron might have a toxic effect on grazers or favor the growth of larger phytoplankton cells indigestible to grazers. The unknown effects of bacterial populations was also thought to influence growth rates.

Buma et al. (1991) studied the factors controlling phytoplankton growth and species composition in the Antarctic Ocean. Iron-enrichment experiments were performed on water from the Weddell and Scotia seas. They always observed iron to stimulate chlorophyll *a* synthesis and nutrient assimilation. In four out of five experiments, particulate organic carbon (POC) levels were also higher in the iron-enriched cultures. The control samples also showed increases in chlorophyll *a* and POC levels. The authors concluded that despite the fact that iron enhances phytoplankton growth, it is not the major controlling factor in the Weddell and Scotia Seas.

In addition to atmospheric deposition and iron-rich sediments in shallow waters, iron may also be supplied to offshore waters by melting sea ice. The bioavailability of iron in sea ice is unknown. Blooms of phytoplankton often occur at the ice-edge in the spring. Such blooms may result from the formation of a shallow, vertically stable upper layer from melting ice. Subsequent seeding by actively growing sea ice microbes may promote bloom formation.

The eastern and central equatorial Pacific Ocean exhibits HNLC conditions. Surface levels of nitrate range from 4 to 12 μmol/kg, while levels of chlorophyll *a* never exceed 0.5 μg/kg (Chavez and Barber, 1987). Natural phytoplankton communities in this region are dominated by picoplankton and nanoplankton, with a combination of high primary production rates and high grazing rates keeping standing stocks of phytoplankton low (Chavez et al., 1991). Atmospheric dust levels in the equatorial Pacific are among the lowest in the world.

Price et al. (1991) undertook iron enrichment experiments along a transect between 9°N and 3°S in the equatorial Pacific Ocean. They found that the addition of 1 nmol/L iron to seawater samples increased the final concentration of chlorophyll *a*, POC, and nitrogen in incubation bottles relative to controls without added iron. Ammonium was the major inorganic nitrogen source for the indigenous phytoplankton community, and little nitrate was taken up despite its high ambient concentration. In the samples with added iron, the phytoplankton switched from using ammonium ion for growth (regenerated production) to nitrate (new production). Accompanying this shift in nitrogen utilization was a change in phytoplankton community structure.

Chavez et al. (1991) also undertook iron enrichment experiments on samples from the equatorial Pacific. They also found significant shifts in phytoplankton community structure upon iron addition (from picoplankton- to diatom-dominated). Measured zooplankton grazing rates were higher in the iron treatments than in the controls, accounting for a large fraction of daily productivity. Growth rates for pennate diatoms appeared to be increased under iron-enriched conditions relative to the controls.

DiTullio et al. (1993) performed nutrient enrichment (N, P, Si, Fe) experiments at 9°N, 147°W in the tropical Pacific Ocean. Specific growth rates for proclorophytes increased upon addition of macronutrients (N, P, and Si). Diatoms, however, were unable to grow without added iron. Zooplankton grazing on prochlorophytes was in approximate balance with prochlorophyte growth. Grazing processes were found to not be as efficient at controlling diatom standing crop.

Martin (1992) reported that iron, when added as natural aerosol particles, showed the greatest productivity increase compared to other forms of iron (Fe(III)-sorbitol or unchelated Fe(III)).

Iron and Climate Change

The discovery that glacial atmospheric CO_2 concentrations are lower than those in interglacial time has led to many hypotheses. Several explanations for this change deal with variations in southern ocean phytoplankton productivity and the related utilization of major nutrients. It is suggested that if the abundant nutrients in the Southern Ocean were utilized, the resulting bloom may have resulted in a draw-down of CO_2 into the ocean from the atmosphere.

Martin (1990, 1992) extended his iron-limitation hypothesis to suggest that iron may have been more available to Southern Ocean phytoplankton during glacial times. Martin used dust data obtained from the Vostok ice core (De Angelis et al., 1987) to compare iron input to the glacial and interglacial Southern Ocean. The dust concentrations in the ice cores showed that dust loads were 50 times higher during the last glacial maximum because of stronger winds and larger tropical arid areas during glacial time. He suggested that "iron-rich dust blew into the Antarctic, the phytoplankton bloomed, the biological pump turned on, and CO_2 was withdrawn from the atmosphere" (Martin, 1990).

However, evidence for increased productivity in the Southern Ocean during glacial periods is still very controversial. Sedimentary records of silicate preservation in the southern ocean have been interpreted as showing decreased glacial productivity due to the lower abundance of opal (Berger and Wefer, 1991). This interpretation, however, is complicated by the nonpreservation of $CaCO_3$ in southern ocean waters and the fact that many major southern ocean species of plankton contain no opal (Martin, 1992). Non-sea-salt sulfur as found in the Vostok ice core (Legrand et al, 1991), when used as a proxy for DMS (dimethyl sulfide) indicates an increase in DMS productivity during glacial times.

Glacial-age productivity in the equatorial Pacific and in eastern boundary upwelling systems was enhanced, probably due to stronger glacial winds, promoting increased upwelling of nutrients (Berger and Wefer, 1991). Prahl (1992) presented organic geochemical evidence in support of the iron hypothesis for the eastern tropical Pacific. He showed that the deposition of total organic carbon (primarily derived from marine productivity) was highly correlated with the input of eolian dust. He also found that biomarkers for prymnesiophyte productivity preceded the total organic carbon maximum by 4–6 kyr. This, Prahl suggests, is evidence for "an ecological change in phytoplankton community during the last glacial transition as a consequence of systematic variation in the supply of eolian dust and associated bioavailable iron."

An In-situ Iron Enrichment Experiment on the Mesoscale

Owing to the problems exposed by using sample bottles to perform iron enrichment studies, plans were developed to enlarge the system under investigation. Feasibility studies were made for a mesoscale *in situ* iron-enrichment experiment to be carried out in the equatorial Pacific (US JGOFS Planning Report #15, 1992; Watson et al, 1991). Martin (1990) proposed that CO_2 could be removed from the atmosphere by fertilizing the ocean with iron, which might amplify the biological uptake of CO_2 from surface waters. By such a mechanism, the enhanced "greenhouse" effect from increased anthropogenic carbon emissions may be reduced. This idea (dubbed the Geritol solution to greenhouse warming) ignited a storm of controversy over the ethics of deliberately perturbing healthy ecosystems (Roberts, 1991; Baum, 1990).

The mesoscale experiment nonetheless took place in the equatorial Pacific Ocean in October 1993 (Martin et al., 1994). Surface seawater at the study site was inoculated with 15,600 L of Fe(III) solution pumped into the propeller wash of a research ship during a 24-hr period, adding a total of 450 kg Fe to the surface water. At the same time, SF_6 was added as a tracer for the movement of the Fe-enriched patch of water. Measurements of Fe and SF_6 concentrations, as well as those of nutrients and chlorophyll pigments, were performed in a grid pattern across the study site for several days after the iron enrichment. Dissolved Fe concentrations as high as 6.2 nmol/L were measured in the core of the patch within 4 hr of fertilization. These decreased rapidly on the first day as a result of nighttime convective mixing, with the highest values on the subsequent day being 3.6 nmol/L. Thereafter, concentrations in the core of the patch decreased by about 15% per day. Contour plots of both dissolved Fe and the SF_6 tracer were in good agreement 3 days after the fertilization.

Photosynthetic energy conversion efficiency (relative fluorescence) was the first biological response observed after the fertilization (Kolber at al., 1994). Relative fluorescence dramatically increased during the first transect across the patch, indicating a large physiological response during the first 24 hr, the time required to add the Fe(III). After fertilization, ^{14}C-productivity increased 3–4 times and chlorophyll concentrations increased about threefold (Martin et al., 1994). Nutrient measurements showed little change in measured nitrate, phosphate, or silicate concentrations in the patch; however, any changes in nutrients expected on the basis of extra chlorophyll production were calculated to be within the limits of analytical precision. Much larger utilization of these nutrients would have been expected on the basis of single-bottle incubation studies. Significant changes in the distribution of ammonium ion were observed, and measurable increases in both particulate dimethylsulfaniopropionate (DMSP) and methyl iodide were recorded, inside the patch.

Measurements of CO_2 fugacity and total CO_2 were significantly lower inside the patch than outside (Watson et al., 1994). However, as with the nutrient results, the decreases were much less than those expected on the basis of single-bottle incubation studies using comparable inoculations of Fe(III). These key results therefore do not support the idea that iron fertilization would significantly affect atmospheric CO_2 concentrations.

Peng and Broecker (1991) applied a box model to examine the dynamical aspects of a possible iron fertilization in Antarctic waters. From their model they concluded that even after 100 years of totally successful fertilization, the CO_2 content of the atmosphere would only be reduced $10 \pm 5\%$ below what it was projected to be without fertilization. They identified that the rate of vertical mixing in Antarctic waters was too slow to create a significant decrease in the CO_2 content of the atmosphere. Other studies using different box models also concluded that iron fertilization of the Southern Ocean would only reduce atmospheric CO_2 levels slightly over a 100 year fertilization program. This in all cases was due to dynamical limitations (Joos et al., 1991a,b; Kurz and Maier-Reimer, 1993).

CONCLUSION

In this review, we have sought to demonstrate that research over the last decade has revolutionized current thinking about the roles of trace metals in aquatic biological processes. In particular, it has become clear that even in pristine environments, phytoplankton use highly specific ligands to control the availability of both toxic and essential metal ions. This ability has probably been very important in their genetic evolution.

So far, we have a gradually clearing picture of how these processes work in the comparatively simple environment of the open ocean. Much remains to be done with respect to the application of the same principles to near-shore, estuarine, and freshwater environments where other factors that affect trace metal chemistry are operative. We should see, over the next decade, many exciting and interesting research findings. In particular, it will be extremely interesting if examples of the strongly complexing ligands produced by phytoplankton such as cyanobacteria can be isolated and structurally characterized. Chemists have been very successful at making useful coordinating ligands for metal ions, but none with the very high selectivity displayed by these phytoplankton-derived substances.

Even without further work, it is clear that the use of natural waters as cesspools for the discharge of unwanted metals is a criminally negligent activity. Cyanobacteria have survived since the early near-anaerobic conditions of the primordial ocean by evolving the ability to exude ligands that detoxify their environment. The limits of this detoxifying mechanism are set by the concentrations of ligand normally found in natural waters, typically a few nmol/L. In this way, cyanobacteria cope with most (but not all) natural variations in the supply of toxic metals to their environment. Clearly the assimilative capacity of natural waters with respect to metal-sensitive phytoplankton is strictly limited.

References

Ahlers, W. W., Reid, M. R., Kim, J. P., and Hunter, K. A., 1990, Contamination-free sample collection and handling protocols for trace elements in natural freshwaters. *Aust. J. Mar. Freshwater Res.*, 41, 713–720.

Anderson, M. A., and Morel, F. M. M., 1982, The influence of aqueous iron chemistry on the uptake of iron by the coastal diatom *Thalassiosira weissflogii*. *Limnol. Oceanogr.*, 27, 789–813.

Banse, K., 1990, Does iron really limit phytoplankton production in the offshore Arctic Pacific? *Limnol. Oceanogr.*, 35, 772–775.

Banse, K., 1991a, Iron, nitrate uptake by phytoplankton and mermaids. *J. Geophys. Res.*, 96, 20701.

Banse, K., 1991b, Iron availability, nitrate uptake and exportable new production in the sub-Arctic Pacific. *J. Geophys. Res.*, 96, 741–748.

Baum, R., 1990, Adding iron to ocean makes waves as way to cut greenhouse CO_2. *Chem. Eng. News*, July 2, 21–24.

Bender, M. L., Klinkhammer, G. P., and Spencer, D. W., 1977, Manganese in seawater and the marine manganese balance. *Deep-Sea Res.*, 24, 799–812.

Berger, W. H., and Wefer, G., 1991, Productivity of the glacial ocean: discussion of the iron hypothesis. *Limnol. Oceanogr.*, 36, 1899–1918.

Boyle, E. A., 1984, Cadmium in foraminifera and abyssal hydrography: Evidence for a 41 kyr obliquity cycle. In: J. E. Hansen and T. Takahashi, eds., *Climate Processes and Climate Sensitivity*, American Geophysical Union, Washington D.C., pp. 360–368.

Boyle, E. A., 1986, Paired carbon isotope and cadmium data from benthic foraminifera: implications for changes in oceanic phosphorus, oceanic circulation and oceanic carbon dioxide. *Geochim. Cosmochim. Acta*, 50, 265–276.

Boyle, E. A., and Edmond, J. M., 1975, Copper in surface waters south of New Zealand. *Nature*, 253, 107–109.

Boyle, E. A., and Keigwin, L. D., 1982, Deep circulation of the North Atlantic over the last 200,000 years: geochemical evidence. *Science*, 218, 784–787.

Boyle, E. A., and Keigwin, L. D., 1985, Comparison of Atlantic and Pacific paleochemical records for the last 250,000 years: changes in deep ocean circulation and chemical inventories. *Earth Planet. Sci. Lett.*, 76, 135–150.

Boyle, E. A., and Keigwin, L. D., 1987, North Atlantic thermohaline circulation during the last 20,000 years linked to high latitude surface temperature. *Nature*, 330, 35–40.

Boyle, E. A., Sclater, F. R., and Edmond, J. M., 1976, On the marine geochemistry of cadmium. *Nature*, 263, 42–44.

Boyle, E. A., Sclater, F., and Edmond, J. M., 1977, The distribution of dissolved copper in the Pacific. *Earth Planet. Sci. Lett.*, 37, 38–54.

Boyle, E. A., Huested, S. S., and Jones, S. P., 1981, On the distribution of copper, nickel and cadmium in the surface waters of the North Atlantic and North Pacific Oceans. *J. Geophys. Res.*, 86, 8048–8066.

Brand, L. E., 1991, Minimum iron requirements of marine phytoplankton and the implications for the biogeochemical control of new production. *Limnol. Oceanogr.*, 36, 1756–1771.

Brand, L. E., Sunda, W. G., and Guillard, R. R. L., 1983, Limitation of marine phyto plankton reproductive rates by zinc, manganese and iron. *Limnol. Oceanogr.*, 28, 1182–1195.

Brand, L. E., Sunda, W. G., and Guillard, R. R. L., 1986, Reduction of marine phytoplankton reproduction rates by copper and cadmium. *J. Exp. Mar. Biol. Ecol.*, 96, 225–250.

Bruland, K. W., 1980, Oceanographic distributions of cadmium, zinc, nickel and copper in the North Pacific. *Earth Planet. Sci. Lett.*, 47, 176–198.

Bruland, K. W., 1983, Trace elements in seawater. In: J. P. Riley and R. Chester, eds., *Chemical Oceanography*, Vol. 8, Academic Press, London, pp. 157–215.

Bruland, K. W., 1989, Complexation of zinc by natural organic ligands in the central north Pacific. *Limnol. Oceanogr.*, 32, 269–285.

Bruland, K. W., 1992, Complexation of cadmium by natural organic ligands in the central north Pacific. *Limnol. Oceanogr.*, 37, 1008–1017.

Bruland, K. W., and Franks, R. P., 1983, Mn, Ni, Cu, Zn and Cd in the western North Atlantic. In: C. S. Wong, E. A. Boyle, K. W. Bruland, and J. D. Burton, eds., *Trace Metals in Seawater*, Plenum Press, New York, pp. 395–414.

Bruland, K. W., Knauer, G. A., and Martin, J. H., 1978a, Cadmium in northeast Pacific waters. *Limnol. Oceanogr.*, 23, 618–625.

Bruland, K. W., Knauer, G. A., and Martin, J. H., 1978b, Zinc in northeast Pacific waters. *Nature*, 271, 741–742.

Bruland, K. W., Franks, R. P., Knauer, G. A., and Martin, J. H., 1979, Sampling and analytical methods for the determination of copper, cadmium, zinc and nickel at the nanogram per liter level in seawater. *Anal. Chim. Acta*, 105, 233–245.

Bruland, K. W., Donat, J. R., and Hutchins, D. A., 1991, Interactive influences of bioactive trace metals on biological production in oceanic waters. *Limnol. Oceanogr.*, 36, 1555–1577.

Buma, A. G. J., De Baar, H. J. W., Nolting, R. F., and van Bennekom, A. J., 1991, Metal enrichment experiments in the Weddell-Scotia Seas: effect of iron and manganese of various plankton communities. *Limnol. Oceanogr.*, 36, 1865–1878.

Byrne, R. H., and Kester, D. R., 1976, Solubility of hydrous ferric oxide and iron speciation in seawater. *Mar. Chem.*, 4, 255–274.

Byrne, R. H., Kump, L. R., and Cantrell, K. J., 1988, The influence of temperature and pH on trace metal speciation in seawater. *Mar. Chem.*, 25, 163–181.

Chavez, F. P., and Barber, R. T., 1987, An estimate of new production in the equatorial Pacific. *Deep-Sea Res.*, 34, 1229–1243.

Chavez, F. P., Buck, K. R., Coale, K. H., Martin, J. H., DiTullio, G. R., Welschmeyer, N. A., Jacobson, A. A., and Barber, R. T., 1991, Growth rates, grazing, sinking and iron limitation of equatorial Pacific phytoplankton. *Limnol. Oceanogr.*, 26, 1816–1833.

Coale, K. H., and Bruland, K. W., 1988, Copper complexation in the northeast Pacific. *Limnol. Oceanogr.*, 33, 1084–1101.

Coale, K. H., and Bruland, K. W., 1990, Spatial and temporal variability in copper complexation in the North Pacific. *Deep-Sea Res.*, 47, 317–336.

Cullen, J. J., 1991, Hypotheses to explain high-nutrient conditions in the open sea. *Limnol. Oceanogr.*, 36, 1578–1599.

Danielsson, L. G., 1980, Cadmium, cobalt, copper, iron, lead, nickel and zinc in Indian Ocean water. *Mar. Chem.*, 8, 199–215.

Danielsson, L. G., Magnusson, B., and Westerlund, S., 1978, An Improved metal extraction procedure for the determination of trace metals in sea water by atomic absorption spectrometry with electrothermal atomization. *Anal. Chim. Acta*, 89, 47–58.

Danielsson, L. G., Magnusson, B., and Westerlund, S., 1985, Cd, Cu, Fe, Ni and Zn in the northeast Atlantic Ocean. *Mar. Chem.*, 17, 23–41.

Davis, J. A., and Leckie, J. O., 1978, Effect of adsorbed complexing ligands on trace metal uptake by hydrous oxides. *Environ. Sci. Techol.*, 12, 1309–1315.

De Angelis, M., Barkov, N. I., and Petrov, V. N., 1987, Aerosol concentrations over the last climatic cycle (160 kyr) from an Antarctic ice core. *Nature*, 325, 318–321.

DiTullio, G. R., Hutchins, D. A., and Bruland, K. W., 1993, Interaction of iron and major nutrients controls phytoplankton growth and species composition in the tropical North Pacific. *Limnol. Oceanogr.*, 38, 495–508.

Donat, J. R., and Bruland, K. W., 1990, A comparison of two voltammetric techniques for determining zinc speciation in Northeast Pacific ocean waters. *Mar. Chem.*, 28, 301–323.

Duce, R. A., 1986, The impact of atmospheric nitrogen, phosphorus and iron species on marine biological productivity. In: P. Buat-Menard, ed., *The Role of Air-Sea Exchange in Geochemical Cycling*, Reidel, Dordrecht, pp. 497–529.

Duce, R. A., and Tindale, N. W., 1991, Atmospheric transport of iron and its deposition in the ocean. *Limnol. Oceanogr.*, 36, 1715–1726.

Dugdale, R. C., and Wilkerson, F. P., 1990, Iron addition experiments in the Antarctic: a re-analysis. *Global Biogeochem. Cycles*, 4, 13–19.

Entsch, B., Sim, R. G., and Hatcher, B. G., 1983, Indication from photosynthetic components that iron is a limiting nutrient in primary producers on coral reefs. *Mar. Biol.*, 73, 17–30.

Fitzwater, S. E., Knauer, G. A., and Martin, J. H., 1982, Metal contamination and its effect on primary production measurements. *Limnol. Oceanogr.*, 27, 544–551.

Frew, R. D., 1995, Antarctic bottom water formation and the global cadmium to phosphorus relationship. *Geophys. Res. Lett.*, 22, 2349–2352.

Frew, R. D., and Hunter, K. A., 1992, Influence of Southern Ocean waters on the cadmium-phosphate properties of the global ocean. *Nature*, 360, 144–146.

Frew, R. D., and Hunter, K. A., 1995, Cadmium–phosphate properties in the Subtropical Convergence south of New Zealand. *Mar. Chem.*, 51, 223–237.

Gordon, R. M., Martin, J. H., and Knauer, G. A., 1982, Iron in northeast Pacific waters. *Nature*, 299, 611–612.

Goyne, E. R., and Carpenter, E. J., 1974, Production of iron-binding compounds by marine microorganisms. *Limnol. Oceanogr.*, 19, 840–842.

Gran, H. H., 1931, On the conditions for production of plankton in the sea. *Rapp. Proces-Verbaux Reunions Cons. Int. l'Explor. Mer*, 75, 37–46.

Hart, T. J., 1934, On the phytoplankton of the Southwest Atlantic and the Bellinghausen Sea 1929–1931. *Discovery Rep.*, 8, 1–268.

Harvey, H. W., 1937a, The supply of iron to diatoms. *J. Mar. Biol. Assoc. (UK)*, 22, 221–225.

Harvey, H. W., 1937b, Note on colloidal ferric hydroxide in sea water. *J. Mar. Biol. Assoc. (UK)*, 22, 205–219.

Hecky, E. E., and Kilham, P., 1988, Nutrient limitation of phytoplankton in freshwater and marine environments: a review of recent evidence on the effects of enrichment. *Limnol. Oceanogr.*, 33, 796–822.

Hester, K. and Boyle, E. A., 1982, Water chemistry control of the cadmium content of recent benthic foraminifera. *Nature*, 298, 260–262.

Hider, R. C., 1984, Siderophore-mediated absorption of iron. *Structure and Bonding*, 58, 26–87.

Ho, F. W. T., 1988, *On the Marine Chemistry of Copper, Nickel and Cadmium*. Ph.D. thesis, University of Otago, Dunedin, New Zealand.

Hong, H., and Kester, D. R., 1986, Redox state of iron in offshore waters of Peru. *Limnol. Oceanogr.*, 31, 512–524.

Hudson, R. J. M., and Morel, F. M. M., 1989, Distinguishing between extra- and intra-cellular iron in marine phytoplankton. *Limnol. Oceanogr.*, 34, 1113–1120.

Hudson, R. J. M., and Morel, F. M. M., 1990, Iron transport in marine phytoplankton: kinetics of cellular and medium coordination reactions. *Limnol. Oceanogr.*, 35, 1002–1020.

Huizenga, D. L., and Kester, D. R., 1983, The distribution of total and electrochemically available copper in the northwestern Atlantic Ocean. *Mar. Chem.*, 13, 281–291.

Hunter, K. A., 1982, The marine geochemistry of trace elements: a perspective. *Chem. N.Z.*, 46, 28–32.

Hunter, K. A., and Ho, F. W. T., 1991, Phosphorus–cadmium cycling in the Northeast Tasman Sea 35–40°S. *Mar. Chem.*, 33, 279–298.

Jaques, G., 1989, Primary production in the open Antarctic ocean during the austral summer: a review. *Vie Milieu*, 39, 1–17.

Joos, F., Sarmiento, J. L., and Siegenthaler, U., 1991a, Estimates of the effect of Southern Ocean iron fertilization on atmospheric CO2 concentrations. *Nature*, 349, 772–775.

Joos, F., Siegenthaler, U., and Sarmiento, J. L., 1991b, Possible effects of iron fertilization in the Southern Ocean on atmospheric CO_2 concentration. *Global Biogeochem. Cycles*, 5, 135–150.

Klinkhammer, G. P., 1980, The distribution of manganese in the Pacific Ocean. *Earth Planet. Sci. Lett.*, 49, 81–101.

Knauer, G. A., and Martin, J. H., 1981, Phosphorus–cadmium cycling in northeast Pacific waters. *J. Mar. Res.*, 39, 65–76.

Kolber, Z. S., Barber, R. T., Coale, K. H., Fitzwater, S. E., Greene, R. M., Johnson K. S., Lindley, S., and Falkowski, P. G., 1994, Iron limitation of phytoplankton photosynthesis in the equatorial Pacific Ocean. *Nature*, 371, 145–149.

Kulaev, I. S., 1979, *The Biochemistry of Inorganic Polyphosphates*. John Wiley & Sons, New York.

Kuma, K., Nakabayashi, S., Suzuki, Y., and Matsunaga, K., 1992, Dissolution rate and solubility of colloidal hydrous ferric oxide in seawater. *Mar. Chem.*, 38, 133–143.

Kurz, K. D., and Maier-Reimer, E., 1993, Iron fertilization of the austral ocean—the Hamburg Model assessment. *Global Biogeochem. Cycles*, 7, 229–244.

Landing, W. M., and Bruland, K. W., 1981, The vertical distribution of iron in the Northwest Pacific. *EOS: Trans. Am. Geophys. Union*, 62, 906.

Landing, W. M., and Bruland, K. W., 1987, The contrasting biogeochemistry of iron and manganese in the Pacific Ocean. *Geochim. Cosmochim. Acta*, 51, 29–43.

Landing, W. M., and Westerlund, S., 1988, The solution chemistry of iron(II) in Framvaren Fjord. *Mar. Chem.*, 23, 329–343.

Legrand, M., Feniet-Saigne, C., Saltzmann, E. S., Germain, C., Barkov, N. I., and Petrov, V. N., 1991, Ice-core record of oceanic emissions of dimethylsulfide during the last climate cycle. *Nature*, 350, 144–146.

Lewin, J., and Chen, C. H., 1971, Available iron: a limiting factor for marine phytoplankton. *Limnol. Oceanogr.*, 16, 670–675.

Lewis, B. L., and Landing, W. M., 1991, The biogeochemistry of manganese and iron in the Black Sea. *Deep-Sea Res.*, 38, S773–S803.

Mackey, D. J., 1983, Metal-organic complexes in seawater-an investigation of naturally occurring complexes of Cu, Zn, Fe, Mg, Ni, Cr, Mn, and Cd using HPLC. *Mar. Chem.*, 13, 169–180.

Martin, J. H., 1990, Glacial–interglacial CO2 change: the iron hypothesis. *Paleoceanogr.*, 5, 1–13.

Martin, J. H., 1992, Iron as a limiting factor in oceanic productivity. In: P. G. Falkowski and A. D. Woodhead, eds., *Primary Productivity and Biogeochemical Cycles in the Sea*, Plenum Press, New York, pp. 123–137.

Martin, J. H., and Fitzwater, S. E., 1988, Iron deficiency limits phytoplankton growth in the north east Pacific subarctic. *Nature*, 331, 341–343.

Martin, J. M., and Gordon, R. M., 1988, Northeast Pacific iron distributions in relation to phytoplankton productivity. *Deep-Sea Res.*, 35, 177–196.

Martin, J. H., Bruland, K. W., and Broenkow, W. W., 1976, Cadmium transport in the California Current. In: H. L. Windom and R. A. Duce, eds., *Marine Pollutant Transport*, Heath, Lexington, MA, pp. 159–184.

Martin, J. H., Gordon, R. M., Fitzwater, S., and Broenkow, W. W., 1989, VERTEX: phytoplankton/iron studies in the gulf of Alaska. *Deep-Sea Res.*, 36, 649–680.

Martin, J. H., Broenkow, W. W., Fitzwater, S. E., and Gordon, R. M., 1990a, Yes it does: a reply to the comment by Banse. *Limnol. Oceanogr.*, 35, 775–777.

Martin, J. H., Fitzwater, S. E., and Gordon, R. M., 1990b, Iron deficiency limits phytoplankton growth in Antarctic waters. *Global Biogeochem. Cycles*, 4, 5–12.

Martin, J. H., Gordon, R. M., and Fitzwater, S. E., 1990c, Iron in Antarctic waters. *Nature*, 345, 156–158.

Martin, J. H., Gordon, R. M., and Fitzwater, S. E., 1991a, The case for iron. *Limnol. Oceanogr.*, 36, 1793–1802.

Martin, J. H., Fitzwater, S. E., and Gordon, R. M., 1991b, We still say iron deficiency limits phytoplankton growth in the subarctic Pacific. *J. Geophys. Res.*, 96, 20699–20700.

Martin, J. H., Fitzwater, S. E., Gordon, R. M., Hunter, C. N., and Tanner, S. J., 1993, Iron, primary production and carbon–nitrogen flux studies during the JGOFS North Atlantic Bloom Experiment. *Deep-Sea Res.*, 40, 115–134.

Martin, J. H., and 43 others, 1994, Testing the iron hypothesis in the ecosystems of the equatorial Pacific Ocean. *Nature*, 371, 123–129.

McKnight, D. M., and Morel, F. M. M., 1979, Release of weak and strong copper-complexing agents by algae. *Limnol. Oceanogr.*, 24, 823–837.

Menzel, D. W., Hulburt, E. M., and Ryther, J. H., 1963, The effects of enriching Sargasso Sea water on the production and species composition of the phytoplankton. *Deep-Sea Res.*, 10, 209–219.

Meybeck, M., 1981, Pathways of major elements from land to ocean through rivers. In: *River Inputs to Ocean Systems*, United Nations, New York, pp. 18–30.

Moffett, J. W., and Zika, R. G., 1987, Solvent extraction of copper acetylacetone in studies of copper(II) speciation in seawater. *Mar. Chem.*, 21, 301–313.

Moffett, J. W., Zika, R. G., and Brand, L. E., 1990, Distribution and potential sources and sinks of copper chelators in the Sargasso Sea. *Deep-Sea Res.*, 37, 27–36.

Moore, R. M., 1978, The distribution of dissolved Cu in the eastern Atlantic ocean. *Earth Planet. Sci. Lett.*, 41, 461–468.

Morel, F. M. M., 1983, *Principles of Aquatic Chemistry*, Wiley-Interscience, New York.

Morel, F. M. M., and Hudson, R.J.M., 1985, The geobiological cycle of trace elements in aquatic systems: Redfield revisited. In: W. Stumm, ed., *Chemical Processes in Lakes*, John Wiley & Sons, New York, pp. 251–281.

Morel, F. M. M., Reinfelder, J. R., Roberts, S. B., Chamberlain, C. P., Lee, J. G., and Yee, D., 1994, Zinc and carbon co-limitation of marine phytoplankton. *Nature*, 369, 740–742.

Morley, N. H., Staham, P. J., and Burton, J. D., 1993, Dissolved trace metals in the southwestern Indian Ocean. *Deep-Sea Res.*, 40, 1043–1062.

Muller, F. L. L., and Kester, D. R., 1991, Voltammetric determination of the complexation parameters of zinc in marine and estuarine waters. *Mar. Chem.*, 33, 71–90.

Murphy, T. P., Lean, D. R. S., and Nalewajko, C., 1976, Blue-green alga. their excretion of iron-selective chelators enables them to dominate other algae. *Science*, 192, 900–902.

Neilands, J. B., 1981, Microbial iron compounds, *Annu. Rev. Biochem.*, 50, 715.

Nolting, R. N., De Baar, H. J. W., Masson, A., and Van Bennekom, A. J., 1991, Cadmium, copper and iron in the Weddell Sea and Weddell/Scotia Confluence (Antarctica). *Mar. Chem.*, 35, 219–243.

O'Sullivan, D. W., Hanson, A. K., Miller, W. L., and Kester, D. R., 1991, Measurement of Fe(II) in surface water of the equatorial Pacific. *Limnol. Oceanogr.*, 36, 1727–1741.

Patterson, C. C., and Settle, D. M., 1976, The reduction of orders of magnitude errors in lead analysis of biological materials and natural waters by controlling external sources of industrial Pb contamination introduced during sample collection, handling and analysis. In: D. M. La Fleur, ed., *Accuracy in Trace Analysis*, National Bureau of Standards, Washington, D.C., pp. 321–323.

Peng, T.-H., and Broecker, W. S., 1991, Factors limiting the reduction of atmospheric CO_2 by iron fertilization. *Limnol. Oceanogr.*, 36, 1919–1927.

Prahl, G., 1992, Prospective use of molecular paleontology to test for iron limitation on marine primary productivity. *Mar. Chem.*, 39, 167–185.

Price, N. M., Andersen, L. F., and Morel, F. M. M., 1991, Iron and nitrogen nutrition of equatorial Pacific plankton. *Deep-Sea Res.*, 38, 1361–1378.

Price, N. M., and Morel, F. M. M., 1990, Cadmium and cobalt substitution for zinc in a marine diatom. *Nature*, 344, 658–660.

Redfield, A. C., 1934, On the proportions of organic derivatives in sea water and their relation to the composition of plankton. In: *James Johnstone Memorial Volume*, Liverpool University Press, Liverpool, UK, pp. 176–192.

Redfield, A. C., Ketchum, B. H., and Richards, F. A., 1963, The influence of organisms on the composition of seawater. In: M. N. Hill, ed., *The Sea*, Vol. 2, John Wiley & Sons, New York, pp. 26–27.

Reid, R. T., and Butler, A., 1991, Investigation of the mechanism of iron acquisition by the marine bacterium *Alteromonas luteoviolaceus*: characterization of siderophore production. *Limnol. Oceanogr.*, 36, 1783–1792.

Rich, H. W., and Morel, F. M. M., 1990, Availability of well-defined iron colloids to the marine diatom *Thalassiosira weissflogii. Limnol. Oceanogr.*, 35, 652–662.

Roberts, L., 1991, Report nixes "Geritol" fix for global warming. *Science*, 253, 1490–1491.

Rueter, J. G., and Morel, F. M. M., 1981, The interaction between zinc deficiency and copper toxicity as it affects the silicic acid uptake mechanisms in *Thalassiosira pseudonana. Limnol. Oceanogr.*, 26, 67–73.

Saager, P. M., De Baar, H. J. W., and Burkill, P. H., 1989, Manganese and iron in Indian Ocean waters. *Geochim. Cosmochim. Acta*, 53, 2259–2267.

Saager, P. M., De Baar, H. J. W., and Howland, R. J., 1992, Cd, Zn, Ni and Cu in the Indian Ocean. *Deep-Sea Res.*, 39, 9–35.

Saager, P. M., Schijf, J., and De Baar, H. J. W., 1993, Trace metal distributions in seawater and anoxic brines in the eastern Mediterranean Sea. *Geochim. Cosmochim. Acta*, 57, 1419–1432.

Sclater, F. F., Boyle, E. A., and Edmond, J. M., 1976, On the marine geochemistry of nickel. *Earth Planet Sci Lett*, 31, 119–128

Shen, G. T., and Boyle, E. A., 1987, Lead in corals: reconstruction of historical industrial fluxes to the surface ocean. *Earth Planet. Sci. Lett.*, 82, 289–304.

Sherrell, R. M., and Boyle, E. A., 1988, Zinc, chromium, vanadium and iron in the Mediterranean Sea. *Deep-Sea Res.*, 35, 1319–1334.

Sunda, W. G., 1989, Trace metal interactions with marine phytoplankton. *Biol. Oceanogr.*, 6, 411–442.

Sunda, W. G., and Hanson, A. K., 1987, Measurement of the free cupric ion concentration in seawater by a ligand competition technique involving copper sorption onto C18 SEP-PAK cartridges. *Limnol. Oceanogr.*, 32, 537–551.

Sunda, W. G., and Huntsman, S. A., 1992, Feedback interactions between zinc and phytoplankton in seawater. *Limnol. Oceanogr.*, 37, 25–40.

Sunda, W. G., and Huntsman, S. A., 1995, Regulation of copper concentration in the oceanic nutricline by phytoplankton uptake and regeneration cycles. *Limnol. Oceanogr.*, 40, 132–137.

Sunda, W. G., Swift, D. G. and Huntsman, S. A., 1991, Low iron requirement for growth in oceanic phytoplankton. *Nature*, 351, 55–57.

Symes, J. L., and Kester, D. R., 1985, The distribution of iron in the northwest Atlantic. *Mar. Chem.*, 17, 57–74.

Thomas, W. H., Holm-Hansen, O., Siebert, D. L. R., Azam, F., Hodson, R., and Takahashi, M., 1977, Effects of copper on phytoplankton standing crop and productivity: controlled ecosystem pollution experiment. *Bull. Mar. Sci.*, 27, 34–43.

Trick, C. G., 1989, Hydroxamate-siderophore production and utilization by marine eubacteria. *Curr. Microbiol.*, 18, 375–378.

Trick, C. G., Andersen, R. J., Gillam, A., and Harrison, P. J., 1983a, Prorocentrin: an extracellular siderophore produced by the marine dinoflagellate *Prorocentrum minimum*. *Science*, 219, 306–308.

Trick, C. G., Andersen, R. J., Price, N. M., Gillam, A. and Harrison, P. J., 1983b, Examination of hydroxamate-siderophore production by neritic eukaryotic marine phytoplankton. *Mar. Biol.*, 75, 9–17.

Turekian, K. K., 1977, The fate of metals in the oceans. *Geochim. Cosmochim. Acta*, 41, 1139–1144.

Turner, D. R., Whitfield, M., and Dickson, A. G., 1981, The equilibrium speciation of dissolved components in freshwater and seawater at 25°C and 1 atm pressure. *Geochim. Cosmochim. Acta*, 45, 855–881.

Tuschall, J. R., and Brezonik, P. L., 1981, Evaluation of the copper anodic stripping voltammetry complexometric titration for complexing capacities and conditional stability constants. *Anal. Chem.*, 53, 1986–1989.

Van den Berg, C. M. G., 1982, Determination of copper complexation with natural organic ligands in seawater by equilibration with MnO_2. Part 1: Theory; Part 2: Experimental procedures and application to surface seawater. *Mar. Chem.*, 11, 307–342.

Van den Berg, C. M. G., 1984, Organic and inorganic speciation of copper in the Irish Sea. *Mar. Chem.*, 14, 201–212.

Van den Berg, C. M. G., 1985, Determination of the zinc complexing capacity in seawater by cathodic stripping voltammetry of zinc-APDC complex ions. *Mar. Chem.*, 15, 1–18.

Van den Berg, C. M. G., Merks, A. G. A., and Duursma, E. K., 1987, Organic complexation and its control of the dissolved concentrations of copper and zinc in the Scheldt Estuary. *Estuar. Coast. Shelf Sci.*, 24, 785–797.

Waite, T. D., and Morel, F. M. M., 1983, Characterization of complexing agents in natural waters by copper(II)/copper(I) amperometry. *Anal. Chem.*, 55, 1268–1274.

Walsh, R. S., and Hunter, K. A., 1992, Influence of phosphorus storage on the uptake of cadmium by marine algae and implications for cadmium biogeochemistry. *Limnol. Oceanogr.*, 37, 1361–1369.

Watson, A. J., Laws, C. S., Van Scoy, K. A., Millero, F. J., Yao, W., Friedrich, G. E., Liddicoat, M. I., Wanninkhof, R. H., Barber, R. T., and Coale, K. H., 1994, Minimal effect of iron fertilization on sea-surface carbon dioxide concentrations. *Nature*, 371, 143–145.

Watson, A. M., Liss, P. S., and Duce, R. A., 1991, Design of a small-scale in situ iron fertilization experiment. *Limnol. Oceanogr.*, 36, 1960–1965.

Webb, M., 1979, *The Metallothioneins*, Elsevier/North-Holland, Amsterdam, 195 pp.

Wells, M. L., 1989, The availability of iron in seawater: a perspective. *Biol. Oceanogr.*, 6, 463–476.

Wells, M. L., and Mayer, L. M., 1991, Variations in the chemical lability of iron in estuarine, coastal and shelf waters and its implications for phytoplankton. *Mar. Chem.*, 32, 195–210.

Wells, M. L., Zorkin, N. G., and Lewis, A. G., 1983, The role of colloid chemistry in providing a source of iron to phytoplankton. *J. Mar. Res.*, 41, 747–768.

Wells, M. L., Mayer, L. M., and Guillard, R. L. L., 1991, A chemical method for estimating the availability of iron to phytoplankton in seawater. *Mar. Chem.*, 33, 23–40.

Westerlund, S., and Ohman, P., 1991, Cadmium, copper, cobalt, nickel, lead and zinc in the water column of the Weddell sea, Antarctic. *Geochim. Cosmochim. Acta*, 55, 2127–2146.

Yeats, P. A., and Campbell, J. A., 1983, Nickel, copper, cadmium and zinc in the northwest Atlantic Ocean. *Mar. Chem.*, 12, 43–58.

Young, R. W., Carder, K. L., Betzer, P. R., Costello, D. K., Duce, R. A., DiTullio, G. R., Tindale, N. W., Laws, E. A., Uemastu, M., Merrill, J. T., and Feely, R. A., 1991, Atmospheric iron inputs and primary productivity: phytoplankton responses in the North Pacific. *Global Biogeochem. Cycles*, 5, 119–134.

Zhuang, G., Duce, R. A., and Kester, D. R., 1990, The dissolution of atmospheric iron in the surface seawater of the open ocean. *J. Geophys. Res.*, 95, 16207–16216.

Zhuang, G., Zhen, Y., Duce, R. A., and Brown, P. R., 1992, Chemistry of iron in marine aerosols. *Global Biogeochem. Cycles*, 6, 161–173.

Atmosphere–Water–Rock Interactions; As Observed in Alpine Lakes

LAURA SIGG, JERALD L. SCHNOOR, AND WERNER STUMM

ABSTRACT. Alpine lakes are well suited to study the interaction of atmospheric constituents with rocks. Such studies indicate that alpine lakes in crystalline areas are especially sensitive to acidic deposition. The chemical composition of the water depends on the type of rocks present in the catchment area. The pH of different lake waters vary between about 5 and 7, depending on whether they are located in purely crystalline area or in an area including some carbonate schist and dolomite. Weathering rates can be estimated from the chemical composition of the lake waters using a mass balance approach. These rates, observed in the field, are compared with the dissolution rate of minerals in the laboratory. In the proton-promoted dissolution of aluminum silicate minerals by acid rain, hydrogen ions attack the bridging oxygen in the $\equiv Si-O-Al \equiv$ and cause the subsequent breaking of the bond. The metals Pb(II) and Zn(II) in these lakes show a clear pH dependence in their concentrations. This can be explained by the pH dependence of the metal adsorption reaction which in the simplest case can be characterized by a reaction of the type $S-OH + Me^{2+} \rightleftharpoons S-OMe^+ + H^+$. In the case of Cu(II) the adsorption is less apparent because of complex formation in solution, and in the case of Al(III) the pH dependence can be accounted for by the pH dependence of Al(III) solubility.

INTRODUCTION

Studies on small alpine lakes provide some perspectives of the major geochemical cycles and the anthropogenic effects of atmospheric deposition. Atmospheric constituents react with the crust of the earth through the process of rock weathering. The general term *weathering* encompasses a variety of processes by which parent rocks are broken down mechanically or are chemically dissolved. In these processes, water occupies a central role, serving both as a reactant and as a transporting agent of suspended and dissolved material. The atmosphere provides a reservoir of weak acids (CO_2 and some organic acids) and oxidants (oxygen, ozone, radicals) and may contain anthropogenic pollutants (strong acids, heavy metals, organic compounds). During chemical weathering, rocks and primary minerals undergo chemical reactions with water (and its acids and oxidants), thereby becoming transformed to solutes and soils and eventually to sediments and sedimentary rocks.

Alpine lakes occur uppermost in the river basin; they receive first the products of weathering and they are the first system to experience changes in atmospheric deposition. Thus, they reflect most directly the close linkage between natural and introduced atmospheric constituents and rock weathering. Understanding the role of alpine lakes and their catchment areas in neutralizing acid deposition and regulating toxic metal pollutants is the key to assessing impacts of civilization on remote areas and generalizing on the effects elsewhere. Furthermore, because alpine lakes are small and poorly buffered, they respond rapidly to changes in the atmosphere and may serve as early warning systems for changes that may occur at lower elevations at a later time.

For the past 10 years, we have been studying (Schnoor and Stumm, 1986; Zobrist et al., 1987; Giovanoli et al., 1988; Sulzberger et al., 1990; Barry et al., 1994) a few alpine lakes to assess the effect of atmospheric deposition of acids and heavy metals on lake-water composition, mainly, to establish the relationship between rock-weathering and resulting water chemistry. We deduced plausible weathering reactions and estimated their rates from quantitative analysis of precipitation, of rocks (including rock model analysis, X-ray fluorescence, electron microscopy and X-ray spectrometry) and water chemistry.

The Alpine Lakes

The alpine lakes which we have been investigating are from an area in the southern part of the Swiss Alps (upper Val Maggia), in which the predominant rocks are crystalline and consist mostly of gneiss, containing variable amounts of quartz, plagioclase, biotite, K-feldspar, muscovite, amphibole, and epidote (Günthert et al., 1976). In some parts of this area, schists and dolomites are also present, so that a variety of geochemical environments are encountered within a few kilometers.

Unaltered rocks are exposed in most of this area, and only very thin soils are present (Figure 19.1, photograph of *L. Cristallina*). The lakes examined here are small (0.01–0.05 km^2); their catchment areas are also rather small (0.01–0.05 km^2); they are situated at altitudes between 1800 and 2500 m and are thus ice-covered during 6–8 months of the year. The catchments receive an average of 1.9–2.4 m of precipitation each year. They lie above the tree line and are characterized by steep slopes. There is little soil or at most thin patches of immature soils with few vascular plants. In the absence of soils and macroscopic vegetation, chemical weathering of the exposed bare rocks with the water derived from atmospheric depositions (mostly snow) is virtually the only process by which acid deposition can be neutralized. The contact time of water with the rocks is usually very short and varies over the year from a few days to weeks (snow-melt versus summer rains).

ACID DEPOSITION AND CHEMICAL WEATHERING

Acidic Precipitation

Most of the major and minor components of the atmosphere (e.g., N_2, O_2, CO_2, NO, NO_2, SO_2, CH_4) participate in elemental cycles that are governed by oxida-

Figure 19.1. Lake Cristallina is situated in the southern part of Switzerland at an altitude of 2400 m in an area with crystalline rocks (gneiss, granite). Since there is no river inflow, this lake acts as a snow and rain collector.

tion–reduction reactions, mostly of biological origin. Photosynthesis and respiration are major reactions in these cycles. These oxidation and reduction processes produce and consume these gases of the atmospheric reservoir. In oxidation–reduction processes, electron transfers are coupled with the transfer of protons (charge balance). Every modification of the redox balance elicits a modification of the acid–base balance. The oxidation of carbon, sulfur, and nitrogen during the combustion of fossil fuels and released also by other industrial processes increases the concentrations of sulfur oxides, nitrogen oxides, and carbon dioxide in the atmosphere. The oxides are converted to sulfuric acid, nitric acid, and carbonic acid and become associated with aerosols, rain drops, snow flakes, and fog. The compounds mainly responsible for acid rain and snow are the strong acids sulfuric acid and nitric acid; hydrochloric acid is also often present, from the incineration of chloride-bearing polymers. These acids are also present in nonpolluted atmospheres, but at significantly lower concentrations. These strong acids are usually only partially neutralized by bases, mostly ammonia, of natural (e.g., manure) or industrial origin (Stumm and Morgan, 1996). Sulfur dioxide has a maximal residence time in the atmosphere of a few days. It is removed mostly by oxidation to sulfuric acid, dissolution in water droplets, and deposition of rain, snow, or aerosols. During this time the sulfur dioxide may be transported several hundred kilometers. Nitric acid is formed more rapidly from nitrogen oxides, resulting in a more restricted spatial dispersion (Berner and Berner, 1987).

Chemical Weathering

Alpine weathering rates have been estimated from the chemical composition of the lake waters. The alpine rocks which interact with atmospheric depositions (dilute solutions of carbon dioxide and mineral acids from rain and melted snow) are built of a variety of minerals that interact at different rates with the hydrogen ions of these acids (Table 19.1). Calcium carbonate dissolves at least 100 times faster than aluminum silicates in slightly acidic waters (pH ~5) (Chou and Wollast, 1989). We used a mass balance approach which implies that the quantities of the various elements (e.g., Ca, K, Na, Si, Al) in the lake or the runoff (corrected for the ions that have been deposited by the precipitation) equals the quantities of these elements dissolved from the minerals (Likens, 1985).

Lake Cristallina

At the highest altitude, in the catchment area of Lake Cristallina, the hydrogen ion deposition amounts to about 20 mmol m^{-2} yr^{-1}. The composition of the rain or snow, as given in Figure 19.2, is characterized by a hydrogen ion concentration of ~18 μmol/L (pH ~4.75). Table 19.2 illustrates some plausible weathering reactions (Giovanoli et al., 1989). The 18 μmol/L react with 4 μmol plagioclase (a calcium-feldspar consisting mostly of anorthite), 2 μmol epidote (a calcium aluminum/silicate containing some Fe(III)), 1 μmol biotite (a three-layer mineral containing some Mg and Fe(II)), 1 μmol of orthoclase (a K-feldspar), and 1 μmol of CaCO$_3$ (calcite). The latter is not present in the bedrock but from wind-blown dust. For every mole of H$^+$ neutralized (used) in the chemical weathering reaction, one equivalent of base cations goes into solution. The lake water formed is still slightly acid and contains ~6 μmol/L of H$^+$ (pH ~5.2) (Figure 19.2).

TABLE 19.1
Some Important Dissolution Reactions

1. $CaCO_3$ (calcite), the major component of limestone
   (a) $CaCO_3(s) + H^+ \rightleftharpoons Ca^{2+} + HCO_3^-$
   (b) $CaCO_3(s) + H_2CO_3^* \rightleftharpoons Ca^{2+} + 2HCO_3^-$ [a]
2. $CaMg(CO_3)_2$ (dolomite)
   (a) $CaMg(CO_3)_2(s) + 2H^+ \rightleftharpoons Ca^{2+} + Mg^{2+} + 2HCO_3^-$
   (b) $CaMg(CO_3)_2(s) + 2H_2CO_3^* \rightleftharpoons Ca^{2+} + Mg^{2+} + 4HCO_3^-$
3. SiO_2 (silica, quartz)
   $SiO_2(s) + 2H_2O \rightleftharpoons H_4SiO_4$
4. $Al(OH)_3$ (aluminum hydroxide), often present as gibbsite
   (a) $Al(OH)_3(s) + H^+ \rightleftharpoons Al(OH)_2^+ + H_2O$ [b]
   (b) $Al(OH)_3(s) + 3H^+ \rightleftharpoons Al^{3+} + 3H_2O$
   (c) $Al(OH)_3 + 3H_2CO_3^* \rightleftharpoons Al^{3+} + 3HCO_3^- + 3H_2O$
5. $NaAlSi_3O8$ (albite, a sodium feldspar or sodium plagioclase)
   (a) $NaAlSi_3O_8 + H^+ + 4\frac{1}{2} H_2O \rightleftharpoons Na^+ + 2 H_4SiO_4 + \frac{1}{2}Al_2Si_2O_5(OH)_4(s)$ (kaolinite)
   (b) $NaAlSi_3O_8 + H_2CO_3^* + 4\frac{1}{2} H_2O \rightleftharpoons Na^+ + HCO_3^- + 2H_4SiO_4 + \frac{1}{2} Al_2Si_2O_5(OH)_4(s)$ (kaolinite) [c,d]

[a] $H_2CO_3^*$ is the sum of $CO_2(aq)$ and H_2CO_3.
[b] Dissolved Al can occur as Al^{3+} and as hydroxo species $Al(OH)_n^{3-n}$.
[c] The dissolution of feldspars occurs in such a way that Na^+ and SiO_2 (as H_4SiO_4) are released into the water while Al is preserved in the form of the common clay mineral kaolinite.
[d] The weathering process of aluminum silicates may also lead directly to aluminum hydroxide (or $Al(OH)_n^{3-n}$ ions) and silicic acid; for example,
$$KAlSi_3O_8 + H^+ + 7H_2O \rightleftharpoons K^+ + 3H_4SiO_4 + Al(OH)_3(s)$$

Weathering Rates. The dissolution reactions given (Table 19.2) contain some kinetic information on the rate of weathering if we know the runoff and the area of the catchment of the lake. The drainage basin is $\sim$17 ha (1.7×10^5 m^2), and runoff is $\sim$1.2 m yr^{-1} (1.2 m^3 water per square meter and year). In each liter of water that leaves the catchment of Lake Cristallina, 18 μmol of H^+ have been neutralized and 18 μeq of cations and 23 μmol of H_4SiO_4 have been released. In the reaction with a hypothetical aluminosilicate we have

$$Ca_5Mg_{0.5}Na_3K_2Al_{12}Fe_3Si_{23}O_{75} + CaCO_3 + 18H^+ + 0.75O_2 + 60.5H_2O \rightarrow$$
$$6Ca^{2+} + 0.5\ Mg^{2+} + 3Na^+ + 2K^+ + 23H_4SiO_4 + 12\ Al(OH)_3 + 3Fe(OH)_3 + H_2CO_3^*$$

For every square meter of catchment area, this is 22 meq of cations and 2.8×10^{-2} mol H_4SiO_4 yr^{-1}.

About half of the $Al(OH)_3$ is redissolved and appears as inorganic aluminum species in discharge waters. The extent of chemical denudation per m^2 yr^{-1} amounts to 3.4 g mineral (34 kg ha^{-1} yr^{-1}). For the composition of the Swiss alpine rivers, weathering rates of 20–500 meq m^{-2} yr^{-1} (milliequivalents acid neutralized per square meters of land area per year) can be estimated (Drever and Zobrist, 1992). That corresponds to a yearly dissolution (chemical denudation) of 7–100 g of silicate rock per square meter per year. If there were no mechanical erosion, only about 5–35 cm^3 of rock would be lost per square meter of catchment area; this corresponds to a layer of only about 5–35 micrometers per year. Mechanical weathering is the fragmentation by primarily physical processes into small grain particles; it includes wind abrasion and rock fragmentation by the freezing of water.

Water chemistry and geology

The chemical composition of the water in the alpine lakes depends on the geochemistry (type of rocks present) in each catchment area (compare left side and right side of Figure 19.2). The inputs of hydrogen ions and heavy metal ions from atmospheric depositions are expected to be uniform over the area considered (about 10 km apart). The pH of different lake waters varies between about 5 and 7, depending on whether they are located in a purely crystalline area or in an area including some carbonate-bearing schist and dolomite. As we have seen, weathering reactions consume protons and produce alkalinity (for definitions see Figure 19.3). The acid–base balance is considered with respect to the composition of water in equilibrium with carbon dioxide only as a reference point (Figure 19.3). Acidic deposition contains strong acidity with respect to this reference. On the opposite side, basic waters contain alkalinity, the sum

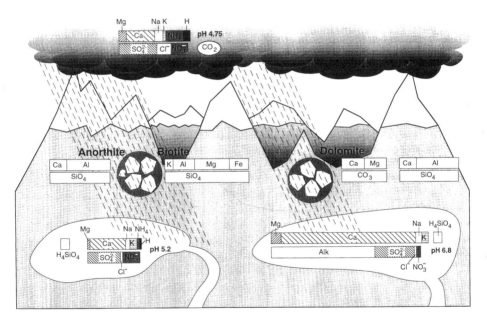

Figure 19.2. Acidic rain and snow interact with rocks. Weathering products of gneiss (composed of anorthite, biotite, and other feldspartic minerals) are calcium, magnesium, potassium, sodium, silicate, aluminum, and iron. These weathering reactions are slow, however, so that acids are only partly neutralized and the indicated composition of slightly acidic lake waters results (Lake Cristallina, pH 5.2; see left side of figure). In one hectare of catchment area within a year, 220 mol of hydrogen ions (acid deposition) react with 34 kg of rocks of the stoichiometric composition $Ca_5Mg_{0.5}Na_3K_2Al_{12}Fe_3Si_{23}O_{75} \cdot CaCO_3$ to give the water composition indicated in the figure. If dolomite is present together with aluminum silicates in the catchment area, its dissolution is much faster and the resulting lake water contains higher concentrations of calcium, magnesium, and bicarbonate and is nearly neutral (Lake Val Sabbia, pH 6.8, see right side of figure).

TABLE 19.2
Chemical Weathering in the Lake Cristallina Catchment[a]

Rock-Forming Mineral	Chemical Weathering Reaction $(\mu mol/L)$
Calcite	$CaCO_3 + 2H^+ \rightarrow Ca^{2+} + H_2CO_3$
Plagioclase	$4Na_{0.75}Ca_{0.25}Al_{1.25}Si_{2.75}O_8 + 5H^+ + 27H_2O \rightarrow 3Na^+ + Ca^{2+} + 11H_4SiO_4 +$ $5Al(OH)_3$
Epidote	$2Ca_2Al_{2.5}Fe_{0.5}Si_3O_{12}(OH) + 8H^+ + 16H_2O \rightarrow 4Ca^{2+} + 6H_4SiO_4 + 5Al(OH)_3 +$ $Fe(OH)_3$
Biotite	$KMg_{0.5}Fe_{2.5}AlSi_3O_{10}(OH)_2 + 7H^+ + 3H_2O \rightarrow K^+ + 0.5Mg^{2+} + 3H_4SiO_4 +$ $Al(OH)_3 + 2.5Fe^{2+}$ $2.5Fe^{2+} + 0.625O_2 + 6.25H_2O \rightarrow 2.5Fe(OH)_3 + 5H^+$
Orthoclase (K-feldspar)	$KAlSi_3O_8 + H^+ + 7H_2O \rightarrow K^+ + 3H_4SiO_4 + Al(OH)_3$

[a]Minerals that undergo chemical weathering in the Lake Cristallina (CH) catchment area. In addition, some H^+ ions are produced when Fe^{2+} oxidizes and precipitates as $Fe(OH)_3$ and are consumed when $Al(OH)_3$ is dissolved as $Al(OH)_x^{(3-x)}$

Summary: Interaction of acid with hypothetical aluminosilicate
$$Ca_5Mg_{0.5}Na_3K_2Al_{12}Fe_3Si_{23}O_{75} + CaCO_3 + 18H^+ + 0.75O_2 + 60.5H_2O \rightarrow$$
$$6Ca^{2+} + 0.5Mg^{2+} + 3Na^+ + 2K^+ + 23H_4SiO_4 + 12Al(OH)_3 + 3Fe(OH)_3 + H_2CO_3$$

Other reactions occurring (to a lesser extent than the reactions given above):

$$KAl_3Si_3O_{10}(OH)_2(s) + H^+ + 9H_2O \rightarrow K^+ + 3H_4SiO_4 + 3Al(OH)_3$$
muscovite

$$Al(OH)_3 + 3H^+ \rightarrow Al^{3+} + 3H_2O$$
(noncrystalline)

$$FeOOH_{(s)} + 3H^+ \rightarrow Fe^{3+} + 2H_2O$$
goethite

$$SiO_{2(s)} + 2H_2O \rightarrow H_4SiO_4$$
quartz

of bases, which consists here mostly of bicarbonate and carbonate ions. In areas with more readily weatherable carbonate minerals, the strong acidity from depositions is neutralized to a larger extent, so that the resulting lake water bears alkalinity, and the pH and concentrations of base cations (calcium, potassium, sodium, magnesium) are higher.

Critical loads of acidity—that is, allowable acid loading that will not acidify lakes and forest soils—can be developed from areal rates of weathering (= H^+ ion-neutralization). Models have been developed by Gherini et al. (1985); Cosby et al. (1985); Nikolaidis et al. (1985); Furrer et al. (1989, 1990); Sverdrup et al. (1992), and Warfvinge et al. (1992).

MECHANISM AND KINETICS OF MINERAL DISSOLUTION

In the past 10 years we have studied the dissolution of minerals under controlled conditions in the laboratory (Furrer and Stumm, 1986; Wieland and Stumm, 1992; Stumm

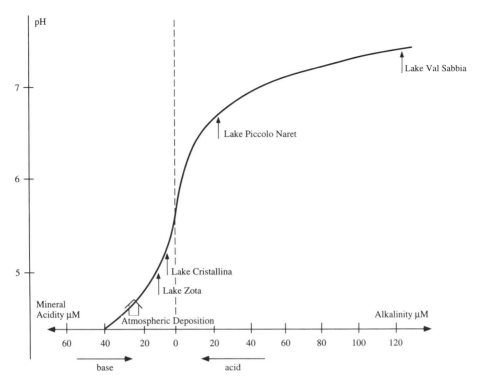

Figure 19.3. The curve gives the acid–base balance in the waters of the rain (snow) and of four lakes and is defined with respect to H_2O in equilibrium with carbon dioxide. Weathering reactions consume protons and produce alkalinity (see Table 19.1); that is, the composition of the water during weathering moves along the curve from left to right, from low pH to high pH, and from acidity to alkalinity. One has to distinguish between the H^+ concentration as an intensity factor and the availability of H^+, the H^+-ion reservoir as given by the base-neutralizing capacity (BNC), or the H-acidity [H-Acy]. The BNC relates to the alkalinity [Alk] or acid neutralizing capacity (ANC) by

$$H\text{-}Acy = -[Alk]$$
$$\text{Base neutralizing capacity} = -\text{Acid neutralizing capacity}$$

The BNC can be defined by a net proton balance with regard to a reference level—the sum of the concentrations of all the species containing protons in excess of the reference level, less the concentrations of the species containing protons in deficiency of the proton reference level. For natural waters, a convenient reference level (corresponding to an equivalence point in alkalimetric titrations) includes H_2O and H_2CO_3. The acid-neutralizing capacity (ANC) or alkalinity [Alk] is related to [H-Acy] by

$$-[H\text{-}Acy] = [Alk] = [HCO_3^-] + 2\,[CO_3^{2-}] + [OH^-] - [H^+]$$

and Wollast, 1990; Stumm, 1992). Among the first questions was whether the dissolution reaction is controlled by a transport step (e.g., the transport of a reactant or weathering product controls the reaction rate) or by a surface reaction (a chemical reaction step at the surface is slow in comparison to the transport step). Our batch studies confirmed earlier results of Holdren and Berner (1979) and Berner and Schott (1982) that the dissolution reaction is typically controlled by reactions at the solid–aqueous solution interface. The surfaces of oxides and aluminum silicate minerals in the presence of water are characterized by amphoteric surface hydroxyl groups whose oxygen donor atoms can coordinate with hydrogen ions and metal ions (Chapter 1, this volume). A

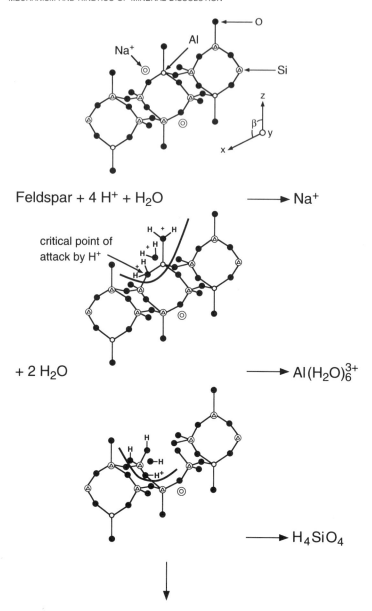

Figure 19.4. Mechanistically the dissolution reaction is controlled by the interaction of hydrogen ions (H^+) which first replace any cation attached to the aluminum silicate and then attack at the surface of the oxygen or hydroxide ions which bridge the Al and Si atoms in the lattice. This is shown for a feldspar structure in Figure 19.4a and for a mineral such as mica or muscovite that is built up by a layer of octahedrally coordinated Al-oxide interconnected with a layer of tetrahedrally coordinated Si-oxide in Figure 19.4b. The detachment of the Al-center is usually the rate- determining step followed by the release of Si from the same surface site. Subsequent dissolution follows in the same sequence. (Adapted from Wieland and Stumm, 1992.)

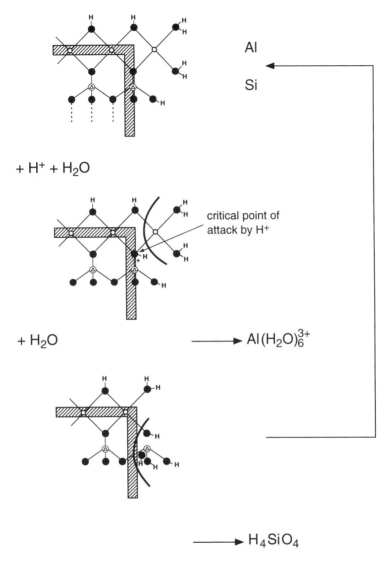

Figure 19.4. Continued

direct link between the dissolution rate and the amount of adsorbed H^+, based on surface speciation modeling and surface titration experiments has been established (Furrer and Stumm, 1986; Blum and Lasaga, 1988; Wieland et al., 1988; Caroll-Webb and Walther, 1988; Brady and Walther, 1989; Stumm and Wollast, 1990; Casey and Sposito, 1992, and Stumm, 1992). In a series of steps involving geometric and chemical principles at surface defects (kinks, edges, surface dislocations), a detachment of an activated surface-coordinated complex is the rate-determining step.

The scheme illustrated in Figures 19.4a and 19.4b makes it plausible that in the dissolution of feldspars and layer silicates the hydrogen ion of the precipitation (snow, rain) causes surface protonation; the H^+ ion attacks the bridging oxygen of the

Si—O—Al bridging bond (Hellmann et al., 1990; Wieland and Stumm, 1992) and causes a coupled release of Al and Si. Schematically,

$$\equiv Si—O—Al\equiv + H^+ \rightleftharpoons \equiv Si—OH—Al\equiv$$

A fast surface protonation (which weakens these bonds) is followed by a slow hydrolysis reaction:

$$\equiv Si—OH—Al\equiv \xrightarrow{H_2O} \equiv SiOH + \equiv AlOH_2^+$$

This is in accord with recent *ab initio* quantum mechanical studies by Xiao and Lasaga (1994).

   In laboratory experiments a zero-order dissolution rate—that is, a dissolution rate that is independent of time—is typically observed, thus implying that the surface is continuously reconstituted (as shown in the scheme of Figure 19.4) and that the surface area and morphology of the surface grains remains at steady state (Schnoor, 1990). Therefore, the neutralization of acid rain by crystalline minerals is kinetically controlled. The rate does not depend linearly on the hydrogen ion concentration, because the protons are not adsorbed in linear dependence of their concentration, and the adsorbed rather than the dissolved protons determine the dissolution rate. An increase in the proton concentration by a factor of 10 (a decrease in pH by one unit) accelerates the dissolution rate by a factor of ~2.6. Laboratory data on the dissolution rate at pH = 5 are on the order of 7 μmol m^{-2}yr^{-1} (quartz) to 400 μmol m^{-2} yr^{-1} (biotite, plagioclase). A rough calculation shows that this corresponds to the dissolution of, at best, only a few atomic layers per year. Mechanical erosion has been estimated to be at least 10 times larger.

   The laboratory data on actual rates of weathering (dissolution) of rocks on a per square meter mineral surface area basis may be compared with the field data on the weathering rate on a per-square-meter catchment area basis. Within the catchment area of Lake Cristallina such a comparison, assuming the same weathering rates in the field and in the laboratory, indicates that for a square meter of geographic area approximately 100 m^2 of mineral grains are active in weathering (Figure 19.5). Drever and Zobrist (1992) have shown more generally for a region underlying Lake Cristallina (having the same type of rocks) that the weathering rates increase significantly with decreasing catchment elevation (from a rate of ca 20 meq m^{-2}yr^{-1} at 2400 m altitude to a rate of about 500 meq m^{-2}yr^{-1} at 250 m altitude). This difference is due to different factors, particularly on the higher surface area of active mineral grains available with increasing soil depth. Furthermore, respiration activities under these conditions are much higher and vascular plants exude through their roots organic acids that may increase weathering rates. Finally, temperature may have a (small) effect.

   Chemical weathering is an important feature of the global hydrogeochemical cycle of elements. We may compare the CO_2 consumption for the weathering of carbonates with that of silicates in the following simplified reactions:

$$CaCO_3(s) + CO_2 + H_2O \rightarrow Ca^{2+} + 2HCO_3^-$$

$$CaSiO_3(s) + 2CO_2 + 3H_2O \rightarrow Ca^{2+} + 2HCO_3^- + H_4SiO_4$$

Both reactions produce Ca^{2+} and HCO_3^-, which in the sea eventually produce again $CaCO_3$:

$$Ca^{2+} + 2HCO_3^- \rightarrow CaCO_3(s) + CO_2 + H_2O$$

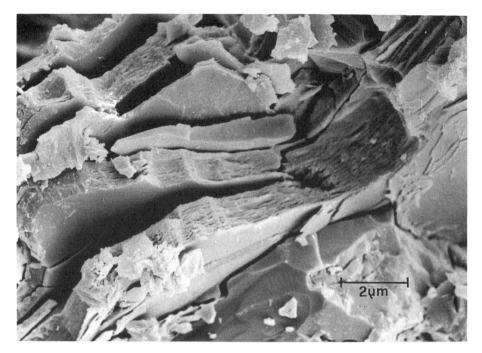

Figure 19.5. Scanning electron micrograph (made by R. Giovanoli of the University of Berne) of weathered gneiss, showing packs of mica splitting into sheets. Attack was probably from edges and propagated through interlayer cleavage. Sheet surfaces appear clean and unattacked.

But, in the silicate weathering we consume CO_2 stoichiometrically and only half of the CO_2 consumed is released and returned to the atmosphere upon $CaCO_3$ formation. Thus, silicate weathering results in a net loss of atmospheric CO_2 (Berner and Lasaga, 1989).

FATE OF HEAVY METALS

The atmosphere is also an important conveyor belt for *heavy metals*. The geochemical cycle of many heavy metals such as lead, zinc, cadmium, mercury, and copper have been substantially perturbed by civilization, and the effects of perturbation can be demonstrated at least in large parts of the Northern Hemisphere. For many of these trace elements, the natural fluxes from rocks to the atmosphere and to water have been drastically increased by human activities. Especially the fluxes to the atmosphere have increased due to combustion of fossil fuels, waste combustion, use of lead as a gasoline additive, industrial exhausts, and so on. Although the alpine lakes considered here are relatively remote from direct sources, atmospheric deposition to the area provides significant inputs of metal ions. Contributions by the weathering of minerals is negligible in comparison to atmospheric inputs; weathering is, however, predominant for Al(III) and Fe(II, III).

Thus, we have been able to compare residual metal concentrations in these lakes that differ in pH values due to somewhat different geological background but for which the metal inputs are very similar (Figure 19.6). The concentrations of metal ions in the surface waters of these lakes result from complex chemical, biological, and physical processes in the catchment and lake. Above all, the bridging of metals to the surface of the settling particles is a prerequisite for their transport to the sediments, resulting in removal from the water column. The binding is governed by the interaction with the functional groups on the surface of the suspended solids in line with a reaction of the type $S—OH—Me^{2+} \rightleftharpoons S-OMe^+ + H^+$. Thus, binding is favored at high pH (Figure 19.6).

On the other hand, biological factors also play a role for the speciation (chemical forms) of the metal ions. Metal ions may also be bound by organic compounds in solution. This process is in competition with the binding to solid particles and is, in a similar manner, pH-dependent, insofar as the tendency for binding to organic compounds also decreases at lower pH. In addition, organic compounds produced by the biota in these lakes may be significant. The biological productivity thus affects the concentration of organic matter in the lakes and in turn the extent of binding of metal ions by organic compounds. In addition, algae produced in these lakes will also bind some metal ions and ultimately contribute to their removal from the water column by sedimentation.

The removal of metal ions to the sediments thus depends on the pH, on the sedimentation rate of solid particles, and on the presence of ligands in the water. These interactions can be included in a simple steady-state model for the concentration of metal ions in a lake. Input by atmospheric depositions and removal by sedimentation and by outflow are considered to be in a steady state. The fraction removed by sedimentation, for given inflow and outflow conditions and sedimentation rate, depends then on the fraction of the metal bound to particles. In a simple quantitative model the complex chemical interactions, which determine the fraction bound to sediments, are summarized by a distribution coefficient, K_d (ratio of concentration in particles to concentration in water). This distribution coefficient depends on various factors (pH, ligands, particle composition). It can then be shown that the fraction of metal ions removed by sedimentation depends on K_d. The steady-state concentration in water is thus higher for low distribution coefficients such as those found in acidic alpine lakes (see Figure 19.6).

The data shown in Table 19.3 illustrate that the concentrations of Zn, Cd, Pb, and, to a lesser extent, Cu are higher in the more acidic Lakes Cristallina and Zotta than in Lake Val Sabbia. This pH-dependence of the scavenged-type metals Zn, Cd, and Pb is in accord with the findings of other authors (Borg et al., 1989; Linhurst et al., 1988; Tessier et al., 1989). In the more eutrophic Lakes Constance and Zurich, which are also affected by larger anthropogenic inputs, the concentrations of Cd, Pb and Zn are lower than those in the remote alpine lakes, because of more efficient removal of the metals to sediments. Atmospheric transport is less important for Cu; its concentration is therefore low in the alpine lakes.

With respect to the interactions with biota, the concentrations of "free" metal ions are more relevant than the total dissolved concentrations; "free" means that the cations in solution are surrounded only by water molecules. Numerous ions and molecules act

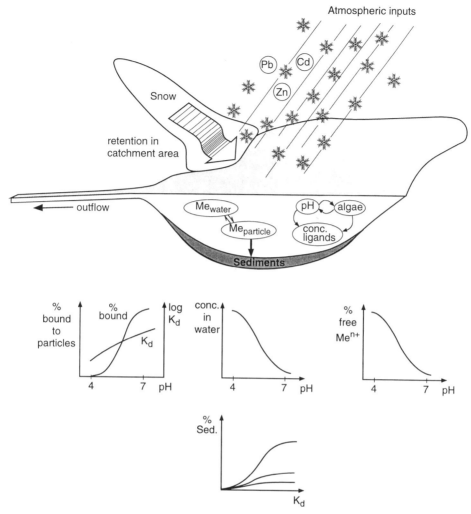

Figure 19.6. Heavy metal ion inputs (lead, cadmium, zinc) from atmospheric depositions are subject to various interactions in the lake. Removal from the water column occurs by sedimentation and depends on the fraction of metal ions bound to solid particles. Binding to particle surfaces is strongly pH-dependent. The ratio of metal ions in particles to metal ions in solution can be described by a distribution coefficient (K_d). The resulting concentrations in water are higher at lower pH. Different fractions of metal ions removed to sediments (% Sed.) result as a function of K_d for different overall sedimentation rates.

as ligands, this means that they enter the coordination sphere of the metal ion, replace some of the water molecules, and form chemical bonds. The most important of such ions and molecules in natural waters are carbonate, hydroxide, and small organic acids. Binding reactions of metal ions by ligands are also pH-dependent. In the alpine lakes, low pH, low concentrations of ions, and low concentrations of organic compounds favor free metal ions over complexes. Figure 19.7 illustrates the data on Zn(II) in the alpine lakes and shows how the concentration of total Zn(II) decreases with increasing pH. Due to the low productivity of these waters, biologically produced organic compounds are only scarcely present. This means that the free metal ions are more

TABLE 19.3
Concentrations of Trace Metals in Alpine Lakes and in More Eutrophic Lakes
(Total Concentrations of Cu, Zn, Cd, Pb, and Dissolved Concentrations of Al)

Lake	Date	pH	Cu (nM)	Zn (nM)	Cd (nM)	Pb (nM)	Al (Dissolved) (μM)
Cristallina[a]	Aug. 92	5.9	5	42	0.4	3.4	0.65
Zotta[a]	Aug. 92	5.7	5	50	0.5	3.0	1.04
Val Sabbia[a]	Aug. 92	6.9	4	11	0.2	1.5	0.24
Constance	1981/82	7.8–8.5	5–15	10–30	0.05–0.1	0.2–0.5	—
Zurich	1983/84	7.5–8.5	6–12	5–45	0.04–0.1	0.05–1	—

[a]Small alpine lakes in the southern part of the Swiss Alps.

available to the biota, especially the phytoplankton, and may thus be more toxic (Xue and Sigg, 1993; Sigg et al., 1994).

A special case is aluminum, which occurs naturally in large amounts in the rocks. Its concentration in natural waters is limited by its solubility as an oxide or hydroxide, which is also strongly pH-dependent. Its concentration in water strongly increases at pH values lower than 6 (Table 19.3). Large differences in the dissolved aluminum concentrations are thus encountered in lakes in the range of pH 5–7. Dissolved aluminum is toxic, especially to fish; it has been demonstrated in different studies that the dissolved aluminum concentrations in acidic waters are detrimental to sensitive fish species and affect their ability to reproduce (Schofield, 1976; Schindler, 1987).

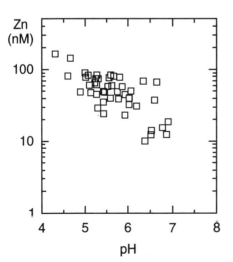

Figure 19.7. Total concentrations of Zn(II) in slightly acidic alpine lakes, collected 1983–1992.

References

Barry, R. C., Schnoor, J. L., Sulzberger, B., Sigg, L., and Stumm, W., 1994, Iron oxidation kinetics in acidic alpine lakes. *Water Res.*, 28, 323–333.

Berner, E. K., and Berner, R. A., 1987, *Global Water Cycle Geochemistry and Environment*, Prentice-Hall, New York.

Berner, R. A., and Schott, J., 1982, Mechanism of pyroxene and amphibole weathering. II. Observations of soil grains. *Am. J. Sci.*, 282, 1214–1231.

Berner, R. A., and Lasaga, A. C., 1989, Modeling the geochemical carbon cycle. *Sci. Am.*, 260, 74–81.

Blum, A. E., and Lasaga, A. C., 1988, Role of surface speciation in the low-temperature dissolution of minerals. *Nature*, 331, 431–433.

Borg, H., Andersson, P., and Johansson, K., 1989, Influence of acidification on metal fluxes in Swedish forest lakes. *Sci. Total Environ.*, 87/88, 241–253.

Brady, P. V., and Walther, J. V., 1989, Controls on silicate dissolution rates in neutral and basic pH solutions at 25°C. *Geochim. Cosmochim. Acta*, 53, 2823–2830.

Caroll-Webb, S. A., and Walther, J. V., 1988, A Surface Complex Reaction Model for the pH-dependence of corundum and kaolinite. *Geochim. Cosmochim. Acta*, 52, 2609–2623.

Casey, W. H., and Sposito, G., 1992, On the temperature dependence of mineral dissolution rates. *Geochim. Cosmochim. Acta*, 56, 3825–2830.

Chou, L., and Wollast, R., 1989, Is the exchange-reaction of alkali feldspars reversible? *Geochim. Cosmochim. Acta*, 53(2), 557–558.

Cosby, B. J., Hornberger, G. M., Galloway, J. N., and Wright, R. F., 1985, Modeling the effects of acid deposition: assessment of a lumped parameter model of soil water and streamwater chemistry. *Water Resour. Res.*, 21, 51–63.

Drever, J. L., and Zobrist, J., 1992, Chemical weathering of silicate rocks as a function of elevation in the southern Swiss Alps. *Geochim. Cosmochim. Acta*, 56, 3209–3216.

Furrer, F., and Stumm, W., 1986, The coordination chemistry of weathering, I. Dissolution kinetics of δ-Al_2O_3 and BeO. *Geochim. Cosmochim. Acta*, 50, 1847–1860.

Furrer, R., Westall, J., and Sollins, P., 1989, The study of soil chemistry through quasi-steady-state models: I. Mathematical definition of model. *Geochim. Cosmochim. Acta*, 53, 595–601.

Furrer, R., Sollins, P., and Westall, J., 1990, The study of soil chemistry through quasi-steady-state models: II. Acidity of soil solution. *Geochim. Cosmochim. Acta*, 54, 2363–2374.

Gherini, S. A., Mok, L., Hudson, J. M., Davis, G. F., Chen, C. W., and Goldstein, R. A., 1985, The ILWAS model: formulation and application. *Water Air Soil Pollut.*, 26(4), 425–460.

Giovanoli, R., Schnoor, J. L., Sigg, L., Stumm, W., and Zobrist, J., 1988, Chemical weathering of crystalline rocks in the catchment area of Acidic Ticino Lakes, Switzerland. *Clays Clay Min.*, 36, 521–529.

Günthert, A., Stern, W. B., and Schwander, H., 1976, Isochemische Granitgneiss Bildung im Maggia-Lappen. *Schweiz. Min. Petrogr. Mitt.*, 56, 105–143.

Hellmann, R., Eggleston, C. M., Hochella, M. F., and Crerar, D. A., 1990, The formation of leaded layers on albite surfaces during dissolution under hydrothermal conditions. *Geochim. Cosmochim. Acta*, 54(5), 1267–1281.

Holdren, G. R., Jr., and Berner, R. A., 1979, Mechanism of feldspar weathering. *Geochim. Cosmochim. Acta*, 43, 1161–1187.

Likens, G. E., 1985, *An Ecosystem Approach to Aquatic Ecology*, Springer, New York.

Linhurst, R. A., Landers, D. H., Eilers, J. M., Brakke, D. F., Overton, W. S., Meier, E. P., and Crowe, R. E., 1988, *Characteristics of lakes in the eastern United States*, EPA 600/4-86/007a, U.S. Environmental Protection Agency, Washington, D.C.

Nikolaidis, N. P., Rajaram, H., Schnoor, J. L., and Georgakakos, K. P., 1988, A generalized soft water acidification model. *Water Resour. Res.*, 24, 1983–1996.

Schindler, D. W., 1987, Detecting ecosystem responses to anthropogenic stress. *Can. J. Fish. Aquat. Sci.*, 44, 6–29.

Schnoor, J. L., 1990, Kinetics of chemical weathering: a comparison of laboratory and field weathering rates. In: W. Stumm, ed., *Aquatic Chemical Kinetics*, Wiley-Interscience, New York, pp. 475–504.

Schnoor, J. L., and Stumm, W., 1986, The role of chemical weathering in the neutralization of acidic deposition. *Schweiz. Z. Hydrologie*, 48(2), 171–195.

Schofield, C. L., 1976, Acid precipitation: effect on fish. *Ambio*, 5, 228–230.

Sigg, L., Kuhn, A., Xue, H., Kiefer, E., and Kistler, D., 1994, Cycles of trace elements (copper and zinc) in a eutrophic lake. *ACS Adv. Chem. Series*, 244, 177–194.

Stumm, W., 1992, *Chemistry of the Solid-Water Interface*, Wiley-Interscience, New York.

Stumm, W., and Morgan, J. J., 1996, *Aquatic Chemistry; Equilibria and Rates in Natural Waters*, Wiley-Interscience, New York.

Stumm, W., and Wollast, R., 1990, Coordination chemistry of weathering: kinetics of the surface-controlled dissolution of oxide minerals. *Rev. Geophys.*, 28(1), 53–69.

Sulzberger, B., Schnoor, J. L., Giovanoli, R., Hering, J. G., and Zobrist, J., 1990, Biogeochemistry of iron in an acidic lake. *Aquatic Sci.*, 52, 56–74.

Sverdrup, H., Warfinge, P., Frogner, T., Haoya, A. O., Johansson, M., and Andersen, B., 1992, Critical loads for forest soils in the Nordic countries. *Ambio*, 21, 348–355.

Tessier, A., Carignan, R., Dubreuil, B., and Rapin, F., 1989, Partitioning of zinc between the water column and the oxic sediments in lakes. *Geochim. Cosmochim. Acta*, 53, 1511–1522.

Warfvinge, P., Holmberg, M., Posch, M., and Wright, R. F., 1992, The use of dynamic models to set target loads. *Ambio*, 21, 369–376.

Wieland, E., and Stumm, W., 1992, Dissolution kinetics of kaolinite in acid aqueous solutions at 25°C. *Geochim. Cosmochim. Acta*, 56, 3339–3355.

Wieland, E., Wehrli, B., and Stumm, W., 1988, The coordination chemistry of weathering: III. A generalization on the dissolution rates of minerals. *Geochim. Cosmochim. Acta*, 52, 1969–1981.

Xiao, Y., and Lasaga, A. L., 1994, *Ab initio* quantum mechanical studies of the kinetics and mechanisms of silicate dissolution: H^+ catalysis. *Geochim. Cosmochim. Acta*, 58, 5379–5400.

Xue, H. B., and Sigg, L., 1993, Free cupric ion concentration and Cu(II) speciation in a eutrophic lake. *Limnol. Oceanogr.*, 38, 1052–1059.

Zobrist, J., Sigg, L., Schnoor, J. L., and Stumm, W., 1987, Buffering mechanism in acidified alpine lakes. In: H. Barth, ed., *Reversibility of Acidification*, Elsevier, Barking, U.K., pp. 95–103.

Passive Bioremediation of Metals and Inorganic Contaminants

THOMAS WILDEMAN AND DAVID UPDEGRAFF

ABSTRACT. There has recently been considerable interest in the use of constructed wetlands and/or passive bioreactors to remove metal contaminants from water, especially acid-mine drainage. This chapter considers some chemical, geochemical, and engineering design features which govern the effectiveness of constructed wetlands. The discussion focuses on a particular attempt to design a passive bioreactor to treat acid-mine drainage, while introducing general principals and chemical interactions which govern the operation of such systems. Aerobic bioreactor systems operate best when the pH of the influent stream is greater than 5.5 and are most effective at the removal of Fe and Mn as $Fe(OH)_3$ and MnO_2 precipitates. In anaerobic systems, consortia of sulfate-reducing microorganisms are used to increase the pH of the influent stream and precipitate metals as sulfides.

Constructed wetlands/bioreactors are capable of treating waters with a pH as low as 3 and heavy-metal concentrations in excess of 1000 mg/L. Microbially mediated precipitations dominate the metal-removal processes. This fact facilitates process-level research through a staged design process that is quite similar to the design of other bioremediation systems. The result is that constructed wetlands are viewed as passive bioreactors as opposed to natural ecosystems, which greatly facilitates research and innovation in the bioremediation of metal/inorganic-contaminated waters.

INTRODUCTION

From our childhood, many of us can remember stepping in the wrong place in a swamp and finding ourselves knee deep in muck. We were immediately repulsed by the foul nature of that environment. Ecologists have long understood that soils in wetlands are often foul because they naturally accumulate and treat contaminants. The methods by which a wetland treats contaminants include:

1. Filtering of suspended and colloidal material from the water
2. Uptake of contaminants into the roots and leaves of live plants
3. Adsorption or exchange of contaminants onto inorganic soil constituents, organic solids, dead plant material, or algal material

4. Neutralization and precipitation of contaminants through the generation of HCO_3^-, H_2S, and NH_3 by bacterial decay of organic matter

5. Destruction or precipitation of contaminants catalyzed by the activity of aerobic or anaerobic bacteria

In the last decade, engineers began to use wetlands for the removal of contaminants from water (Hammer, 1989; Reed et al., 1988). In some instances, natural wetlands were used. However, a natural system will accommodate all the above removal processes and probably will not operate to maximize a desired process. If a wetland is constructed (e.g., Figure 20.1), it can be designed to maximize a specific process suitable for the removal of certain contaminants from water. Engineering as well as ecological reasons lead to the choice of constructing a wetland for contaminant removal rather than using an existing natural ecosystem.

Because of the remote location of many mines and constant flow of effluent, constructed wetlands have particular appeal in treating acid-mine drainage (AMD) (Klusman and Machemer, 1991). Early wetland designs treating coal mine drainage generally included peat or compost substrate, cattails, and limestone gravel within a surface flow system (Brodie, 1991; Brodie et al., 1989a,b, 1991; Eger and Lapakko, 1989; Hiel and Kerins, 1988; Stark et al., 1988). Recently, microbial SO_4^{2-} reduction in wetlands has been used to treat acid-mine drainage from coal mines in the eastern United States (Hammack and Hedin, 1989; Hammack and Edenborn, 1991; Hedin et al., 1988, 1989). In the western United States, metal-mine drainages commonly have low pH and high heavy metal concentrations (Wildeman et al., 1974). One of the first wetlands to be built in Colorado to treat water from such a metal mine was the Big Five pilot wetland in Idaho Springs. The wetland was designed to passively remove heavy metals from metal mine waters and raise the pH (Howard et al., 1989; Wildeman et al., 1993a). The mine drainage flowing from the Big Five Tunnel was chosen for study due to its low pH (3.0), high metal concentrations, moderate flow rate, and accessibility.

It was during the Big Five Project that it was realized that in the anaerobic zone, metals treatment is dominated by bacterial activity. Particularly important for metals precipitation is the sulfate-reducing bacterial (SRB) consortium (Machemer et al., 1993;

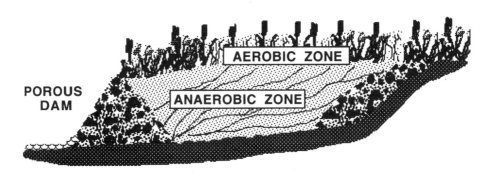

TYPICAL CONSTRUCTED WETLAND

Figure 20.1. Cross-sectional view of a wetland where inflow water can flow down through the substrate instead of across the surface.

Reynolds et al., 1991). In this type of treatment, macroscopic plants become less important (Wildeman et al., 1993b), and the constructed system resembles a flow-through bioreactor (Wildeman et al., 1993a). This bioreactor concept has also been extended to aerobic treatment through the use of algae. In this case, photosynthesis by algae adds oxygen and alkalinity to the water, providing two important reactants for oxidation and hydrolysis reactions (Duggan et al., 1992). It is the purpose of this review to develop and present examples of the biogeochemical principles and engineering methods underlying aerobic and anaerobic passive treatment.

MICROBIAL GUIDELINES FOR METALS REMEDIATION

With an expenditure of energy, metals have been extracted from ores found at the surface or near surface of the earth. Unlike organic compounds, metals generally cannot be degraded to gaseous products or otherwise destroyed. However, just as organic compounds can be mineralized to CO_2 and H_2O, metal contaminants can be mineralized back to the compounds found in ore deposits. Because these inorganic minerals generally represent chemical forms that persisted over geologic time, it is probable that they represent a very stable chemical form for the metal in geologic environments (Wildeman and Laudon, 1989). Consequently, the objective of metals treatment is often to return mobile contaminants to their stable, immobile mineral forms. Many of these minerals, such as FeS_2 and MnO_2, are formed from water solution in a sedimentary environment, and the reactions are catalyzed by bacteria. Thus, the bioremediation of metal contaminants involves optimizing what has been naturally occurring throughout geologic time.

When treating to remove metal contaminants in water, the objectives usually require adjusting the pH to about 7 and removal of the metals as sulfide, hydroxide, or carbonate precipitates. These objectives generate one of the primary guidelines for metals bioremediation. Because S^{2-}, CO_3^{2-}, and OH^- are common products of bacterial activity, enzymatic uptake of the metal into bacteria is not necessary for remediation. In nature, S^{2-}, CO_3^{2-}, and OH^- are usually generated by a consortium of bacterial species that are working to break down complex organic material. This situation generates a second important guideline for metals bioremediation. This is that inocula of natural bacterial consortia rather than a special, single species are best for effective metals remediation. For the bioremediation of metals, these two guidelines can be combined into the one design criterion of choosing bacterial consortia that are found in typical aquatic environments. This area of study is often called microbial ecology.

Treatment by Traditional Microbial Methods

For this example, consider the reduction of chromate to Cr(III) and its subsequent precipitation as an hydroxide. A traditional treatment study would be initiated by sampling chromium-contaminated site materials. These materials probably contain microbial consortia that include chromate-reducing bacteria. Various colony types from these materials would be transferred to fresh media containing chromate in order to isolate organisms that reduce chromate. Standard bacteriological procedures involving cultural, morphological, and biochemical tests would be employed to identify the isolates. Such a study has already been accomplished by Komori's group (Komori et al., 1990; Wang et al., 1989). Next, the chromate-reducing bacterial species would be utilized to explore the effects of experimental variables such as:

1. Added inorganic nutrients
2. Competing electron acceptors (including sulfate, nitrate, ferric iron, and Mn (II))
3. Use of different electron donors in the form of various organic compounds
4. Amendment with cellulose and inexpensive plant materials such as hay and alfalfa
5. Supplementation with accessory growth factors including amino acids and vitamins

In addition, using the isolated bacteria, laboratory studies would be designed and accomplished to determine the effect of pH, ionic strength, temperature, and chromate concentration on the rate of reduction of chromate in both contaminated site materials and synthetic culture media. Some cultures would be grown in sealed serum bottles plugged with butyl rubber stoppers, so that the effect of different gaseous atmospheres, including air, nitrogen, hydrogen, and carbon dioxide, could be evaluated. Consideration would be given to the development of biomarkers, such as gene probes, for the principal organisms so that the specific isolates could be readily identified. Analyses would be carried out on contaminated site materials to detect other organic and inorganic co-contaminants, as well as their fate during the chromate reduction process.

Finally after all of the these studies were completed, including adaptation mechanisms (if any) of the isolated organisms, the design of a practical treatment system would be considered. Then, the design would be constructed and operated on a bench- and then pilot-scale level.

Treatment Approach Using Microbial Ecology

The traditional approach emphasizes isolating specific bacterial species and studying how to optimize its removal capabilities. In the microbial ecology approach, emphasis is less on specific species and more on the consortium of bacteria in an aquatic environment that might render inorganic contaminants immobile. As stated above, for most metals, the stable mineral (and presumably thermodynamic) forms are sulfides, hydroxides, or carbonates, which can, under the right conditions, be made to form as precipitates in the bioreactor. The reactants needed to form these precipitates often come from the products of microbial activity, and enzymatic uptake of the metal may not be necessary.

The example of chromate removal can also be discussed from this microbial ecology perspective. In addition to being effective in a natural environment, any chromate-reducing bacterial species isolated by traditional methods would have to be able to survive within a consortium of bacteria indigenous to a specific environment. Also, that specific species will probably have to depend on the consortium to supply the simple organic compounds that it best uses for electron donors (Wildeman and Updegraff, 1994). In this scenario, it is often difficult to determine whether chromate is reduced by direct microbial activity or whether other readily available reduced constituents such as Fe(II), Mn(II), S(−II), or reactive organic compounds would abiotically reduce Cr(VI). If an aquatic environment is designed to ensure that it remains suboxic and neutral, then Cr may be reduced and removed even if the specific chromate-reducing bacteria are not present.

For this method, soils and scums from different aquatic environments (including a chromate-contaminated site) would be collected. The samples would be exposed to chromate-contaminated waters, and removal progress would be monitored over the course of a month. The tests would be run under aerobic and anaerobic conditions and in sunlight and darkness to see how the microbial systems respond to general external conditions. Systems and conditions that show removal promise would be examined further to determine the nature of removal and possible nutrients, electron donors, and electron acceptors that enhance removal. With this information, a practical treatment system is designed and tested on a bench- and pilot-scale level.

In the end, the treatment system design could be the same regardless of the approach. Using the microbial ecology approach may lead to a more rapid decision on the proper design and a more robust treatment system. This is because removal might not depend upon the activity of one bacterial species.

This microbial ecology approach can, of course, be used to design treatment systems for other contaminants. Several of these will be described in later sections of this review. First, however, a review of important microbially mediated reactions will be presented as an aid to understanding these examples (Wildeman and Updegraff, 1994).

APPLICATION OF MICROBIAL ECOLOGY TO WETLANDS AND BIOREACTORS

Microbial processes in aerobic environments are very different from those in anaerobic environments. Aerobic conditions are effective in removing metals whose oxides, hydroxides, or carbonates are relatively insoluble. These include Fe(III) and Mn(IV). Anaerobic processes, including sulfate reduction, are effective in removing metals that form insoluble sulfides. These include Cu, Zn, Cd, Pb, Ag, and Fe(II). Both aerobic and anaerobic processes can neutralize acids (increasing the pH) and add alkalinity to water in the form of HCO_3^-. Consequently, in either environment it is possible to remove Al and Cr(III) as hydroxides, or Zn and Cu as carbonates.

Aerobic Processes

The most important aerobic biological processes in wetlands are iron oxidation and photosynthesis. Both are autotrophic processes in which carbon dioxide is the source of carbon for the organisms concerned. Photosynthesis, carried out by blue-green bacteria, algae, and plants, consumes carbonic acid and bicarbonate and produces hydroxyl ions:

$$6HCO_3^- \ (aq) \ + \ 6H_2O \ \rightarrow \ C_6H_{12}O_6 \ + \ 6O_2 \ + \ 6OH^-$$

In this case, plants and aquatic organisms are making organic matter by taking up dissolved bicarbonate to produce dissolved oxygen and hydroxide ions (Wetzel, 1983).

The oxidation of iron pyrite by aerobic autotrophic bacteria of the genus *Thiobacillus* is the cause of acid-mine drainage, as summarized by the following reactions from Stumm and Morgan (1981):

$$FeS_2(s) + 7/2O_2 + H_2O \rightarrow Fe^{2+} + 2SO_4^{2-} + 2H^+$$

$$Fe^{2+} + 1/4O_2 + H^+ \rightarrow Fe^{3+} + 1/2H_2O$$

$$Fe^{3+} + 3H_2O \rightarrow Fe(OH)_3 + 3 H^+$$

$$FeS_2 + 14Fe^{3+} + 8H_2O \rightarrow 15Fe^{2+} + 2SO_4^{2-} + 16H^+$$

Note that H^+ is produced by the oxidation of bisulfide and by the precipitation of $Fe(OH)_3$. Manganese oxidation and precipitation also releases H^+:

$$2H_2O + Mn^{2+} \rightarrow MnO_2 + 3H^+ + 2e^-$$

Finally, oxidation of organic matter, a half-reaction used by the majority of heterotrophic bacteria, produces H^+:

$$H_2O + \text{``}CH_2O\text{''} \rightarrow CO_2 + 4H^+ + 4e^-$$

Here, "CH_2O" represents organic matter such as cellulose and other carbohydrates. The reduction potential for the above reaction is lower than that for the formation of H_2 (Stumm and Morgan, 1981). Consequently, organic matter in aquatic environments acts as a strong reducing agent.

Anaerobic Processes

Under anaerobic conditions in wetlands, five general types of microbial processes are of importance:

1. Hydrolysis of biopolymers by extracellular bacterial enzymes. An example is the hydrolysis of cellulose, the most abundant organic material in plants, to glucose:

$$(C_6H_{11}O_5)_n + nH_2O \rightarrow nC_6H_{12}O_6$$

2. Fermentation. Examples are the formation of ethanol and pyruvic acid:

$$C_6H_{12}O_6 \rightarrow 2C_2H_5OH + 2CO_2$$

$$C_6H_{12}O_6 \rightarrow 2C_3H_4O_3 + 4H_2$$

Note that hydrogen is produced in the second reaction. It is a common product of fermentation.

3. Methanogenesis:

$$CO_2 + 4H_2 \rightarrow CH_4 + 2H_2O$$

4. Sulfate reduction:

$$2H^+ + SO_4^{2-} + 2\text{``}CH_2O\text{''} \rightarrow H_2S + 2H_2CO_3$$

Note that organic matter is used as the reducing agent.

5. Iron reduction:

$$Fe^{3+} + e^- \rightarrow Fe^{2+}$$

Again, organic matter could be used as the reducing agent.

Fermentation often produces organic acids, decreasing pH, while sulfate reduction consumes H^+ and increases pH. Methanogenesis also consumes hydrogen ions. Proton-reducing bacteria, which are symbionts with methanogenic bacteria, convert H^+ to H_2, and this is used by the methanogenic bacteria to reduce CO_2 to CH_4. In our anaerobic wetland environments, sulfate reduction and methanogenesis proceed together. Because Postgate (1979) reports that the activity of sulfate-reducing bacteria (SRB) is severely limited below pH 5, organic materials in a constructed wetland environment must be chosen so that fermentation does not dominate over sulfate reduction.

Overall Guidelines

Although still not completely understood, the principles outlined above appear to be the predominant removal mechanisms in the treatment of mine drainage and other metal-contaminated waters by constructed wetlands (Hammer, 1989; Wildeman et al., 1993a). In the early 1980s, aerobic removal processes were emphasized and the precipitation of $Fe(OH)_3$ was an important objective (Hammer, 1989; Brodie, 1991). Because precipitation of $Fe(OH)_3$ produces H^+ ions, iron was removed but the pH of the effluent often was around 3. Brodie (1991) has had success with metals treatment using aerobic constructed wetlands as long as the pH of the influent was above 5.5 and carried some alkalinity in the form of dissolved bicarbonate. Around 1987, groups from the U.S. Bureau of Mines (Hedin et al., 1989) and the Colorado School of Mines (Wildeman and Laudon, 1989; Wildeman et al., 1993a) began to investigate the role of anaerobic processes, particularly sulfate reduction, in treating acid-mine drainage.

The microbial guidelines presented above have resulted from these early studies. Application of these guidelines to aerobic wetlands leads to the following four guiding principles for success:

1. Aerobic removal processes are successful when the pH of the effluent water is above 5.5 and dissolved bicarbonate is present.

2. Any practice, such as use of anoxic limestone drains (Brodie et al., 1991), that will raise the pH and add alkalinity should be used.

3. Precipitation of iron and manganese oxyhydroxides is a primary removal process, and other metal contaminants are removed by adsorption onto these precipitates or by precipitating as carbonates.

4. Plants are essential to success because photosynthesis is a primary process for raising pH, adding oxygen to the water, and supplying organic nutrients.

The role of photosynthesis was discussed in the section entitled "Aerobic Processes." At about pH 5.5, significant amounts of bicarbonate can be present in aqueous solution (Stumm and Morgan, 1981), and photosynthesis can occur through uptake of bicarbonate from the water in preference to the uptake of CO_2 from the air. This appears to be why aerobic constructed wetlands require plants and are effective at adding alkalinity when the pH is above 5.5.

Application of the guidelines to anaerobic removal processes is much more direct, because plants can be absent and the system is dominated by microbial processes. The following four practices lead to success:

1. Wetland substrates are formulated so that organic material necessary for metabolism is in high abundance and the wetland soil can provide acid buffering capacity at a pH above 7.

2. Microbial processes that transform strong acids such as H_2SO_4 into weak acids such as H_2S should be promoted.

3. The products of these reactions are used to precipitate metal contaminants as sulfides (CuS, ZnS, PbS, CdS), hydroxides ($Al(OH)_3$, $Cr(OH)_3$), and carbonates ($MnCO_3$).

4. To remain effective, the reactions that consume H^+ must predominate over the reactions that produce H^+.

In a natural wetland, most water courses across the surface and the anaerobic subsurface remains somewhat isolated. For best application of these guidelines, design of the wetland has to ensure that water goes through the substrate. Hence, the system can look like a plug-flow bioreactor. These passive anaerobic systems have been successful at raising the pH of metal mine drainages from below 3 to above 6 and significantly reducing metal concentrations (Wildeman et al., 1993a). The design of metal-remediation systems using these principles will be the focus of the rest of this chapter.

STAGED DESIGN OF METAL-TREATMENT SYSTEMS

In our studies on treatment of metal-mine drainage by constructed wetlands (Wildeman et al., 1993a), when it was determined that precipitation of metals by sulfide generated from SRB is an important process, it was realized that establishing and maintaining the proper microbial environment in the substrate is the key to success for removal (Wildeman and Updegraff, 1994). If this is the case, then construction of large pilot cells is not necessary to optimize the anaerobic bacterial processes needed for removal (Reynolds et al., 1991).

Consequently, study of wetland processes and design of optimum systems can proceed from laboratory experiments, to bench-scale studies, and then to the design and construction of actual cells. We call this "staged design of wetland systems." Although staged design is best carried out on anaerobic substrates, it has also been used with success on design of aerobic systems. Algal photosynthesizers are excellent generators of oxygen and alkalinity in water, and they can be readily used in laboratory and bench-scale studies of aerobic treatment (Duggan et al., 1992).

Example Laboratory Studies

In early laboratory studies, culture bottle experiments were used for studies on how to establish tests to determine the production of sulfide by bacteria and also to determine what substrate will provide the best initial conditions for growth of SRB

(Reynolds et al., 1991). Recently, laboratory studies have concentrated on the practical aspects of wetland design. In particular, great emphasis is placed on testing local organic and soil materials to determine what mix provides the best environment for sustained SRB activity. Also, a number of studies have been conducted to "prove-in-principle" that treatment of metals and inorganics is possible and to provide an indication of what reduction in concentration is possible. Examples of these studies are described below. Other examples are described in the literature (Wildeman et al., 1994a,b).

Measurement of sulfate reduction activity

In an extensive laboratory study, a series of culture bottles was sealed and incubated at 18°C to determine the activity of SRB and whether metal removal in the laboratory was comparable to that in a demonstration anaerobic reactor (Reynolds et al., 1991). For the laboratory study, 20 g of substrate and 70 mL of mine drainage, whose chemistry is shown in Table 20.1, were sealed into 120-mL serum bottles. The substrate came from an active anaerobic cell from the Big Five Pilot Wetlands in Idaho Springs, Colorado (Wildeman et al., 1993a). On the day of collection, the mine drainage and the effluent from Cell B-Upflow had the chemistry shown in Table 20.1. The results for the metals removal on the first day and after 35 days are also shown in Table 20.1.

In the baseline bottles, measured 1 day after adding the mine drainage, the pH was significantly higher and metal concentrations were significantly lower than in the mine drainage. As stated in the wetlands guidelines, substrates are formulated to immediately raise the mine drainage pH to neutral conditions. Several processes could have contributed to the immediate removal of metals, including precipitation due to the increase in pH and adsorption onto organic materials (Machemer and Wildeman, 1992). Over the course of 35 days, the pH continued to rise until it was the same as in the effluent from the anaerobic reactor. The concentration of sulfate gradually decreased until, by day 35, sulfate concentration was lower than that in the reactor effluent. At 25 days, the concentration of sulfate matched that in the reactor effluent. Based on this observation, laboratory-scale tests are conducted for at least 4 weeks to simulate conditions in the field. For Fe, the majority was removed by the first day. We believe this is caused by organic adsorption (Machemer and Wildeman, 1992). Then, the Fe concentration continues to decrease until, by day 35, it is at the concentration of the cell

TABLE 20.1
Comparison of pH, Sulfate, and Metal Concentrations in Mine Drainage, Wetland Output, and Serum Bottles[a]

Sample	pH	SO_4^{2-} (mg/L)	Cu (mg/L)	Fe (mg/L)	Mn (mg/L)	Zn (mg/L)
Mine drainage	3.0	1720	0.57	39	31	8.6
Reactor effluent	6.7	1460	<0.05	0.64	15.8	0.07
Serum bottles						
(one-day baseline)[b]	6.1	1680	<0.05	10.5	15.5	0.04
Serum bottles						
(35 days)[c]	6.7	1240	<0.05	4.51	10.6	0.16

[a]Adapted from Reynolds et al. (1991).
[b]Values for these samples are the average of four replicates.
[c]Values for these samples are the average of three replicates.

effluent. The differences in how SO_4^{2-} and Fe change indicate that more than one process is operating in the treatment of the mine drainage.

Because the serum bottles were sealed, all volatile products were retained. In particular, all retained sulfide species could be titrated and the amount of production gives an estimate of the activity of sulfate reduction. Figure 20.2 shows how that rate of sulfate reduction changes with time. For the first 40 days, this rate was 1.2 μmol sulfide/g substrate/day (Reynolds et al., 1991), and then after 40 days the rate dropped to 0.75 μmol/g/day. In design calculations for bench- and pilot-scale reactors, a sulfide production rate of 300 nmol/g/day is used.

Poisoned control bottles were prepared by adding sodium azide until the final concentration of azide was 0.5%. Acid volatile sulfides in these bottles were determined at three times during the course of the experiment, and the results are shown in Figure 20.2. The rate of sulfate reduction in these bottles is negligible compared to those bottles that were not poisoned. This adds to the evidence that bacterial sulfate reduction is indeed responsible for the decrease in sulfate and sulfide precipitation for removal of metals from the mine drainage (Machemer et al., 1993).

Anaerobic removal of cyanide

A certain company was interested in whether cyanide concentrations typical of milling-waste effluents could be treated in an anaerobic reactor or whether the cyanide would kill the SRB (Filas and Wildeman, 1992). The composition of the spent tailings solution is shown in Table 20.3. It contains appreciable free cyanide and also bound CN^-, most likely in the form of Fe and Cu complexes.

To typical culture bottles were added various amounts of substrate, milling efflu-

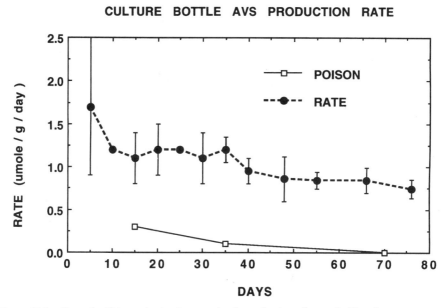

Figure 20.2. Rate of sulfide production in serum bottles and poisoned controls. Error bars represent one standard deviation. Data points without error bars are the average of two samples.

TABLE 20.2
Protocol for the Cyanide Microbiology Experiments Incubated at 27°C

| Culture No. | O_2 | Nutrients | | | Bacterial Inoculation (amt in mL) | Tailings (amt in mL) |
		Compost (gm)	Lactate (g/mL)	Sulfate (g/mL)		
5	No	10	None	None	10	140
7	Yes	10	None	None	10	140
8	Yes	10	None	None	10	140
11	Yes	10	6.8	None	10	135
12	Yes	10	6.8	None	10	135
13	No	10	6.8	2.1	None	140
17	No	10	None	None	10	135
18	No	30	None	None	10	125
19	Yes	20	6.8	None	10	125

ent, and inoculum. The laboratory protocol is summarized in Table 20.2. For those bottles containing 10 g of substrate, it was hypothesized that this was a minimal amount of soil necessary to provide organic nutrients for the SRB. To test this, extra amounts of lactate and sulfate were added to some bottles. Also there was concern that the concentration of cyanide would be toxic to the bacteria, so, for three hours, oxygen was bubbled through a subsample of the milling effluent to oxidize cyanide and lower the pH. A group of culture bottles was also prepared using this oxygenated milling effluent (labeled "yes" in the O_2 column of Table 20.2). The bottles were incubated for 50 days and then the solutions in the bottles were analyzed. The analytical results are shown in Table 20.3.

From the results in Table 20.3, it can be seen that cyanide decreased in all the bottles. A decrease in the concentration of sulfate is a good measure of the activity of SRB. In most bottles there was little or no decrease. However, in bottle 11 there was appreciable decrease in the sulfate concentration caused by the activity of SRB. In this bottle there was also a great decrease in total cyanide. It appears that, given the proper conditions, cyanide degradation as well as sulfate reduction can occur in the anaero-

TABLE 20.3
Results of Analyses of the Solutions from Cultures That Showed Decreased Concentrations of Cyanide[a]

Culture No.	Fe	Cu	Sulfate	Total Cyanide	Free Cyanide	pH
Original Solution	470	78	298	330	24	9.0
5	85	98	310	57	60	8.9
7	150	90	320	56	49	8.8
8	8.2	9	570	56	54	8.1
11	21	10.4	<5	2.3	<5	7.3
12	5.4	6.0	460	2.2	<5	7.3
13	65	142	4470	120	72	8.4
17	120	38	360	56	21	8.6
18	60	87	520	2.0	<5	8.4
19	23	58	790	57	42	7.0

[a]All concentrations are in milligrams per liter.

bic zone of a wetland substrate. The cyanide disappearance could have occurred by a number of processes, including:

(i) Volatilization of cyanide by lowering the solution pH
(ii) Metabolic use of the cyanide by microbes
(iii) Nonmetabolic degradation of cyanide to cyanate

It is probable that all three are occurring since the inoculum is a consortium of bacteria (Reynolds et al., 1991) and the substrate is complex organic material (Wildeman et al., 1993a). In any event, the laboratory experiment proved in principle that cyanide disappearance, sulfate reduction, and heavy-metal removal are all simultaneously possible in the anaerobic zone of a constructed wetland.

Aerobic removal of manganese

Manganese, a common contaminant in mine drainages, is difficult to remove from solution due to the high pH required to form insoluble manganese(II) hydroxides, carbonates, and sulfides (Stumm and Morgan, 1981; Watzlaf and Casson, 1990). In addition, although MnO_2 should form at a pH as low as 4, the kinetics of oxidation of Mn(II) to Mn(IV) are slow at a pH below 8 (Stumm and Morgan, 1981; Wehrli and Stumm, 1989). As can be seen in Table 20.1 for the Big Five system, even though the pH of the effluent has risen to above 6.7, manganese is still only reduced by about 50%. Thus, a polishing stage system utilizing an alternate microbial process seemed necessary to remove high concentrations of manganese.

A laboratory study that utilized the principles of microbial ecology was conducted to determine an optimum system that would remove manganese from waters where the pH is above 6. Having some alkalinity in the water allowed aerobic experiments to be conducted. A mixture of water, soil, and scum was taken from various ecosystems listed in Table 20.4, including a mining environment where MnO_2 was forming. Ap-

TABLE 20.4
Individual Inocula Used for Manganese Removal Studies

Ecosystem Sample and Ecosystem	pH	Mn Concentration (mg/L)	
		Initial	Final
Algae growing in acid-mine drainage	3	220	310
Brown precipitate in effluent pipe at Big Five wetland	3	220	392
Black mud in effluent of a wastewater treatment plant	8	190	147
Black manganese oxlde (?) solid in acid mine drainage stream	6.5	200	109
Pond scum from freshwater stream	7.5	190	<0.3

proximately 75 mL of each sample was mixed with a Big Five mine drainage adjusted to pH = 7 (Table 20.1) in Erlenmeyer flasks exposed to the atmosphere. To the flasks were added 10 mL of a Mn(II) solution sufficiently concentrated to increase the concentration in the flasks to approximately 200 mg Mn/L. The flasks were incubated in sunlight at laboratory temperatures for 40 days.

The removal results are quite dramatic. Only the pond scum system containing green and blue-green algae (predominantly Cladophora) was able to reduce manganese below 100 mg/L, and dissolved Mn was undetectable in this flask. Later studies have shown that the dissolved oxygen and hydroxide ions produced from photosynthesis by the algae are responsible for catalyzing Mn oxidation and precipitation of MnO_2. Adsorption and metabolic uptake of Mn by the algae are minor removal processes (Duggan and Wildeman, 1994). The study confirms the importance of photosynthesis in aerobic wetland systems and establishes the use of algae and soil bacteria for use in laboratory studies of aerobic wetland processes.

Example Bench-Scale Studies

Anaerobic removal of heavy metals

For anaerobic bench-scale studies, plastic garbage cans were used to conduct experiments to provide answers necessary for the design of a subsurface cell (Bolis et al., 1991). Typical design parameters include the optimum loading factor, substrate, cell configuration, and the hydraulic conductivity of the substrate. A recent study used Big Five mine drainage to determine whether an anaerobic system could raise the pH to above 7 where it was found that $MnCO_3$ would precipitate (Bolis et al., 1991, 1992).

Four reactors using two types of substrate were used. Two reactors contained an organic substrate composed of 75% composted cow manure and 25% planter soil by volume. This mix had been found to be effective in a previous bench-scale reactor study (Bolis et al., 1991). To this was added about 10% by weight of inoculum from currently active anaerobic cells at the Big Five pilot wetland. The second set of reactors contained a primarily inorganic substrate composed of approximately 77% by volume limestone rock, 14% alfalfa, and 9% of the same inoculum used in the first reactors. For both sets of reactors, one was soaked with mine drainage for one week prior to operation and the other was left dry. The flow rate was maintained at approximately 10 mL/min. This value was determined by using the value of 0.3 mol of sulfide generated per cubic meter of substrate per day, evaluating the concentration of heavy metals in the Big Five drainage (Table 20.1), and limiting the flow such that the amount of heavy metals flowing through the system did not exceed the amount of sulfide generated by the substrate (Machemer et al., 1993; Reynolds et al., 1991; Wildeman et al., 1993a). The reactors were run from July through November.

Figure 20.3 shows the pH and the concentration ratios of SO_4^{2-}, Fe, and Mn in the reactor effluent to that in the original mine drainage. For the manure reactors, pH behaved as expected, maintaining a value above 7 for most of the experiment but dropping toward the end when the ambient temperature dropped significantly. For the limestone reactors, the pH remained below 6 for over 30 days. During this time, the reactors were emitting nasty smells indicative of low-molecular-weight organic acids. Apparently, the soaked alfalfa was undergoing anaerobic fermentation. Analysis of the sulfate concentration ratios gives a good measure of how strongly sulfate reduction

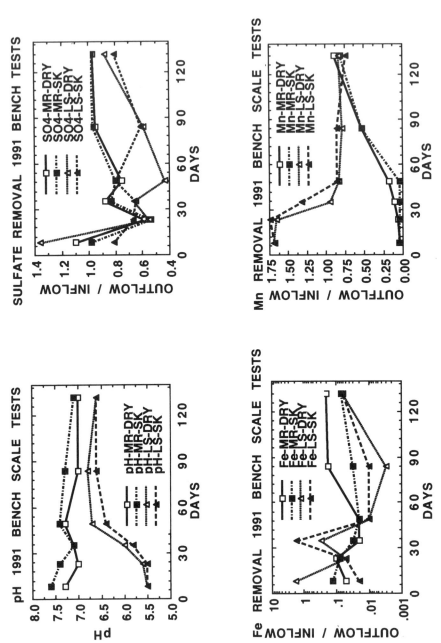

Figure 20.3. pH, sulfate removal, iron removal, and manganese removal versus time for the anaerobic bench-scale study. Removal is the ratio of concentration in reactor outflow concentration in the mine drainage.

was occurring. For all systems in the first month, sulfate was significantly reduced. This is because the low-molecular-weight organic acids that are the primary nutrients for SRB are readily available. For the manure in the later stages of the experiment, sulfate reduction is much less apparent. We assume this is because all the readily available nutrients for the SRBs have been used up and further nutrients are produced by the breakdown of complex organic material. For the alfalfa, sulfate reduction is still active in the later stages. Perhaps this implies that breakdown of organic material is more facile in the alfalfa system.

Copper and zinc were completely removed throughout the entire course of the study. The removal patterns for Fe and Mn are more instructive in terms of the processes occurring in the reactors. For the manure reactors, Fe removal was excellent in the first half of the experiment. Figure 20.3 shows Fe concentration ratios on a log scale so the smaller and larger ratios can be more easily compared. In these systems, iron precipitates as FeS (Machemer et al., 1993), which is somewhat soluble. Consequently, as sulfate reduction decreases, the concentration of Fe in the effluent increases. For the limestone reactors in the early stages of the experiment, Fe is released, probably being dissolved from the limestone. In the later stages of the experiment, the limestone–alfalfa reactors are more efficient at retaining Fe than the manure systems. This is attributed to better sulfate reduction. For Mn, good removal only occurs when the pH is above 7. This is attributed to the formation of $MnCO_3$. Because it appears that continuous maintenance of reactor pH above 7 is difficult, we consider that consistent anaerobic removal of Mn is speculative.

Aerobic removal of manganese

The manganese removal results using algae were encouraging enough that bench-scale studies were conducted (Duggan et al., 1992). In this case, the reservoirs were constructed from small plastic wading pools, approximately 1.1 m in diameter. Each of the two pools initially contained 97 L of effluent from the Big Five Wetland and 5 L of scum comprised primarily of Cladophora from a local pond. The Big Five cell effluent used here contained approximately 32 mg/L of manganese and had a pH of 5.8. The only difference between the two reservoirs was that one reservoir also contained 12 kg of limestone. This reservoir is referred to as "reservoir LS," and the reservoir that did not contain limestone is denoted "reservoir NoLS." The reservoirs were placed outside to have full exposure to the environment.

The four-month duration of the experiment was from August to December, during which the reservoirs were exposed to a wide range of weather conditions. The reservoirs were static for the first two months of the experiment, with water being added occasionally to account for water loss due to evaporation. The weather was typically warm and sunny during this portion of the experiment. A flow system was installed during the last two months of the experiment to determine approximate loading and removal rates. This was accomplished using a peristaltic pump to monitor flow from a feed tank into the reservoirs, along with an outlet tube 7 cm above the bottom of each pool. The weather during this portion of the experiment was typically cold and snowy, and the reservoirs froze several times. A high concentration manganese solution was twice added to the reservoirs during the study period, once during the static portion and once during the flow portion of the experiment. The purpose of these additions was to study the effect of high Mn concentrations on the effectiveness of the algae in Mn removal. This 100-mg/L manganese solution was prepared using manganese sulfate and deionized water.

For reservoir LS, the concentration of manganese and the pH over the static portion of the study is shown in Figure 20.4. As for the anaerobic system, there is a strong correlation between pH and Mn removal. In this case it is for a different reason. Algal photosynthesis raises the pH to values where oxidation of Mn(II) is rapid (Wehrli and Stumm, 1989) and provides the dissolved oxygen for the oxidation. Addition of

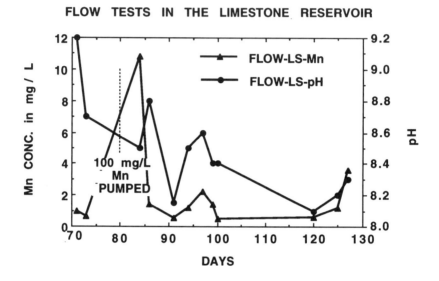

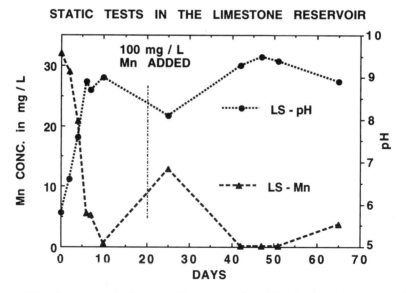

Figure 20.4. Manganese and pH results of the aerobic, static and flow bench-scale study for the limestone reservoir (LS).

the limestone improves removal. This is not due to limestone dissolution because, at the near-neutral pHs in the reservoirs, dissolution is slow. It is more likely that the rock provides surfaces for adsorption of manganese oxides and these sites provide auto-catalysis for further Mn oxidation (Stumm and Morgan, 1981). The 40 L of 100-mg/L manganese solution had a pH of 4.9. Its addition put dual stresses of lower pH as well as high Mn concentrations on the system. Recovery of the system occurred within a few days.

During the static portion of the experiment, the algal biomass had grown exten-sively and appeared healthy, and black precipitates could be seen in the algal mat. Therefore, a flow system was designed to simulate a possible pilot scale wetland and determine the efficiency of the reservoirs at different flow rates. The pump system was installed in mid-October with flow rates varying between 3 and 5 mL/min. During the course of the experiment, the Big Five cell effluent in the feed tank was replaced with the same 100 mg/L manganese solution as was used in the static experiment and was pumped into the reservoirs for 5 days.

Outflow manganese concentrations and pH for the LS reservoir during the flow portion of the study are shown in Figure 20.4. Removal in the NoLS reservoir was less efficient (Duggan et al., 1992). Both systems recovered from the addition of the 100-mg/L Mn solution. This tolerance to high manganese concentrations is important be-cause Mn concentrations as low as 10–20 mM have been shown to greatly inhibit man-ganese oxidizing microbes (Nealson et al., 1988).

The severe weather conditions present during much of the pump flow study had a visible effect on the health of the biomass. During the static experiment, the thick algal mat was bright green and floated on the surface due to the large number of oxy-gen gas bubbles produced during photosynthesis. Throughout the pump flow experi-ment, the algae lost much of its bright green color and most of the algae sank below the water surface. Gas bubbles were still observed during the sunlight hours; this in-dicated that photosynthesis was occurring, but to a much lesser extent than during the warmer, sunnier months. At the completion of the experiment in mid-December, the reservoirs had frozen several times and the algal biomass did not appear very healthy. The fact that the reservoirs performed so well even under these adverse conditions im-plies that other removal processes besides photosynthesis are operating. Analyses of the resulting solids points to adsorption of Mn on the limestone as the next most im-portant process (Duggan and Wildeman, 1994). The results from the bench-scale stud-ies have been used to construct a pilot-scale algal pond that has shown to be very suc-cessful for manganese removal (Wildeman et al., 1993c).

Example Pilot-Scale Studies

Big five pilot wetlands

Pilot systems are built to test out new processes with the anticipation that config-urations may change if the first designs prove unsatisfactory. The Big Five pilot wet-land was an excellent example of this purpose (Wildeman et al., 1993a). The initial cell configuration was primarily a surface flow system. The final configuration that produced the best results was a reactor bed configuration. In between these two con-figurations, a horizontal plug-flow system was tried and found to be inadequate. Two identical reactor beds were built: one to run upflow, and the other downflow (Machemer and Wildeman, 1992; Machemer et al., 1993; Wildeman et al., 1993a). The cells were filled with mushroom compost to approximately 3 m × 3 m × 1 m deep. The system was started in September 1989. For the first six months, the flow through the system was 3–4 L/min. After that, a flow was maintained at 1 L/min for the rest of the two-year operation.

The two-year removal results for the upflow cell are shown in Figure 20.5. The results for the first six months were poor, and this is shown by the erratic pattern in Figure 20.5. After that, complete removal of Cu and almost complete removal of Zn is maintained. Removal of Fe and reduction of Mn and SO_4^{2-} vary with the seasons. At about six months, conditions had been established which eventually led to the understanding necessary to establish the principles given in the section entitled "Overall Guidelines" for the operation of an anaerobic reactor.

A key factor for reactors that depend upon sulfate reduction is the maintenance of an optimum microenvironment for sulfate-reducers. Besides adequate sulfate concentrations, the most important environmental conditions are reducing conditions and a pH of around 7 (Postgate, 1979). Because the wetland cell is receiving mine drainage of pH below 3 and Eh of above 700 mV (Wildeman et al., 1993a), the water can easily overwhelm the microenvironment established by the anaerobic bacteria. This leads to the limiting reactant concept for determining how much water can be treated, and it is our primary guide for loading anaerobic reactors.

Consider the following precipitation reaction, where the sulfide is generated within the reactor substrate and the Fe is delivered to the system by the mine drainage:

$$Fe^{2+} + S^{2-} \rightarrow FeS$$

At high flow rates of mine drainage through the substrate, sulfide will be the limiting reactant, the microbial environment will be under stress to produce more sulfide, the

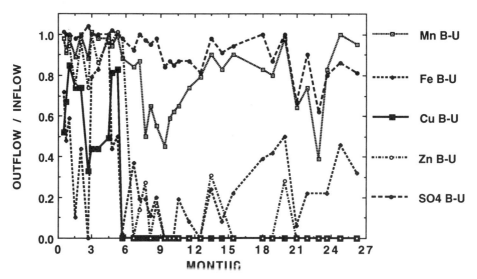

Figure 20.5. Removal (ratio of concentration in reactor outflow over concentration in the mine drainage) over two years for cell B upflow at the Big Five pilot wetland. The system was started in September 1989.

pH of the microbial environment will drop, and removal will be inconsistent. At low flow rates of mine drainage through the substrate, iron will be the limiting reactant, the excess sulfide will ensure a reducing environment and a pH near 7, the microbial population will remain healthy, and removal of the metal contaminants will be consistent and complete. Using this idea, loading factors should be set such that the heavy-metal contaminants are always the limiting reactants. The goal then is to determine how much sulfide a colony of sulfate-reducing bacteria can produce per cubic centimeter of substrate per day. As previously described, the measurement of sulfate reduction activity is critical to this estimation process.

The results shown in Figure 20.2 show that between 750 and 1200 nmol of sulfide were generated per gram of substrate per day. Other studies (Hammack and Hedin, 1989; Hedin et al., 1989; Dvorak et al., 1991) suggest that a reasonable figure for sulfide generation in a wetland is 300 nmol of sulfide per cubic centimeter of substrate per day. This number, the volume of the wetland cell, and the metals concentrations in the mine drainage are used to set the flow of mine drainage through the wetland cell. This is how the flow of 1 L/min was determined for the flow through Cell B-Upflow. In Figure 20.5, the results of applying this limiting reactant guideline are obvious.

PROSPECTS FOR THE FUTURE

Rather than providing the usual summary, it may be more useful to speculate on the future of research designed to improve the performance and prospects for constructed wetland treatment systems. The first issue is how to get off the applied research and development track and into the study of fundamentals. Most of the results described above were from applied projects. Any fundamental questions had to be addressed as side projects. It appears that the treatment systems are working even though there are numerous aquatic chemistry and geomicrobiology questions that have been left unanswered. Because the aquatic chemistry and microbiology have been readily transferable, that has sufficed. However, to develop greater loading capacities and tolerances for contaminants, scientific investigations are imperative. Two questions are most pressing:

1. Within each removal process, there is a rate-limiting step that controls how fast treatment can be accomplished. For the removal processes that have been developed, determining this rate-limiting step more specifically would greatly help in devising design modifications that would speed the treatment processes. Because the kinetics involve abiotic and biotic components, investigations are necessarily interdisciplinary.

2. Most of the precipitates formed are amorphous solids (Machemer et al., 1993; Wildeman et al., 1993a), which makes them almost impossible to completely characterize scientifically. This characteristic also makes the precipitates quite reactive, which is generally detrimental from a technical design point of view. Research into how to form crystalline precipitates and into the nature of the precipitates formed is therefore an important research need. For example, amorphous FeS rather than pyrite is formed in anaerobic reactors (Machemer et al.,

1993). The questions as to why FeS forms and whether pyrite will eventually form are both technically important and scientifically intriguing.

To this point in time, only metals with a comparatively simple aqueous chemistry have been treated. Contaminants with more complex aqueous chemistries may prove to be more challenging. Two important and currently interesting examples are As and Se. The multiple oxidation states exhibited by both of these elements makes the questions associated with the design of a system that provides long-term removal quite complex.

Besides the consideration of metal or metalloid removal, wetlands have also been used to a limited extent to remove organic contaminants from wastewater flows (Hammer, 1989). Since it is almost certain that this type of treatment will be increasingly important in the future, increased efforts to understand organic removal processes in wetlands is important. There are two general types of questions important in consideration of organic processes in wetlands:

1. If there are spills of organics into natural wetlands, what stresses will be put on the ecosystem and what can be done to ensure that the organics will be immobilized and/or destroyed?

2. What are the possibilities for using constructed wetlands to treat various types of organic contaminants and mixed wastes?

Research directed toward answering these questions is virtually nonexistent at the present time, and, therefore these questions represent fertile areas for future research considerations.

The topic of this review has been *passive* treatment of metals using bioreactors. Such systems are designed and constructed with great margins of safety so that maintenance can be kept to a minimum. Nevertheless, the same principles can be applied to an active system that is designed to provide optimum treatment at the price of greater maintenance and control. If active treatment systems for metals removal are investigated, then the questions posed above concerning determination of the rate-limiting step and how it can be accelerated become quite important.

Research in this field has been rewarding because of the close link between scientific principles and their application. In addition, the area is highly interdisciplinary and progress critically depends on cooperative problem solving. It is certain that this field will grow, and it is hoped that 10 years from now the ideas developed in this review will be appear to be rudimentary.

ACKNOWLEDGMENTS

This research has been carried forward by many graduate and undergraduate students. Those most prominent in the research can be found as authors on the articles cited in the text. Colleagues in the Chemistry/Geochemistry and Environmental Sciences and Engineering Departments at the Colorado School of Mines, most notably Ron Klusman, Ron Cohen, and John Emerick, have also contributed substantially to this effort. Outside support for the research has come from the EPA SITE Program, where Ed Bates was the technical supervisor; the USBM AML Program, where Valois Shea-Albin was the technical supervisor; and from the Edna Bailey Sussman Fund.

References

Bolis, J. L., Wildeman, T. R., and Cohen, R. R, 1991, The use of bench scale parameters for preliminary analysis of metal removal from acid mine drainage by wetlands. In: *Proceedings of the 1991 National Meeting of the American Society of Surface Mining and Reclamation*, American Society of Surface Mining & Reclamation, Princeton, WV, pp. 123–135.

Bolis, J. L., Wildeman, T. R., and H. E. Dawson, 1992, Hydraulic conductivity of substrates used for passive acid mine drainage treatment. In: *Mining and Reclamation*, American Society of Surface Mining and Reclamation, Princeton, WV, pp. 79–89.

Brodie, G. A., 1991, Achieving compliance with staged, aerobic, constructed wetlands to treat acid drainage. In: *Proceedings of the 1991 National Meeting of the American Society of Surface Mining and Reclamation*, American Society of Surface Mining & Reclamation, Princeton, WV, pp. 151–174.

Brodie, G. A., Hammer, D. A., and Tomljanovich, D. A., 1989a, Treatment of acid mine drainage with a constructed wetland at the Tennessee Valley Authority 950 coal mine. In: Hammer, D. A., ed., *Constructed Wetlands for Wastewater Treatment*, Lewis Publishers, Chelsea, MI, pp. 201–209.

Brodie, G. A., Hammer, D. A., and Tomljanovich, D. A., 1989b, Constructed wetlands for treatment of ash pond seepage. In: Hammer, D. A., ed., *Constructed Wetlands for Wastewater Treatment*, Lewis Publishers, Chelsea, MI, pp. 211–219.

Brodie, G. A., Britt, C. R., Tomaszewski, T. M., and Taylor, H. N., 1991, Use of passive anoxic limestone drains to enhance performance of acid drainage treatment wetlands. In: *Proceedings of the 1991 National Meeting of the American Society of Surface Mining and Reclamation*, American Society of Surface Mining & Reclamation, Princeton, WV, pp. 211–228.

Duggan, L. A., and Wildeman, T. R., 1994, Processes contributing to the removal of manganese from mine drainage by an algal mixture. Preprints of papers presented at the 27th ACS National Meeting, American Chemical Society, Environ. Chem. Div., Vol. 34, 2, pp 488–492.

Duggan, L. A., Wildeman, T. R., and Updegraff, D. M., 1992, The aerobic removal of manganese from mine drainage by an algal mixture containing Cladophora. In, *Proceedings of the 1992 National Meeting of the American Society of Surface Mining and Reclamation*, American Society of Surface Mining & Reclamation, Princeton, WV, pp. 241–248.

Dvorak, D. H., Hedin, R. S., Edenborn, H. M., and McIntire, P. E., 1991, Treatment of metal contaminated water using bacterial sulfate reduction: results from pilot scale reactors. In: *Proceedings of the 1991 National Meeting of the American Society of Surface Mining and Reclamation*, American Society of Surface Mining & Reclamation, Princeton, WV, pp. 109–122.

Eger, P., and Lapakko, K., 1989, Use of wetlands to remove nickel and copper from mine drainage. In: Hammer, D. A., ed., *Constructed Wetlands for Wastewater Treatment*, Lewis Publishers, Chelsea, MI, pp. 780–787.

Filas, B. A., and Wildeman, T. R., 1992. The use of wetlands for improving water quality to meet established standards. In: *Successful Reclamation: What Works*, Nevada Mining Association Annual Reclamation Conference, Sparks, NV, pp. 157–176.

Hammack, R. W., and Hedin, R. S., 1989, Microbial sulfate reduction for the treatment of acid mine drainage: a laboratory study. In: *Proceedings of the Conference "Reclamation, A Global Perspective,"* Alberta Land Conservation and Reclamation Council Report No. RRTAC 89-2, pp. 673–680.

Hammack, R. W., and Edenborn, H. M., 1991, The removal of nickel from mine waters using bacterial sulfate reduction. In: *Proceedings of the 1991 National Meeting of the American Society of Surface Mining and Reclamation*, American Society of Surface Mining & Reclamation, Princeton, WV, pp. 97–107.

Hammer, D. A., ed., 1989, *Constructed Wetlands for Wastewater Treatment*, Lewis Publishers, Chelsea, MI, 831 pp.

Hedin R. S., Hyman, D. M., and Hammack, R. W., 1988, Implications of sulfate-reduction and pyrite formation processes for water quality in a constructed wetland: preliminary observations. In: *Proceedings of a Conference on Mine Drainage and Surface Mine Reclamation. V. I: Mine Water and Mine Waste*, U.S. Department of the Interior, Bureau of Mines Information Circular IC-9183, pp. 382–388.

Hedin, R. S., Hammack, R. W., and Hyman, D. M., 1989, Potential importance of sulfate reduction processes in wetlands constructed to treat mine drainage. In Hammer, D. A., ed., *Constructed Wetlands for Wastewater Treatment*, Lewis Publishers, Chelsea, MI, pp. 508–514.

Hiel M. T., and Kerins F. J., 1988, The Tracy wetlands: a case study of two passive mine drainage treatment systems in Montana. In: *Proceedings of a Conference on Mine Drainage and Surface Mine Reclamation. V. I: Mine Water and Mine Waste*, U.S. Department of the Interior, Bureau of Mines Information Circular IC-9183, pp. 352–358.

Howard, E. A., Emerick, J. E., and Wildeman, T. R., 1989, Design and construction of a research site for passive mine drainage treatment in Idaho Springs, Colorado. In: Hammer, D. A., ed., *Constructed Wetlands for Wastewater Treatment*, Lewis Publishers, Chelsea, MI, pp. 761–764.

Klusman, R. W., and Machemer, S. D., 1991, Natural processes of acidity reduction and metal removal from acid mine drainage. In: Peters, D. C., ed., *Geology in Coal Resource Utilization*, Tech Books, Fairfax, VA, pp. 513–540.

Komori, K., Rivas, A., Toda, K., and Ohtake, H., 1990, A method for removal of toxic chromium using dialysis-sac cultures of a chromate-reducing strain of *Entobacter cloacae. Appl. Microbiol. Biotechnol*, 33, 117–119,

Machemer, S. D., and Wildeman, T. R. 1992, Adsorption compared with sulfide precipitation as metal removal processes from acid mine drainage in a constructed wetland. *J. Contam. Hydrol*, 9, 115–131.

Machemer, S. D., Reynolds, J. S., Laudon, L. S., and Wildeman, T. R., 1993, Balance of S in a constructed wetland built to treat acid mine drainage, Idaho Springs, Colorado, U.S.A. *Appl. Geochem*, 8, 587–603.

Nealson, K. H., Tebo, B. M., and Rosson, R. A., 1988, Occurrence and mechanisms of microbial oxidation of manganese. *Adv. Appl. Microbiol.*, 33, 279–318.

Postgate, J. R., 1979, *The Sulfur-Reducing Bacteria*, Cambridge University Press, New York, 151 pp.

Reed, S. C., Middlebrooks, E. J., and Crites, R. W., 1988, *Natural Systems for Waste Management and Treatment*, McGraw-Hill, New York, 308 pp.

Reynolds, J. S., Machemer, S. D., Wildeman, T. R., Updegraff, D. M., and Cohen, R. R., 1991, Determination of the rate of sulfide production in a constructed wetland receiving acid mine drainage. In: *Proceedings of the 1991 National Meeting of the American Society of Surface Mining and Reclamation*, American Society of Surface Mining & Reclamation, Princeton, WV, pp. 175–182.

Stark L. R., Kolbash R. L., Webster H. J., Stevens S. E., Jr., Dionis K. A., and Murphy E. R., 1988, The Simco #4 wetland: biological patterns and performance of a wetland receiving mine drainage. In: *Proceedings of a Conference on Mine Drainage and Sur-*

face Mine Reclamation. V. 1: Mine Water and Mine Waste, U.S. Department of the Interior, Bureau of Mines Information Circ. IC-9183, pp. 332–344.

Stumm, W., and Morgan, J. J., 1981, *Aquatic Chemistry*, 2nd ed. John Wiley & Sons, New York, 780 pp.

Wang, P. C., et al., 1989, Isolation and characterization of an *Entobacter cloacae* strain that reduces hexavalent chromium under anaerobic conditions. *Appl. Environ. Microbiology*, 55, 1665–1669.

Watzlaf, G. R., and Casson, L. W., 1990, Chemical stability of manganese and iron in mine drainage treatment sludge: effects of neutralization chemical, iron concentration, and sludge age. In: *Proceedings of the 1990 National Meeting of the American Society of Surface Mining and Reclamation*, American Society of Surface Mining & Reclamation, Princeton, WV, pp. 3–9.

Wehrli, B., and Stumm, W, 1989, Vanadyl in natural waters: Adsorption and hydrolysis promote oxygenation. *Geochim Cosmochim. Acta*, 53, 69–77.

Wetzel, R. G., 1983, *Limnology*, 2nd ed., Saunders College Publishers, Philadelphia, 767 pp.

Wildeman, T. R., and Laudon, L. S., 1989, The use of wetlands for treatment of environmental problems in mining: non-coal mining applications. In: Hammer, D. A., ed., *Constructed Wetlands for Wastewater Treatment*, Lewis Publishers, Chelsea, MI, pp. 221–329.

Wildeman, T. R., and Updegraff, D. M., 1994, Passive bioremediation of metals from water using reactors or constructed wetlands. In: J. L. Means and R. E. Hinchee eds., *Emerging Technology for Bioremediation of Metals*, Lewis Publishers, Boca Raton, FL, pp. 13–24.

Wildeman, T. R., Cain, D. A., and Ramirez, A. J., 1974, The relation between water chemistry and mineral zonation in the Central City Mining District. In: Hadley, R. F., and Snow, D. T., eds., *Water Resources Problems Related to Mining*, American Water Resources Association Proceedings No. 18, pp. 219–229.

Wildeman, T. R., Brodie, G. A., and Gusek, J. J., 1993a, *Wetland Design for Mining Operations*, Bitech Publishing Co. Vancouver, BC, Canada, 300 pp.

Wildeman, T. R., Duggan, L. A., Updegraff, D. M., and Emerick, J. C., 1993b, The role of macrophytes and algae in the removal of metal contaminants in wetland processes. In: *Proceedings Annual Meeting Air & Waste Management Association*, Pittsburgh, Paper A690.

Wildeman, T. R., Duggan, L. A., Updegraff, D. M., and Emerick, J. C., 1993c, Passive treatment methods for manganese: preliminary results from two pilot sites. In: *Proceedings of the 1993 National Meeting of the American Society of Surface Mining and Reclamation*, American Society of Surface Mining and Reclamation, Princeton, WV, pp. 665–677.

Wildeman T. R., Filipek, L. H., and Gusek, J. J., 1994a, Proof-of-principle studies for passive treatment of acid rock drainage and mill tailing solutions from a gold operation in Nevada. In: *Proceedings of the International Land Reclamation and Mine Drainage Conference*, Vol. 2, U.S. Bureau of Mines Special Publication SP 06B-94, pp. 387–394.

Wildeman, T. R., Cevaal, J., Whiting, K., Gusek, J. J., and Scheuring J., 1994b, Laboratory and pilot-scale studies on the treatment of acid rock drainage at a closed gold-mining operation in California. In: *Proceedings of the International Land Reclamation and Mine Drainage Conference*, Vol. 2, U.S. Bureau of Mines Special Publication SP 06B-94, pp. 379–386.

Index

Page references to figures are in italics